LIST OF
SERIAL PUBLICATIONS
IN THE
BRITISH MUSEUM
(NATURAL HISTORY) LIBRARY

SECOND EDITION

VOLUME III   R - Z

LONDON : 1975

Publication No. **778**

© Trustees of the British Museum (Natural History) 1975

ISBN No. 0 565 00778 5

TITLE	SERIAL No.

RAOU Newsletter. Melbourne.
    RAOU Newsl. 1970 →   T.B.S 7001 B

R.G.S. Research Series. Royal Geographical Society. London.
    R.G.S.Res.Ser. 1948 →   S. 211 L

R.G.S. Technical Series. Royal Geographical Society. London.
    R.G.S.tech.Ser. 1920-1933.   S. 211 I

RIC Reviews. London.
    RIC Rev. 1968-1971.
    Replaced by Chemical Society Reviews. London.   M.S 129

RIVON - Mededeling. Rijksinstituut voor Veldbiologisch Onderzoek
    ten behoeve van het Natuurbehoud. Bilthoven.
    Published in Verslagen van de Werkzaamheden. Rijksinstituut
    voor Veldbiologisch Onderzoek ten behoeven van het
    Natuurbehoud.   S. 619

R. Ruggles Gates Memorial Papers. Edinburgh.
    R. Ruggles Gates Meml Pap. 1963 →   P.A.S 17 A

Rabotȳ Dono-Kubanskoĭ Nauchnoĭ Rybokhozyaĭstvennoĭ Stantsii.
    Rostov-Don.
    Rab.dono-kuban.nauch.rybokhoz.Sta. 1934-1937.   Z.S 1838 A

Rabotȳ Geologicheskago Otdêleniya Imperatorskago Obshchestva
    Lyubiteleĭ Estestvoznaniya, Antropologii i Etnografii. Moskva.
    Rab.Geol.Moskva 1917   P.S 524

Rabotȳ iz Laboratorīi Zoologicheskago Kabineta Imperatorskago
    Varshavskago Universiteta.
    Rab.Lab.zool.Kab.imp.varsh.Univ. 1896, 1898, 1909,
       1911, 1913.   Z.O 72Q o W

Rabotȳ Murmanskoĭ Biologicheskoĭ Stantsii. Murmansk.
    Rab.murmansk.biol.Sta. 1925-1929.
    Continued as Trudy Murmanskoĭ Biologicheskoĭ
    Stantsii. Moskva.   S. 1855 D

Rabotȳ Okskoĭ Biologicheskoĭ Stantsiĭ v Gor. Murome.
    (later v Nizhni-Novgorode).
    Rab.oksk.biol.Sta.Murom 1921-1931.   S. 1843

Rabotȳ Parazitologicheskoĭ Laboratorii. Moskovskii
    Gosudarstvennȳĭ Universitet.
    Rab.parazit.Lab.mosk.gos.Univ. 1926.   S. 1844 D

Rabotȳ Severo-Kavkazskoĭ Gidrobiologicheskoĭ Stantsii pri
    Gorskom Sel'sko-Khozyaĭstvennom Institute. Vladikavkaz.
    Rab.sev.kavk.gidrobiol.Sta.gorsk.sel'-Khoz.Inst.
    Vol.1 - Vol.4 No.3, 1925-1945.   Z.S 1835 A

Rabotȳ Tyan'-Shan'skoĭ Vysokogornoĭ Fiziko-Geograficheskoĭ
    Stantsii. Frunze.
    Rab.Tyan'sh.fiz.-geogr.Sta. 5 → 1962 →   P.S 73 A.o.F

Rabotȳ Volzhskoĭ Biologicheskoĭ Stantsīi. Saratov.
    Rab.volzh.biol.Sta. Vol.3 No.2 - Vol.10, 1906-1929.
    Until 1926, formed part of Trudy Saratovskago Obshchestva
    Estestvoispȳtateleĭ i Lyubiteleĭ Estestvoznaniya.
    Formerly Otchet Volzhskoi Biologicheskoĭ Stantsīi.
    Continued as Trudy Saratovskogo Otdeleniya.   Z.S 1840

Raccolte Planctoniche fatte della R. Nave "Liguria" Firenze.
    See Pubblicazioni del R. Istituto di Studi Superiori
    Pratici e di Perfezionamento in Firenze. Sezione di
    Scienze Fisiche e Naturali.   S. 1122

Rad Fitopatološkog Zavoda u Sarajevu.
    Rad.fitopat.Zav. 1928-1933 (imp.)
    Formerly Izvjestaj o Radu Drzavnoga Fitopatoloskoga Zavoda.   E.S 1882

TITLE	SERIAL No.

Rad Jugoslavenske Akademije Znanosti i Umjetnosti. u Zagrebu.
 Rad.jugosl.Akad.Znan.Umjetn. 1867 →   S. 1705 A

Radiocarbon. New Haven.
 Radiocarbon 1961 →
 Formerly American Journal of Science. Radiocarbon Supplement.   P.S 810

Radioekologiya Vodnykh Organizmov. Riga.
 Radioekol.vod.Orgm 1972 →   S. 1852 e F

Raksti. Geologijas un Geografijas Instituts.
 Latvijas PSR Zinatnu Akademija. Riga.
 See Trudỹ Instituta Geologii. Akademiya Nauk Latviĭskoĭ
 SSR. Riga.   P.S 559

Raksti. Latvijas Augstskola.
 See Latvijas Augstskolas Raksti.   S. 1851 a

Raksti. Latvijas Universität.
 See Latvijas Universitates Raksti.   S. 1851 a

Raksti. Latvijas Universitātes Botaniskā Darba.
 See Acta Horti Botanici Universitatis Latviensis.   B.S 1500

Rameau de Sapin. Neuchâtel.
 Rameau Sapin 1866-1944.   S. 1242

Random Notes on Natural History. Providence, R.I.
 Random Notes nat.Hist. 1884-1886.   S. 2355 a

Range Science Series. Colorado State University.
 See Science Series. Colorado State University Range Science
 Department. Fort Collins.   S. 2334 a

Rapport d'Activité du Laboratoire de Géologie du Quaternaire.
 Bellevue.
 Rapp.Act.Lab.Geol.Quatern. 1966-1967 →   P.S 1209

Rapport d'Activité. Office de la Recherche Scientifique et
 Technique Outre Mer. Paris.
 Rapp.Activ.O.R.S.T.O.M. 1965 →   S. 953 E

Rapport d'Activité de la Station Zoologique de Naples. Naples.
 Rapp.Activ.Stn zool.Naples 1962-1964 →
 Formerly Activity Report of the Zoological Station
 of Naples.   Z.S 1131 A

Rapport Administratif. Conseil Permanent International pour
 l'Exploration de la Mer. Charlottenlund.
 Rapp.adm.Cons.perm.Int.Explor.Mer. 1958 →
 (From 1902-1957 in Rapport et Procès-Verbaux des Réunions.
 Conseil Permanent International pour l'Exploration
 de la Mer.)   Z.S 2715 M

Rapport sur l'Administration du Muséum d'Histoire Naturelle
 de Genève. Genève.
 Rapp.Adm.Mus.Hist.nat.Genève 1939, 1952-1967.
 (Includes Compte Rendu de l'Assemblée Générale de la
 Société Auxiliaire du Muséum d'Histoire Naturelle.
 1940, 1953-1968.)
 Continued as Rapport du Muséum d'Histoire Naturelle de Genève. S. 1206

Rapport Annuel du Conseil de l'Université. Rennes.
 Rapp.a.Cons.Univ.Rennes 1903-1921.   S. 958 a B
                                          (O.B. Store)

Rapport Annuel de la Direction des Mines et de la Géologie.
 République de la Côte d'Ivoire.
 Rapp.a.Dir.Mines Geol.Rép.Côte Ivoire 1966 →   P.S 1179 S

TITLE	SERIAL No.

Rapport Annuel sur le Fonctionnement Technique de l'Institut
    Pasteur du Viet-Nam. Saigon.
        Rapp.a.Fonct.tech.Inst.Pasteur Viet-Nam 1959-1964.   S. 1922 C

Rapport Annuel. Fonds National de la Recherche Scientifique.
    Bruxelles.
        Rapp.a.Fonds natn.Rech.scient.Brux. 1927 →   S. 702 a

Rapport Annuel. Inspection Générale des Mines et de la Géologie
    de la France et d'Outre Mer. Paris.
        Rapp.a.Insp.gén.Min.Géol.Fr. 1955.   P.S 1246

Rapport Annuel de l'Institut Géologique de Hongrie. Budapest.
    See Evi Jelentés a Magyar Kir Foldtani Intézet. Budapest.   P.S 1366

Rapport Annuel. Institut de Médecine Tropicale Prince Léopold.
    Antwerpen.
        Rapp.a.Inst.Méd.trop.Prince Léopold 1967/1968 →   S. 776

Rapport Annuel. Institut National pour l'Etude Agronomique
    du Congo Belge. Bruxelles.
        Rapp.a.Inst.natn.Etude agron.Congo belge 1938, 1944-1959.   B.S 353 b

Rapport Annuel. Institut pour la Recherche Scientifique en
    Afrique Centrale. Bruxelles.
        Rapp.a.Inst.Rech.scient.Afr.cent. 1948-1959.
        Continued as Rapport. Institut pour la Recherche Scientifique
    en Afrique Centrale. Butare.   S. 709 a

Rapport Annuel. Laboratoire de Geologie de la Faculté des Sciences
    de l'Université de Dakar.
        Rapp.a.Lab.Geol.Fac.Sci.Univ.Dakar No.13 → 1966 →   P.S 1239

Rapport Annuel de la Section de Géologie, de Minéralogie
    et de Paléontologie du Musée Royal de l'Afrique Centrale
    (et de la Commission de Géologie du Ministère des Affaires
    Africaines.)
        Rapp.a.Sect.Géol.Miner.Paléont.Mus.r.Afr.centr. 1960 →
        Formerly Rapport Annuel de la Section de Géologie,
    de Minéralogie et de Paléontologie du Musée Royal du
    Congo Belge et de la Commission de Géologie du Ministère
    du Congo Belge et du Ruanda-Urundi.   P.S 248 a

Rapport Annuel de la Section de Géologie, de Minéralogie
    et de Paléontologie du Musée Royal du Congo Belge et de la
    Commission de Géologie du Ministère (des Colonies)
    du Congo Belge et du Ruanda-Urundi.
        Rapp.a.Sect.Geol.Miner.Paleont.Mus.r.Congo belge 1957-1959.
        Continued as Rapport Annuel de la Section de Géologie,
    de Minéralogie et de Paléontologie du Musée Royal de
    l'Afrique Centrale et de la Commission de Géologie
    du Ministère des Affaires Africaines.   P.S 248 a

Rapport Annuel. Service de Botanique du Muséum d'Histoire
    Naturelle. Paris.
        Rapp.a.Serv.bot.Mus.Hist.Nat. 1910-1914.   B.S 413 e

Rapport Annuel du Service Géologique, Tunis.
        Rapp.a.Serv.géol.Tunis 1954   P.S 1236

Rapport Annuel du Service Géologique de l'Afrique Equatoriale
    Francaise.
        Rapp.a.Serv.géol.Afr.équat.fr. 1954.   P.S 1235 A

Rapport Annuel du Service Géologique, Territoire du Cameroun. Paris.
        Rapp.a.Serv.géol.Territ.Cameroun 1954, 1959 →   P.S 1235 a C

Rapport Annuel. Service des Mines, Nouvelle - Calédonie et
    Dependances. Nouméa.
        Rapp.a.Serv.Mines Nouv.-Caledonie 1965 →   P.S 1243 B

TITLE	SERIAL No.

Rapport Annuel. Société Provancher d'Histoire Naturelle du Canada.
   See Report of the Provancher Society of Natural History of Canada. Quebec.   S. 2636

Rapport Annuel. Station de Biologie Marine. Grande-Rivière, Québec.
   See Rapport. Station de Biologie Marine de Grande-Rivière. Québec.   Z.S 2656 A

Rapport Annuel. Station Fédérale d'Essais Viticoles et Arboricoles à Lausanne. Berne.
   Rapp.a.Stn féd.Essais vitic.arboric.Lausanne 1898-1900.   E.S 1211

Rapport Annuel sur les Travaux de la Société d'Histoire Naturelle de l'Ile Maurice. Port Louis.
   Rapp.a.Trav.Soc.Hist.nat.Ile Maurice 1830-1836; 1850-1851.   S. 2091 A

Rapport Annuel. Union Internationale pour la Conservation de la Nature et de ses Ressources.
   See Report. International Union for Conservation of Nature and Natural Resources. Morges, Switzerland.   S. 2715 H

Rapport Annuel. Université Officielle du Congo. Lubumbashi.
   Rapp.a.Univ.off.Congo 1965-1966 →   S. 2058 B

Rapport et Bilans d'Exercise. Comité Spécial du Katanga. Bruxelles.
   Rapp.Bilans Exerc.Com.spéc.Katanga 1951 (1952).   74 K.o.C

Rapport. Bosbouwproefstation. Bogor.
   Rapp.BosbProesfstn, Bogor. 1949 → (imp.)   B. 581.9(595.p40) BOS

Rapport. Congrès Géologique International. London.
   See International Geological Congress.   M.S 3025

Rapport. Congrès International de la Mer.
   Rapp.Congr.int.Mer 3rd 1946.   S. 2722

Rapport. Congrès International pour la Protection de la Nature.
   Rapp.Congr.int.Prot.Nat. 1923 & 1931.   S. 2730

Rapport du Conseil. Section Regionale Ouest Palearctique, Organisation Internationale de Lutte Biologique. Wageningen.
   Rapp.Cons.SROP 1971-1972.
     Continued in Bulletin. Section Regionale Ouest Palearctique, Organisation Internationale de Lutte Biologique.   E.S 2569

Rapport. Danmarks Geologiske Undersøgelse. København.
   Rapp.Danm.geol.Unders. 1968 →   P.S 1395 A

Rapport sur le Fonctionnement des Services de l'Institut Pasteur d'Algérie. Alger.
   Rapp.Fonct.Servs Inst.Pasteur Algér. 1910-1928.   S. 2029 C

TITLE	SERIAL No.

Rapport sur le Fonctionnement Technique de l'Institut Pasteur
de la Guyane Francaise et du Territoire de l'Inini. Cayenne.
  Rapp.Fonct.tech.Inst.Pasteur Guyane 1942 →   S. 2205 b B

Rapport Général sur la Mission de Délimitation, Afrique
  Equatoriale Française-Cameroun. (1912-14). Paris.
  Rapp.gén.Miss.Délim.Afr.Equat.Franc.-Cameroun
    Tom 3. 1916.   74 L.o.F

Rapport Général sur les Travaux de l'Académie des Sciences,
  Arts et Belles-Lettres de la Ville de Caen.
  Rapp.gén.Acad.Sci.Caen 1811-1815.   S. 838 A

Rapport Généraux des Travaux de la Société Philomathique
  de Paris.
  Rapp.gén.Soc.philom.Paris 1788-1799.   S. 943 A

Rapport. Institut Océanographique de Monaco.
  Rapp.Inst.océanogr. 1937-1938.   S. 906 D

Rapport de l'Institut Pasteur d'Algérie.
  See Rapport sur le Fonctionnement des Services de l'Institut
  Pasteur d'Algérie. Alger.   S. 2029 C

Rapport de l'Institut Pasteur Guyane Francaise.
  See Rapport sur le Fonctionnement Technique de l'Institut
  Pasteur de la Guyane Française et du Territoire de
  l'Inini. Cayenne.   S. 2205 b B

Rapport. Institut pour la Recherche Scientifique en Afrique
  Centrale. Butare.
  Rapp.Inst.Rech.scient.Afr.cent. 1960-1964 →
  Formerly Rapport Annuel. Institut pour la Recherche
  Scientifique en Afrique Centrale. Bruxelles.   S. 709 a

Rapport. Institutt for Marin Biologi, Universitetet i Oslo.
  Rapp.Inst.mar.Biol.Univ.Oslo No.2 → 1971 →
  (Wanting No.4 & 5.)   S. 550 G

Rapport inzake Nederlands Nieuw-Guinea. (The Hague.)
  Rapp.ned.Nieuw-Guinea 1951-1952, 1955-1961.   S. 639

Rapport. Musées d'Histoire Naturelle de Lausanne.
  Rapp.Musées Hist.nat.Lausanne 1887 →   S. 1234

Rapport. Musées d'Histoire Naturelle de Lyon.
  Rapp.Musées Hist.nat.Lyon 1871-1885.   S. 891 A

Rapport du Muséum d'Histoire Naturelle de Genève. Genève.
  Rapp.Mus.Hist.nat.Genève 1968 →
  (Includes Compte Rendu de l'Assemblée Générale des Amis
  du Muséum. 1969 →)
  Formerly Rapport sur l'Administration du Muséum d'Histoire
  Naturelle de Genève.   S. 1206

TITLE	SERIAL No.

Rapport Norsk Institutt for Tang- og Tareforskning. Oslo.
   Rapp.Norsk Inst.Tang- Tareforsk. 1951 →   B.A.S 22

Rapport de l'Office de Biologie, Ministère de la Chasse
   et des Pêcheries. Province de Québec.
   Rapp.Off.Biol.Québ. 1943-1955.   S. 2638 A

Rapport. Office de la Recherche Scientifique et Technique
   Outre Mer. Centre de Noumea.
   Rapp.O.R.S.T.O.M.Centre Noumea Oceanographie 1966-1969.   S. 953 G

Rapport sur la Paléobotonique dans le Monde. Utrecht.
   See World Reports on Palaeobotany. Utrecht.   P.S 50 o.I

Rapport Particulier. Centre de Recherches du Service de Santé
   des Armées. (Paris.)
   Rapp.part.Cent.Rech.Serv.Santé Armées No.10 → 1967 →   S. 908

Rapport sur les Pêcheries d'Egypte. Le Caire.
   Rapp.Pêch.Egypte 1931-1933 & 1935.
   Formerly Report on the Fisheries of Egypt.   Z.O 74B q E

Rapport sur les Pêcheries (Pêches) du Québec. Québec.
   Rapp.Pêch.Québ. 1962/1963 →   Z.S 2656 C

Rapport et Procès-Verbaux des Réunions de la Commission
   Internationale pour l'Exploration Scientifique de la
   Mer Méditerranée. Paris.
   Rapp.P.-v.Réun.Commn int.Explor.scient.Mer Méditerr. 1926 →
   Formerly Bulletin de la Commission Internationale
   pour l'Exploration Scientifique de la Mer Méditerranée.
   Paris.   Z.S 2720

Rapport et Procès-Verbaux des Réunions du Conseil Permanent
   International pour l'Exploration de la Mer. Copenhague.
   Rapp.P.-v.Réun.Cons.perm.int.Explor.Mer 1902 →   Z.S 2715 D

Rapport Scientifiques. Institut Polytechnique Ukrainien.
   Augsburg.
   See Naukoviy̆ı Byuleten. Ukrayins'ky̆yi Tekhnichno-Hospodarsky̆i
   Instȳtut. Augsburg, etc.   S. 1315

Rapport du Service de la Faune. Québec.
   Rapp.Serv.Faune Québec 1962 →   Z.S 2602

Rapport de Session. Conseil Général des Pêches pour la Méditerranée.
   Organisation des Nations Unies pour l'Alimentation et
   l'Agriculture. Rome.
   See Session Report. General Fisheries Council for the
   Mediterranean. Rome.   Z.S 2713 D

| TITLE | SERIAL No. |

Rapport Société Entomologique de Genève.
   See Mitteilungen der Schweizerischen Entomologischen
   Gesellschaft. Lausanne.   E.S 1205

Rapport Société Vaudoise d'Entomologie.
   See Mitteilungen der Schweizerischen Entomologischen
   Gesellschaft. Lausanne.   E.S 1205

Rapport. Station Agronomique de Guadeloupe. Pointe-à-Pitre.
   Rapp.Stn agron.Guadeloupe 1918-1920.   E.S 2389

Rapport. Station de Biologie Marine de Grande-Rivière. Québec.
   Rapp.Stn Biol.mar.Grande-Rivière 1961 →
   Previously issued as part of Contributions du
   Département des Pêcheries, Quebec.   Z.S 2656 A

Rapport. Station Biologique du St. Laurent à Trois Pistoles. Québec.
   Rapp.Stn biol.St.Laurent 1931-1949.   S. 2637 A

Rapport sur les Travaux de Recherches Effectués. Service Botanique
   et Agronomique de Tunisie. Tunis.
   Rapp.Trav.Rech.effect.Serv.bot.agron.Tunis 1952 →
   Formerly Bulletin du Service Botanique et Agronomique
   de Tunisie. Tunis.   B.S 2318

Rapport. Université Officielle du Congo a Lubumbashi.
   Rapp.Univ.off.Congo Lubumbashi 1965/1966.   S. 2058 B

Rapport och Uppsatser. Institutionen för Skogsentomologi. Stockholm.
   Rapp.Upps.Inst.Skogsent.Stockholm 1964 →   E.S 511

Rapport en Verhandelingen uitgegeven door het Rijksinstituut
   voor Visscherijonderzoek. s'Gravenhage.
   Rapp.Verhand.Rijksinst.Vissch.-Onderz. Deel 1, Afl.1-4, 1913-1919.
   Continued as Verhandelingen en Rapporten Uitgegeven
   door het Rijksinstituut voor Visscherijonderzoek.
   s'Gravenhage.   Z. 22 q H

Rapport. Visscherij en de Industrie van Zeeprodukten in de
   Kolonie Curacao. s'Gravenhage.
   Rapp.Vissch.Ind.Zeeprod.Curacao 1-2, 1907-1919.   Z.O 75F q C

Rapporten, Rapporter & Rapports.
   See Rapport.

Rasprave Geološkog Instituta Kraljovine Jugoslavije. Beograd.
   See Mémoires du Service Géologique du Royaume de
   Yougoslavie. Béograd.   P.S 1464

Rasprave Zavoda za Geološko i Geofizičko Istraždvanje N.R. Serbije.
   Beograd.
   See Mémoires du Service Géologique et Geophysique de la R.P.
   de Serbie. Beograd.   P.S 1464

TITLE	SERIAL No.

Rassegna Bibliografica Zoologica di Lavori Pubblicati
in Italia. Torino.
Rass.biblfica zool.Lav.pubbl.Ital. 1954 →   Z.S 1180

Rassegna Faunistica. Roma.
Rass.Faun. 1934-1939.   Z.S 1137

Rassegna delle Scienze Geologiche in Italia. Roma.
Rass.Sci.geol.Ital. 1891-1892 (imp.)   P.S 605

Rastenie i Sreda. Moskva.
See Trudy Laboratorii Evolyutsionnoĭ i Ekologicheskoĭ
Fiziologii im. B.K. Kellera. Moskva.   B.S 1381

Rastenievudni Nauki. Akademiya na Selskostopanskite Nauki. Sofiya.
Rasten.Nauki Vol.2 → No.2 → 1965 → (imp.)   B.S 1531 a

Rastitel'noe Sȳr'e. Leningrad, Moskva.
See Trudy Botanicheskogo Instituta Akademii
Nauk SSSR. Leningrad, Moskva. Ser. 5.   B.S 1377 d

Rastitel'nost'Latviĭskoĭ SSR. Riga.
See Trudy Instituta Biologii. Akademiya Nauk
Latviĭskoĭ SSR.   S. 1852 e C

Rastitel'nye Resursȳ. Akademiya Nauk. Moskva, Leningrad.
Rastit.Resursȳ Tom 2 → 1966 →   B.S 1403 i

Rastitelnȳe Resursȳ Sibiri i Dal'nego Vostoka: Tekushchii
Ukazatel' Literaturȳ. Novosibirsk.
Rastit.Resurȳ Sibiri dal.V. 1974 →   B.S 1415 a

Rat en Muis. Mededelingen Betreffrende de Bostrijding
van Ratten en Muizen. Wageningen.
Rat Muis 1953 →   Z. Mammal Section

Raymond Dart Lectures. Institute for the Study of Man in Africa.
Johannesburg.
Raymond Dart Lect. 1964 →   P.A.S 506 A

Raymondiana. Lima.
Raymondiana 1968 →   B.S 4037

Razprave Matematično-Prirodoslovnega Razreda, Akademije
Znanosti in Umetnosti v Ljubljana.
Razpr.mat.-prir.Razr.Akad.Znan.Umet.Ljubl. 1940-1942.
Continued as Razprave. Slovenska Akademija Znanosti in
Umetnosti. Ljubljana.   S. 1746 E

Razprave. Slovenska Akademija Znanosti in Umetnosti. Ljubljana.
Razpr.Slov.Akad.Znan.Umet. Razred Mat., Prirod.Med.Tech.
Prirod.Odsek. Vol.4. 1949.
Class IV Hist.Nat.et Med. Pars Hist.Nat. 1951 →
Formerly Razprave Matematicno-Prirodoslovnega Razreda.
Akademije Znanosti in Umetnosti v Ljubljane.   S. 1746 E

TITLE	SERIAL No.

Reading Naturalist. Reading.
    Reading Nat. 1949 →
    Formerly Quaestiones Naturales, Reading.   S. 339 B

Reading Ornithological Club Reports. Reading.
    Reading orn.Club Rep. 1947 →   T.B.S 262

Reading University Geological Reports. Reading.
    Reading Univ.geol.Rep. 1967 →
    (No.1 also published as Computation in Sedimentology.
    Report, No.5)   P.S 152 B

Réalités Scientifiques et Techniques Francaises. Paris.
    Réalités scient.tech.fr. No.52 → 1971 →   S. 887

Recent Advances in Botany. Toronto.
    Recent Adv.Bot. 1961.   B 58 C.I.B

Recent Advances in Phytochemistry. Amsterdam.
    Recent Adv.Phytochem. 1968 →   B.S 299

Recent Advances in Zoology in India. Delhi.
    Recent Adv.Zool.India 1963 →   Z.S 1913

Recent Geographical Literature, Maps and Papers (Photographs).
    Royal Geographical Society. London.
    Recent geogr.Lit.Maps Pap.R.geogr.Soc. 1918-1941.
    (Supplement to the Geographical Journal.)
    Continued as New Geographical Literature and Maps.
    Royal Geographical Society. London.   S. 211 B

Recent Polar Literature. Cambridge.
    Recent polar Lit. 1973 →
    Formerly included in Polar Record.   S. 6b B

Recherche. Paris.
    Recherche Vol.4 → 1973 →
    (1973 → includes Science Progrès Découverte. Paris.)   S. 996 b

Recherches Agronomiques. Québec.
    Rech.agron. 1957 →   E.S 2530 A

Recherches Géologiques en Afrique. Paris.
    Rech.géol.Afr. 1972 →   P.S 193 A

Recherches d'Hydrobiologie Continentale. Paris.
    Rech.Hydrobiol.cont. 1969.
    Continued as Annales d'Hydrobiologie. Paris.   S. 944 a

Recherches Marines. Constanta.
    See Cercetari Marine. Constanta.   S. 1894 e

Record and Abstracts Birmingham & Midland Institute
    Scientific Society.
    Rec.Abstr.Bgham Midl.Inst.scient.Soc.
    Record 1872-1934. Abstracts 1927-1935.
    Formerly Record & Proceedings. Birmingham & Midland
    Institute Scientific Society.   S. 17 b c

| TITLE | SERIAL No. |

Record of Agricultural Research Station. Rehovot. Palestine.
    See Ktavim.   E.S 2036 c

Record of the Albany Museum. Grahamstown.
    Rec.Albany Mus. 1903-1935.   S. 2021 A & T.R.S 4010

Record of the Auckland Institute and Museum.
    Rec.Auckland Inst.Mus. 1930 →   S. 2185 A

Record of the Australian Academy of Science. Canberra City.
    Rec.Aust.Acad.Sci. 1966 →   S. 2148

Record of the Australian Museum. Sydney.
    Rec.Aust.Mus. 1890 →   S. 2126 A & T.R.S 7206 B

Record of Bare Facts. Caradoc and Severn Valley Field Club. Shrewsbury.
    Rec.bare Facts Caradoc Severn Vall.Fld Club 1894-1942.
    Formerly Caradoc Record of Bare Facts.   S. 360 A

Record of the Botanical Survey of India. Calcutta.
    Rec.bot.Surv.India 1893 →   B.S 1601

Record of the Canterbury Museum. Christchurch. N.Z.
    Rec.Canterbury Mus. 1907 →   S. 2171

Record of the Department of Geology of Travancore. Trivandrum.
    Rec.Dep.Geol.Travancore 1921 (1922).   P.S 1116

Record of Department of Mineralogy. Ceylon. Colombo. Professional Paper.
    Rec.Dep.Miner.Ceylon 1943-1945.   P.S 1116 A & M.S 1920
    See also Professional Paper. Department of Mineralogy. Ceylon.

Record of the Dominion Museum. Wellington, N.Z.
    Rec.Dom.Mus.Wellington 1942 →
    From 1946-1953 See Dominion Museum Records in Entomology and Zoology.   S. 2165 C

Record of the Egyptian Government School of Medicine. Cairo.
    Rec.Egypt.Govt.Sch.Med. 1901-1905.
    Continued as Record of the School of Medicine. Cairo.   Z.S 2020

Record in Entomology. Dominion Museum.
    See Dominion Museum Records in Entomology.   E.S 2264

Record of the Fiji Museum. Suva.
    Rec.Fiji Mus.Suva 1965 →   S. 2180

TITLE	SERIAL No.

Record of General Science. London.
    Rec.gen.Sci. 1835-1836.   S. 461

Record of the Geological Committee of the Russian Far East.
    Vladivostock.
    See Materialȳ po Geologii i Poleznȳm Iskopaemȳm Dal'nyago
    Vostoka. Vladivostock.   P.S 1540

Record of the Geological Department of the State of Mysore.
    Bangalore.
    Rec.geol.Dep.St.Mysore 1894-1935.   M.S 1912

Record of the Geological Survey. Bechuanaland.
    See Record of the Geological Survey Department. Bechuanaland.
    (Pietermaritzburg.)   P.S 1175 E

Record. Geological Survey of British Guiana. Georgetown.
    Rec.geol.Surv.Br.Guiana 1962 →   P.S 1168 B

Record of the Geological Survey. Department of Mines.
    Victoria, Melbourne.
    See Record of the Geological Survey of Victoria.   M.S 2417

Record of the Geological Survey. Federation of Malaya.
    See Economic Bulletin. Geological Survey. Federation
    of Malaya.   P.S 1117 C

Record of the Geological Survey Department. Bechuanaland.
    (Pietermaritzburg.)
    Rec.geol.Surv.Dep.Bechuan. 1957/58-1961/62.   P.S 1175 E

Record of the Geological Survey of India. Calcutta.
    Rec.geol.Surv.India 1868 →   P.S 1110

Record of the Geological Survey of Malawi. Zomba.
    Rec.geol.Surv.Malawi 1961 →
    Formerly Record of the Geological Survey of Nyasaland.
    Zomba.   P.S 1179 B

Record of the Geological Survey of New South Wales. Sydney.
    Rec.geol.Surv.N.S.W. 1889-1922.   P.S 1124

Record of the Geological Survey of Nigeria.
    Rec.geol.Surv.Nigeria 1954-1959   P.S 1179 K A

Record of the Geological Survey. Northern Rhodesia, Lusaka.
    Rec.geol.Surv.Nth.Rhod. 1956, 1958, Vol.9 (1963).
    Continued as Record of the Geological Survey, Zambia.
    Lusaka.   P.S 1172 H

Record of the Geological Survey of Nyasaland. Zomba.
    Rec.geol.Surv.Nyasald 1959-1960.
    Continued as Record of the Geological Survey of Malawi.
    Zomba.   P.S 1179 B

| TITLE | SERIAL No. |

Record of the Geological Survey of Pakistan. Karachi.
    Rec.geol.Surv.Pakist. 1948 →   P.S 1098

Record of the Geological Survey of Queensland. Brisbane.
    Rec.geol.Surv.Qd 1904-1905.
    (Publications. Geological Survey of Queensland.
       No. 190 & 196.)   P.S 1145

Record of the Geological Survey of Tanganyika. Dar Es Salaam.
    Rec.geol.Surv.Tanganyika 1951 →   P.S 1179 A

Record. Geological Survey. Tasmania. Hobart.
    Rec.geol.Surv.Tasm. 1913-1919   P.S 1154

Record of the Geological Survey (Department) Uganda. Entebbe.
    Rec.geol.Surv.Uganda 1950 →
    Formerly in Report of the Geological Survey Department,
    Uganda Protectorate, Entebbe.   P.S 1179 c

Record of the Geological Survey of Victoria. Melbourne.
    Rec.geol.Surv.Vict. 1902-1928.   P.S 1133
              Vol.1 pt.2 - Vol.2, pt.4. 1903-1908.   M.S 2417

Record of the Geological Survey, Zambia. Lusaka.
    Rec.geol.Surv.Zambia Vol.10 → 1966 →
    Formerly Record of the Geological Survey, Northern Rhodesia.
    Lusaka.   P.S 1172 H

Record of the Geology and of the Mineral Resources of East-Siberia.
    Irkutsk.
    See Materialy po Geologii i Poleznȳm Iskopaemȳm Vostochnoi Sibiri.
    Irkutsk.   P.S 1530

Record of the Geology of the West Siberian Region. Tomsk.
    See Materialy po Geologii Zapadno-Sibirskogo Kraya. Tomsk.   P.S 1532

Record of the Indian Museum. Calcutta.
    Rec.Indian Mus. 1907-1962.   T.R.S 3010 B &
    Continued as Record of the Zoological Survey of India.   Z.S 1910 C

Record of Investigations. Department of Agriculture, Uganda. Entebbe.
    Rec.invest.Dep.Agric.Uganda 1948 →   E.S 2179 b

Record of Lectures & Addresses. Haslemere Microscope and
    Natural History Society.
    Rec.Halsemere Microsc.nat.Hist.Soc. 1893-1898.
    Continued as Report of the Haslemere Microscope and
    Natural History Society.   S. 117 A

Record of the London and West Country Chamber of Mines. London.
    Rec.Lond.W.Ctry Chamb.Mines 1901-1912   M.S 159

Record of the Malaria Survey of India. Calcutta.
    Rec.Malar.Surv.India 1929-1937.
    Continued as Journal of the Malaria Institute of India.
    Calcutta.   E.S 1993

TITLE	SERIAL No.

Record of the Mines of South Australia, Adelaide.
    Rec.Mines S.Aust. 1898-1908   M.S 2410 A

Record. Ministry of Commerce and Communications, Siam, Bangkok.
    (Technical & Scientific Supplements only.)
    Rec.Minist.Commerce Commun.Siam Nos.2-3, 6-7 & 14. 1926-1930. B.S 1760

Record. Mysore Geological Department. Bangalore.
    See Record of the Geological Department, State of Mysore.   M.S 1912

Record. New Zealand Oceanographic Institute. Wellington.
    See NZOI Records. New Zealand Oceanographic Institute. Wellington.   S. 2160 a C

Record of Observations. Scripps Institution of Oceanography. Berkeley, California.
    Rec.Obsns Scripps Instn Oceanogr. 1942-1947.   S. 2319 P

Record of Oceanographic Works in Japan. National Research Council. Tokyo.
    Rec.oceanogr.Wks Japan 1928-1941: 1953 →   S. 1994 C

Record of the Otago Museum. Dunedin.
    Rec.Otago Mus. Zoology. 1964 →   Z.S 2185
                  Anthropology. 1964 →   P.A.S 880

Record of the Papua and New Guinea Museum. Port Moresby.
    Rec.Papua New Guin.Mus. 1970 →   S. 2142 a B

Record and Proceedings. Birmingham & Midland Institute Scientific Society.
    Rec.Proc.Bgham Midl.Inst.scient.Soc.
    Record 1872-1923. Proceedings 1922-1924.
    Continued as Record and Abstracts. Birmingham & Midland Institute Scientific Society.   S. 17 b C

Record of Proceedings. Cleveland Naturalists' Field Club. Middlesbrough.
    Rec.Proc.Cleveland Nat.Fld Club 1889-1902.
    Continued as Proceedings of the Cleveland Naturalists' Field Club. Middlesbrough.   S. 80

Record of the Queen Victoria Museum. Launceston.
    Rec.Queen Vict.Mus. 1942 →   S. 2155

Record of Researches. Faculty of Agriculture, University of Tokyo.
    Rec.Res.Fac.Agric.Univ.Tokyo 1950 →   E.S 1920b

Record of the Royal Institution of Great Britain. London.
    Rec.R.Instn Gt Br. 1949-1952, 1961-1968.   S. 213 E

Record of the Royal Society of London.
    Rec.R.Soc. 1897, 1901, 1912, 1940.   S. 3 D

Record of the Russian Mineralogical Society.
    See Zapiski Vserossiiskogo Mineralogicheskogo Obshchestva.   M.S 1820

| TITLE | SERIAL No. |

Record of the School of Medicine. Cairo.
    Rec.Sch.Med., Cairo 1911
    <u>Formerly</u> Record of the Egyptian Government School
of Medicine.   Z.S 2020

Record of the School of Mines and of Science applied to the Arts.
London.
    Rec.Sch.Mines 1851-1852.   M.S 307 & P.S 1030

Record. Scottish Plant Breeding Station. Pentlandfield, Roslin.
    Rec.Scott.Pl.Breed.Stn 1962-1965.
    <u>Formerly and Continued as</u> Report. Scottish Plant
Breeding Station.   B.S 131

Record of the South Australian Museum. Adelaide.
    Rec.S.Aust.Mus. 1918 →   S. 2109

Record of the Survey of India. Calcutta.
    Rec.Surv.India 1909-1934.   S. 1916

Record of the Western Australian Museum. Perth.
    Rec.West Aust.Mus. 1910-1939.   S. 2156 A

Record of Zoological Literature. London.   REF.
    Rec.zool.Lit. 1864-1869.   Z. R97 o Z
                                    T.R.Ref, T.B.S 9999 & E. REF.
    <u>Continued as</u> Zoological Record. London.

Record of the Zoological Survey of India. Delhi.
    Rec.zool.Surv.India 1963 →
    <u>Formerly</u> Record of the Indian Museum.   T.R.S 3010 B & Z.S 1910 C

Record. Zoological Survey of Pakistan. Karachi.
    Rec.zool.Surv.Pakist. 1969 →   Z.S 1929 A

Record in Zoology. Dominion Museum.
    <u>See</u> Dominion Museum Records in Zoology.   Z.S 2160

Recorders' Reports. Bradford Natural History and Microscopical
    Society.
    Rec.Rep.Bradford nat.Hist.& microsc.Soc. 1906.   S. 32

Records.
    <u>See</u> Record.

Recreative Science. London.
    Recreative Sci. 1860-1862.
    Succeeded by "The Intellectual Observer".   S. 462 A

Recueil des Actes de l'Académie Nationale des Sciences,
    Belles-Lettres et Arts de Bordeaux.
    <u>See</u> Actes de l'Académie Royale (Nationale) des Sciences,
Belles-Lettres et Arts de Bordeaux.   S. 829 C

Recueil des Actes de la Séance Publique de l'Académie
    Imperiale des Sciences de St.Petersbourg.
    Recl Act.Acad.Sci.St.Petersb. 1827-1848.   S. 1802 A

| TITLE | SERIAL No. |

Recueil des Actes de la Société de Santé de Lyon.
  Recl Act.Soc.Santé Lyon 1798. — S. 897

Recueil Géobotanique. Kiev.
  See Heobotanichnȳyi Zbirnȳk. Kyyiv. — B.S 1371 a

Recueil de l'Institut Botanique 'Léo Errera'. Bruxelles.
  Recl Inst.bot.'Léo Errera' 1902-1913. — B.S 352

Recueil de l'Institut Zoologique Torley-Rousseau. Bruxelles.
  Recl Inst.zool.Torley-Rousseau 1927-1942.
  Formerly Annales de Biologie Lacustre. — Z.S 712

Recueil de Mémoires, ou Collection de Pièces Académiques. Dijon.
  Recl Mém.Coll.Piec.acad.Dijon 1754-1769.
  Continued as Collection Académique, etc. — S. 983

Recueil de Mémoires, etc. Société Académique d'Aix.
  Recl Mém.Soc.acad.Aix 1819-1827.
  Continued as Mémoires de l' Académie des Sciences,
  Agriculture, Arts et Belles-Lettres. Aix. — S. 807

Recueil des Mémoires et des Travaux de la Société Botanique
  du Grand-Duché de Luxembourg.
  Recl Mém.Trav.Soc.bot.Luxemb. 1874-1903. — B.S 384

Recueil Périodique de Société Géologique. Suisse.
  See Eclogae Geologicae Helvetiae Lausanne. — P.S 705

Recueil des Pièces qui ont remporte le Prix de l'Académie
  Royale des Sciences. Paris.
  Recl Pièc.Acad.Sci.Paris 1720-1772. — S. 804 C

Recueil des Procès-Verbaux de la Conférence Internationale
  pour la Protection de la Nature.
  Rec.P.v.Conf.int.Prot.Nat. 1913. — S. 2729

Recueil des Travaux Botaniques Néerlandais. Société Botanique
  Néerlandaise. Nimègue.
  Recl Trav.bot.néerl. 1904-1951.
  Replaced by Acta Botanica Neerlandica. — B.S 302

Recueil de Travaux. Centre de Nouméa, Office de la Recherche
  Scientifique et Technique Outre-Mer. Nouméa.
  Recl Trav.Cent.Nouméa O.R.S.T.O.M. 1969 → — S. 953 L

Recueil des Travaux des Facultés de Géologie et de Mines. Beograd.
  See Zbornik Geološkog i Rudarskog Fakulteta. — P.S 457

Recueil des Travaux. Institut Biologique. Beograd.
  See Zbornik Radova. Biološki Institut N.R. Srbije.
  Beograd. — S. 1888 b B

Recueil des Travaux de l'Institut Botanique de l'Université
  de Montpellier.
  Recl Trav.Inst.bot.Univ.Montpellier Fasc.1 & 4, 1944-1949.
  Continued as Recueil des Travaux des Laboratoires
  de Botanique, Géologie et Zoologie de la Faculté des
  Sciences de l'Université de Montpellier. Série Botanique. — B.S 405

| TITLE | SERIAL No. |

Recueil des Travaux. Institut d'Ecologie et de Biogéographie.
Beograd.
See Zbornik Radova. Institut za Ekologiju i Biogeografiju.
Srpska Akademija Nauka. Beograd.   S. 1888 H

Recueil des Travaux. Institut de Géologie. Beograd.
See Zbornik Radova. Geoloshki Institut N.R. Srbije. Beograd.   P.S 456 A

Recueil des Travaux de l'Institut de Géologie "Jovan Zujović". Beograd.
See Zbornik Radova Geoloshkog Instituta "Jovan Zhujovich".
Beograd.   P.S 456 A

Recueil de Travaux du Laboratoire Boerhaave. Leiden.
Recl Trav.Lab.Boerhaave 1899.   S. 631 E

Recueil des Travaux des Laboratoires de Botanique, Géologie et
Zoologie de la Faculté des Sciences de l'Université
de Montpellier.
Recl Trav.Labs bot.Géol.Zool.Univ.Montpellier
Série Botanique, Fasc.6-7, 1953-1955.   B.S 405
Formerly Recueil des Travaux de l'Institut Botanique
de l'Université de Montpellier.
Continued as Naturalia Monspeliensia. Montpellier.
Série Botanique.   B.S 405
Série Zoologique   Z.S 911

Recueil de Travaux. Musée de la Ville de Skopje.
See Zbornik. Muzej na Grad Skopje.   P.S 112

Recueil des Travaux des Professeurs à l'Université d'Etat
à Irkoutsk.
See Sbornik Trudov Professorov i Prepodavatelei
Gosudarstvennogo Irkutskogo Universiteta. Irkutsk.   S. 1831 A

Recueil des Travaux des Sciences Medicales au Congo Belge.
Leopoldville.
Recl Trav.Sci.méd.Congo belge 1942-1945.   S. 2056

Recueil des Travaux Scientifiques. Institut de Médecine Tropicale
Prince Leopold. Antwerpen.
Recl Trav.scient.Inst.Méd.trop.Prince Leopold.
Vol.12 → 1972 →   S. 776 B

Recueil des Travaux de la Société (d'Amateurs) des Sciences,
de l'Agriculture et des Arts à Lille.
Recl Trav.Soc.Sci.Agric.Lille 1819-1827.
Continued as Mémoires de la Société (Royale) des Sciences,
de l'Agriculture et des Arts à Lille.   S. 881

Recueil. Travaux de la Station Biologique du Lac d'Oredon. Toulouse.
Rec.Trav.Stn biol.Lac Oredon Nos.1-2. 1962-1963.
Replaced by S.970 B. Annales de Limnologie. Station
Biologique du Lac d'Oredon. Toulouse.   S. 970 A

Recueil des Travaux. Station Hydrobiologique. Ohrid.
See Zbornik na Rabotite. Hidrobiološki Zavod. Ohrid.   S. 1764 b A

Recueil des Travaux de la Station Marine d'Endoume. Marseille.
Recl Trav.Stn mar.Endoume 1949-1969   S. 899 C
Fasc.Hors Série Suppl. 1962-1970   S. 899 D

TITLE   SERIAL No.

Recueil de Voyages et de Mémoires de la Société de
   Géographie. Paris.
     Rec.Soc.géogr.Paris 1824-1864.   S. 938 C

Recueil Zoologique Suisse. Geneve.
     Recl zool.suisse 1884-1892.   T.R.S 1208 & Z.S 1280

Redia. Firenze.
     Redia 1903 →   E.S 1105

Redogörelse. Kungliga Universitets i Uppsala.
     See Kungliga Universitetets i Uppsala Redogörelse.   S. 596 A

Redwing. St. Andrew's College Natural History Society. Grahamstown.
     Redwing 1954 →   S. 2022

Referati. Kongres Geologa Jugoslavije.
     See Kongres Geologa Jugoslavije. Sarajevo.   P.S 459

Referativnyĭ Zhurnal. Moskva.
     Referat.Zh. Biologiya (current year only.)   REF.ABS.95
                  Geologiya 1956 (imp.) 1957 →   P. REF.
                  Biologiya, K. Zooparasit. 1963 →   Z. Asch. Section
     Nauchnaya Tekhnicheskaya Informatsiya (English
     translation) See Abstract Journal. Scientific &
     Technical Information. Moscow.

Referatȳ Nauchno-Issledovatel'skikh Rabot. Akademiya Nauk SSSR. Moskva.
   Otdel.Geol.-Geogr.Nauk.
     Ref.nauchno-issled.Rab.Akad.Nauk SSSR 1945. (1947).   P.S 1542 A

Referatȳ Nauchnȳkh Rabot Instituta Biologiya Morya. Vladivostok.
     Ref.nauchnȳkh Rabot Inst.Biol.Morya Vladivostok 1969.
       Continued as Nauchnȳe Soobshcheniya Instituta
     Biologii Morya. Vladivostock.   Z.S 1821 A

Reference Lists. Natural History Division, Provincial Museum
   and Archives of Alberta. Edmonton.
     Ref.List nat.Hist.Div.prov.Mus.Arch.Alberta 1968 →   S. 2609 D

Regionalna Geologia Polski. Polski Towarzystwo Geologiczne. Kraków.
     Reg.Geol.Pol. 1951.   P.S 566

Regional'naya Stratigrafiya Kitaya, Moskva.
     Reg.Stratigr.Kitaya 1960 →   P.S 74 H.q.S

Regional'naya Stratigrafiya SSSR. Akademiya Nauk SSSR. Moskva.
     Reg.Stratigr.SSSR 4 → 1960 →   P.S 1500 c

Registro de Datos Oceanograficos y Meteorologicos. Cumaná.
     Registro Datos oceanogr.met. 1971 →   S. 2202 f E

Regnum Vegetabile. Utrecht.
     Regnum veg. 1952 →   B. See Author Catalogue

Regulus. Bulletin de la Ligue Luxembourgeoise pour l'Etude
   et la Protection des Oiseaux, Luxembourg.
     Regulus 1957 → (imp.)   T.B.S 781

| TITLE | SERIAL No. |

Reichenbachia. Staatliches Museum für Tierkunde in Dresden.
    Reichenbachia 1962 →   E.S 1370

Reimpresion Serie Geologia. Facultad de Ciencias Exactas
y Naturales. Universidad de Buenos Aires.
    Reimpr.Ser.geol.Fac.Cienc.exact.nat.Univ.B.Aires 1964 →   P.S 961

Reinwardtia. Kebrun Raya.
    Reinwardtia 1950 →
    Formerly Bulletin of the Botanic Gardens, Buitenzorg.   B.S 1800 b

Reise in Ostafrika.
    See Wissenschaftliche Ergebnisse. Reise in Ostafrika
    1903-1905. Stuttgart.   70.q.V

Relations Annuae Instituti Geologici Publici Hungarici. Budapest.
    See Evi Jelentés a Magyar Kir Földtani Intézet. Budapest.   P.S 1366

Relatório Anual do Departamento de Botânica do Estado. Sao Paulo.
    Relat.a Dep.Bot., S.Paulo 1939-1941
    Continued as Relatorio Anual do Instituto de Botânica.
    São Paulo.   B.S 3072

Relatório Anual. Divisão de Geologia e Mineralogia. Brasil.
Rio de Janeiro.
    Relat.a.Div.geol.miner.Bras. 1939 →
    Formerly Relatório Anual Servico Geologico e Mineralogico
    do Brasil. Rio de Janeiro.   P.S 2013

Relatório Anual do Instituto de Botanica. São Paulo.
    Relat.a Inst.Bot., S.Paulo 1942-1951.
    Formerly Relatório Anual do Departamento de Botânica
    da Estado. São Paulo.   B.S 3072

Relatório Anual. Museu Nacional. Rio de Janeiro.
    Relat.a.Mus.nac.Rio de J. 1956-1958, 1962-1963.
    (1956 was published as: Publicacões Avulsas No.19.)   S. 2213 F

Relatório Anual do Serviço Geologico e Mineralogico do Brasil.
Rio de Janeiro.
    Relat.a.Serv.Geol.Min.Bras. 1922-1938.
    Continued as Relatório Anual. Divisão de Geologia e
    Mineralogia. Brasil. Rio de Janeiro.   P.S 2013

Relatório. Commissão de Linhas Telegraphicas Estrategicas
de Matto Grosso ao Amazonas. Rio de Janeiro.
    See Publicacoes. Commissão de Linhas Telegraphicas
    Estrategicas de Matto Grosso ao Amazonas. Rio de Janeiro.   76 D.o & q.B

Relatório. Conselho Nacional de Pesquisas.
    Relat.Cons.nac.Pesq. 1964 →   S. 2213 a B

Relatório. Museu Nacional. Rio de Janeiro.
    Relat.Mus.nac.Rio de J. 1919-1920.   S. 2213 G

Relatórios e Comunicacões do Instituto de Investigacão Cientifica
de Angola. Luanda.
    Relat.Comuncões Inst.Invest.cient.Angola 1962 →   S. 2090 a C

| TITLE | SERIAL No. |

Relazioni e Monografie Agrario-Coloniali. Firenze.
  Relaz.Monogr.agr.-colon. Nos.5, 7, 12-17 1915-1930.   S. 1124

Relazioni Scientifiche della Spedizione Italiana de Filippi,
  nell'Himalaia, Caracorum e Turchestan Cinese (1913-1914).
  Bologna.
  Relaz.scient.Sped.Ital.Filippi Himalaia 1922-1934.   73.q.K

Remote Sensing of Environment. New York.
  Remote Sens.Environ. 1969 →   S. 2503

Rendiconti. Accademia Nazionale dei XL. Roma.
  Rc.Accad.naz.XL 1950 →
  Formerly Rendiconti. Societá Italiana delle Scienze detta
  Accademia dei XL. Roma.   S. 1142 A

Rendiconti dell' Accademia delle Scienze Fisiche e Matematiche.
  Napoli.
  Rc.Accad.Sci.fis.mat.Napoli 1842 →   S. 1106 A

Rendiconti. Accademia delle Scienze dell'Istituto di Bologna.
  See Rendiconti delle Sessioni dell'Accademia delle Scienze
  dell'Istituto di Bologna, and Atti dell'Accademia delle
  Scienze dell'Istituto di Bologna. Classe di Scienze
  Fisiche, Rendiconti.   S. 1103 A-B

Rendiconti dell'Istituto Lombardo di Scienze e Lettere. Milano.
  Rc.Ist.lomb.Sci.Lett. 1864 →
  Classe di Scienze Matematiche e Naturali: Series I, II & III.
  From 1957 onwards issued in sections. A. Scienze Matematiche,
  Fisiche, Chimiche e Geologiche.   S. 1104 E

Rendiconti e Memorie dell'Accademia di Scienze,Lettere ed
  Arti degli Zelanti. Acireale.
  Rc.Mem.Accad.Sci.Lett.Arti Zelanti 1901-1936.
  From 1927-1929 Styled Memorie della Reale Accademia
  di Scienze, Lettere ed Arti degli Zelanti. Acireale.
  Formerly Atti e Rendiconti della Reale Accademia di Scienze
  Lettere ed Arti degli Zelanti. Acireale.   S. 1113

Rendiconti e Memoire della R. Accademia di Scienze Lettere ed Arti
  degli Zelanti. Acireale.
  See Rendiconti e Memorie dell'Accademia di Scienze,Lettere
  ed Arti degli Zelanti. Acireale.   S. 1113

Rendiconti della R. Accademia d'Italia.
  See Atti della Reale Accademia d'Italia. Rendiconti. Roma.   S. 1107 C

Rendiconti. R. Accademia (Nazionale) dei Lincei. Roma.
  See Atti della Reale Accademia (Nazionale) dei Lincei.   S. 1107 C & D

Rendiconti della R. Accademia delle Scienze Fisiche e
  Matematiche. Napoli.
  See Rendiconti dell'Accademia delle Scienze Fisiche e
  Matematiche. Napoli.   S. 1106 A

Title	Serial No.
Rendiconti delle Sessioni dell'Accademia delle Scienze dell'Istituto di Bologna. Rc.Sess.Accad.Sci.Ist.Bologna 1851-1907: Classe di Scienze Fisiche 1907-1954. (From 1837-1851 published in Nuovi Annali delle Scienze Naturali. Bologna at S. 1184.)	S. 1103 A
Rendiconti. Società Italiana delle Scienze detta Accademia dei XL. Roma. Rc.Soc.ital.Sci.Accad.XL 1948-1949. Formerly Memorie della Società Italiana delle Scienze detta dei XL. Roma. Continued as Rendiconti. Accademia Nazionale dei XL. Roma.	S. 1142 A
Rendiconti della Società Mineralogica Italiana. Pavia. Rc.Soc.miner.,ital. 1941 →	M.S 1102
Rendiconti. Società Nazionale di Scienze, Lettere ed Arti in Napoli. See Rendiconti dell'Accademia delle Scienze Fisiche e Matematiche. Napoli.	S. 1106 A
Rendiconti. Società Reale Borbonica. See Rendiconti dell'Accademia delle Scienze Fisiche e Matematiche. Napoli.	S. 1106 A
Rendiconti Società Reale di Napoli. See Rendiconti dell'Accademia delle Scienze Fisiche e Matematiche. Napoli.	S. 1106 A
Rendiconti delle Tornate dell'Accademia Pontaniana. Napoli. Rc.Accad.pontan. 1853-1875.	S. 1150 B
Rendiconti. Unione Zoologica Italiana. See Bollettino di Zoologia. Pubblicato dall' Unione Zoologica Italiana.	Z.S 1180 A
Rendiconto. See Rendiconti.	
Renner Research Report, Hoblitzelle Agricultural Laboratory, Texas Research Foundation. Renner. Renner Res.Rep. 1968-1972.	E.S 2490 E
Répertoire de Chimie Appliquée. Paris Rép.Chim.appl. 1859-63	M.S 908
Répertoire de Chimie Pure.. Paris. Rép.Chim.pure 1858-62	M.S 908
Repertorio Italiano per la Storia Naturale. Bologna. Reprio ital.Stor.nat. 1854.	S. 1182 a
Repertorium für Anatomie und Physiologie. Berlin. Reprium Anat.Phys.Berlin 1836-1845.	S. 1625
Repertorium Entomologicum. Berlin. Reprium ent. 1924-1933. Formerly Entomologische Literaturblätter. Berlin.	E.S 1308

TITLE	SERIAL No.

Repertorium zum Neuen Jahrbuch (und Zentralblatt) für Mineralogie,
   Geologie und Palaeontologie.
    <u>Reprium Neuen Jb.Min.Geol.Palaeont.</u> 1870 →   M.S 1304 & P. REF.
    <u>Formerly</u> Allgemeines Repertorium der Mineralogie,
   Geognosie, Geologie und Petrefakten-Kunde.

Repertorium des Neuesten und Wissenswürdigsten aus der
   gesammten Naturkunde. Berlin.
    <u>Reprium Neuest.WissWürd.ges.Naturk.</u> 1811-1813.   S. 1626

Repertorium Novarum Specierum Regni Vegetabilis. Berlin.
    <u>Reprium nov.Spec.Regni veg.</u> 1905-1909.
      Beihefte - <u>See</u> Beihefte zum Novarum Specierum Regni
      Vegetabilis. Berlin.
      Sonderhefte - <u>See</u> Sonderhefte zum Repertorium Novarum
      Specierum Regni Vegetabilis. Berlin.
    <u>Continued as</u> Repertorium Specierum Novarum Vegetabilis.
   Berlin.   B.S 904

Repertorium Plantarum Succulentarum. Leeds, Utrecht.
    <u>Reprium Pl.succul.</u> 1950 →
   (From 1956 published as part of Regnum Vegetabile.)   B.S 23

Repertorium Specierum Novarum Regni Vegetabilis. Berlin.
    <u>Reprium Spec.nov.Regni veg.</u> Bd.8-51, 1910-1942.
    <u>Formerly</u> Repertorium Novarum Specierum Regni Vegetabilis.
   Berlin.
    <u>Continued as</u> Feddes Repertorium Specierum Novarum
   Vegetabilis. Berlin.   B.S 904

Report & Abstracts of Proceedings of the Croydon
   Microscopical Club.
    <u>Rep.Abstr.Proc.Croydon microsc.Club</u>   1871-1877.
    <u>Continued as</u> Proceedings and Transactions of the Croydon
   Microscopical & Natural History Club.   S. 28

Report. Aberdeen University Ghana Expeditions to Mole National Park.
   Aberdeen.
    <u>Rep.Aberdeen Univ.Ghana Exped.Mole natn.Park</u> 1974 →   S. 97 C

Report. The Academy of Natural Sciences of Philadelphia.
    <u>Rep.Acad.nat.Sci.Philad.</u> 1920-1922.
    <u>Formerly contained in</u> Proceedings. Academy of Natural
   Sciences of Philadelphia.
    <u>Continued as</u> Yearbook. The Academy of Natural Sciences
   of Philadelphia.   S. 2305 E

Report of the Acclimatisation Society of New South Wales. Sydney.
    <u>Rep.Acclim.Soc.N.S.W.</u> Nos.1, 3, 4, 6, 1862-1867.   Z.S 2134

Report & Accounts. Herbert Whitley Trust. Paignton.
    <u>Rep.Accts H.Whitley Trust</u> 1957-1968.   S. 316
    <u>Continued as</u> Report. Herbert Whitley Trust. Paignton.   S. 316 B
    <u>and</u> Report & Accounts. Paignton Zoological and
   Botanical Gardens. Paignton.   S. 316 C

Report & Accounts. Paignton Zoological and Botanical Gardens.
   Paignton.
    <u>Rep.Accts Paignton zool.bot.Gdn</u> 1969 →   S. 316 C
    <u>Formerly included in</u> Report & Accounts. Herbert Whitley
   Trust. Paignton.   S. 316

Report and Accounts. Wildlife and Nature Protection Society
   of Ceylon. Colombo.
    <u>Rep.Accts Wildl.Nat.Prot.Soc.Ceylon</u> 1971/72 →
    <u>Formerly</u> Report and Accounts. Wildlife Protection Society
   of Ceylon. Colombo.   Z.S 1943 A

| TITLE | SERIAL No. |

Report and Accounts. Wildlife Protection Society of Ceylon. Colombo.
    Rep.Accts Wildl.Prot.Soc.Ceylon 1958/59, 1967/68 - 1970/71.
    Continued as Report and Accounts. Wildlife and Nature
    Protection Society of Ceylon. Colombo.   Z.S 1943 A

Report. Acting Secretary for Mines, Tasmania. Hobart. (1939 only)
    See Report. Secretary for Mines, Tasmania. Hobart.   M.S 2444

Report & Addresses. Worcester Natural History Society.
    Rep.Worcs.nat.Hist.Soc. 1833-1866.   S. 400

Report. Adhesives Research Committee. D.S.I.R. London.
    Rep.Adhes.Res.Comm.D.S.I.R. 1922-1932.   S. 205 B

Report on Administration of Chosen.
    Rep.Admin.Chosen 1930-1932.
    Formerly Report on Reforms and Progress in Chosen.   S. 1977 b

Report of the Administrator of Agricultural Research, U.S.
    Department of Agriculture. Washington.
    Rep.Admr agric.Res.U.S. 1943-1953   E.S 2458r

Report of the Advisory Committee on Sand and Gravel. Ministry
    of Town and Country Planning. London.
    Rep.advis.Comm.Sand & Gravel 1948-1954.   P.S 1031

Report of the Advisory Committee for the Tropical Diseases
    Research Fund. London.
    Rep.advis.Comm.trop.Dis.Res.Fd Lond. 1906-1914.   O. 72 Aa.O.G

Report of the Advisory Council of the Science Museum. London.
    Rep.advis.Coun.Sci.Mus. 1930-1938.
    Formerly Report on the Science Museum. London.   S. 241 E

Report of the Advisory Council of Scientific und Industrial
    Research of Alberta. Edmonton.
    Rep.advis.Coun.scient.ind.Res.Alberta 1920.
    Continued as Report of the Scientific and Industrial Research
    Council of Alberta. Edmonton.   S. 2611

Report. Advisory Council of Scientific and Industrial
    Research, Canada. Ottawa.
    Rep.advis.Coun.Scient.ind.Res.Can. 1918-1925
    Continued as Report of the National Research Council
    of Canada. Ottawa.   S. 2602 D

Report of the Advisory Council for Scientific and Industrial
    Research of Canada. Ottawa. Annual Report.
    Rep.advis.Coun.scient.ind.Res.Can.a.Rep. 1917-1924.
    Continued as Report of the National Research Council of
    Canada. Ottawa. Annual Report.   S. 2602 B

Report of the Advisory Council on Scientific Policy. London.
    Rep.advis.Coun.scient.Policy 1947-1964.   S. 172 a

Report. Africana Museum. Johannesburg.
    Rep.afr.Mus. 1938 → (imp.)   S. 2008 B

Report of the Agricultural Deparment, Antigua. St. John.
    Rep.agric.Dep.Antigua 1915-1939.   E.S 2382

Report of the Agricultural Department, British Virgin Islands.
    Tortola.
    Rep.agric.Dep.Br.Virgin Isl. 1914-1923.   E.S 2381f

Report. Agricultural Department, Dominica.
    See Report. Agricultural and Forestry Department,Dominica.E.S 2381 b

TITLE	SERIAL No.

Report of the Agricultural Department. Gold Coast. Accra.
   Rep.agric.Dep.Gold Cst 1927-1954.
   Continued as Report of the Department of Agriculture, Ghana. E.S 2174a

Report on the Agricultural Department, Grenada. St. George.
   Rep.agric.Dep.Grenada 1912-1940 (imp.).   E.S 2381 c

Report on the Agricultural Department, Montserrat. Plymouth.
   Rep.agric.Dep.Montserrat 1917-1942.   E.S 2381e

Report on the Agricultural Department, Nigeria. Lagos.
   Rep.agric.Dep.Nigeria 1921 →   E.S 2172

Report on the Agricultural Department, St. Kitts & Nevis. Bridgetown.
   Rep.agric.Dep.St.Kitts & Nevis 1914-38
   Formerly Report on the Botanic Stations, St.Kitts & Nevis.   E.S 4120

Report. Agricultural Department, St. Lucia.
   Rep.agric.Dep.St.Lucia 1911-1945.   E.S 2381

Report. Agricultural Department, St. Vincent, Kingstown.
   Rep.agric.Dep.St.Vincent 1911 →   E.S 2381 a

Report. Agricultural Experiment Station, California.
   See Report of the California College of Agriculture and Agricultural Experiment Station. Berkeley.   E.S 2467 b

Report. Agricultural Experiment Station, Cornell University.
   See Report. Cornell University College of Agriculture and Experiment Station. Ithaca.   E.S 2470 a

Report of the Agricultural Experiment Station, Hawaii.
   See Report of the Hawaii Agricultural Experiment Station. Washington, Honolulu.   E.S 2276 a

Report. Agricultural Experiment Station, Kansas State College of Agriculture.
   See Report of the Kansas Agricultural Experiment Station, Topeka.   E.S 2475 a

Report. Agricultural Experiment Station, New Hampshire.
   See Report of the New Hampshire Agricultural Experiment Station. Durham.   E.S 2486

Report. Agricultural Experiment Station. Rio Piedras.
   See Report. Porto Rico Insular Agricultural Experiment Station. Rio Piedras.   E.S 2393

Report. Agricultural Experiment Station, University of Florida.
   See Report of the Florida Agricultural Experiment Station. Gainesville.   E.S 2471 b

Report. Agricultural Experiment Station, University of Minnesota.
   See Report of the Minnesota Agricultural Experiment Station. St. Paul.   E.S 2482 a

| TITLE | SERIAL No. |

Report. Agricultural Experiment Station, University of Wyoming.
    See Report of the Wyoming Agricultural Experiment
    Station. Laramie.   E.S 2496 c

Report on Agricultural Experiment Stations. United States
    Department of Agriculture. Washington.
    Rep.agric.Exp.Stns U.S.Dep.Agric. 1913-1942 (imp.)   E.S 2448a

Report of the Agricultural Extension Division, Ministry of
    Agriculture, Forests and Wildlife, Tanzania. Dar es Salaam.
    Rep.agric.Ext.Div.Minist.Agric.Forests Wildl.Tanzania
    1963 →   E.S 2176 A

Report. Agricultural and Forestry Department, Dominica.
    Rep.agric.For.Dep.Dominica 1912 →   E.S 2381b

Report of the Agricultural & Horticultural Research Station,
    University of Bristol. Long Ashton.
    Rep.agric.hort.Res.Stn Univ.Bristol 1913 →   E.S 71

Report of the Agricultural Research Council. London.
    Rep.agric.Res.Coun. 1931-1935.   E.S 69

Report. Agricultural Research Institute. Peshawar.
    Rep.agric.Res.Inst.Peshawar 1966-1967 →   E.S 2021

Report of the Agricultural Research Institute and College,
    Pusa. Calcutta.
    Rep.agric.Res.Inst.Coll.Pusa 1907-1916.
    Continued as Scientific Reports Agricultural Research
    Institute, Pusa. Calcutta.   E.S 1996

Report on the Agricultural Work in the Botanical Gardens.
    Georgetown.
    Rep.Agric.Wk bot.Gdns Georgetown 1890-1891/1892   B.S 3090

Report. Air Pollution Research Group. Council for Scientific and
    Industrial Research. Pretoria.
    Rep.Air Poll.Res.Grp C.S.I.R.Pretoria 1970/71 →   S. 2064

Report. Alabama Museum of Natural History. Alabama.
    Rep.Ala Mus.nat.Hist. 1965.   S. 2512 B

Report of Alaska Agricultural Experiment Stations. Washington.
    Rep.Alaska agric.Exp.Stns 1914-1926.   E.S 2461a

Report. Alaska (Board) Department of Fish & Game. Jeneau, Alaska.
    Rep.Alaska Dep.Fish Game No.10-11, 1958-1960.
    Continued as Progress Report. Alaska (Board) Department
    of Fish & Game. Juneau.   Z.S 2596

Report of the Albany Museum. Grahamstown.
    Rep.Albany Mus. 1882-1913.   S. 2021 C

Report. Allan Hancock Atlantic Expedition. Los Angeles.
    Rep.Allan Hancock Atlant.Exped. 1942-1964.   S. 2404 b B

TITLE	SERIAL No.

Report. Allan Hancock Foundation of the University of
    Southern California. Los Angeles.
    Rep.Allan Hancock Fdn 1952-1957.   S. 2404 b F

Report. Allan Hancock Pacific Expeditions. Los Angeles.
    See Allan Hancock Pacific Expeditions.   S. 2404 b E

Report. Alvecote Pools Nature Reserve, Warwickshire. Birmingham.
    Rep.Alvecote Pools Nat.Reserve 1959, 1961 →   S. 387 a

Report of the American Association of Museums.
    Washington, D.C.
    Rep.Am.Ass.Mus. 1925-1935.
    Continued as Summary of Reports. American Association
    of Museums. Washington, D.C.   S. 2302 D

Report. American Malacological Union (and) AMU Pacific Division.
    Buffalo.
    Rep.Am.malac.Un.AMU Pacif.Div. No.27-37, 1960-1970.
    Continued as Bulletin. American Malacological Union.
    (Seaford, N.Y.)   Z. Mollusca Section

Report of the American Museum of Natural History. New York.
    Rep.Am.Mus.nat.Hist. 1870 →   S. 2356 A
                         No.28-56, 1897-1925.   T.R.S 5112 C

Report. Animal Research Laboratories, Commonwealth Scientific and
    Industrial Research Organization, Australia. Melbourne.
    Rep.Anim.Res.Lab.CSIRO 1961-1966.
    Continued as Report. Division of Animal Health, Division
    of Animal Physiology and Research Report, Divison of
    Nutritional Biochemistry, CSIRO, Australia.   Z.S 2112 C

Report of the Anniversary of the Microscopical Society
    of London.
    Rep.Anniv.microsc.Soc.Lond. 1841 & 1852.   S. 415 A

Report of Annual Meeting. New Zealand Ecological Society.
    Wellington.
    Rep.a.Meet.N.Z.ecol.Soc. 1953.
    Formerly Report of Ecological Conference. New Zealand
    Ecological Society. Wellington.
    Continued in Proceedings of the New Zealand Ecological
    Society. Wellington.   S. 2166

Report. Anti-Locust Research Centre. London.
    Rep.Anti-Locust Res.Centre 1961/1964 - 1970/1971.   E.S 42 b

Report of the Antiqua Agricultural Department.
    See Report of the Agricultural Department, Antigua.   E.S 2382

Report on the Antiquities Service and Museums. Government of Sudan.
    Khartoum.
    Rep.Antiq.Serv.Mus.Sudan 1938-1957.
    Formerly Report of the Archaeological and Museums Board.
    Anglo-Egyptian Sudan. Khartoum.   S. 2045 a

Report. Applied Scientific Research Corporation of Thailand. Bangkok.
    Rep.appl.scient.Res.Corp.Thailand 1964 →   S. 1914 c B

TITLE	SERIAL No.

Report of the Archaeological and Museums Board. Anglo-Egyptian
    Sudan. Khartoum.
        Rep.archaeol.Mus.Bd Sudan 1934.
        Continued as Report of the Antiquities Service and Museums.
    Government of Sudan. Khartoum.   S. 2045 a

Report of the Archives Department. Mauritius. Port Louis.
        Rep.Arch.Dep.Mauritius 1971 →   S. 2092 a

Report. Arctic Institute of North America. Montreal.
        Rep.Arctic Inst.N.Am. 1968/1969 →   S. 2640 B

Report. Area Museums Service for South Eastern England.
    Hemel Hempstead.
        Rep.Area Mus.Serv.s.-east.Engl. 1971/1972 →   S. 115 a

Report. Argentine-Chilean Boundary. London.
        Rep.Argentine-Chilean Bound. 1900.   76.q.A

Report of the Arkansas Geological Survey. Little Rock.
        Rep.Ark.geol.Surv.   1888-1892 (imp.)   P.S 1875

Report. Art, Historical and Scientific Association.
    Vancouver, B.C.
        Rep.Art Hist.Sci.Ass.Vancouver 1939: 1945-1958.   S. 2659 B

Report. Arthropod-Borne Virus Epidemiology Unit, Microbiological
    Research Establishment. Porton Down.
        Rep.Arthrop.-Borne Virus Epidem.Unit microbiol.
        Res. Establ. 1966-1967 →   E.S 96

Report. Ashmolean Natural History Society of Oxfordshire.
        Rep.Ashmol.nat.Hist.Soc. 1901-1907.
        Formerly Report. Oxfordshire Natural History Society
    & Field Club.
        Continued as Proceedings & Report. Ashmolean Natural
    History Society of Oxfordshire.   S. 313 A

Report. Asiatic Society. Calcutta.
        Rep.Asiat.Soc. 1971 →   S. 1902 M

Report of the Association of American Geologists and Naturalists. Boston.
        Rep.Ass.Am.Geol.& Nats 1840-1842. (1843)   P.S 854

Report of the Association of Public School Science Masters.
        Rep.Ass.Publ.Sch.Sci.Masters 1909-1918.
        Continued as Report of Science Masters' Association.   S. 2 a

Report of the Astrakhan Ichthyological Laboratory.
        See Trudy Astrakhanskoi Nauchnoĭ Rybokhozyaĭstvennoĭ
    Stantsii.   Z.S 1830

Report of the Astrakhan Scientific Fishery Station.
        See Trudy Astrakhanskoi Nauchnoĭ Rybokhozyaĭstvennoĭ
    Stantsii.   Z.S 1830

TITLE	SERIAL No.

Report. Attenborough Ringing Group. Nottingham.
    Rep.Attenbor.Ring.Grp 1968 →
    Formerly Attenborough Ringing Report. Nottingham.   T.B.S 267

Report. Auckland Institute and Museum.
    Rep.Auckland Inst.Mus. 1918 →   S. 2185 C

Report. Australasian Association for the Advancement of Science. Sydney.
    Rep.Austalas.Ass.Advmt Sci. 1888-1928.
    Continued as Report. Australian and New Zealand Association for the Advancement of Science. Sydney.   S. 2101 A

Report of the Australian Academy of Science. Canberra.
    Rep.aust.Acad.Sci. 1957 →
    (wanting No.2, 4, 5 and 6)   S. 2148 B

Report. Australian Conservation Foundation. Eastwood, N.S.W.
    Rep.Aust.Conserv.Found. 1967-1968 →   S. 2122 a D

Report of the Australian Museum. Sydney.
    Rep.Aust.Mus. 1863 →   S. 2126 F

Report. Australian National Antarctic Research Expeditions. Melbourne.
    See A.N.A.R.E. Reports   80.q.A

Report. Australian and New Zealand Association for the Advancement of Science. Sydney.
    Rep.Aust.N.Z.Ass.Advmt Sci. 1930-1954.
    Formerly Report. Australasian Association for the Advancement of Science. Sydney.
    Continued in Australian Journal of Science.   S. 2101 A

Report. Avon Biological Research. University College, Southampton
    Rep.Avon.biol.Res. 1932-1938.   Z.S 128

Report. B.A.N.Z. Antarctic Research Expedition 1929-1931. Adelaide.
    Rep.B.A.N.Z.Antarctic Res.Exped. 1937 →   80.q.B

Report. B.I.O.S. (British Intelligence Objectives Sub-Committee). London.
    See B.I.O.S. Survey. British Intelligence Objectives Sub-Committee. Final Report. London.   M.S 145

Report & Balance Sheets. Huddersfield Naturalist, Photographic & Antiquarian Society.
    Rep.Huddersfield Nat.photogr.Soc. 1891-1901, 1918-1919.
    (The Reports for 1891-1901 were published in the Monthly Magazine.)   S. 122 C

Report. Balham & District Antiquarian and Natural History Society.
    Rep.Balham nat.Hist.Soc. Nos.8-9, 13-18. 1905-1916.   S. 10 A

Report of the Banding Committee, Ornithological Society of New Zeland. Wellington, N.Z.
    Rep.Banding Comm.orn.Soc.N.Z. No.9 → 1959 →
    Formerly included in Notornis.   Z. Bird Section

TITLE	SERIAL No.

Report. Bardsey Bird and Field Observatory. Eglwysfach.
    Rep.Bardsey Bird Fld Obs. 1953 →   Z.S 363

Report. Barrow Naturalists' Field Club.
    See Report and Proceedings. Barrow Naturalists' Field Club.   S. 9

Report. Bath Royal Literary and Scientific Institution.
    Rep.Bath lit.scient.Instn 1925-1938.   S. 12 a

Report. Beaudette Foundation for Biological Research.
    Solvang. California.
    Rep.Beaudette Fdn biol.Res. 1958-1963.   S. 2390 B

Report of the Belfast Natural History and Philosophical Society.
    Rep.Belfast nat.Hist.phil.Soc. 1837-1878. (imp.)   S. 14 D

Report. Belfast Naturalists' Field Club.
    Rep.Belfast Nat.Fld Club No.3-10, 1865-1873.
    Continued as Report & Proceedings of the Belfast
    Naturalists' Field Club.   S. 15

Report. Berkshire, Buckinghamshire and Oxfordshire Naturalists'
    Trust. Oxford.
    Rep.Berks.Bucks.Oxfs.Nat.Trust 1961 →   S. 315 a B

Report. Bernice Pauahi Bishop Museum. Honolulu.
    Rep.Bernice Pauahi Bishop Mus. 1899 →
    (From 1899-1921 published in Occasional Papers,
    1922-1953 in Bulletin.)   S. 2175 A, D-E.

Report. Biological Board of Canada. Ottawa.
    Rep.biol.Bd Can. 1930-1937.
    Continued as Report of the Fisheries Research Board
    of Canada.   Z.S 2628 F

Report of the Biological Bureau, Game and Fisheries Department,
    Québec.
    See Rapport de l'Office de Biologie, Ministère de la
    Chasse et des Pêcheries, Province de Québec.   S. 2638 A

Report. Biological Study of Estuarine and Marine Waters of
    Louisiana. Zoology Department, Tulane University.
    Rep.biol.Study estuar.mar.Wat.La No.11-14, 1955-1956.
    Formerly Report. Biological Study of Lake Pontchartrain.
    Zoology Department, Tulane University.   Z.S 2415 A

Report. Biological Study of Lake Pontchartrain. Zoology Department,
    Tulane University.
    Rep.biol.Study Lake Pontchartrain No.1-10, 1953-1955.
    Continued as Report. Biological Study of Estuarine and
    Marine Waters of Louisiana. Zoology Department,
    Tulane University.   Z.S 2415 A

Report. Biology Department, Brookhaven National Laboratory.
    Upton, N.Y.
    Rep.Biol.Dep.Brookhaven natn.Lab. 1957-1970.   S. 2413 a C
    Replaced by Brookhaven Highlights. Springfield, Va.   S. 2413 a D

TITLE   SERIAL No.

Report of the Bird Banding Committee, Ornithological Society
    of New Zealand. Wellington, N.Z.
    Rep.Banding Comm.orn.Soc.N.Z. No.9 → 1959 →
    Formerly included in Notornis.   T.B.S 7301 B

Report. Bird Observatory and Field Study Centre, Gibraltar
    Point, Lincs. Lincoln.
    Rep.Bird Obs.Fld Study Cent.Gibraltar Pt 1949-1953.
    Continued in Transactions of the Lincolnshire
    Naturalists' Union.   T.B.S 271

Report on Birds. Lancashire & Cheshire Fauna Committee. Burnley.
    Rep.Birds Lancs.Cheshire Fauna Comm. 1958-1959.
    Formerly Ornithological Report. Lancashire & Cheshire
    Fauna Committee.
    Continued as Lancashire Bird Report.   T.B.S 272

Report. Birds of Leicestershire and Rutland. Leicester.
    Rep.Birds Leicestersh.& Rutl. 1948 →
    Formerly Report on the Wild Birds of Leicestershire
    and Rutland.   T.B.S 260

Report of the Birds of Warwickshire, Worcestershire and
    South Staffordshire.
    Rep.Birds Warwicksh.Worcs.S.Staffs. No.13-14, 1946-1947.
    Continued as West Midland Bird Report.   T.B.S 258

Report. Birmingham and Midland Institute.
    Rep.Bgham Midl.Inst. 1906.   S. 17 b B

Report. Birmingham Natural History and Philosophical Society.
    Rep.Bgham nat.Hist.Phil.Soc. 1893-1949. (imp.)   S. 17 a A

Report. Birmingham and West Midland Bird Club.
    See Report. Birds of Warwickshire, Worcestershire
    and South Staffordshire.   T.B.S 258

Report. Bishop Museum. Honolulu.
    See Report. Bernice Pauahi Bishop Museum. Honolulu.   S. 2175 A, D-E.

Report of the Blakeney Point Committee of Management.
    Rep.Blakeney Pt Comm.Mgmt 1913, 1915-1919.   85 o.B

Report of the Board of Agriculture and Department of Public
    Gardens and Plantations, Jamaica. Kingston.
    Rep.Bd Agric.Jamaica 1901-1907.
    Formerly Report. Public Gardens & Plantations, Jamaica. Kingston.
    Continued as Report on the Department of Agriculture,
    Jamaica. Kingston.   B.S 4100 a

Report of the Board of Agriculture & Forestry, Hawaii. Division
    of Entomology. Honolulu.
    Rep.Bd agric.For.Hawaii. 1900-22 (imp.)   E.S 2277a

Report of the Board of Agriculture for Scotland. London.
    Rep.Bd Agric.Scotl. 1913-1928 (imp.)
    Continued as Report of the Department of Agriculture, Scotland. E.S 78

TITLE	SERIAL No.

Report of the Board of Irrigation, Survey & Experiment.
    Topeka. Kansas.
        Rep.Bd Irrig.Surv.& Exp. 1895-1896.   S. 2409

Report of the Board of Regents of the Smithsonian Institution.
    Washington.
        Rep.Smithson.Instn   1847-1964.   S. 2426 A
                              1888-1964.   T.R.S 5145 A
        Continued as   Smithsonian Year. Washington.

Report of the Board of Visitors. Dublin Museum of Science and Art.
        Rep.Bd Vis.Dubl.Mus.Sci.Art   1905-1907.
        Continued as   Report of the Board of Visitors. National
Museum of Science and Art and Royal Botanic Gardens, Dublin.   S. 48 a B

Report of the Board of Visitors. National Museum of Science and Art
    and Royal Botanic Gardens, Dublin.
        Rep.Bd Vis.natn.Mus.Sci.Art R.bot.Gdns Dubl.   1907-1928.
        Formerly   Report of the Board of Visitors. Dublin Museum
of Science and Art.   S. 48 a B

Report. Bodleian Library. Oxford.
        See   Report of the Curators. Bodleian Library. Oxford.   S. 310 C c

Report of the Bose Research Institute. Calcutta.
        Rep.Bose Res.Inst.   1952 →   B.S 1607 a

Report. Botanic Gardens Department, Colony of Singapore.
        Rep.bot.Gdns Dep.Singapore   1948 →
        Formerly   Report of the Botanic Gardens, Singapore
and Penang.   B.S 1746 a

Report of the Botanic Gardens and Government Herbarium. Cape Town.
        Rep.bot.Gdns Govt.Herb.Cape Tn   1884-1902.   B.Botanic Gardens,
                                                                               Vol.5, 58.006

Report of the Botanic Gardens, Singapore and Penang.
        Rep.bot.Gdns Singapore Penang   1889-1938 (imp.)
        Continued as   Report. Botanic Gardens Department,
Colony of Singapore.   B.S 1746 a

Report on the Botanic Stations, St. Kitts & Nevis.
        Rep.bot.Stns St.Kitts Nevis   1900-1914.
        Continued as   Report on the Agricultural Department,
St.Kitts & Nevis. Bridgetown.   E.S 4120

Report. Botanical Exchange Club. London.
        Rep.botl Exch.Club   1877-1878.
        Formerly   Report of the Curator. Botanical Exchange Club. London.
        Continued as   Report. Botanical Exchange Club of the
British Isles. Manchester.   B.B.H.S 97

Report. Botanical Exchange Club of the British Isles. Manchester.
        Rep.botl Exch.Club Br.Isl.   1879-1900.
        Formerly   Report. Botanical Exchange Club. London.
        Continued as   Report. Botanical Exchange Club and Society of
the British Isles. Manchester, &c.   B.B.H.S 97

| TITLE | SERIAL No. |

Report. Botanical Exchange Club and Society of the British Isles.
Manchester, &c.
Rep.botl Exch.Club Soc.Br.Isl.   1901-1913.
Formerly Report. Botanical Exchange Club of the British Isles.
Manchester.
Continued as Report. Botanical Society and Exchange Club of
the British Isles. Arbroath.   B.B.H.S 97

Report of the Botanical Garden in Tiflis.
See Trudy Tiflisskago Botanicheskago Sada. Tiflis.   B.S 1430

Report of the Botanical Office, British Columbia. Victoria.
Rep.botl Off.Br.Columb 1913-1915.   B.S 4500

Report. Botanical Society and Exchange Club of the British Isles.
Arbroath.
Rep.botl Soc.Exch.Club Br.Isl.   1914-1947.
Formerly Report. Botanical Exchange Club and Society of the
British Isles. Manchester, &c.
Replaced by Watsonia. Arbroath.   B.B.H.S 97

Report of the Botanical Survey of India. Calcutta.
Rep.botl Surv.India 1894 → (imp.)   B.S 1601 a

Report of Botany Branch and Queensland Herbarium. Queensland
Department of Primary Industries. Brisbane.
Rep.Bot.Brch Qd Herbm 1973/74 →
Formerly Report of Botany Section. Government Botanist.
Queensland.   B.S 2409 a

Report of Botany Section, Government Botanist, Queensland.
Brisbane.
Rep.Bot.Sect.Govt Bot.Qd 1966.
Continued as Report of Botany Branch and Queensland Herbarium.
Queensland Department of Primary Industries. Brisbane.   B.S 2409 a

Report of the Bournemouth Society of Natural Science
(Natural Science Society).
Rep.Bournemouth Soc.nat.Sci. 1904-1914.   S. 18 A

Report. Bradfield College Natural History Society.
Rep.Bradfield Coll.nat.Hist.Soc. 1955-1958/1959.   S. 32 a

Report. Brighton Free Library, Museum and Picture Gallery,
Royal Pavilion.
Rep.Brighton Free Libr.Mus.Pict.Gall. 1874.
Continued as Report. Brighton Public Museum.   85.o.B

Report. Brighton and Hove Natural History and Philosophical
Society.
Rep.Brighton Hove nat.Hist.phil.Soc. 1938-1954.
(Not issued 1941-1948 & 1950-1953.)
Formerly Abstracts of Papers. Brighton and Hove
Natural History and Philosophical Society.   S. 19

TITLE	SERIAL No.

Report. Brighton Public Library, Museums and Art Galleries.
    Rep.Brighton publ.Libr.Mus.Art Gall. 1904, 1911.
    Formerly Report. Brighton Public Museum.   85.o.B

Report. Brighton Public Museum.
    Rep.Brighton publ.Mus. 1891-1895
    Formerly Report. Brighton Free Library, Museum and Picture Gallery, Royal Pavilion.
    Continued as Report. Brighton Public Library, Museums and Art Galleries.   85.o.B

Report (and Abstracts of Proceedings and Papers) of the Brighton and Sussex Natural History and Philosophical Society.
    Rep.Brighton nat.Hist.Soc. 1855-1887.
    Continued as Abstracts of Papers. Brighton and Sussex Natural History and Philosophical Society.   S. 19

Report. Bristol Museum and Library (and Art Gallery.)
    Rep.Bristol Mus.Libr. 1891-1930 imp.   S. 20 a

Report of the British Association for the Advancement of Science. London.
    Rep.Br.Ass.Advmt Sci. 1831-1938.   S. 1 A
    1892, 1911-1915, 1933.   T.R.S 10
    Continued as Advancement of Science.   S. 1 A

Report (Annual). British Association for the Advancement of Science. London.
    Rep.a.Br.Ass.Advmt Sci. 1959/1960 →   S. 1 M

Report. British Bryological Society. London.
    Rep.Br.bryol.Soc. 1923-1946.
    Continued as Transactions British Bryological Society.
    Formerly Report. Moss Exchange Club.   B.B.S 4

Report of the British Columbia Department of Agriculture. Victoria.
    Rep.Br.Columb.Dep.Agric. 1959 →   E.S 2526c

Report. British Columbia Department of Recreation and Conservation. Victoria, B.C.
    Rep.Br.Columb.Dep.Recr.Conserv. 1970 →   S. 2658 A

Report. British Columbia. Minister of Mines.
    See Report. Minister of Mines, British Columbia. Victoria, B.C.   M.S 2713

Report. British Columbia Provincial Museum. Victoria, B.C.
    See Report. Provincial Museum (of Natural History and Anthropology), Province of British Columbia. Victoria, B.C.   S. 2658 A

Report. British Commonwealth Geological Liaison Office. London.
    Rep.Br.Commonw.geol.liaison Off. 57 (2) → 1957 →   P.S 172 A

| TITLE | SERIAL No. |

Report of the British Council. London.
    Rep.Br.Coun. 1944/1945 →   S. 173

Report. British Guiana Geological Survey.
    See Report of the Geological Survey Department,
    British Guiana.   P.S 1168 A

Report of the British Guiana Museum (and Zoo). Georgetown.
    Rep.Br.Guiana Mus. Nos.2 & 4-13. 1954-1957.
    Continued as Journal of the British Guiana Museum
    and Zoo. Georgetown.   S. 2205 B

Report of the British Guiana Museums and Georgetown Public
    Free Library.
    Rep.Br.Guiana Mus.Georgetown Publ.Free Libr. 1939-1940.   S. 2205 a A

Report. The British Library. London.
    Rep.Br.Lib. 1973/74 →   S. 273 a C

Report. British Mosquito Control Institute. Hayling Island.
    Rep.Br.Mosq.Control Inst. 1930-1932.
    Formerly Report. Hayling Mosquito Control.   E.S 40 a

Report. British Museum and British Museum (Natural History). London.
    Rep.Br.Mus.Br.Mus.nat.Hist. 1922-1938.
    Formerly Return. British Museum. London.
    Continued as Report of the Trustees. British Museum
    & Report on the British Museum (Natural History). London.   B.M.o

Report on the British Museum (Natural History). London.
    Rep.Br.Mus.nat.Hist. 1963-1965 →   SBM.4, P.S 186 E
                                                         & T.R.S 1 C
    For previous Reports See Report. British Museum &
    British Museum (Natural History). London.   SBM.4 & P.S 186 E

Report on British New Guinea. Melbourne.
    See British New Guinea, Annual Report.   S. 2143 B

Report. British Records Association. London.
    Rep.Br.Rec.Assoc. 1973/74 →   Archivist Room

Report. British Schools Exploring Society. London.
    Rep.Br.Sch.Explor.Soc. 1948 →   E.S 44 & S. 186

Report. British Scientific Instrument Research Association. London.
    Rep.Br.scient.Instrum.Res.Ass. No.2-3, 1919-1921.   83 o B

Report. British Standards Institution. London.
    Rep.Br.Stand.Instn 1969 →   S. 220 a B

Report. The British Trawlers' Federation. London.
    Rep.Br.Trawlers' Fed. 1964 →   Z.S 330

Report. British Trust for Ornithology. Oxford.
    Rep.Br.Trust Orn. 1935 →   T.B.S 103 B

Report. Brookhaven National Laboratory. Biology Department.
    Upton, N.Y.
    See Report. Biology Department, Brookhaven National Laboratory.
    Upton, N.Y.   S. 2413 a C

Report. Brooklyn Botanic Garden.
    Rep.Brooklyn bot.Gdn 1958/61 →   B.S 4306 d

TITLE	SERIAL No.

Report. Burnley Literary and Philosophical Society.
    Rep.Burnley lit.phil.Soc. 1895 & 1899.   S. 22 a A

Report of the Burton-on-Trent Natural History and
    Archaeological Society.
    Rep.Burton-on-Trent nat.Hist.archaeol.Soc. 1884-1892.   S. 23 A

Report. Bury Natural History Society.
    Rep.Bury nat.Hist.Soc. 1868-1871.   S. 23 a

Report. Bryanston School Natural History Society. Blandford
    Rep.Bryanston Sch.nat.Hist.Soc. 1946 →   S. 17 e

Report. Buffalo Society of Natural Sciences.
    Rep.Buffalo Soc.nat.Sci. 1927-1969.   S. 2326 C
    (The Reports for 1921-1926 published in a condensed
    form in "Hobbies".)
    Continued in Science on the March. Buffalo, New York.

Report and Bulletin. The Nyasaland Museum. Blantyre.
    Rep.Nyasaland Mus. 1960 →   S. 2053

Report and Bulletin of the Geological Survey Department, Uganda
    Protectorate. Annual Report. Entebbe.
    See Report of the Geological Survey Department, Uganda
    Protectorate. Annual Report. Entebbe.   P.S 1179 c

Report of the Bureau of American Ethnology. Smithsonian
    Institution. Washington.
    Rep.Bur.Am.Ethnol. 1879/1880 - 1963/1964.
    (Volume 16 wanting.)   P.A.S 800 A

Report of the Bureau of Animal Industry. U.S. Department
    of Agriculture. Washington, D.C.
    See Report of the United States Bureau of Animal Industry.
    Washington.   Z.S 2523 A

Report. Bureau of Animal Population. Oxford.
    Rep.Bur.Anim.Popul.Oxf. 1935-1938.   Z.S 129

Report Bureau of Applied Entomology. Leningrad.
    See Izvestiya Otdela Prikladnoĭ Entomologii. Leningrad.   E.S 1915

Report of the Bureau of Commercial Fisheries. Fish & Wildlife
    Service, U.S. Department of the Interior, Washington, D.C.
    Rep.Bur.comm.Fish. 1957 →   Z.S 2510 G

Report. Bureau of Economic Geology. University of Texas. Austin.
    Rep.Bur.econ.Geol.Univ.Tex. 1960 →   P.S 1978

Report (annual). Bureau of Mineral Resources, Geology and
    Geophysics. Canberra.
    Rep.a.Bur.miner.Resour.Geol.Geophys. 1971 →   P.S 1128 F

Report. Bureau of Mineral Resources. Geology and Geophysics.
    Commonwealth of Australia. Melbourne.
    Rep.Bur.miner.Resour.Geol.Geophys.Aust. 1947 →   P.S 1128 A

| TITLE | SERIAL No. |

Report of the Bureau of Mines. Ontario, Toronto.
    See Report of the Ontario Department of Mines. Toronto.   M.S 2708

Report of the Bureau of Mines, Quebec. Quebec.
    Rep.Bur.Mines, Quebec   1929-1930 (imp.)   M.S 2710

Report of the Bureau of Mines, Republic of the Philippines, Manila.
    Rep.Bur.Mines Repub.Philipp.   1952 →   M.S 2003

Report. Bureau of Mines, United States, Washington.
    Rep.Bur.Mines U.S.   1911-1915.   M.S 2620

Report of the Bureau of Science, Philippine Islands. Manila.
    Rep.Bur.Sci.Philipp.Isl.   Nos.5-36.   1905-1938.   S. 1976 C
                                            Nos.7-18.   1908-1919.   T.R.S 6520
    Formerly Report of the Superintendent of Government
    Laboratories in the Philippine Islands. Manila.   S. 1976 C

Report. Buxton Field Club. Buxton.
    Rep.Buxton Fld Club   1972 →   S. 30 c

Report. C.S.I.R. Australia.
    See Report of the Council for Scientific and Industrial
    Research. Commonwealth of Australia. Melbourne.   S. 2113 F

Report. CSIRO Division of Entomology. Canberra.
    Rep.CSIRO Div.Ent.   1965/66 →   E.S 2241

Report on Cacao Research. Imperial College of Tropical Agriculture,
    University of the West Indies. St. Augustine, Trinidad.
    Rep.Cacao Res.   1964 →   E.S 2378 b

Report. California Academy of Sciences. San Francisco.
    Rep.Calif.Acad.Sci.   1946 → (imp.)   S. 2401 H

Report of the California Archaeological Survey. Berkeley.
    Rep.Calif.archaeol.Surv.   1948-1949.
    Continued as Report of the University of California
    Archaeological Survey.   P.A.S 776 C

Report of the California College of Agriculture and Agricultural
    Experiment Station. Berkeley.
    Rep.Calif.Coll.Agric.   1895-1938 (imp.)   E.S 2467 b

Report. California Cooperative Oceanic Fisheries Investigations.
    (Sacramento).
    Rep.Calif.coop.oceanic.Fish.Invest.   7 → 1958-1959 →
    Formerly Progress Report. California Cooperative
    Oceanic Fisheries Investigations.   Z.S 2481

Report. Camberley Natural History Society.
    Rep.Camberley nat.Hist.Soc.   1958-1965.   S. 34 a

Report. Cambridge Bird Club.
    Rep.Camb.Bird Club   1930 →
    Formerly Report. Cambridge Ornithological Club.   T.B.S 256

Report. Cambridge Ornithological Club.
    Rep.Camb.orn.Club   1928-1929.
    Continued as Report. Cambridge Bird Club.   T.B.S 256

TITLE	SERIAL No.

Report of the Canadian Arctic Expedition 1913-1918. Ottawa.
    Rep.Can.arct.Exped.   1918-1928.   Z. 71 o C

Report of the Canadian Institute.
    Rep.Canad.Inst.   1886-1894.   S. 2605 C

Report. Canterbury College and Museum. Christchurch, N.Z.
    Rep.Canterbury Coll.Mus.   No.49-72   1921-1944 (imp.)   S. 2172

Report. Canterbury Museum. Christchurch.
    Rep.Canterbury Mus.   1948 →   S. 2171 B

Report of the Canterbury Philosophical and Literary Institution.
    Rep.Canterbury phil.lit.Instn   1827-1832.   S. 24 a A

Report. Cape Clear Bird Observatory. Cape Clear Island.
    Rep.Cape Clear Bird Obs.   No.2 →   1960 →   T.B.S 511

Report. Cape Department of Nature Conservation. Cape Town.
    See   Report. Department of Nature Conservation, Union of
    South Africa. Cape Town.   Z.S 2031

Report. Cardiff Naturalists' Society.
    See   Report and Transactions of the Cardiff Naturalists'
    Society.   S. 25 & T.R.S 463

Report. Caribbean Commission.
    Rep.Caribbean Commn
    West Indian Conference. 3-5 Session.   1949-1952.
      United States Section. 1948.   75 F.o.C

Report Caribbean Geological Conference.
    See   Report of the Meetings. Caribbean Geological Conference.   P.S 964

Report. Carlisle Natural History Society.
    Rep.Carlisle nat.Hist.Soc.   1948-1949.   S. 37 B

Report of the Carnegie Institute. Pittsburgh.
    Rep.Carnegie Inst.   1910-1911; 1914-1918.   S. 2373 F

Report of the Carnegie Museum. Pittsburgh, Pa.
    Rep.Carnegie Mus.   1898-1972.
    Continued as   Report. Carnegie Museum of Natural History.   S. 2373 A

Report. Carnegie Museum of Natural History (Pittsburgh, Pa.)
    Rep.Carnegie Mus.nat.Hist.   1973 →
    Formerly   Report of the Carnegie Museum.   S. 2373 A

Report. Carnegie United Kingdom Trust. Edinburgh &c.
    Rep.Carnegie U.K. Trust   1915 →   S. 68 A

Report of the Castle Museum Committee to the (Town) Council. Norwich.
    Rep.Castle Mus.   1894-1928.
    Continued as   Report of the Museum Committee to the
    City Council. Norwich.   S. 298 A

Report. Cawthron Institute for Scientific Research. Nelson, N.Z.
    Rep.Cawthron Inst.scient.Res.   1941-1961.
    Continued as   Biennial Report. Cawthron Institute.
    Nelson, N.Z.   S. 2182 A

Report. Center for Short-lived Phenomena. Smithsonian
    Institution. Cambridge, Mass.
    See   Report. Smithsonian Institution Center for Short-lived
    Phenomena. Cambridge, Mass.   P. REF

| TITLE | SERIAL No. |

Report of the Central Fisheries Department. Pakistan.
    Rep.cent.Fish.Dep.Pakist. 1951 →   Z.S 1931

Report. Centre for Overseas Pest Research. London.
    Rep.Centre overseas Pest Res. 1971/1972 →   E.S 42 B

Report on Cetacea Stranded on the British Coasts. London.
    Rep.Cetacea Br.Csts 1913 →   BM.Eb.17 I.q
                                                            Z.O 72Aa q B
                                                            & Cetacea Room
                                                              P.B.M.Eb

Report from the Ceylon Marine Biological Laboratory. Colombo.
    Rep.Ceylon mar.biol.Lab. 1905-1912.   S. 1920 a

Report. Challenger Society for the Promotion of the Study of
    Oceanography. London.
    Rep.Challenger Soc. 1903-1966.
    Continued as Report and Proceedings. Challenger Society.   S. 167 A

Report. Chamber of Mines Precambrian Research Unit. Department
    of Geology. University of Cape Town.
    Rep.Chamber Mines Precambr.res.Unit 1963 →   P.S 191 A

Report of the Cheltenham College Natural History Society.
    Rep.Cheltenham Coll.nat.Hist.Soc. 1895-1952.
      (Wanting Report for 1900 & 1903.)   S. 26

Report. Cheltenham and District Naturalists' Society. Cheltenham.
    Rep.Cheltenham Distr.Nat.Soc. 1951-1956
    Continued as Report. North Gloucestershire Naturalists'
    Society.   S. 40 a B

Report of the Chemical Branch, Mines Department, Western Australia,
    Perth.
    Rep.Chem.Brch Mines Dep.West.Aust. 1923-1940.
    Formerly & Continued as Report of the Government
    Mineralogist, Analyst and Chemist, Department of Mines,
    Western Australia.   M.S 2407 C

Report of the Chemical Laboratories, Department of Mines,
    Western Australia, Perth.
    See Report of the Government Mineralogist, Analyst
    and Chemist, Department of Mines, Western Australia.   M.S 2407 C

Report of the Chester Society of Natural Science
    Literature and Art.
    Rep.Chester Soc.nat.Sci. No.3-27. 1871-1898.
    Continued as Report and Proceedings Chester Society
    of Natural Science, Literature and Art.   S. 27 A

Report. Chicago Academy of Sciences.
    Rep.Chicago Acad.Sci. 1895-1897: 1938-1944.   S. 2329 G

Report. Chicago Natural History Museum.
    Rep.Chicago nat.Hist.Mus. 1943-1964.
    Formerly Publication. Field Museum of Natural History.
    Chicago. Report Series.
    Continued as Report. Field Museum of Natural History.
    Chicago.   S. 2330 A & T.R.S 5125 B

| TITLE | SERIAL No. |

Report. Chicago Zoological Society. Chicago.
    Rep.Chicago zool.Soc. 1968 →   Z.S 2331 A

Report. Chichester and West Sussex Natural History and
    Microscopical Society.
    Rep.Chichester nat.Hist.microsc.Soc. 1877-1882.
    Continued as Transactions of the Chichester and West Sussex
Natural History and Microscopical Society.   S. 27 a A

Report of the Chief Game Guardian Province of Saskatchewan. Regina.
    Rep.chf Game Guard.Sask. 1913-1914, 1916-1923.
    Continued as Report of the Game Commissioner.
Saskatchewan, Regina.   Z.O 75B o S

Report of the Chief Inspector of Mines, Mysore. Madras.
    Rep.chf Insp.Mines Mysore   1899-1914.
    Continued as Report. Mysore Department of Mines and
Geology (and Explosives). Bangalore.   M.S 1915

Report by the Chief Secretary on the Fisheries in New South Wales.
    Sydney.
    Rep.chf Secr.Fish.N.S.W. 1949-1952, 1954 →
    Formerly Report of the Department of Fisheries of
New South Wales.   Z.S 2106

Report of the Christ's Hospital (West Horsham) Natural
    History Society. Horsham.
    Rep.Christ's Hosp.nat.Hist.Soc. 1903-1945. (imp.)   S. 168 A

Report on the City Industrial Museum, Kelvingrove Park, Glasgow.
    Rep.Cy ind.Mus.Glasgow 1876, 1878-1879.
    Continued as Report on the Kelvingrove Museum and the
Corporation Galleries of Art, Glasgow.   S. 95

Report of the City of London College Science Society.
    Rep.Cy Lond.Coll.sci.Soc. 1885-1888.
    Continued as Journal of the City of London
College Science Society.   S. 169

Report. City Museums and Mappin Art Gallery. Sheffield.
    Rep.Cy Mus.Mappin Art Gall. 1932-1934.
    Formerly Report. Public Museums and Mappin Art Gallery.
City of Sheffield.
    Continued as   Report. Sheffield City Museums.   S. 358 a

Report of the Civil Service Department. London.
    Rep.Civil Serv.Dep. 1969 →   S. 204 a

Report of the Clifton Scientific Society. Clifton.
    Rep.Clifton Coll.scient.Soc. 1908-1914; 1926-1933;
1947-1949.   S. 27 b B

Report. Cocoa Research Institute, Ghana Academy of Sciences. Tafo.
    Rep.Cocoa Res.Inst.Ghana 1965-1966 →   B.S 2315 a

Report of the Cocoa Research Institute of Nigeria. Ibadan.
    Rep.Cocoa Res.Inst.Nigeria 1965/1966 →
    Formerly Report of the West African Cocoa Research
Institute. (Nigeria). Ibadan.   B.S 2316

TITLE   SERIAL No.

Report. Coffee Research and Experimental Station. Lyamungu, Moshi.
    Tanganyika Territory. Dar-es-Salaam.
    Rep.Coff.Res.exp.Stn Lyamungu 1955-1956.   E.S 2175 b

Report of the Colchester and District Natural History Society
    and Field Club.
    See Report and Records of the Colchester and District
    Natural History Society and Field Club.   S. 37 a

Report. Colchester and Essex Museum. Colchester.
    Rep.Colchester Essex Mus. 1967-1968 →   S. 41 a

Report on the Collections made by the British Ornithologists'
    Union Expedition and the Wollaston Expedition in Dutch New
    Guinea, 1910-13. London.
    Rep.B.O.U.Exped.Dutch New Guinea 1916.   77 Ba.q.G

Report of the College of Liberal Arts, University of Iwate. Morioka.
    Rep.Coll.Lib.Arts Univ.Iwate Vol.21, Pt.3 - Vol.25, Pt.3,
    1963-1965.
    Formerly Report of the Gakugei Faculty of Iwate
    University.
    Continued as Report of the Faculty of Education, University
    of Iwate.   S. 1996 d

Report of the Colonial (Museum) and Laboratory. Geological Survey
    of New Zealand. Wellington.
    Rep.Colon.Lab.N.Z. No.3-40. 1868-1907   P.S 1160

Report of the Colorado Biological Association. Washington, D.C.
    Rep.Colo biol.Ass. 1888-1889.   S. 2333 A

Report. Colorado Geological Survey. Denver.
    See Report on the Geological Survey, Colorado.   M.S 2637 A

Report of the Colorado Museum of Natural History. Denver.
    Rep.Colo.Mus.nat.Hist. 1913-1947.
    Continued as Report of the Denver Museum of Natural
    History. Denver.   S. 2338 B

Report of the Columbia Basin Commission. State of Washington. Olympia.
    Rep.Columbia Basin Commn 1933-1935.   P.S 1988 c

Report of the Commission for the Exhibition of 1851.
    Rep.Comm.Exhibit.1851 No.9 1935.   S. 409

Report of the Commission for the Investigation of Mediterranean
    Fever. London.
    Rep.Commn Invest.Mediterr.Fever 1905-1907.   S. 3 H5

Report. Commission for the Preservation of Natural and Historical
    Monuments and Relics. Lusaka.
    Rep.Commn Pres.nat.hist.Mon.Relics Lusaka 1968 →   S. 2073

| TITLE | SERIAL No. |

Report of the Commissioner for Fisheries, Province of
    British Columbia. Victoria, B.C.
        Rep.Commnr Fish.Br.Columb. 1914-1932.   Z.S 2690

Report of the Commissioners of Inland Fisheries. Providence, R.I.
        Rep.Commnr inl.Fish., Providence, R.I.
            No. 32, 33, 35, 39, 40, 56-57. 1902-1927.   Z.S 2377

Report of the Commissioners of the State Reservation at Niagara.
    Albany, New York.
        Rep.Commnrs St.Reserv.Niagara   1885-1906.
            (Wanting Nos. 4, 16 & 19.)   S. 2365
                             No.11,   1893-1894.   P.S 1961

Report of the Commissioners on the Zoological Survey of
    the State (of Massachusetts). Boston.
        Rep.Commnrs Zool.Surv.Mass. 1838-1846.   75 C.o.M

Report of the Committee on Bird Sanctuaries in the Royal Parks.
    London.
        Rep.Comm.Bird Sanct.R.Pks   1922 →   Z. 85 B o G

Report. Committee of British Palaeobotantists. London.
        Rep.Comm.Br.Palaeobot. 1959 →   P.S 121

Report of the Committee of Control. South African Central Locust
    Bureau. Cape Town.
        Rep.Comm.Control S.Afr.cent.Locust Bur. 1907-1910.   E.S 2168 c

Report. Committee on the Deterioration of Structures of
    Timber-Metal and Concrete exposed to the Action of
    Sea Water. London.
        Rep.Comm.Deterior.Struct.Timb.-Metal Concr.Sea-Wat. 1920-1940. S. 205 C

Report of the Committee on Foreign Scientific Research.
    Kyushu University. Fukuoka.
        Rep.Comm.for.scient.Res.Kyushu Univ. 1963-1964.   S. 1998 e A

Report of the Committee on Locust Control, Economic Advisory
    Council. London.
        Rep.Comm.Locust Control   1929-1934.   E.S 68

Report of the Committee on Marine Ecology as related to
    Paleontology. Washington, D.C.
        Rep.Comm.mar.Ecol.Paleont.,Wash.   1941-1946.   S. 2419 E
                                        1941-1942.   P.S 884 A
        Formerly   Report of the Subcommittee on the Ecology
        of Marine Organisms. Washington, D.C.
        Continued as   Report of the Committee on a Treatise
        on Marine Ecology and Paleoecology. Washington, D.C.

Report of the Committee on the Measurement of Geologic Time.
    Washington.
        Rep.Comm.Meas.geol.Time Wash.   1932-1952.   P.S 884
                                  1933-1949 (imp.)   M.S 2675

| TITLE | SERIAL No. |

Report of the Committee on Paleoecology. Washington, D.C.
    Rep.Comm.Paleoecol.,Wash. 1935-1937.
    Continued as Report of the Subcommittee on the Ecology
    of Marine Organisms. Washington, D.C.   S. 2419 E

Report of the Committee of the Privy Council for Scientific
    and Industrial Research. London.
    Rep.Comm.Privy Coun.scient.ind.Res. 1915-1927.
    Continued as Report of the Department of Scientific
    and Industrial Research.   S. 205 P

Report of the Committee on Research. Allan Hancock Foundation.
    See Report. Allan Hancock Foundation of the University
    of Southern California. Los Angeles.   S. 2404 b F

Report of the Committee on Sedimentation. Washington.
    Rep.Comm.Sedim.,Wash. 1924-1949 (imp.)   P.S 884 B

Report of the Committee on Submarine Configuration and
    Oceanic Circulation. Washington, D.C.
    Rep.Comm.submar.Config.oceanic Circul.,Wash. 1931.   S. 2419 F

Report of the Committee on a Treatise on Marine Ecology
    and Paleoecology. Washington, D.C.
    Rep.Comm.Treatise mar.Ecol.Paleoecol., Wash. 1946-1951.
    Formerly Report of the Committee on Marine Ecology as
    related to Paleontology. Washington, D.C.   S. 2419 E

Report. Commonwealth Entomological Conference. London.
    Rep.Commonw.ent.Conf. 1948 →
    Formerly Report on the Imperial Entomological
    Conference. London.   E.S 2555

Report. Commonwealth Forestry Institute, University of Oxford.
    Rep.Commonw.For.Inst.Univ.Oxf. 1961 →
    Formerly Report. Imperial Forestry Institute
    University of Oxford.   B.S 90 a

Report. Commonwealth Institute.
    Rep.Commonw.Inst.Lond. 1959 →
    Formerly Report. Imperial Institute.   S. 187 B

Report on the Commonwealth Mycological Conference. London, Kew.
    Rep.Commonw.mycol.Conf. 1948.
    Formerly Report on the Imperial Mycological Conference.
    London, Kew.   B.S.M 5 h

Report of the Commonwealth Scientific and Industrial Research
    Organization. Australia, Canberra.
    Rep.Commonw.scient.ind.Res.Org.Aust. 1949 →
    Formerly Report. Council for Scientific and Industrial
    Research. Commonwealth of Australia. Melbourne.   S. 2113 F

Report. Computation in Sedimentology.
    See Computation in Sedimentology. Report.   P.S 152

| TITLE | SERIAL No. |

Report of Conference of North Central States Entomologists.
    Lafayette, Indiana.
    <u>See</u> Report of the North Central States Entomologists'
    Conference.   E.S 2421 c

Report Connecticut State Board of Fisheries and Game,
    Lake & Pond Survey Unit. Hartford.
    <u>Rep.Conn.St.Bd.Fish.Game</u> 1959 →   Z. 75C o C

Report of the Connecticut State Entomologist. New Haven.
    <u>Rep.Conn.St.Ent.</u> 1901-1939 (imp.)   E.S 2468 c

Report. Conservation Commission Department. New York.
    <u>See</u> Report. New York State Conservation (Commission)
    Department.   S. 2364

Report of the Conservation Commission of Maryland. Baltimore.
    <u>Rep.Conserv.Commn Maryld</u> 1908-1909   P.S 1929 A

Report. Cornell University College of Agriculture and
    Experiment Station. Ithaca.
    <u>Rep.Cornell Univ.Coll.Agric.Exp.Stn</u> 1879-1890.   E.S 2470 a

Report of the Council. Cornwall Naturalists' Trust Ltd. Penzance.
    <u>Rep.Coun.Corn.Nat.Trust</u> 1969 →   S. 321 a

Report of the Council on Environmental Quality. Washington.
    <u>See</u> Environmental Quality. Council on Environmental Quality.
    Washington.   S. 2428 a

Report of the Council. Essex Naturalists' Trust Ltd. Felsted.
    <u>Rep.Coun.Essex Nat.Trust</u> 1970 →   S. 74 B

Report of Council. Federation of Zoological Gardens of Great Britain
    and Ireland. London.
    <u>Rep.Coun.Fed.zool.Gdns G.B.Ir.</u> 1973 →   Z.S 2 A

Report of the Council of the League of Nations on the Administration
    of the Territory of New Guinea. Canberra.
    <u>Rep.Coun.League Nat.Admin.Terr.New Guinea</u> 1929-1936.
    <u>Continued as</u> Report to the General Assembly of the United
    Nations on the Administration of the Territory of New Guinea.   S. 2146

Report of the Council of the Leicester Literary &
    Philosophical Society.
    <u>Rep.Leices.lit.phil.Soc.</u> 1852-1886.   S. 151 A

Report of the Council. Museums Association. London.
    <u>Rep.Mus.Ass.</u> 1966 →   S. 2 K

Report. Council for Nature. London.
    <u>Rep.Coun.Nature</u> 1965 →   S. 229 C

Report. Council for the Preservation of Rural Wales. Aberystwyth, etc.
    <u>Rep.Coun.Pres.rur.Wales</u> 1947-1961 (imp.)
    <u>Continued as</u> Report. Council for the Protection of Rural Wales. S. 6 d

Report of the Council & Proceedings of the Hampstead
    Scientific Society. London.
    <u>See</u> Report of the Hampstead Scientific Society.   S. 114 a A

| TITLE | SERIAL No. |

Report. Council for the Promotion of Field Studies. London.
    Rep.Coun.Promot.Fld Stud. 1946-1954.
    Continued as Report Field Studies Council.   S. 182 A

Report. Council for the Protection of Rural Wales. Machynlleth.
    Rep.Coun.Prot.rur.Wales 1961-1964.
    Formerly Report. Council for the Preservation of
    Rural Wales.   S. 6 d

Report of the Council of the Ray Society. London.
    See Report of the Ray Society. London.   S. 203

Report of the Council for Scientific and Industrial Research.
    Commonwealth of Australia. Melbourne.
    Rep.Coun.scient.ind.Res.Aust. 1927-1948.
    Formerly Report. Institute of Science and Industry.
    Commonwealth of Australia. Melbourne.
    Continued as Report of the Commonwealth Scientific and Industrial
    Research Organization. Canberra.   S. 2113 F

Report of the Countryside Commission. London.
    Rep.Ctryside Commn 1969 →
    Formerly Report of the National Parks Commission.   S. 225 a

Report. Countryside Commission for Scotland. Edinburgh.
    Rep.Ctryside Commn Scotland 1968 →   S. 87

Report. Cranbrook Institute of Science, Bloomfield Hills, Mich.
    Rep.Cranbrook Inst.Sci. 1930 → (imp.)   S. 2316 a C

Report. Crichton Royal Institution, Dumfries.
    Rep.Crichton R.Instn Nos.83-84, 1922-1923.   S. 49

Report. Croydon Microscopical Club.
    See Report & Abstracts of Proceedings of the Croydon
    Microscopical Club.   S. 28

Report of the Curator. Botanical Exchange Club. London.
    Rep.Cur.botl Exch.Club 1869-1876.
    Formerly Report of the Curators. London Botanical Exchange Club.
    Continued as Report. Botanical Exchange Club. London.   B.B.H.S 97

Report of the Curator of the University Museum to the Board
    of Regents, University of Michigan. Ann Arbor.
    Rep.Cur.Univ.Mus.Univ.Mich. 1903-1912.
    Continued as Report of the Director of the Museum of
    Zoology to the Board of Regents, University of Michigan.   Z.S 2425 F

Report of the Curators. Bodleian Library. Oxford.
    Rep.Cur.Bodleian Libr. 1949 →   S. 310 C c

Report of the Curators. London Botanical Exchange Club.
    Rep.Cur.Lond.botl Exch.Club 1866-1868.
    Formerly Curator's Report. Botanical Exchange Club. Thirsk.
    Continued as Report of the Curator. Botanical Exchange
    Club. London.   B.B.H.S 97

TITLE	SERIAL No.

Report. Cyprus Ornithological Society.
   Rep.Cyprus orn.Soc. 1957-1959.
   Continued as Bird Report. Cyprus Ornithological
Society.   T.B.S 1951 B

Report of the Czechoslovak Botanical Society at Prague.
   See Preslia. Casopis Ceskeslovenské Botanické
Spolecnosti. Praha.   B.S 1232

Report of the Danish Biological Station to the Board of
Agriculture (Ministry of Fisheries) Copenhagen.
   Rep.Dan.biol.Stn No.3-54, 1893-1952.   S. 525

Report on the Danish Oceanographical Expeditions 1908-1910
to the Mediterranean and adjacent Seas. Copenhagen.
   Rep.Dan.oceanogr.Exped.Mediterr. 1912-1939.   72 P.q.D

Report. Darlington and Teesdale Naturalists' Field Club. Darlington.
   Rep.Darlington Teesdale Nat.Fld Club 1969 →   S. 337

Report. Dauntsey's School Natural History Society. West Lavington.
   Rep.Dauntsey's Sch.nat.Hist.Soc. 1953-1960/1961.
   Formerly School House Natural History Society Report.
Dauntsey's School.   S. 393 a B

Report. De Beers Consolidated Mines, Ltd., Kimberley.
   Rep.De Beers Mines. 1889 →   M.S 2104

Report of the Delegates of the University Museum. Oxford.
   Rep.Univ.Mus.Oxf. 1888-1944 (imp.)   S. 310 F

Report of the Denver Museum of Natural History. Denver.
   Rep.Denver Mus.nat.Hist. 1948 →
   Formerly Report of the Colorado Museum of Natural History.
Denver.   S. 2338 B

Report of the Department of Agriculture, Alberta. Edmonton.
   Rep.Dep.Agric.Alberta 1905-1919 (imp.).   E.S 2527 c

Report of the Department of Agriculture, Antigua.
   See Report of the Agricultural Department, Antigua.   E.S 2382

Report of the Department of Agriculture, Barbados. Bridgetown.
   Rep.Dep.Agric. Barbados 1899-1928.
   Continued as Report of the Department of Science and
Agriculture, Barbados, Bridgetown.   E.S 2384

Report Department of Agriculture, Basutoland. Bloemfontein.
   Rep.Dep.Agric.Basutold 1936-1949   E.S 2169 b

Report of the Department of Agriculture, British Columbia. Victoria.
   See Report of the British Columbia Department of Agriculture,
Victoria.   E.S 2526 c

Report of the Department of Agriculture, British East Africa. Nairobi.
   Rep.Dep.Agric.Br.E.Afr. 1908-1920.
   Continued as Report. Department of Agriculture, East Africa
Protectorate. Nairobi.   E.S 2171

TITLE	SERIAL No.

Report. Department of Agriculture, British Guiana. Georgetown.
   See Administration Report. Director of Agriculture,
   British Guiana. Georgetown.   E.S 2369

Report. Department of Agriculture, British Honduras. Belize.
   Rep.Dep.Agric.Br.Hond. 1935 →   E.S 2395 a

Report. Department of Agriculture, Canada. (e) Entomology. Ottawa.
   Rep.Dep.Agric.Can. 1897-1918 (imp.)   E.S 2524 a

Report of the Department of Agriculture, Cape of Good Hope. Cape Town.
   Rep.Dep.Agric.Cape Good Hope 1908-1909.   E.S 2162

Report of the Department of Agriculture, Ceylon. Colombo.
   Rep.Dep.Agric.Ceylon 1924-1926, 1937 →   E.S 2020
                        1911/1912, 1916, 1927.   B.S 1720 b

Report of the Department of Agriculture, Cyprus.
   See Report of the Director of Agriculture, Cyprus. Nicosia. E.S 2035 c

Report. Department of Agriculture, East Africa Protectorate. Nairobi.
   Rep.Dep.Agric.E.Afr.Prot. 1920-1921.
   Formerly Report of the Department of Agriculture, British
   East Africa. Nairobi.
   Continued as Report. Department of Agriculture,
   Kenya. Nairobi.   E.S 2171

Report. Department of Agriculture, Fiji. Suva.
   Rep.Dep.Agric.Fiji 1915 →   E.S 2270 a

Report. Department of Agriculture, Formosa, Taihoku.
   See Report. Department of Agriculture, Government Research
   Institute, Formosa, Taihoku.   E.S 1921

Report of the Department of Agriculture, Gambia. London.
   Rep.Dep.Agric.Gambia 1923 →   E.S 2172 c

Report of the Department of Agriculture, Ghana. Accra.
   Rep.Dep.Agric.Ghana 1955-1956.
   Formerly Report of the Agricultural Department.
   Gold Coast.   E.S 2174 a

Report. Department of Agriculture. Government Research
   Institute. Formosa.
   Rep.Dep.Agric.Govt res.Inst.Formosa 1922-1939 (imp.)
   Continued as Report. Government Research Institute.
   Department of Agriculture. Taiwan.   E.S 1957 a

Report of the Department of Agriculture, Irish Free State. Dublin.
   See Annual General Report (Dept.of Agriculture, &c.
   for Ireland). Dublin.   E.S 82

| TITLE | SERIAL No. |

Report on the Department of Agriculture, Jamiaca. Kingston.
    Rep.Dep.Agric.Jamaica 1913-1922.
        Formerly Report of the Board of Agriculture & Department
        of Public Gardens & Plantations, Jamaica. Kingston.
        Continued as Report of the Department of Science and
        Agriculture, Jamaica. Kingston.   E.S 2375

Report. Department of Agriculture, Kenya. Nairobi.
    Rep.Dep.Agric.Kenya 1921-1938.
        Formerly Report. Department of Agriculture, East Africa
        Protectorate. Nairobi.   E.S 2171

Report. Department of Agriculture, Madras.
    Rep.Dep.Agric.Madras 1899-1932.   E.S 2013

Report of the Department of Agriculture, Malaya. Singapore.
    Rep.Dep.Agric.Malaya 1937-1939   E.S 1971

Report of the Department of Agriculture, Malta. Valetta.
    Rep.Dep.Agric.Malta 1919 → (imp.)   E.S 1120

Report of the Department of Agriculture, Mauritius. Port Louis.
    Rep.Dep.Agric.Mauritius 1935-1966.
        Continued as Report of the Ministry of Agriculture
        and Natural Resources, Mauritius.   E.S 2183 a

Report of the Department of Agriculture Mysore. Bangalore.
    Rep.Dep.Agric.Mysore 1922-1923   E.S 2015

Report. Department of Agriculture, New South Wales. (Sydney.)
    Rep.Dep.Agric.N.S.W. 1891, 1893, 1900.   E.S 2248

Report. Department of Agriculture, New Zealand. Wellington.
    Rep.Dep.Agric.N.Z. 1906-1919.   E.S 2263 a

Report of the Department of Agriculture, North-West
    Territories. Regina.
    Rep.Dep.Agric.N.-W.Terr. 1902.   E.S 2528 c A

Report. Department of Agriculture, Northern Rhodesia. Lusaka.
    Rep.Dep.Agric.Nth.Rhod. 1926-1959.   E.S 2177 a

Report. Department of Agriculture, Nyasaland. Zomba.
    Rep.Dep.Agric.Nyasald 1925-1946   E.S 2175 a

Report. Department of Agriculture, Queensland.
    See Report of the Department of Agriculture & Stock,
    Queensland. Brisbane.   E.S 2257 c

Report. Department of Agriculture, St.Lucia.
    See Report Agricultural Department. St.Lucia.   E.S 2381

Report. Department of Agriculture, St.Vincent.
    See Report Agricultural Department, St.Vincent. Kingtown.   E.S 2381 a

Report of the Department of Agriculture, Saskatchewan. Regina.
    Rep.Dept.Agric.Saskatchewan 1905-1906.   E.S 2528 c C

| TITLE | SERIAL No. |

Report of the Department of Agriculture, Scotland.
   <u>Rep.Dep.Agric.Scotl</u>. 1930-1950. (imp.)
   <u>Formerly</u> Report of the Board of Agriculture for Scotland.
London.   E.S 78

Report. Department of Agriculture, Seychelles Islands. Victoria.
   <u>Rep.Dep.Agric.Seychelles</u> 1923-1932.   E.S 2181

Report of the Department of Agriculture, Sierra Leone. Freetown.
   <u>Rep.Dep.Agric.Sierra Leone</u> 1929 →   E.S 2173

Report of the Department of Agriculture, South Australia. Adelaide.
   <u>Rep.Dep.Agric.S.Aust</u>. 1929-1945.   E.S 2260 a

Report. Department of Agriculture, Tanganyika Territory.
London and Dar-es-Salaam.
   <u>Rep.Dep.Agric.Tanganyika</u> 1923 →   E.S 2176

Report. Department of Agriculture, Transvaal. Pretoria.
   <u>Rep.Dep.Agric.Transv</u>. 1903-1904.
   <u>Continued as</u> Report of the Government Veterinary
Bacteriologist. Union of South Africa. Pretoria.   Z.S 2081

Report of the Department of Agriculture, Trinidad and Tobago.
Port of Spain.
   <u>Rep.Dep.Agric.Trin</u>. 1910-1925 (imp.)
   <u>Continued as</u> Administration Report of Director of
Agriculture, Trinidad & Tobago.   E.S 2379

Report of the Department of Agriculture, Uganda. Entebbe.
   <u>Rep.Dep.Agric.Uganda</u> 1913-1949   E.S 2179

Report of the Department of Agriculture, Union of South Africa.
Pretoria.
   <u>Rep.Dep.Agric.Un.S.Afr</u>. 1910-1920.   E.S 2168

Report of the Department of Agriculture, Victoria, Melbourne.
   <u>Rep.Dep.Agric.Vict</u>. 1874-1910 (imp.).   E.S 2259 b

Report of the Department of Agriculture, Zanzibar.
   <u>Rep.Dep.Agric.Zanzibar</u> 1901; 1924-1938 (imp.).   E.S 2176 b

Report. Department of Agriculture and Fisheries, Palestine.
   <u>See</u> Report. Department of Agriculture and Forests.
Jerusalem, Palestine.   E.S 2036

Report of the Department of Agriculture & Forests. Jerusalem,
Palestine.
   <u>Rep.Dep.Agric.Forests Palest</u>. 1927-1946.   E.S 2036

Report. Department of Agriculture and Forests, New South Wales.
(Sydney.)
   <u>Rep.Dep.Agric.For.N.S.W</u>. 1894.   E.S 2248 a

Report of the Department of Agriculture and Stock,
Queensland. Brisbane.
   <u>Rep.Dep.Agric.Stk Qd</u> 1892-1900.   E.S 2257c

Report. Department of Agriculture, Stock and Fisheries, Territory
of Papua and New Guinea.
   <u>Rep.Dep.Agric.Stk Fish.Papua New Guinea</u> 1959 →   E.S 2271a

| TITLE | SERIAL No. |

Report of the Department of Animal Health, Gold Coast. Accra.
    Rep.Dep.Anim.Hlth Gold Cst   1931-1932 →   Z. 74M f G
                                    1931-1934.   E.S 2174 b
    Formerly Report of the Veterinary Department, Gold Coast.

Report of the Department of Animal Health.
    Northern Rhodesia. Livingstone.
    Rep.Dep.Anim.Hlth, Nth.Rhod. 1929-1933.
    Formerly & Continued as Report of the Veterinary
    Department of Northern Rhodesia.   Z. 74D f R

Report. Department of Botanical Research, Carnegie Institution.
    Washington.
    Rep.Dep.bot.Res.Carnegie Instn 1909-1921.   B.S 4412

Report on the Department of Conservation and Development.
    State of New Jersey. Trenton, N.J.
    See Report of the New Jersey Department of Conservation
    and Development.   S. 2412 a

Report. Department of Energy, Mines and Resources. Ottawa.
    Rep.Dep.Energy Mines Resour. 1965 →
    Formerly Report of the Department of Mines and Technical
    Surveys, Canada.   M.S 2703

Report of the Department of Fisheries, Baroda State. Baroda.
    Rep.Dep.Fish.Baroda   1938-1949.   Z.S 1908

Report of the Department of Fisheries, Maharashtra State. Bombay.
    Rep.Dep.Fish.Maharashtra State 1959-1960 →   Z.S 1927

Report of the Department of Fisheries of New South Wales. Sydney.
    Rep.Dep.Fish.N.S.W.   1946-1948.
    Formerly Report on the Fisheries of New South Wales.
    Continued as Report by the Chief Secretary on the
    Fisheries in New South Wales.   Z.S 2106

Report. Department of Fisheries and Forestry, Canada. Ottawa.
    Rep.Dep.Fish.For.Can. 1968/1969 →   E.S 2537
                          1969/1970 →   S. 2601 a
    Formerly Report. Department of Forestry (and Rural Development)
    of Canada. Ottawa.   E.S 2537

Report. Department of Forestry (and Rural Development) of Canada.
    Ottawa.
    Rep.Dep.For.rur.Dev.Can. 1965/1966 - 1967/1968.
    Continued as Report. Department of Fisheries and Forestry,
    Canada. Ottawa.   E.S 2537

Report. Department of Forestry, Government Research Institute,
    Formosa. Taihoku.
    Rep.Dep.For.Res.Inst.Formosa No.9, 1930.   B. 58.00p5 Tai SAS

Report of the Department of Forestry. Republic of South Africa.
    See Jaarverslag van die Departement van Bosbou.
    Republiek van Suid-Afrika.   B.S 2304 b

| TITLE | SERIAL No. |

Report. Department of Geological Survey, Bechuanaland.
    (Parow, Cape.)
       Rep.Dep.geol.Surv.Bechuan. 1953-1965. (imp.)
       Continued as Report of the Geological Survey Department,
    Botswana. Gaberones.   P.S 1175 C

Report. Department of Hop Research, Wye College, Ashford, Kent.
    Rep.Dep.Hop Res., Wye Coll. 1948 → (imp.)   B.S 1

Report. Department of Lands, Mines and Surveys. Kenya.
    See Report. Lands, Mines & Surveys Department,
    Mines Division, Kenya.   M.S 2109 A

Report. Department of Lands and Surveys, Tasmania. Hobart.
    Rep.Dep.Lds Survs Tasm. 1912-1913.   E.S 2262a

Report of the Department of Liberal Arts, the Iwate University.
    Morioka.
       Rep.Dep.Lib.Arts Iwate Univ. 1949.
       Continued as Report of the Gakugei Faculty of Iwate
    University.   S. 1996 d

Report of the Department of Mines, Canada, Ottawa.
    Rep.Dep.Mines Can. 1920-1936
       Continued as Report. Department of Mines and Resources,
    Canada, Ottawa.   M.S 2703

Report of the Department of Mines, New South Wales. Sydney.
    Rep.Dep.Mines N.S.W. 1877-1880.   P.S 1120
                         1875 → (imp.)   M.S 2403

Report. Department of Mines, Ontario, Toronto.
    See Report. Ontario Department of Mines. Toronto.   M.S 2708

Report of the Department of (Development and) Mines,
    Queensland. Brisbane.
       Rep.Dep.Mines Qd 1898, 1904 →   M.S 2401

Report of the Department of Mines, Western Australia, Perth.
    Rep.Dep.Mines West Aust. 1896 →   M.S 2407

Report of the Department of Mines, Agriculture and Resources.
    St. John's. Province of Newfoundland & Labrador.
       Rep.Dep.Mines Agr.Resour.St.John's 1965 →   S. 2644

Report. Department of Mines and Natural Resources, Manitoba.
    Winnipeg.
       Rep.Dep.Mines nat.Resour. 1961 →   P.S 1093 A

Report of the Department of Mines and Resources, Canada. Ottawa.
    Rep.Dep.Mines Resour.Can. 1936-1943.
       Formerly Report. Department of Mines, Canada. Ottawa.
       Continued as Report of Mines, Forests and Scientific
    Services Branch, Canada, Ottawa.   M.S 2703

TITLE   SERIAL No.

Report. Department of Mines and Resources. Newfoundland. St.John's.
    See Report. Geological Survey, Province of Newfoundland.
St. John's.   P.S 1096 A

Report of the Department of Mines and Technical Surveys, Canada.
    Rep.Dep.Mines tech.Survs Can. 1949-1965.
    Formerly Report of Mines, Forests and Scientific Services
Branch, Canada.
    Continued as Report. Department of Energy, Mines and
Resources. Ottawa.   M.S 2703

Report. Department of Natural Resources, Ohio.
    See Report of the Director of the Ohio Department of
Natural Resources. Columbus.   S. 2332 c A

Report. Department of Nature Conservation, Union of
South Africa. Cape Town.
    Rep.Dep.Nat.Conserv.Un.S.Afr. No.9 → 1952 →
    Formerly Report. Inland Fisheries Department.
Union of South Africa.   Z.S 2031

Report of the Department of the Naval Service. Ottawa.
    Rep.Dept.Naval Serv.Ottawa 1913-1917.   71.o.C

Report of the Department of Science & Agriculture, Barbados,
Bridgetown.
    Rep.Dep.Sci.Agric.Barbados   1928-1957.
    From 1932-1940 styled Agriculture Journal. Department
of Science & Agriculture, Barbados.
    Formerly Report. Department of Agriculture, Barbados.   E.S 2384

Report of the Department of Science & Agriculture, Jamaica, Kingston.
    Rep.Dep.Sci.Agric.Jamaica 1923-1956.
    Formerly Report on the Department of Agriculture,
Jamaica. Kingston.
    Continued as Report on the Ministry of Agriculture &
Lands, Jamaica.   E.S 2375

Report of the Department of Science and Art. London.
    Rep.Dep.Sci.Art, Lond. 1854-1899.   O. 72 Aa.o.G

Report of the Department of Scientific and Industrial Research.
London.
    Rep.Dep.scient.ind.Res.,Lond. 1927-1938; 1947-1956. (imp.)
    Formerly Report of the Committee of the Privy Council for
Scientific & Industrial Research.
    Continued as Report of the Research Council, D.S.I.R. London. S. 205 P

Report of the Department of Veterinary Research of the
Federation of Nigeria. Lagos.
    Rep.Dep.vet.Res.Fed.Nigeria 1954-1955 - 1956-1957.
    Formerly Report of the Veterinary Department, Nigeria.
    Continued as Report of the Federal Department of Veterinary
Research.   Z. 74M f N

Report of the Department of Veterinary Science and Animal
Husbandry, Tanganyika. London.
    Rep.Dep.vet.Sci.Anim.Husb.Tanganyika 1922-1953.
    Continued as Report of the Veterinary Department,
Tanganyika.   Z. O. 74Db f T

| TITLE | SERIAL No. |

Report. Department of Veterinary Services, Kenya.
    Rep.Dep.vet.Servs Kenya 1947 →
    Formerly Report. Veterinary Department. Colony &
Protectorate of Kenya.   Z. Mammal Section

Report. Department of Veterinary Services, Northern
    Rhodesia. Lusaka.
    Rep.Dep.vet.Servs Nth.Rhod. 1949-1958.
    Formerly Report. Veterinary Department, Northern Rhodesia.
    Continued as Report. Department of Veterinary & Tsetse
Control Services. Northern Rhodesia. Lusaka.   Z. 74D f R

Report. Department of Veterinary Services and Animal Industry,
    Malawi. Zomba.
    Rep.Dep.vet.Servs Anim.Ind.Malawi 1962 →
    Formerly Report of the Department of Veterinary Services
and Animal Industry, Nyasaland.   Z. 74D f N

Report of the Department of Veterinary Services and Animal
    Industry, Nyasaland. Zomba.
    Rep.Dep.vet.Servs Anim.Ind.Nyasald 1949-1961.
    Formerly Report. Veterinary Department. Nyasaland.
    Continued as Report of the Department of Veterinary
Services & Animal Industry, Malawi.   Z. 74D f N

Report of the Department of Veterinary Services and Animal
    Industry, Uganda Protectorate. Entebbe.
    Rep.Dep.vet.Servs Anim.Ind.Uganda 1953 →
    Formerly Report of the Veterinary Department. Uganda.   Z. 74Da f U

Report. Department of Veterinary and Tsetse Control Services,
    Northern Rhodesia. Lusaka.
    Rep.Dep.vet.Tsetse Control Servs Nth.Rhod. 1959-1963.   Z. 74D f R
                                            1963.   E.S 2177 b
    Formerly Report. Department of Veterinary Services,
Northern Rhodesia. Lusaka.

Report. Department of Zoology, University of Singapore. Singapore.
    Rep.Dep.Zool.Univ.Singapore 1964-1965 →   Z.S 1945 a

Report of the Derby Free Library and Museum and Art Gallery.
    Rep.Derby Free Libr.& Mus. No.15-56, 1885-1927 imp.   S. 30 b

Report on the Desert Locust Investigation. Lyallpur, etc.
    Rep.Desert Locust Invest. 1931-1934.   E.S 2000 c

Report of the Development Commissioners. London.
    Rep.Dev.Commnrs 1910-1939.   S. 207

Report. Devon Bird-Watching and Preservation Society. Exeter.
    Rep.Devon Bird-Watch.Preserv.Soc. 33rd → 1960 →   T.B.S 286

Report of the Director of Agriculture, Cyprus. Nicosia.
    Rep.Dir.Agric.Cyprus 1901-1904, 1929 →   E.S 2035 c

TITLE	SERIAL No.

Report. Director of Agriculture, Hong Kong.
    See Annual Departmental Report. Director of Agriculture,
Fisheries and Forestry. Hong Kong.   Z.S 1946 A

Report of the Director of Agriculture, Fisheries and
    Forestry. Hong Kong.
    See Annual Departmental Report. Director of Agriculture,
Fisheries and Forestry. Hong Kong.   Z.S 1946 A

Report of the Director of Agriculture and Forestry. Hong Kong.
    See Annual Departmental Report. Director of Agriculture
& Forestry, Hong Kong.   Z.S 1946 A

Report of the Director of the Extension Service, U.S. Department
    of Agriculture. Washington.
    Rep.Dir.Ext.Serv.U.S.Dep.Agric. 1918-1930 (imp.).
    Continued as Report on Extension Work in Agriculture &
Home Economics, U.S. Department of Agriculture. Washington.   E.S 2458c

Report of the Director of Fisheries. Hong Kong
    Rep.Dir.Fish.Hong Kong 1949-1950.
    Formerly Report of the Officer i/c Fisheries. Hong Kong.
    Continued as Annual Departmental Report. Director of
Agriculture, Fisheries and Forestry. Hong Kong.   Z.S 1946 A

Report of the Director of Fishery Research. Ministry of
    Agriculture, Fisheries and Food. London.
    Rep.Dir.Fish.Res.   1966 →   Z.O 72Aa f G

Report of the Director of Geological Survey. Ghana.
    See Report of the Geological Survey, Ghana.   P.S 1179 E

Report of the Director of Geological Survey. Gold Coast.
    See Report of the Geological Survey, Gold Coast.   P.S 1179 E

Report. Director Geophysical Laboratory, Carnegie Institution,
    Washington.
    Rep.Dir.geophys.Lab.Carnegie Instn 1919-1940 (imp.); 1941 → M.S 2622

Report of the Director of Government Chemical Laboratories.
    Western Australia, Perth.
    See Report of the Government Chemical Laboratories,
Western Australia.   M.S 2407 C

Report of the Director for Mines, Tasmania, Hobart.
    Rep.Dir.Mines Tasm. 1940 →
    Formerly Report. Secretary for Mines, Tasmania. Hobart.   M.S 2444

Report of the Director of Mines and Goverment Geologist.
    South Australia.
    Rep.Dir.Mines Govt.Geol.S.Aust. 1916 →   P.S 1138
                            1916-1939.   M.S 2410
    Formerly Report of the Government Geologist,
South Australia.

TITLE	SERIAL No.

Report of the Director of the Museum of Zoology to the Board of Regents, University of Michigan. Ann Arbor.
Rep.Dir.Mus.Zool.Univ.Mich. 1912-1954 (imp.)
Formerly Report of the Curator of the University Museum to the Board of Regents, University of Michigan.   Z.S 2425 F

Report. Director of National Parks and Wild Life Management. Rhodesia.
See Report of the National Parks Advisory Board and Director of National Parks and Wild Life Management. Ministry of Lands and Mines. Rhodesia.   Z.S 2065 A

Report of the Directorate of Botanical & other Public Gardens, West Bengal. Alipore.
Rep.Direct.Bot.& Publ.Gard., W.Bengal 1955 →
Formerly Report of the Indian Botanic Garden & the Gardens in Calcutta (Parks & Gardens in Cooch Behar) and the Lloyd Botanic Garden, Darjeeling. Calcutta.   B.S 1602 a

Report. Directorate of Fisheries Research. Department of Agriculture and Fisheries for Scotland. Edinburgh.
See Directorate of Fisheries Research Report. Department of Agriculture and Fisheries for Scotland. Edinburgh.   Z.S 483

Report. Directorate of Overseas Surveys. London.
Rep.Direct.overs.Survs. 1964 →   S. 171 b

Report of the Directors of the Museum of the California Academy of Sciences. San Francisco.
Rep.Mus.Calif.Acad.Sci. 1912.   S. 2401 E

Report of the Directors of the Ohio Department of Natural Resources. Columbus.
Rep.Dir.Ohio Dep.nat.Resour. 1950-1964.   S. 2332 c A

Report of the Directors of the Peabody Museum. New Haven.
Rep.Peabody Mus. 1928-1934.   S. 2352 Q

Report. "Discovery" Expedition. London.
Rep."Discovery" Exped. 1926.
Continued as Report. Discovery Investigations. London.   S. 172 B

Report."Discovery"Investigations. London.
Rep.Discovery Invest. 1927-1928.
Formerly Report "Discovery" Expedition.
Continued as Report on the Progress of the"Discovery" Committee's Investigation.   S. 172 B

Report. Division of Animal Health. Commonwealth Scientific and Industrial Research Organisation, Australia. Melbourne.
Rep.Div.anim.Hlth 1966 →
Formerly in Report. Animal Research Laboratories. C.S.I.R.O. Australia.   Z.S 2112 C b

Report of the Division of Animal Industry, Board of Agriculture and Forestry. Hawaii.
See Report of the Board of Agriculture and Forestry, Hawaii. Division of Entomology.   E.S 2277 a

TITLE	SERIAL No.

Report. Division of Animal Physiology. Commonwealth Scientific
and Industrial Research Organization, Australia. Sydney.
Rep.Div.anim.Physiol. 1966 →
Formerly in Report. Animal Research Laboratories.
C.S.I.R.O. Australia.   Z.S 2112 C c

Report of the Division of Apiculture, State College of
Washington. Olympia.
Rep.Div.Apic.St.Coll.Wash. 1921-1925   E.S 2495 a

Report of the Division of Entomology, Board of Agriculture
and Forestry, Hawaii.
See Report of the Board of Agriculture and Forestry,
Hawaii. Division of Entomology.   E.S 2277 a

Report of the Division of Entomology of the Department of
Agriculture, Union of South Africa. Pretoria.
Rep.Div.Ent.Dep.Agric.Un.S.Afr. 1911-1913.   E.S 2168a

Report. Division of Fisheries. Union of South Africa. Pretoria.
Rep.Div.Fish.Un.S.Afr. No.19-30, 1947-1959.
Formerly Report. Fisheries & Marine Biological Survey,
Union of South Africa.
Continued as Report. Division of Sea Fisheries.
Republic of South Africa. Pretoria.   Z.S 2030 C

Report. Division of Fisheries and Oceanography. C.S.I.R.O.
Australia. Cronulla.
Rep.Div.Fish.Oceanogr.CSIRO 1963 →   Z.S 2112 D

Report. Division of Fisheries Research Newfoundland.
Rep.Div.Fish.Res.Newfoundld Fauna Ser. 1936.
Published in Research Bulletin. Division of Fishery
Research. Department of Natural Resources,
Newfoundland.   Z.S 2670 B

Report. Division of Forest Research, Zambia. Kitwe.
Rep.Div.For.Res.Zambia Sect.2, 1968 →   B.S 2313 a

Report. Division of Sea Fisheries. Republic of South Africa.
Pretoria.
Rep.Div.Sea Fish.Rep.S.Afr. No.31 → 1959 →
Formerly Report. Division of Fisheries. Union of
South Africa. Pretoria.   Z.S 2030 C

Report. Division of Wildlife Research, C.S.I.R.O., Australia.
Melbourne.
Rep.Div.Wildl.Res.C.S.I.R.O.Aust. 1961 →
Formerly Report. Wildlife Survey Section, C.S.I.R.O.,
Australia.   Z.S 2112 H

Report of the Dominion Museum, N.Z. Wellington.
See Report. National Art Gallery and Dominion Museum.
Wellington, N.Z.   S. 2165 G

Report of the Doncaster Municipal Art Gallery & Museum.
Doncaster.
Rep.Doncaster Mus. 1910-1914.   S. 82

TITLE	SERIAL No.

Report of the Don-Kuban Station.
   See Raboty Dono-Kubanskoĭ Nauchnoi Rybokhozyaĭstvennoĭ Stantsii.   Z.S 1838 A

Report. Dorman Museum and Municipal Art Gallery. County Borough of Middlesbrough.
   Rep.Dorman Mus. 1966 →   S. 294

Report. Dorset Naturalists' Trust. (Poole.)
   See News Letter. Dorset Naturalists' Trust.   S. 41 b

Report. Dove Marine Laboratory. Cullercoats. Northumberland. Newcastle-on-Tyne.
   Rep.Dove mar.Lab. 1912 →
   Formerly Report on the Scientific Investigations. Northumberland Sea Fisheries Committee.   Z.S 300

Report of the Dover Museum.
   Rep.Dover Mus. 1931-1938.   S. 42 C

Report. Dublin Institutions of Science and Art.
   Rep.Dublin Instns Sci.Art 1900-1908.
   Continued as Report on the National Museum of Science and Art. Dublin.   S. 48 a A

Report. Dublin Museum of Science and Art.
   See Report of the Board of Visitors. Dublin Museum of Science and Art.   S. 48 a B

Report of the Dublin Naturalists' Field Club.
   Rep.Dublin Nat.Fld Cl. 1898, 1905-1916. (imp.)   S. 47

Report of the Dulwich College Science Society.
   Rep.Dulwich Coll.sci.Soc. 1878-1885. (imp.)
   Continued as Science Society Report. Dulwich College.   S. 171

Report. Dundee Free Library.
   Rep.Dundee Free Libr. 1870-1927 imp.   S. 52 b

Report. Dundee Museums and Art Galleries.
   Rep.Dundee Mus.Art Gall. 1971/3 →   S. 52 d

Report of the Dundee Naturalists' Society.
   Rep.Dundee nat.Soc. 4-10 & 12, 1876-1885.   S. 52 A

Report of the Durban Museum.
   Rep.Durban Mus. 1910-1917, 1923 → (imp.)   S. 2013 B

Report from the E.M. Museum of Geology and Archaeology. Princeton, N.J.
   Rep.E.M.Mus.Geol.Archaeol. 1882-1885.   P.S 1958

Report of the Ealing Microscopical and Natural History Society.
   Rep.Ealing microsc.nat.Hist.Soc. 3-17, 1880-1893.
   Continued as Report of the Ealing Natural Science & Microscopical Society.   S. 53

TITLE   SERIAL No.

Report of the Ealing Natural Science and Microscopical Society.
    Rep.Ealing nat.Sci.microsc.Soc. 18-27, 1894-1904.
    Formerly Report of the Ealing Microscopical and
    Natural History Society.
    Continued as Report of the Ealing Scientific &
    Microscopical Society.   S. 53

Report of the Ealing Scientific and Microscopical Society.
    Rep.Ealing scient.microsc.Soc. 1904-1935.
    Formerly Report of the Ealing Natural Science and
    Microscopical Society.   S. 53

Report on Earth Science. Department of General Education,
    Kyushu University. Fukuoka.
    Rep.Earth Sci.Dep.gen.Educ.Kyushu Univ. 1955 → (imp.)   P.S 1768 B

Report. East African Agricultural Research Institute, Amani.
    Dar-es-Salaam.
    Rep.E.Afr.agric.Res.Inst. 1928-1947.
    Continued as Report. East African Agriculture & Forestry
    Research Organisation. Nairobi.   E.S 2170a

Report. East African Agriculture & Forestry Research Organisation.
    Nairobi.
    Rep.E.Afr.Agric.For.Res.Org. 1948 →
    Formerly Report. East African Agriculture Research Institute,
    Amani. Dar-es-Salaam.   E.S 2170 a

Report. East African Fisheries Research Organization. Jinja.
    Rep.E.Afr.Fish.Res.Org. 1953-1959.
    Formerly Report. East African Inland Fisheries Research
    Organization. Jinja.
    Continued as Report. East African Freshwater Fisheries
    Research Organization. Jinja.   Z.S 2056

Report. East African Freshwater Fisheries Research Organization.
    Jinja.
    Rep.E.Afr.Freshwat.Fish.Res.Org. 1960 →
    Formerly Report. East African Fisheries Research
    Organization. Jinja.   Z.S 2056

Report. East African Inland Fisheries Research Organization.
    Jinja.
    Rep.E.Afr.inld Fish.Res.Org. 1948-1949
    Continued as Report. East African Fisheries Research
    Organization. Jinja.   Z.S 2056

Report of East African Institute of Malaria and Vector-borne
    Diseases. Mwanza.
    Rep.E.Afr.Inst.Malar. 1953 →   E.S 2176 c

Report of the East African Marine Fisheries Research
    Organization. Nairobi.
    Rep.E.Afr.mar.Fish.Res.Org. 1952 →   Z.S 2055

| TITLE | SERIAL No. |

Report. East African Trypanosomiasis Research Organization. Nairobi.
    Rep.E.Afr.Trypan.Res.Org. 1956 →
    Formerly Report of the East African Tsetse and Trypanosomiasis
Research and Reclamation Organization. Nairobi.   E.S 2171 c

Report of the East African Tsetse and Trypanomiasis Research
and Reclamation Organization. Nairobi.
    Rep.E.Afr.Tsetse Trypan.Res.Reclam.Org. 1953-54.
    Continued as Report. East African Trypanomiasis Research
Organization.   E.S 2171 c

Report. East African Veterinary Research Organization. Nairobi.
    Rep.E.Afr.vet.Res.Org. 1951, 1955 →   Z. Mammal Section

Report of East African Virus Research Institute. Entebbe.
    Rep.E.Afr.Virus Res.Inst. 1956 →   E.S 2170 b

Report of the East Kent Natural History Society. Canterbury
    Rep.E.Kent nat.Hist.Soc. 1858-1895.
    Continued as Report and Transactions. East Kent Scientific
and Natural History Society. Canterbury.   S. 24 A

Report. East London Museum.
    Rep.E.Lond.Mus. 1966 →   S. 2014 a

Report of the East Malling Research Station. East Malling.
    Rep.E.Malling Res.Stn 1925 →   E.S 77

Report of the East of Scotland Union of Naturalists' Societies.
    Rep.E.Scot.Un.nat.Socs. 1884.
    Continued as Proceedings of the East of Scotland Union
of Naturalists' Societies.   S. 55

Report of the Eastbourne Natural History Society.
    See Papers(& Annual Report)of the Eastbourne Natural
History Society.   S. 54 B

Report of Ecological Conference. New Zealand Ecological Society.
Wellington.
    Rep.ecol.Conf.N.Z. 1952.
    (Issued in New Zealand Science Review.)
    Continued as Report of Annual Meeting. New Zealand
Ecological Society. Wellington.   S. 2166

Report. Economic Unit, Geological Survey Department, Northern
Rhodesia. Lusaka.
    Rep.econ.Unit.geol.Surv.Dep.Nth.Rhod. 1963-1964.
    Continued as Report. Economic Unit, Geological Survey
Department, Republic of Zambia.   P.S 1162 F

| TITLE | SERIAL No. |

Report. Economic Unit, Geological Survey Department, Republic
of Zambia. Lusaka.
Rep.econ.Unit geol.Surv.Dep.Zambia 1964-1965.
Formerly Report. Economic Unit, Geological Survey
Department, Northern Rhodesia.
Continued as Economic Report. Geological Survey Department,
Republic of Zambia.   P.S 1172 F

Report on Economic Zoology. Department of Zoology, British
Museum (Natural History). London.   BM.Ec.o
Rep.econ.Zool.Br.Mus.nat.Hist. 1903-1904.   E.S 70 b

Report on Economic Zoology. South Eastern Agricultural College. Wye.
Rep.Econ.Zool.S.East.agric.Coll. 1906, 1907, 1911.   E.S 70a

Report. Empire Marketing Board. London.
Rep.Emp.Mktg Bd 1927-1932.   E.S 66

Report of the Entomological Meeting at Pusa.
See Report of Proceedings of Entomological
Meetings at Pusa.   E.S 1995

Report. Entomological Society of Hampshire. Southampton.
Rep.ent.Soc.Hamps. 1923.   E.S 30 a

Report of the Entomological Society of Ontario, Toronto.
Rep.ent.Soc.Ont. 1871-1958.
Continued as Proceedings of the Entomological Society
of Ontario. Toronto.   E.S 2522

Report of the Entomologist. Minnesota University Experiment
Station. Minneapolis.
Rep.Ent.Minn.Univ.Exp.Stn 1895-1903.
Continued as Report of the Minnesota State Entomologist.
St. Paul.   E.S 2480

Report of the Entomologist. United States Department of
Agriculture. Washington.
Rep.Ent.U.S.Dep.Agric. 1880-1928 (imp.)
Continued as Report of the United States Bureau of Entomology
and Plant Quarantine Washington.   E.S 2458 1

Report. Entomology Branch. New South Wales Department of
Agriculture. Sydney.
Rep.Ent.Brch N.S.W.Dep.Agric. 1967 →   E.S 2252 a

Report. Environment Canada. Ottawa.
Rep.Envir.Can. 1971/1972 →   S. 2613

Report of the Epsom College Natural History Society. London.
Rep.Epsom Coll.nat.Hist.Soc. 1-11, 13, 21, 23-26. 1889-1950.   S. 65

Report of the Erith and Belvedere Natural History and Scientific
Society.
Rep.Erith & Belvedere nat.Hist.Soc. 5-6. 1882-1884.   S. 66

TITLE	SERIAL No.

Report. Essex Bird Watching and Preservation Society. Chelmsford.
    See Essex Bird Watching and Preservation Society.
    Chelmsford.   T.B.S 266

Report. Eton College Natural History Society.
    Rep.Eton Coll.nat.Hist.Soc. 1929-1963.
    Replaced by Atom. Eton College.   S. 85 A

Report on European Paleobotany. Paleobotanical Department,
    Swedish Museum of Natural History, Stockholm.
    Rep.Eur.Paleobot. 1939-1949.   P.S 426

Report to the Evolution Committee of the Royal Society. London.
    Rep.Evol.Comm.R.Soc. 1902-1909.   S. 3 H3

Report of the Executive Committee. National Central Library. London.
    Rep.exec.Comm.natn.cent.Libr. 1963 →   S. 497 b

Report received from Experiment Stations, Empire Cotton Growing
    Corporation. London.
    Rep.Exp.Stns Emp.Cott.Grow.Corp. 1923-1933.
    Continued as Progress Report from Experiment Stations.
    Empire Cotton Growing Corporation.   E.S 76

Report on Experiment Stations and Extension Work in the United States.
    See Report on Agricultural Experiment Stations.
    United States Department of Agriculture. Washington.   E.S 2448 a

Report on the Experimental Farms. Department of Agriculture,
    Canada. Ottawa.
    Rep.exp.Fms Can. 1894; 1896.   E.S 2524

Report. Experimental and Research Station, Nursery and Market
    Garden Industries Development Society Ltd., Cheshunt.
    Rep.exp.Res.Stn, Cheshunt 1915-1927.   E.S 72

Report of Explorations and Surveys for a Railroad from the
    Missisippi River to the Pacific Ocean. 1853-55. Washington.
    Rep.Explor.& Surv.1853-54 Wash. 1855-1860.   75 C.q.U

Report on Extension Work in Agriculture and Home Economics,
    U.S. Department of Agriculture. Washington.
    Rep.Ext.Wk Agric.Home Econ.U.S. 1931-1949.
    Formerly Report of the Director of the Extension Service,
    U.S. Department of Agriculture. Washington.   E.S 2458c

Report and Extract Moss Exchange Club.
    See Report Moss Exchange Club.   B.B.S 4

Report of the FAO Mission for Nicaragua. Washington, Rome.
    Rep.FAO Miss.Nicaragua 1950.   75 Ee.o.U

Report of the Faculty of Education, University of Iwate. Morioka.
    Rep.Fac.Educ.Univ.Iwate Vol.26, Pt.3 → 1966 →
    Formerly Report of the College of Liberal Arts, University
    of Iwate.   S. 1996 d

TITLE	SERIAL No.

Report. Faculty of Fisheries of the Prefectural University
of Mie. Otanimachi, Tsu, Mie.
Rep.Fac.Fish.prefect.Univ.Mie 1951 →   Z.S 1953 A

Report. Faculty of Science. Egyptian (Fouad I) University. Cairo.
Rep.Fac.Sci.Egypt.Univ. 1932-1951.   S. 2044 B

Report of the Faculty of Science, Kagoshima University. Kagoshima.
Rep.Fac.Sci.Kagoshima Univ. 1968 →
Formerly Science Reports of Kagoshima University.   S. 1990 b

Report of Faculty of Science, Shizuoka University. Shizuoka.
Rep.Fac.Sci.Shizuoka Univ. 1965 →
Formerly Report. Liberal Arts and Science Faculty,
Shizuoka University. Shizuoka.   S. 1995 e

Report. Fair Isle Bird Observatory. Edinburgh.
Rep.Fair Isle Bird Obs. 1949 →   T.B.S 213 A

Report of the Fan Memorial Institute of Biology. Peiping.
Rep.Fan.meml Inst.Biol. 1928-1937.   S. 1977 A

Report on the Fauna and Flora of Wisconsin. Museum of Natural
History, Wisconsin State University. Stevens Point.
See Report. Museum of Natural History. Wisconsin State
University.   S. 2537

Report. Feasibility Study of Computer - Orientated Systems for
Geological Data Handling. Cambridge.
Rep.feasibil.Study comput.-orient.Syst.geol.
Data Handling 1966 →   P.S 124

Report of the Federal Department of Forest Research,
Nigeria. Ibadan.
Rep.Fed.Dep.For.Res.Nigeria 1965/6 →   B.S 2294 d

Report of the Federal Department of Veterinary Research.
Lagos, Nigeria.
Rep.fed.Dep.vet.Res.Nigeria 1957-1958 →
Formerly Report of the Department of Veterinary Research,
Nigeria.   Z. 74M f N

Report of the Federal Fisheries Service, Nigeria. Lagos.
Rep.fed.Fish.Serv.Nigeria 1955-1956 →   Z. 74M o N

Report of the Federal National Parks Board, Rhodesia and
Nyasaland. Salisbury.
Rep.fed.natn.Pks Bd Rhod. 1961-1962 - 1962-1963.
Continued as Report of the National Parks Advisory
Board and Director of National Parks and Wild Life
Management. Ministry of Lands and Mines. Rhodesia.   Z.S 2065 A

Report. Federated Malay States. Kuala Lumpur.
Rep.Fed.Malay St. 1914-1918.   S. 1925 b

| TITLE | SERIAL No. |

Report of the Felsted School Natural History Society. Chelmsford.
    Rep.Felsted Sch.nat.Hist.Soc. 1886-1893.
    Continued as Report of the Felsted School Scientific Society.
    Formerly Report of the Proceedings of the Felsted School
    Natural Science Society.   S. 83

Report of the Felsted School Scientific Society.
    Rep.Felsted Sch.scient.Soc. 1894-1933.
    Continued as Felsted Bury.
    Formerly Report of the Felsted School Natural History Society.   S. 83

Report. Field Columbian Museum.
    See Publication. Field Columbian Museum. Chicago,
    Report Series.   S. 2330 A

Report. Field Museum of Natural History. Chicago.
    Rep.Fld Mus.nat.Hist. 1965 →   T.R.S 5725 B &
    Formerly Report. Chicago Natural History Museum.   S. 2330 A

Report. Field Naturalists' Club of Victoria.
    Rep.Fld Nat.Club Vict. No.6-8. 1885-1888.   S. 2114 B

Report. Field Studies Council.
    Rep.Fld Stud.Coun. 1954 →
    Formerly Report. Council for the Promotion of Field Studies.   S. 182 A

Report of the Fifth Thule Expedition 1921-1924. Copenhagen.
    Rep.Fifth Thule Exped. 1931-1945.   71.o.D

Report of the First Scientific Expedition to Manchoukuo. 1933.
    Rep.1st scient.Exped.Manchoukuo 1934-1939.   73 H.o.T

Report of the First Session of the State Oceanographical
    Institute. Moscow.
    See Doklady Gosudarstvennogo Okeanograficheskogo Instituta.
    Moskva.   S. 1845 C

Report. Fish Marketing Organization, Hong Kong.
    Rep.Fish Mktg Org.Hong Kong 1970/1971 →   Z.S 1946 a B

Report of the Fisheries Department, Malawi. Zomba.
    Rep.Fish.Dep.Malawi 1971 →   Z. 74 D f M

Report on the Fisheries Department, Straits Settlements and
    Federated Malay States, Singapore.
    Rep.Fish.Dep.Straits Settl. 1931-1933, 1935-1938.   Z. 73G f S

Report. Fisheries Department, Western Australia. Perth.
    Rep.Fish.Dep.West.Aust. 1961 →   Z.S 2115 B

Report on the Fisheries of Egypt. Cairo.
    Rep.Fish.Egypt 1919-1930.
    Continued as Rapport sur les Pêcheries d'Egypte.   Z.O 74B q E

Report on Fisheries Investigations, Nigeria. Lagos.
    Rep.Fish.Invest.Nigeria 1942-1948.   Z. 74M o N

| TITLE | SERIAL No. |

Report. Fisheries & Marine Biological Survey, Union of
   South Africa. Cape Town. (later Pretoria.)
   Rep.Fish.mar.biol.Surv.Un.S.Afr. 1920-1936.
   Continued as Report. Division of Fisheries,
   Union of South Africa.   Z.S 2030 C

Report on Fisheries. Marine Department, New Zealand. Wellington.
   Rep.Fish.N.Z. 1930 →   Z. 77D o N

Report on the Fisheries of New South Wales. Sydney.
   Rep.Fish N.S.W.   1938-1939, 1942, 1945.
   Formerly Report. New South Wales State Fisheries.
   Continued as Report of the Department of Fisheries of
   New South Wales.   Z.S 2106

Report of the Fisheries Research Board of Canada. Ottawa.
   Rep.Fish.Res.Bd Can. 1939 →
   Formerly Report of the Biological Board of Canada.   Z.S 2628 F

Report on the Fisheries of Scotland, Scottish Home Department.
   Edinburgh.
   Rep.Fish.Scotl. 1939 →
   Formerly Report of the Fishery Board for Scotland.   Z.S 380

Report of the Fishery Board for Scotland. Edinburgh.
   Rep.Fishery Bd.Scotl. 1883-1938.
   Continued as Report on the Fisheries of Scotland.
   Scottish Home Department.   Z.S 380

Report. Fishery Board of Sweden. Series Hydrography. Göteborg.
   Rep.Fishery Bd Swed. No.1-26, 1953-1972.
   Formerly in Svenska Hydrografisk-Biologiska Kommissionens
   Skrifter. Göteborg.   M.S 522

Report of the Fishery Research Institute. Newfoundland.
   See Report of the Newfoundland Fishery Research Commission.   S. 2670

Report of Fishery Research Laboratory, Kyushu University. Fukuoka.
   Rep.Fish.Res.Lab.Kyushu Univ. 1971 →   Z.S 1972 B

Report of the Fishery Research Laboratory, Newfoundland.
   St. John's.
   Rep.Fishery Res.Lab.Newfoundld 1934, 1936-1937.
   Formerly Report of the Newfoundland Fishery Research Commission.
   Continued in Economic Bulletin. Department of Natural
   Resources, Newfoundland.   Z.S 2670

Report. Fishery Research Unit, Monkey Bay. Zomba.
   Rep.Fish.Res.Unit Monkey Bay 1969/1971 →   Z. 74 D f M

Report of the Florida Agricultural Experiment Station. Gainesville.
   Rep.Fla agric.Exp.Stn   1908-1943 (imp.)   E.S 2471 b

TITLE	SERIAL No.

Report. Florida State Geological Survey. Tallahassee.
   Rep.Fla St.geol.Surv. 1907-1930   P.S 1850

Report. Fly Fishers' Club. London.
   Rep.Fly Fish.Club Lond. 1906-1911.   Z.S 332 A

Report of the Folkestone Natural History Society.
   Rep.Folkestone nat.Hist.Soc. 1870-1871: 1950-1958.   S. 86 B

Report of the Food Investigation Board. London.
   Rep.Fd Invest.Bd 1918-1957.   S. 205 F

Report on the Forest Administration of Northern Nigeria. Kaduna.
   Rep.Forest Adm.nth.Nigeria 1962-1966/1967.   B.S 2295

Report of the Forest Entomology and Pathology Branch, Canada. Ottawa.
   Rep.For.Ent.Path.Canada 1961 →
   Formerly Report of the Forest Insect and Disease Survey,
   Department of Agriculture, Canada, Ottawa.   E.S 2525

Report of the Forest, Fish and Game Commissioner, New York State.
   Albany.
   Rep.Forest Fish Game Commnr N.Y.St. 1907-1909.   75 C.q.N

Report. Forest and Gardens Department, Mauritius.
   Rep.For.Gard.Dept., Mauritius 1907-1908.   B.S 2297

Report of the Forest Insect and Disease Survey, Department
   of Agriculture, Canada. Ottawa.
   Rep.Forest Insect Dis.Surv.Can. 1952-1961.
   Formerly Report of the Forest Insect Survey, Canada. Ottawa.
   Continued as Report of the Forest Entomology and
   Pathology Branch, Canada. Ottawa.   E.S 2525

Report of the Forest Insect Survey, Canada. Ottawa.
   Rep.Forest Insect Surv.Can. 1940-1950.
   Continued as Report of the Forest Insect and Disease Survey,
   Department of Agriculture, Canada. Ottawa.   E.S 2525

Report on Forest Insects, Quebec.
   Rep.Forest Insects, Queb. 1946-1949.   E.S 2526 a

Report of the Forest Products Research Board, Department of Scientific
   and Industrial Research. London.
   Rep.Forest Prod.Res.Bd 1927-1937.
   Continued as Forest Products Research. Department of
   Scientific and Industrial Research. London.   B.S 92

Report. Forest Products Research Institute. Kumasi.
   Rep.Forest Prod.Res.Inst.Kumasi 1970/1971 →   B.S 2317

Report on Forest Research. Forestry Commission. London.
   Rep.Forest Res., Lond. 1955 →   E.S 62a

Report Forest Research Institute, Bogor.
   See Rapport. Bosbouwproefstation, Bogor.   B.S 581.9(595.p40)BOS

TITLE	SERIAL No.

Report of Forest Research Institute. Wellington, N.Z.
    Rep.Forest Res.Inst.N.Z. 1971 →   E.S 2266 b
                           1972 →   B.S 2482 a

Report of the Forestry Commission. London.
    Rep.For.Commn 1919 →   E.S 63a

Report. Forestry Department, British Solomon Islands Protectorate. Honiara.
    Rep.For.Dep.Br.Solomon Isl. 1961 →   B.S 2469

Report. Forestry Department, Ghana. Accra.
    Rep.For.Dep.Ghana 1957-1958.
    Formerly Report. Forestry Department, Gold Coast. Accra.
    Continued as Report. Forestry Division, Ministry
    of Agriculture, Ghana. Accra.   B.S 2250

Report. Forestry Department. Gold Coast, Accra.
    Rep.For.Dep.Gold Cst 1951-1956 (imp.)
    Continued as Report. Forestry Department. Ghana, Accra.   B.S 2250

Report. Forestry Division, Ministry of Agriculture, Ghana. Accra.
    Rep.For.Div., Min.Agric.Ghana 1959-1960.
    Formerly Report. Forestry Department. Ghana, Accra.   B.S 2250

Report. Franz Theodore Stone Institute of Hydrobiology.
Ohio State University. Columbus.
    Rep.Franz Theodore Stone Inst.Hydrobiol. 1950-1954.
    Formerly Report. Franz Theodore Stone Laboratory.
    Ohio State University. Columbus Ohio.   S. 2332 a D

Report. Franz Theodore Stone Laboratory. Ohio State University. Columbus, Ohio.
    Rep.Franz Theodore Stone Lab.Ohio St.Univ. 1926-1950.
    Continued as Report. Franz Theodore Stone Institute of
    Hydrobiology. Ohio State University. Columbus.   S. 2332 a D

Report of the Free Libraries and Museum Committee, Great Yarmouth.
    Rep.Free Libr.Mus.Comm.Gt.Yarmouth No.27, 1912-1913.   S. 111

Report of the Free Public Museum, Liverpool.
    Rep.Mus.Lpool 1914-1949. (imp.)
    Continued as Report. Libraries, Museums & Arts Committee,
    City of Liverpool.
    Formerly Report of the (Free) Public Library,
    Museum, etc. Liverpool.   S. 160 B

Report. Freshwater Biological Association. Ambleside.
    Rep.Freshwat.biol.Ass. 1930 →   S. 7 b A

Report. Freshwater Biological Association of the British Empire. Ambleside.
    See Report. Freshwater Biological Association. Ambleside.   S. 7 b A

Report. Friends of the National Libraries. London.
    Rep.Friends natn.Libr. 1966 →   S. 223 a

| TITLE | SERIAL No. |

Report. Fruit Research Station, Saharanpur.
    Rep.Fruit Res.Stn Saharanpur   1953 →   B.S 1649

Report of the Gakugei Faculty of Iwate University. Morioka.
    Rep.Gakugei Fac.Iwate Univ.   Vol.2, Pt.2 - Vol.17, Pt.2,
    Vol.18, Pt.3 - Vol.20, Pt.3, 1950-1962.
    Formerly   Report of the Department of Liberal Arts,
    Iwate University.
    Continued as   Report of the College of Liberal Arts,
    University of Iwate.   S. 1996 d

Report of the Game Commissioner, Saskatchewan, Regina.
    Rep.Game Commnr Sask.   1924-1928.
    Formerly   Report of the Chief Game Guardian, Province
    of Saskatchewan.   Z.O 75B.o.S

Report of the Game Department, Colony and Protectorate of
    Kenya. Nairobi.
    Rep.Game Dep.Kenya   1924-1937   Z. Mammal Section

Report of the Game Department, Federated Malay States. Kuala Lumpur.
    Rep.Game Dep.F.M.S.   1937-1938.   Z. 77A o F

Report of the Game Department, Tanganyika. Dar-es-Dalaam.
    Rep.Game Dep.Tanganyika   1955-1956 →   Z.O 74 Db o T

Report of the Game Department, Uganda Protectorate. Entebbe.
    Rep.Game Dep.Uganda   1925-1949.
    Continued as Report of the Game and Fisheries Department,
    Uganda.   Z. Mammal Section

Report of the Game and Fisheries Department, Uganda
    Protectorate. Entebbe.
    Rep.Game Fish.Dep.Uganda   1950 →
    Formerly Report of the Game Department, Uganda.   Z. Mammal Section

Report to the General Assembly of the United Nations on the
    Administration of the Territory of New Guinea. Canberra.
    Rep.gen.Assembly U.N.Admin.Terr.New Guinea   1952 - 1969/1970.
    Formerly   Report of the Council of the League of Nations
    on the Administration of the Territory of New Guinea. Canberra. S. 2146
    Replaced by   Report. Papua New Guinea. Canberra.   S. 2146 b

Report of the General Board of Studies. Cambridge University.
    Rep.Bd.Stud.Camb.   1912-1935.   S. 35 A

Report. General Fisheries Council for the Mediterranean.
    Food and Agriculture Organization of the United Nations. Rome.
    Rep.gen.Fish.Coun.Mediterr.   No.10 → 1970 →
    Formerly   Session Report. General Fisheries Council for
    the Mediterranean.   Z.S 2713 D

Report of the Geological Commission. Cape of Good Hope. Cape Town.
    Rep.geol.Commn Cape Good Hope   1896-1911   P.S 1174

| TITLE | SERIAL No. |

Report of the Geological Department. Uganda Protectorate.
   Annual Report. Entebbe.
     Rep.geol.Dep.Uganda a.Rep. 1920.
     <u>Continued as</u> Report of the Geological Survey Department,
     Uganda Protectorate. Annual Report. Entebbe.
     <u>See also</u> Report. Geological Survey of Uganda. Entebbe.   P.S 1179 c

Report. Geological Division. Department of Lands and Mines.
   Tanganyika Territory. Dar-es-Salaam.
     Rep.geol.Div.Tanganyika 1935-1949.
     <u>Formerly</u> Report. Geological Survey. Tanganyika Territory.
     <u>Continued as</u> Report of the Geological Survey Department.
     Tanganyika.   P.S 1179 A

Report of Geological Explorations. Geological Survey of
   New Zealand. Wellington.
     Rep.geol.Explor.geol.Surv. N.Z. 1870-1893   P.S 1155

Report of the Geological Foundation of the Netherlands.
   <u>See</u> Jaarverslag van de Geologische Stichtung. Heerlen.   P.S 1291 A

Report of the Geological Museum called Peter the Great of the
   Imperial Academy of Sciences.
     <u>See</u> Godovoĭ Otchet' Geologicheskago Muzeya imeni Imperatora
     Petra Velikago Imperatoskoĭ Akademii Nauk'   P.S 504

Report. Geological and Natural History Survey of Minnesota.
   Minneapolis.
     Rep.geol.nat.Hist.Surv.Minn. 1872-1901
        (1884 - 1901)   T.R.S 5128 & P.S 1942

Report. Geological and Natural History Survey of Minnesota.
   Annual Report. Minneapolis.
     Rep.geol.nat.Hist.Surv.Minn.a.Rep. 1872-1898.
        (Wanting No.7.)   P.S 1940

Report. Geological Society of America.
   <u>See</u> Report of the Officers and Committees. Geological
   Society of America.   P.S 865 c

Report. Geological Survey of the Anglo-Egyptian Sudan. Khartoum.
     Rep.geol.Surv.Anglo-Egypt.Sudan 1914-1919
     (Summaries only), 1950-1953.
     <u>Continued as</u> Report. Geological Survey Department, Ministry
     of Mineral Resources, Republic of the Sudan. Khartoum.   P.S 1183

Report of the Geological Survey of Arkansas. Little Rock.
   <u>See</u> Report of the Arkansas Geological Survey. Little Rock.   P.S 1875

Report of the Geological Survey Board. Department of Scientific
   and Industrial Research. London.
     Rep.geol.Surv.Bd 1945-1951.
     <u>Formerly and Continued as</u> Summary of Progress of the
     Geological Survey of Great Britain and the Museum
     of Practical Geology. London.   P.S 1000

TITLE	SERIAL No.

Report of the Geological Survey, Borneo Region, Malaysia. Kuching.
    Rep.geol.Surv.Borneo   1963 →
    Formerly Report of the Geological Survey Department.
British Territories in Borneo. Kuching.   P.S 1118 A

Report. Geological Survey Branch. New Zealand. Wellington.
    Rep.geol.Surv.Br.N.Z.   New Series, 1912-1941.
    Formerly Report. New Zealand Geological Survey Department. Wellington.
    Continued as Report. Geological Survey. New Zealand.
Wellington.   P.S 1163

Report. Geological Survey of Canada. Montreal and Ottawa.
    Rep.geol.Surv.Can.   1846-1847, 1853-1858, 1863-1904.   P.S 1070

Report. Geological Survey, Colony of Fiji. Suva.
    See Report. Geological Survey, Fiji. Suva.   S. 1167 A

Report on the Geological Survey, Colorado, Denver.
    Rep.geol.Surv.Colo.   1908.   M.S 2637 A

Report of the Geological Survey of Connecticut. New Haven.
    Rep.geol.Surv.Conn.   1837   P.S 1885

Report. Geological Survey of Denmark. Copenhagen.
    See Rapport. Danmarks Geologiske Undersøgelse.   P.S 1395 A

Report. Geological Survey Department, Bechuanaland.
    See Report. Department of Geological Survey, Bechuanaland.
(Parow, Cape.)   P.S 1175 C

Report of the Geological Survey Department, Botswana. Gaberones.
    Rep.geol.Surv.Dep., Botswana   1966 →
    Formerly Report. Department of Geological Survey,
Bechuanaland. (Parow, Cape.)   P.S 1175 C

Report of the Geological Survey Department. British Guiana. Georgetown.
    Rep.geol.Surv.Dep.Br.Guiana   1947 →   P.S 1168 A

Report of the Geological Survey Department. British Territories in Borneo. Kuching.
    Rep.geol.Surv.Dep.Br.Terr.Borneo   1949-1962
    Continued as Report of the Geological Survey,
Borneo Region, Malaysia. Kuching.   P.S 1118 A

Report of the Geological Survey Department. Cyprus. Nicosia.
    Rep.geol.Surv.Dep.Cyprus   1955 →   P.S 1179 J

Report of the Geological Survey Department. Federated Malay States Kuala Lumpur.
    Rep.geol.Surv.Dep.F.M.St.   1930-1940
    Formerly Report of the Geologist. Federated Malay States.
    Continued as Report of the Geological Survey Department.
Malayan Union.   P.S 1117

Report of the Geological Survey Department, Federation of Malaya. Kuala Lumpur.
    Rep.geol.Surv.Dep.Fed.Malaya   1948-1949.
    Formerly Report of the Geological Survey Department.
Malayan Union.   P.S 1117

TITLE	SERIAL No.

Report. Geological Survey Department. Fiji, Suva.
    Rep.geol.Surv.Dep.Fiji 1958 →
    Formerly Report. Geological Survey, Fiji. Suva.   P.S 1167 A

Report Geological Survey Department, Jamaica. Kingston.
    Rep.geol.Surv.Dep.Jamaica 1951 →   P.S 1169 B

Report of the Geological Survey Department, Malawi. Zomba.
    Rep.geol.Surv.Dep.Malawi 1963 →
    Formerly Report of the Geological Survey Department, Nyasaland Protectorate. Livingstonia & Zombia.   P.S 1179 B

Report of the Geological Survey Department. Malayan Union. Kuala Lumpur.
    Rep.geol.Surv.Dep.Malay.Un. 1946-1947.
    Formerly Report of the Geological Survey Department. Federated Malay States.
    Continued as Report of the Geological Survey Department. Federation of Malaya.   P.S 1117

Report. Geological Survey. Department of Mines and Resources. Newfoundland. St. John's.
    See Report. Geological Survey, Province of Newfoundland. St. John's.   P.S 1096 A

Report. Geological Survey Department, Ministry of Mineral Resources, Republic of the Sudan. Khartoum.
    Rep.geol.Surv.Dep.Sudan 1953-1957.
    Formerly Report. Geological Survey of the Anglo-Egyptian-Sudan. Khartoum.   P.S 1183

Report. Geological Survey Department. Northern Rhodesia. Annual Report. Lusaka.
    See Report. Geological Survey. Northern Rhodesia. Annual Report. Lusaka.   P.S 1172 D

Report of the Geological Survey Department, Nyasaland Protectorate. Livingstonia & Zomba.
    Rep.geol.Surv.Dep.Nyasald 1923-1962 (imp.)
    Continued as Report of the Geological Survey Department, Malawi. Zomba.   P.S 1179 B

Report of the Geological Survey Department. Sierra Leone. Freetown.
    See Report of the Geological Survey. Sierra Leone. Freetown.   P.S 1179 G

Report of the Geological Survey Department. Somaliland Protectorate.
    Rep.geol.Surv.Dep.Somalild 1957-1960.
    Formerly Report of the Geological Survey. Somaliland Protectorate. Annual Report.
    Continued as Report of the Geological Survey of the Ministry of Commerce and Industry. Somali Republic.   P.S 1179 A

TITLE	SERIAL No.

Report. Geological Survey Department, Sudan.
   See Report. Geological Survey Department, Ministry of
Mineral Resources, Republic of the Sudan. Khartoum.   P.S 1183

Report of the Geological Survey Department. Swaziland.
   Rep.geol.Surv.Dep.Swaziland 1944-1957.
   (From 1945-1946 Styled Progress Report.)
   Formerly Report of the Government Geologist. Swaziland.
   Continued as Report of the Geological Survey and   P.S 1175 A
Mines Department. Swaziland.   & M.S 2106 A

Report of the Geological Survey Department. Tanganyika.
   Dar-es-Salaam.
   Rep.geol.Surv.Dep.Tanganyika 1950-1959.
   Formerly Report. Geological Division. Department of Lands
and Mines. Tanganyika Territory.
   Continued as Report of the Geological Survey Division
Tanganyika.   P.S 1179 A

Report of the Geological Survey Department. Uganda Protectorate.
   Entebbe.
   Rep.geol.Surv.Dep.Uganda 1922 →
   See also Report of the Geological Survey of Uganda.
   Formerly Report of the Geological Department.
Uganda Protectorate. Entebbe.   P.S 1179 c

Report of the Geological Survey Department, Uganda Protectorate.
   Annual Report. Entebbe.
   Rep.geol.Surv.Dep.Uganda a.Rep. 1922-1959.
   Formerly Report of the Geological Department. Uganda
Protectorate. Annual Report. Entebbe.
   Continued as Report of the Geological Survey and Mines
Department, Uganda. Annual Report. Entebbe.
   See also Report. Geological Survey of Uganda. Entebbe.   P.S 1179 c

Report. Geological Survey Department. Zambia. Annual Report. Lusaka.
   See Report. Geological Survey. Zambia. Annual Report.
Lusaka.   P.S 1172 D

Report of the Geological Survey Division. Tanganyika.
   Rep.geol.Surv.Div.Tanganyika 1960 →
   Formerly Report of the Geological Survey Department.
Tanganyika.   P.S 1179 A

Report. Geological Survey. Fiji, Suva.
   Rep.geol.Surv.Fiji 1953-1957.
   Continued as Report. Geological Survey Department.
Fiji, Suva.   P.S 1167 A

Report of the Geological Survey, Ghana. Accra.
   Rep.geol.Surv.Ghana 1956 →
   Formerly Report of the Geological Survey,
Gold Coast. Accra.   P.S 1179 E

| TITLE | SERIAL No. |

Report of the Geological Survey, Gold Coast. Accra.
    Rep.geol.Surv.Gold Cst 1913-1956.   P.S 1179 E
                     1928-1953 (imp.)   M.S 2105
    Continued as Report of the Geological Survey, Ghana. Accra.

Report. Geological Survey of Greenland. Copenhagen.
    Rep.geol.Surv.Greenld 1964 →   P.S 1398

Report. Geological Survey, Illinois.
    See Report of Investigations. Illinois State Geological
    Survey. Urbana.   P.S 1902 B

Report of the Geological Survey of India. Calcutta.
    See General Report on the Work carried out by the
    Geological Survey of India. Calcutta.   P.S 1113

Report of the Geological Survey of India and of the Museum of
    Geology. Calcutta.
    Rep.geol.Surv.India 1861-62 & 1864-1865.   P.S 1113

Report of the Geological Survey of Indiana. Indianapolis.
    See Report of the Indiana Department of Geology and Natural
    Resources. Indianapolis.   P.S 1905

Report of the Geological Survey of Iowa.
    See Report of the Iowa Geological Survey. Desmoines.   P.S 1912

Report Geological Survey of Iran. Teheran.
    Rep.geol.Surv.Iran 1964 →   P.S 1745

Report. Geological Survey of Japan. Tokyo.
    Rep.geol.Surv.Japan No.126 → 1948 →
    Formerly Report. Imperial Geological Survey of Japan. Tokyo. P.S 1766

Report. Geological Survey, Kentucky. Frankfort.
    Rep.geol.Surv.Ky 1856-1957.   M.S 2643
                 1854-1880, 1908-1927.   P.S 1925

Report. Geological Survey of Kenya. Nairobi.
    Rep.geol.Surv.Kenya No.15 → 1948 →
    Formerly Report. Lands, Mines and Surveys Department
    Mines Division. Kenya. Nairobi.   M.S 2109 A

Report. Geological Survey of Kwangtung and Kwangsi. Canton.
    Rep.geol.Surv.Kwantung Kwangsi 1927-1933.   P.S 1790

Report of the Geological Survey of Maryland. Baltimore.
    See Maryland Geological Survey. General Series.   P.S 1928

Report of the Geological Survey of Michigan.
    See Report of the State Board of Geological Survey
    of Michigan. Lansing.   P.S 1936

Report of the Geological Survey and Mines Department. Swaziland.
    Rep.geol.Surv.Mines Dep.Swaziland 1958 →
    Formerly Report of the Geological Survey Department.   P.S 1175 A
    Swaziland.   & M.S 2106 A

TITLE	SERIAL No.

Report of the Geological Survey and Mines Department, Uganda.
    Annual Report. Entebbe.
        Rep.geol.Surv.Mines Dep.Uganda a.Rep. 1967 →
        Formerly Report of the Geological Survey Department,
Uganda Protectorate. Annual Report. Entebbe.   P.S 1179 c

Report. Geological Survey and Mines Department, Uganda. Entebbe.
        Rep.geol.Surv.Mines Dep.Uganda 1968 →
        Formerly Report. Geological Survey of Uganda. Entebbe.   P.S 1179 c

Report of the Geological Survey of the Ministry of Commerce
    and Industry. Somali Republic.
        Rep.geol.Surv.Somali 1960 →
        Formerly Report of the Geological Survey Department.
Somaliland Protectorate.   P.S 1179 A

Report. Geological Survey of Missouri. Jefferson City.
        Rep.geol.Surv.Mo. 1853-1874, 1890-1898.
        Continued as Report. Missouri Bureau of Geology and Mines.
Jefferson City.   P.S 1950 & 1952

Report on the Geological Survey and Museum, the Science Museum,
    and the work of the Solar Physics Committee. London.
        Rep.geol.Surv.Mus., Lond. 1908-1912.
        Continued as Report on the Science Museum and on the
Geological Survey and Museum of Practical Geology. London.   O. 72Aa o G

Report of the Geological Survey of Natal and Zululand.
    Pietermaritzburg.
        Rep.geol.Surv.Natal Zululand 1899-1905   P.S 1175

Report. Geological Survey of New Jersey.
        See Report. New Jersey Geological Survey. Trenton.   P.S 1957

Report Geological Survey New South Wales. Sydney.
        Rep.geol.Surv.N.S.W. 1962 →   P.S 1124 A

Report of the Geological Survey of New Zealand. Wellington.
        Rep.geol.Surv.N.Z. 1941-1947.
        Formerly Report. Geological Survey Branch. New Zealand.
Wellington.   P.S 1163

Report of the Geological Survey of Newfoundland. London & Montreal.
        Rep.geol.Surv.Newfoundld 1864-1880.
        (Re-publd.London, 1881.); 1904-1905.
        Replaced by Report. Geological Survey, Province
of Newfoundland. St.John's.   P.S 1095

Report of the Geological Survey of Nigeria.
        Rep.geol.Surv.Nigeria 1930 →   P.S 1179 K B

Report. Geological Survey. Northern Rhodesia. Annual Report.
    Lusaka.
        Rep.geol.Surv.Nth.Rhod.a.Rep. 1963.
        Continued as Report. Geological Survey. Zambia.
Annual Report. Lusaka.   P.S 1172 D

TITLE	SERIAL No.

Report. Geological Survey. Northern Rhodesia. Ndola.
    Rep.geol.Surv.Nth.Rhod. 1954-1963.
    Continued as Report. Geological Survey. Zambia. Lusaka.   P.S 1172 C

Report of the Geological Survey of Ohio. Columbus.
    Rep.geol.Surv.Ohio 1873-1893   P.S 1967
                  Vol.4, 1882.   75 C o D

Report. Geological Survey of Pennsylvania. Harrisburg.
    Rep.geol.Surv.Pa 1836-1885.   P.S 1970

Report. Geological Survey. Province of Newfoundland. St.John's.
    Rep.geol.Surv.Prov.Newfoundld 1953 →
    Replaces Report of the Geological Survey of Newfoundland,
St. John's.   P.S 1096 A

Report of the Geological Survey of Queensland.
    Rep.geol.Surv.Qd 1879-1904.
    Continued in Publications of the Geological Survey
    of Queensland.
    (For a recent Report of the Geological Survey of Queensland
      No.1, 1963 → See P.S 1144 A.)   P.S 1145

Report. Geological Survey of Queensland. Brisbane.
    Rep.geol.Surv.Qd No.1 → (1963 →)
    (For a former Report of the Geological Survey of
      Queensland 1879-1904 See P.S 1145.)   P.S 1144 A

Report of the Geological Survey. Sierra Leone. Freetown.
    Rep.geol.Surv.Sierra Leone 1918 → (imp.)   P.S 1179 G

Report. Geological Survey. Somaliland Protectorate.
    Rep.geol.Surv.Somalild 1956 →   P.S 1179 H

Report of Geological Survey. Somaliland Protectorate. Annual Report.
    Rep.geol.Surv.Somalild a.Rep. 1952-1957.
    Continued as Report of the Geological Survey Department.
Somaliland Protectorate.   P.S 1179 H

Report. Geological Survey of the South African Republic.
    Rep.geol.Surv.S.Afr. 1897-1898   P.S 1176

Report. Geological Survey of South Australia. Adelaide.
    Rep.geol.Surv.S.Aust. 1912-1916.   P.S 1136

Report. Geological Survey of South Australia.
    Annual Report. Adelaide.
    See Report of the Government Geologist. South Australia.   P.S 1138

Report. Geological Survey. Southern Rhodesia. Bulawayo.
    Rep.geol.Surv.Sth.Rhod. 1911, 1913-1915, 1917-1956.   P.S 1172

Report on the Geological Survey of the State of Ohio. Columbus.
    Rep.geol.Surv.St.Ohio No.2. 1838.   P.S 1965

| TITLE | SERIAL No. |

Report of the Geological Survey. Tanganyika Territory. Dar-es-Salaam.
    Rep.geol.Surv.Tanganyika 1926-1934.
    Continued as Report. Geological Division. Department of
    Lands and Mines. Tanganyika Territory.   P.S 1179 A

Report on the Geological Survey, Tasmania. Hobart.
    Rep.geol.Surv.Tasm. 1910-19   M.S 2447

Report of the Geological Survey of Texas. Austin.
    Rep.geol.Surv.Texas 1889-1892.   P.S 1977

Report. Geological Survey of the Transvaal. Pretoria.
    Rep.geol.Surv.Transv. 1903-1908
    Continued as Report of the Geological Survey. Union of
    South Africa. Pretoria.   P.S 1176

Report. Geological Survey of Uganda. Entebbe.
    Rep.geol.Surv.Uganda 1959-1965.
    Continued as Report. Geological Survey and Mines Department,
    Uganda. Entebbe.
    See also Report of the Geological Survey Department,
    Uganda Protectorate. Annual Report. Entebbe.   P.S 1179 c

Report of the Geological Survey. Union of South Africa. Pretoria.
    Rep.geol.Surv.Un.S.Afr. 1909-1913.
    Formerly Report. Geological Survey of the Transvaal.
    Pretoria.   P.S 1176

Report of the Geological Survey of the United Kingdom and of the
    Museum of Practical Geology. London.
    Rep.geol.Surv.Lond. 1856, 1862-63, 1892-1897, 1901.
    (For further Reports See Mem.Geol.Surv. Summary of Progress.) P.S 1000

Report of the Geological Survey, United States, Washington.
    See Report of the United States Geological Survey,
    Washington.   P.S 1860

Report of the Geological Survey of Victoria. Melbourne.
    See Report of Progress. Geological Survey of Victoria.
    Melbourne.   P.S 1130

Report of the Geological Survey of West Virginia. Morgantown.
    Rep.geol.Surv.W.Va 1899 →   P.S 1986

Report of the Geological Survey. Western Australia. Annual
    Report. Perth.
    Rep.geol.Surv.West.Aust.a.Rep. 1940 →
    Formerly Annual Progress Report of the Geological Survey.
    Western Australia. Perth.   P.S 1140

Report. Geological Survey of Western Australia. Perth.
    Rep.geol.Surv.W.Aust. 1969 →   P.S 1140 A

Report. Geological Survey Wisconsin, Iowa and Minnesota.
    Philadelphia.
    Rep.geol.Surv.Wis.etc. 1852   M.S 2614

| TITLE | SERIAL No. |

Report. Geological Survey. Zambia. Annual Report. Lusaka.
    Rep.geol.Surv.Zambia a.Rep. 1964 →
       Formerly Report. Geological Survey. Northern Rhodesia.
       Annual Report. Lusaka.   P.S 1172 D

Report. Geological Survey. Zambia. Lusaka.
    Rep.geol.Surv.Zambia 1964 →
       Formerly Report. Geological Survey. Northern Rhodesia.
       Ndola.   P.S 1172 C

Report of the Geologist. Federated Malay States.
    Rep.Geol.F.M.S. 1903-1927 (imp.)
       Continued as Report of the Geological Survey Department.
       Federated Malay States.   P.S 1117

Report on the Geology of the State of Maine. Augusta.
    Rep.Geol.St.Maine No.2 - 3 1838-1839.   P.S 1927

Report on the Geology of the State of Vermont. Burlington.
    Rep.Geol.St.Vermont 1845-1847.   P.S 1980

Report. Geophysical Laboratory Carnegie Institution,
    Washington, Washington.
       See Report. Director, Geophysical Laboratory,
       Carnegie Institution, Washington.   M.S 2622

Report. German Hydrographic Institute. Hamburg.
    Rep.Germ.Hydrogr.Inst. 1946-1953.
       Translation of Jahresbericht des Deutschen Hydrographischen
       Institut. Hamburg.   M.S 1383

Report. Giza Zoological Gardens. Cairo.
    See Report. Zoological Gardens. Giza.   Z.O 74B o C

Report. Glasgow Archaeological Society.
    Rep.Glasg.archaeol.Soc. 1892-1893.   P.A.S 15 B

Report. Gloucestershire Trust for Nature Conservation Limited.
    Rep.Gloucest.Trust Nat.Conserv. No.5 → 1965 →   S. 104 a

Report of the Gordon Technical College. Geelong.
    Rep.Gordon Tech.Coll. 1894.   S. 2120

Report. Gorgas Memorial Laboratory, Institute of Tropical
    Medicine, Panama. Washington.
    Rep.Gorgas meml Lab. 1951 →   E.S 2416

Report of the Government Biologist. Department of Agriculture,
    Cape of Good Hope. Cape Town.
    Rep.Govt Biol.Cape Good Hope 1900-1904.
       Formerly Report of the Marine Biologist, Department of
       Agriculture, Cape of Good Hope, Cape Town.   Z.S 2030 B

Report Government Botanist and Curator. Cape Town.
    Rep.Govt Bot.& Cur.Cape Town 1897-1903' imp.   B.S 2264

TITLE	SERIAL No.

Report of the Government Botanist and Director of the Botanic
    and Zoologic Garden. Melbourne.
      Rep.Govt Bot.& Dir.bot.zool.Gdn
        1860-1861, 1868, 1874.   B. 58.006
                                                                     MEL Q

Report of the Government Bureau of Microbiology, New South Wales,
    Sydney.
      Rep.Govt Bur.Microbiol.N.S.W.   1909-1912.
      Continued as Report of the Microbiological Laboratory,
      Department of Public Health, New South Wales, Sydney.   S. 2130 a

Report of the Government Chemical Laboratories, Western
    Australia, Perth.
      Rep.Govt Chem.Labs West.Aust. 1946-1950, 1961 →
      Formerly Report of the Government Mineralogist Analyst
      and Chemist, Department of Mines, Western Australia.   M.S 2407 C

Report of the Government Entomologist, Cape of Good Hope. Cape Town.
      Rep.Govt Ent.Cape Good Hope 1896-1901.   E.S 2162

Report of the Government Entomologist, Natal. Pietermaritzburg.
      Rep.Govt Ent.Natal 1899/1900-1901.   E.S 2168 d

Report of the Government Entomologist, Uganda Protectorate. Entebbe.
      Rep.Govt Ent.Uganda   1909/1910.   E.S 2179 a A

Report of the Government Geologist, South Australia. Adelaide.
      Rep.Govt Geol.S.Aust. 1882-1883, 1893-1894, 1912-1915.   P.S 1138
                             1884, 1915.   M.S 2410
      Continued as Report of the Director of Mines and Government
      Geologist, South Australia. Adelaide.

Report of the Government Geologist, Swaziland.
      Rep.Govt Geol.,Swazild   1942-1943.
      Continued as Report of the Geological Survey Department, M.S 2106 A &
      Swaziland.   P.S 1175 A

Report. (Annual General) Government Geologist.
    Western Australia. Perth.
      Rep.Govt Geol.W.Aust. 1888-90   M.S 2409

Report. Government Geologist's Office, Tasmania. Launceston.
      Rep.Govt Geol.Office Tasm.   1902-1904   M.S 2444 A

Report on the Government Horticultural Gardens, Lucknow. Allahabad.
      Rep.Govt hort.Gdns Lucknow   1886-1916 (imp.)   B.S 1585

Report of the Government Mineralogist, Analyst, and Chemist,
    Department of Mines, Western Australia, Perth.
      Rep.Govt Miner.Analyst Chem.West.Aust. 1922, 1941-1945.
      From 1923-1940 Styled Report of the Chemical Branch,
      Mines Department, Western Australia.
      Continued as Report of the Government Chemical Laboratories,
      Western Australia.   M.S 2407 C

TITLE   SERIAL No.

Report. Government Research Institute. Department of Agriculture.
   Formosa.
   Rep.Govt Res.Inst.Dep.Agric.Formosa 1939-1942 (imp.)
   Formerly Report. Department of Agriculture.
   Government Research Institute. Formosa.
   Continued as Bulletin. Taiwan Agricultural Research
   Institute. Taipeh.   E.S 1921

Report of the Government Sugar Experiment Station Taiwan, Formosa.
   Rep.Govt Sug.Exp.Stn Taiwan 1934-1939.   B.S 1985

Report of the Government Veterinary Bacteriologist. Union of
   South Africa. Pretoria.
   Rep.Govt vet.Bact.Un.S.Afr. 1906-1908.
   Formerly Report. Department of Agriculture, Transvaal.
   Pretoria.
   Continued as Report of Veterinary Research, Department
   of Agriculture, Union of South Africa, Pretoria.   Z.S 2081

Report of the Grain Pests Committee of the Royal Society. London.
   Rep.Grain Pests Comm.R.Soc. 1919-1921.   E. Econ.R.

Report of the Great Barrier Reef Committee. Brisbane.
   Rep.Gt Barrier Reef Comm. 1925-1956.
   (Vol.1 published in Transactions of the Royal Geographical
   Society of Australasia, Queensland Branch.)
   Replaced by University of Queensland Papers. Great Barrier
   Reef Committee. Heron Island Research Station.   S. 2137

Report. Great Lakes Institute. Toronto.
   Rep.Gt Lakes Inst. 1967 →   S. 2650 F

Report on Great Lakes Water Quality. Ottawa.
   Rep.Gt Lakes Wat.Qual. 1972 →   S. 2614

Report of the Greenock Philosophical Society.
   Rep.Greenock phil.Soc. 2-97, 1863-1964 (imp.)   S. 108 A

Report on the Groundwater Investigation Programme. Mines
   Department, Victoria. Melbourne.
   Rep.Groundwat.Invest.Progm.Mines Dep.Victoria 1971 →   P.S 1130 a

Report of the Gresham's School Natural History Society. Holt.
   Rep.Gresham's Sch.nat.Hist.Soc. 1922-1964/1965.
   Continued as Gresham's School Journal of Natural Sciences.   S. 123

Report of the Guam Agricultural Experiment Station. Washington.
   Rep.Guam agric.Exp.Stn 1914-1926 (imp.)   E.S 1965

Report. Hackney Microscopical and Natural History Society.
   Rep.Hackney microsc.nat.Hist.Soc. 1877-1892.   S. 181

Report. Haileybury and Imperial Service College Natural History
   Society. Hertford.
   Rep.Haileybury imp.Serv.Coll.nat.Hist.Soc. 1949-1951.
   (For Continuation of Ornithological Report See Ornithological
   Report. Haileybury and Imperial Service College.)   S. 113 b

TITLE	SERIAL No.
Report. Haileybury Natural Science Society. Rep.Haileybury nat.Hist.Soc. 1873-1875.	S. 113 a
Report. Hamilton Natural History Society. Hamilton. Rep.Hamilton nat.Hist.Soc. 1969 →	S. 116 a
Report. Hampshire and Isle of Wight Naturalists' Trust Ltd. Portsmouth. Rep.Hamps.Isle Wight Nat.Trust Ltd 1967-1968 → 1960/1961 →	S. 333 B E.S 30 a A
Report. Hampstead Naturalists' Club. Rep.Hampstead Nat.Cl. 3-4. 1882-1884.	S. 114
Report of the Hampstead Scientific Society. London. Rep.Hampstead scient.Soc. 1899-1938; 1946-1952.	S. 114 a A
Report. Handling and Preservation of Fish. See Torry Research on the Handling & Preservation of Fish Products, Edinburgh.	Z.S 482
Report of the Harrow School Scientific Society. Rep.Harrow Sch.scient.Soc. 1866-1869.	S. 115
Report. Harvard Forest. Petersham, Mass. Rep.Harv.For. 1966/1967 →	B.S 4238 j
Report. Haslemere Educational Museum. Rep.Haslemere Mus. 1926 →	S. 118 B
Report of the Haslemere Microscope and Natural History Society. Rep.Haslemere Microsc.nat.Hist.Soc. 1897-1905. Formerly Record of Lectures and Addresses. Haslemere Microscope and Natural History Society. Continued as Report of the Haslemere Natural History Society.	S. 117 A
Report of the Haslemere Natural History Society. Rep.Haslemere nat.Hist.Soc. 1907, 1916 → Formerly Report of the Haslemere Microscope and Natural History Society.	S. 117 A
Report. Hastings Natural History Society. Rep.Hastings nat.Hist.Soc. 1933-1948. Formerly Report of the Hastings and St.Leonards Natural History Society.	S. 119 A
Report. Hastings and St. Leonards Museum Association. Rep.Hastings St.Leonards Mus.Ass. 1903-1905; 1917-1940.	S. 118 a
Report of the Hastings and St.Leonards Natural History Society. Hastings. Rep.Hastings St.Leon.nat.Hist.Soc. 1899-1933. Continued as Report. Hastings Natural History Society.	S. 119 A

| TITLE | SERIAL No. |

Report of the Hatch Agricultural Experiment Station. Amherst, Mass.
    Rep.Hatch agric.Exp.Stn 1895-1906.
    Continued as Report of the Massachusetts Agricultural
Experiment Station. Amherst.   E.S 2479 a

Report of the Hawaii Agricultural Experiment Station.
    Washington, Honolulu.
    Rep.Hawaii agric.Exp.Stn 1901-1932 (imp.)   E.S 2276 a

Report. Hawaiian Sugar Planters' Association. Experiment Station.
    Honolulu.
    Rep.Hawaiian Sug.Plr Ass.Exp.Stn 1973 →   E.S 2275 a

Report. Hayling Mosquito Control. Hayling Island.
    Rep.Hayling Mosq.Control 1920-1930.
    Continued as Report. British Mosquito Control Institute.   E.S 40 a

Report. Herbert Whitley Trust. Paignton.
    Rep.H.Whitley Trust 1970 →   S. 316 B
    Formerly Report & Accounts. Herbert Whitley Trust.
Paignton.   S. 316

Report. Herefordshire and Radnorshire Nature Trust Ltd. Hereford.
    Rep.Hereford.Radnor.Nat.Trust 1970 →   S. 119 b

Report. Hertfordshire County Museum. St.Albans.
    Rep.Herts Cty Mus. 1903-1912.
    (Wanting 1909.)   S. 349

Report. Hertfordshire and Middlesex Trust for Nature Conservation
    Limited. St. Albans.
    Rep.Herts.Middx Trust Nat.Conserv. 1970 →   S. 391 a C

Report. Hillingdon Natural History Society. West Drayton.
    Rep.Hillingdon nat.Hist.Soc. 1971/3 →
    Formerly Journal. Hillingdon Natural History Society.   S. 123 a

Report of the Horniman Museum. Forest Hill, London.
    Rep.Horniman Mus. 1901-1914.   S. 191 A

Report. Huddersfield Naturalist, Photographic & Antiquarian
    Society.
    See Report & Balance Sheets. Huddersfield Naturalist,
Photographic & Antiquarian Society.   S. 122 C

Report (Proceedings & Transactions) Hull Literary &
    Philosophical Society.
    Rep.Hull lit.phil.Soc. 1864-1884.
    Formerly Report of the Literary & Philosophical Society
at Kingston-upon-Hull.   S. 125 A

Report of the Hungarian Geological Institute. Budapest.
    See Evi Jelentés a Magyar Kir Foldtani Intézet. Budapest.   P.S 1366

TITLE	SERIAL No.

Report. Huntingdonshire Fauna and Flora Society.
    Rep.Huntingdon.Fauna Flora Soc. 1949 →   S. 107

Report of the Hydrobiological Research Unit. University College
  of Khartoum.
    Rep.hydrobiol.Res.Unit Univ.Khartoum 1953 →   S. 2047 A

Report of the Hydrobiological Station on Sevan Lake. Erevan.
    See Trudy Sevanskoĭ Gidrobiologicheskoĭ Stantsii.   Z.S 1905

Report by the Hydrographer of the Navy. London.
    Rep.Hydrogr.Navy 1963 →   S. 200 B
                1961 → (imp.)   M.S 386

Report of the Ichthyological Laboratory in Astrachan.
    See Trudy Astrakhanskoĭ Ikhtiologicheskoĭ Laboratorii
    pri Upravlennii Kaspiĭsko Volzhskikh Rybnykh i
    Tyulen'ikh Promyslov.   Z.S 1830

Report of the Ichthyological Laboratory in Kertch.
    See Trudy Kerchenskoi Ikhtiologicheskoi Laboratorii.   Z.S 1838

Report. Ilfracombe Museum Committee.
    Rep.Ilfracombe Mus.Comm. 1933 →   S. 133

Report of the Illinois Agricultural Experiment Station. Urbana.
    Rep.Ill.agric.Exp.Stn 1894 →   E.S 2473

Report of the Illinois State Entomologist. Bloomington.
    Rep.Ill.St.Ent. 1867-1916.   E.S 2472

Report of the Illinois State Laboratory of Natural History. Urbana.
    See Biennial Report of the Illinois State Laboratory
    of Natural History. Urbana.   S. 2320 B

Report. Illinois State Museum. Springfield.
    Rep.Ill.St.Mus. 1967/1968 →   S. 2406 F

Report. Illinois State Museum of Natural History. Springfield.
    Rep.Ill.Mus.nat.Hist. 1909-1929. (imp.)
    Formerly Biennial Report of the Illinois State Museum
    of Natural History. Springfield.   S. 2406 A

Report. Imperial Agricultural Bureau. London.
    Rep.imp.agric.Bur. 1937-1948.   E.S 65

Report of Imperial Bureau of Fisheries Scientific
    Investigations. Tokyo.
    Rep.imp.Bur.Fish.scient.Invest.Tokyo No.2, 1913.
    Continued as Report of the Imperial Fisheries Institute. Z.S 1966

Report of the Imperial College of Science and Technology. London.
    Rep.imp.Coll.Sci.Technol.,Lond. 1908-1924; 1965 →   S. 233 A

Report of the Imperial Department of Agriculture, India. Calcutta.
    Rep.Imp.Dep.Agric.India   1904-1907.
    Continued as Report on the Progress of Agriculture in
    India. Calcutta.   E.S 1997

| TITLE | SERIAL No. |

Report on the Imperial Entomological Conference. London.
    Rep.imp.ent.Conf. 1920-1935.
    Continued as Report. Commonwealth Entomological
Conference. London.   E.S 2555

Report of the Imperial Fisheries Institute. Tokyo.
    Rep.imp.Fish.Inst.Tokyo No.4, 1915.
    Formerly Report of Imperial Bureau of Fisheries.
Scientific Investigations. Tokyo.   Z.S 1966

Report. Imperial Forestry Institute, University of Oxford.
    Rep.imp.For.Inst.Univ.Oxf. 1935-1961.
    Continued as Report. Commonwealth Forestry Institute,
University of Oxford.   B.S 90 a

Report. Imperial Geological Survey of Japan. Tokyo.
    Rep.geol.Surv.Japan No.88-120 1922-1937.
    Continued as Report. Geological Survey of Japan. Tokyo.   P.S 1766

Report. Imperial Institute.
    Rep.imp.Inst.,Lond. 1905-1914; 1926-1958 (imp.)
    Continued as Report. Commonwealth Institute.   S. 187 B

Report on the Imperial Mycological Conference. London, Kew.
    Rep.imp.mycol.Conf. 1924-1934.
    Continued as Report on the Commonwealth Mycological
Conference. London, Kew.   B.S.M 5h

Report of the Indian Botanic Garden and the Gardens in Calcutta,
    (Parks & Gardens in Cooch Behar) and the Lloyd Botanic Garden,
    Darjeeling. Alipore.
    Rep.Indian Bot.Gdn, &c. 1950-1955.
    Formerly Report of the Royal Botanical Gardens Calcutta.
    (& other Gardens in Calcutta, & of the Lloyd Botanic Garden,
    Darjeeling).
    Continued as Report of the Directorate of Botanical
& other Public Gardens, West Bengal. Alipore.   B.S 1602 a

Report. Indian Council of Agricultural Research. Calcutta.
    Rep.Indian Coun.agric.Res. 1950 →   E.S 1990 a

Report of the Indian Lac Research Institute. Calcutta.
    Rep.Indian Lac Res.Inst. 1927-1948. (imp.)   E.S 2008

Report. Indian Museum, Natural History Section, Calcutta.
    Rep.Indian Mus.nat.Hist.Sect. 1884-1933.   S. 1917 a

Report. Indiana Department of Conservation. Fort Wayne, Ind.
    Rep.Indiana Dep.Conserv. 1920-1940.
        (Wanting Nos. 1, 15 & 16.)   S. 2379

Report of the Indiana Department of Geology and Natural Resources.
    Indianapolis.
    Rep.Indiana Dep.Geol.nat.Resour. 1869-1926.   P.S 1905

TITLE	SERIAL No.

Report of the Industrial Mineral Survey. Tokyo.
   See Industrial Mineral Survey Report. Imperial
   Geological Survey. Japan, Tokyo.   M.S 1943

Report on Injurious Insects and other Animals observed in the
   Midland Counties. Birmingham.
   Rep.injur.Insects Midl.Cties   1903-1908.   Z. 65B o C

Report. Inland Fisheries Department. Union of South Africa.
   Rep.inld Fish.Dep.Un.S.Afr. 1944-1951.
   Continued as Report. Department of Nature Conservation,
   Union of South Africa.   Z.S 2031

Report of the Inspector (Department) of Agriculture, Malta, Valetta.
   See Report of the Department of Agriculture, Malta. Valetta.   E.S 1120

Report of the Inspectors. Sea Fisheries, England and Wales. London.
   Rep.Insps Sea Fish. 1879, 1901-1902.
   Continued as Report of the Proceedings under Acts
   Relating to the Sea Fisheries, England & Wales.   Z.S 460 F

Report. Institut za Oceanografiju i Ribarstvo, Split.
   See Izvjesca. Institut za Oceanografiju i Ribarstvo. Split.S. 1704 a E

Report. Institute of Agricultural Research, Ethiopia. Addis Ababa.
   Rep.Inst.agric.Res.Ethiopia 1966/1968 →   E.S 2159

Report. Institute for Agricultural Research and Special
   Services. Ahmadu Bello University. Samaru, Zaria. Northern
   Nigeria.
   Rep.Inst.Agric.Res.Samaru 1962-1963 →   S. 2035 B

Report. Institute of Animal Physiology, Babraham. Cambridge.
   Rep.Inst.Anim.Physiol.Babraham 1960 →   Z.S 335

Report. Institute of Archaeology. University of London.
   Rep.Inst.Archaeol. 1937 →   P.A.S 9 A

Report. Institute for Biological Field Research. Arnhem.
   See I T B O N Werkzaamheden. Arnhem.   S. 617

Report. Institute of Botany, Academia Sinica. Taipei.
   Rep.Inst.Bot.Acad.sin. 1972/1973 →   B.S 1988 a

Report of the Institute of Fishery Biology. Taipei.
   Rep.Inst.Fish.Biol.,Taipei Vol.1 No.2 → 1957 →   Z.S 1976

Report. Institute of Freshwater Research, Drottningholm. Stockholm.
   Rep.Inst.Freshwat.Res.Drottningholm 1956 →
   Formerly Report and Short Papers. Institute of Freshwater
   Research. Drottningholm.   Z.S 510

Report. Institute of Geological Sciences. Annual Report. London.
   Rep.Inst.geol.Sci. 1965 →
     Part I. Summary of Progress of the Geological Survey
     of Great Britain and the Museum of Practical Geology.
     Part II. Overseas Geological Surveys.   P.S 1026

| TITLE | SERIAL No. |

Report. Institute of Geological Sciences. London.
    Rep.Inst.geol.Sci. 1969 →   M.S 156 & P.S 1027

Report of the Institute of Jamaica. Kingston
    Rep.Inst.Jamaica 1879-1901: 1931-1951/1955 (imp.)   S. 2291 D

Report. Institute for Marine Environmental Research. Plymouth.
    Rep.Inst.mar.envir.Res. 1971/1973 →   S. 224 a G

Report of the Institute of Marine Research Lysekil. Stockholm.
    Rep.Inst.mar.Res.Lysekil Series Biology 1950 →
    Formerly Svenska Hydrografisk-Biologisk Kommissionens
    Skrifter, Ny Serie-Biologi.   Z.S 515

Report. Institute of Marine Resources. University of California.
    Rep.Inst.mar.Resour.Univ.Calif. 1964 →   S. 2319 L

Report of the Institute for Medical Research, Federation
    of Malaya. Kuala Lumpur.
    Rep.Inst.med.Res.Fed.Malaya 1948 →   E.S 1974 a

Report. Institute of Medical and Veterinary Science,
    South Australia. Adelaide.
    Rep.Inst.med.vet.Sci.S.Aust. No.3 → 1940 →   Z. 77Cb o S

Report. Institute for Nature Conservation Research Bilthoven.
    See Verslagen van de Werkzaamheden. Rijksinstituut voor
    Veldbiologisch Onderzoek ten behoeven van het Natuurbehoud.
    R.I.V.O.N. Bilthoven.   S. 619

Report. Institute of Polar Studies. Ohio State University Research
    Foundation. Columbus.
    Rep.Inst.Polar Stud. 1962 →   S. 2332 a L

Report. Institute of Science and Industry, Commonwealth of
    Australia. Melbourne.
    Rep.Inst.Sci.Ind.Aust. 1921-1922.
    Continued as Report of the Council for Scientific and
    Industrial Research. Commonwealth of Australia. Melbourne.   S. 2113 F

Report of the Institute of Scientific Research, Manchoukuo. Hsinking.
    Rep.Inst.scient.Res.Manchoukuo 1936-1940.   S. 1986 c

Report Institute of Seaweed Research. Edinburgh.
    Rep.Inst.Seaweed Res. 1951-1968.
    Formerly Report Scottish Seaweed Research Association.   B.A.S 1

Report. Institute of Tropical Forestry. Rio Piedras.
    Rep.Inst.trop.For.Rio Piedras 1963 →   B.S 4140 a

Report of the Institute of Zoology and Biology. Kiev.
    See Trudȳ Instȳtutu Zoolohiyi ta Biolohiyi. Kȳyiv.   S. 1834 a F

Report. Institutions of Science and Art. Dublin.
    See Report. Dublin Institutions of Science and Art.   S. 48 a A

TITLE	SERIAL No.

Report. Inter-American Tropical Tuna Commission.
    La Jolla, California.
    Rep.inter-Am.trop.Tuna Commn   1950 →   Z.S 2448 A

Report. International Commission for the Northwest Atlantic
    Fisheries. St. Andrews, N.B., etc.
    Rep.int.Commn NW.Atlant.Fish. 1951-1952; 1972/1973 →
    Published as Annual Proceedings. International Commission
    for the Northwest Atlantic Fisheries.   Z.S 2716 B

Report. International Commission on Whaling. London.
    Rep.int.Commn Whal. 1950 →   Z.S 2729

Report. International Committee for Bird Preservation.
    British Section. London.
    Rep.int.Comm.Bird Preserv.Br.Sect.   1937-1959
    Continued as Report. International Council for
    Bird Preservation, British Section.   Z.S 2722

Report. International Council for Bird Preservation, British Section.
    Rep.int.Coun.Bird Preserv.Br.Sect. 1960 →
    Formerly Report. International Committee for Bird
    Preservation, British Section.   Z.S 2722

Report. International Federation for Documentation. Hague.
    Rep.int.Fed.Docum. 1969 →   S. 637 B

Report of the International Fisheries Commission. Seattle.
    Rep.int.Fish.Commn 1931-1953.
    Continued as Report of the International Pacific
    Halibut Commission. Seattle.   Z.S 2706

Report. International Geological Congress. London.
    See International Geological Congress.

Report of the International Health Commission (Board). New York.
    See Report of the International Health Division
    Rockefeller Foundation. New York.   Z.S 2470 A

Report of the International Health Division Rockefeller
    Foundation. New York.
    Rep.int.Hlth Div.Rockefeller Fdn No.2-8, 1915-1921.   Z.S 2470 A

Report of the International North Pacific Fisheries
    Commission. Vancouver, B.C.
    Rep.int.N.Pacif.Fish.Commn 1955 →   Z.S 2707 A

Report. International Office for the Protection of Nature. Brussels.
    Rep.int.Office Prot.Nat. 1940-1946 & 1949-1950.   S. 2728 C

| TITLE | SERIAL No. |

Report of the International Pacific Halibut Commission.
   Seattle.
     Rep.int.Pacif.Halibut Commn No.21 → 1953 →
     Formerly Report of the International Fisheries Commission.
     Seattle.   Z.S 2706

Report. International Pacific Salmon Fisheries Commission.
   New Westminster, B.C.
     Rep.int.Pacif.Salm.Fish.Commn 1937 →   Z.S 2712

Report. International Subcommission on Stratigraphic Classification.
   Copenhagen, &c.
     Rep.int.Subcommn stratigr.Classif. 1961 →   P. REF

Report. International Union for Conservation of Nature and
   Natural Resources. Morges, Switzerland.
     Rep.int.Un.Conserv.Nature nat.Resour. 1961-1969.
     Continued as I.U.C.N. Yearbook.   S. 2715 H

Report on Introduction of Improvements in Agriculture. Calcutta.
     Rep.Introd.Improv.Agric. 1912-1914.   E.S 2003

Report of Investigations. Bureau of Economic Geology.
   University of Texas. Austin.
     Rep.Invest.Bur.econ.Geol.Univ.Tex. 1946 →   P.S 1978

Report of Investigations. Bureau of Mines. Philippine
   Islands. Manila.
     Rep.Invest.Bur.Mines Philipp.Isl. No.11 → 1954 →   M.S 2002

Report of Investigations Delaware Geological Survey. Newark.
     Rep.Invest.Del.geol.Surv. 2 → 1958 →   P.S 1889 B

Report of Investigations. Department of Mines (Geological Survey),
   South Australia. Adelaide.
     Rep.Invest.Dep.Mines S.Aust. 1954 →   P.S 1139

Report of Investigations. Division of Geological Survey.
   Ohio. Columbus.
     Rep.Invest.Div.geol.Surv.Ohio 1947 →   P.S 1967 A

Report of Investigations. Division of Geology. Department of
   Conservation, State of Tennessee. Nashville.
     Rep.Invest.Div.Geol.Tenn. 1955 →   P.S 1976 A

Report of Investigations. Division of (Mines and) Geology.
   Washington State Department of Conservation and Development.
     Rep.Invest.Div.Geol.Wash.St.Dep.Conserv. 1926 →   P.S 1988 A

Report of Investigations. Division State Geological Survey,
   Illinois. Urbana.
     See Report of Investigations. Illinois State Geological
     Survey. Urbana.   P.S 1902 B & M.S 2641

| TITLE | SERIAL No. |

Report of Investigations. Florida Geological Survey. Tallahassee.
    Rep.Invest.Fla geol.Surv. No.7 → 1951 →   P.S 1890 D

Report of Investigations. Geological Survey of Ohio. Columbus.
    See Report of Investigations. Division of Geological Survey.
    Ohio. Columbus.   P.S 1967 A

Report of Investigations. Geological Survey, South Australia.
    See Report of Investigations. Department of Mines
    (Geological Survey), South Australia. Adelaide.   P.S 1139

Report of Investigations. Illinois State Geological Survey. Urbana.
    Rep.Invest.Ill.St.geol.Surv. 1924 →
    (Nos. 4, 6, 7, 10-13, 16-17, 19-20, 22-32, 34-36 wanting)   P.S 1902 B
    Nos. 43-184, 1936-1955. (imp.)   M.S 2641

Report of Investigations. Illinois State Museum of Natural
    History. Springfield.
    Rep.Invest.Ill.St.Mus.nat.Hist. 1948 →   S. 2406 E

Report of Investigations. Kentucky Geological Survey. Lexington.
    Rep.Invest.Ky geol.Surv. 1949 →   P.S 1925 C

Report of Investigations. Maryland Geological Survey.
    Rep.Invest.Md geol.Surv. 1965 →   P.S 1928 A

Report of Investigations. Michigan Geological Survey. Lansing.
    Rep.Invest.Mich.geol.Surv. No.2 → 1967 →   P.S 1935 D

Report of Investigations. Minnesota Geological Survey. Minneapolis.
    Rep.Invest.Minn.geol.Surv. 1963 →   P.S 1943 A

Report of Investigations. Missouri Geological Survey and
    Water Resources. Rolla.
    Rep.Invest.Mo.geol.Surv. 1945 →   P.S 1952 B

Report of Investigations. North Dakota Geological Survey. Grand Forks.
    Rep.Invest.N.Dak.geol.Surv. No.1 → 1953 → Imp.   P.S 1886 B

Report on Investigations. North Pacific Fur Seal Commission.
    (Seattle, Wash.)
    Rep.Invest.N.Pacif.Fur Seal Commn 1958/1961 →   Z.S 2509

Report of Investigations. Philippines Bureau of Mines, Manila.
    See Report of Investigations. Bureau of Mines.
    Philippine Islands Manila.   M.S 2002

Report of Investigations. South Dakota Geological and Natural
    History Survey. Vermillion.
    Rep.Invest.S.Dak.geol.nat.Hist.Surv.
    Nos.1-3, 6-15. 1930-1933.
    Continued as Report of Investigations. South Dakota
    Geological Survey. Vermillion.   P.S 1887 C

TITLE	SERIAL No.

Report of Investigations. South Dakota Geological Survey.
   Vermillion.
     Rep.Invest.S.Dak.geol.Surv. No.109 → 1973 →
     Formerly Report of Investigations. South Dakota Geological
     and Natural History Survey.   P.S 1887 C

Report of Investigations. State Geological and Natural History Survey
   of Connecticut.
     Rep.Invest.St.geol.nat.Hist.Surv.Conn.
        No. 1 → 1961 →   S. 2343 D

Report of Investigations. State Geological Survey, Illinois
   Urbana.
     See Report of Investigations. Illinois State Geological
     Survey. Urbana.   P.S 1902 B & M.S 2641

Report of Investigations. West Virginia Geological and Economic
   Survey. Morgantown.
     Rep.Invest.W.Va geol.econ.Surv. 1947 →   P.S 1986 A

Report of Investigations. Wyoming Geological Survey. Laramie.
     Rep.Invest.Wyo.geol.Surv. No.2 → 1939 →   P.S 1992 B

Report and Investigators' Summaries. Fisheries Research Board
   of Canada, Arctic Unit. Montreal.
     Rep.Invest.Summ.Fish.Res.Bd Can.Arctic Unit 1959/60 →   Z.S 2628 K

Report of the Iowa Geological Survey. Des Moines.
     Rep.Iowa geol.Surv. 1892 →   P.S 1912

Report. Ipswich Museum Free Library (and Art Gallery.)
     Rep.Ipswich Mus.Free Libr. 1910-1911 & 1921-1922.   S. 135 a

Report of the Iraq Natural History Museum. Baghdad.
     Rep.Iraq nat.Hist.Mus. 1950 →   S. 1923 a C

Report. Irish Ornithologist's Club.
     See Irish Bird Report.   T.B.S 501

Report of the Irish Wildbird Conservancy. Dublin.
     Rep.Ir.Wildbird Conserv. 1969 →   T.B.S 502

Report on the Iron Ores of Missouri. Geological Survey
   of Missouri. Jefferson City.
     Rep.Iron Ores, geol.Surv.Missouri No. 2-7; 1892-94(imp.)   M.S 2645

Report. Isle of Thanet Field-Club. Ramsgate.
     Rep.Isle Thanet Fld Club 1948-1951.   S. 336

Report. Israel Oceanographic and Limnological Research. Haifa.
     Rep.Israel oceanogr.limnol.Res. 1971/72 →   Z.S 1999 A

Report of the Japan Sea Regional Fisheries Research
   Laboratory. Niigata.
     Rep.Japan Sea reg.Fish.Res.Lab. No.3 → 1957 →   Z.S 1957 a B

Report. Jersey Wildlife Preservation Trust. Jersey.
     Rep.Jersey Wildl.Preserv.Trust 1967 →   Z.S 490

TITLE	SERIAL No.

Report. Johannesburg Public Library. Johannesburg.
    Rep.Joburg publ.Libr. 1971/72 →   S. 2008 a

Report. John Innes Horticultural Institution. London, Bayfordbury.
    Rep.John Innes hort.Instn 1926-1959. (imp.)
    Continued as Report. John Innes Institute. Bayfordbury.   B.S 83

Report. John Innes Institute. Bayfordbury.
    Rep.John Innes Inst. 1960 →
    Formerly Report. John Innes Horticultural Institution.
    London, Bayfordbury.   B.S 83

Report. Joint Fisheries Research Organisation Northern Rhodesia
    (& Nyasaland). Lusaka.
    Rep.jt Fish.Res.Org.Nth.Rhod. No.8-11. 1958-1961.
    Formerly Included in (Report of the Department of
    Game and Tsetse Control of Northern Rhodesia & of Nyasaland.)
    (Not in Museum.)
    Continued as Fisheries Research Bulletin. Ministry of
    Lands and Natural Resources. Zambia.   Z. 74D f R

Report of the Kaffrarian Museum. King William's Town.
    Rep.Kaffrarian Mus. 1937 → (imp.)   S. 2012

Report of the Kansas Agricultural Experiment Station. Topeka.
    Rep.Kans agric.Exp.Stn 1889-1907.   E.S 2475 a

Report of the Kansas State Board of Agriculture. Topeka.
    Rep.Kans.St.Bd Agric. 1907-1910.   E.S 2475 c

Report of the Karelia Scientific Research Fishery Station.
    See Trudy Karel'skoĭ Nauchno-Issledovatel'skoĭ
    Rȳbokhozyaĭstvennoĭ Stantsii.   Z.S 1857 A

Report. Kasetsart University. Bangkok.
    See Kasetsart University Research Activities. Bangkok.   S. 1913 c B

Report. Keighley Borough Museum.
    Rep.Keighley Mus. 1906-1907.   S. 140 C

Report on the Kelvingrove Museum and the Corporation Galleries
    of Art, Glasgow.
    Rep.Kelvingrove Mus.,Glasgow 1880-1895.
    Formerly Report on the City Industrial Museum,
    Kelvingrove Park, Glasgow.
    Continued as Report. Museums and Art Galleries,
    Corporation of Glasgow.   S. 95

Report of the Kendal Entomological Society. Kendal, Westmorland.
    Rep.Kendal ent.Soc. 1899-1904.   E.S 3 a

Report. Kent Trust for Nature Conservation. Maidstone.
    Rep.Kent Trust Nat.Conserv. 1969 →   S. 255 b

TITLE	SERIAL No.

Report. Kenya Wild Life Society. Nairobi.
    Rep.Kenya wild Life Soc. 1956-1957.   Z.S 2052

Report from the Kevo Subarctic Research Station. Turku.
    See Annales Universitatis Turkuensis.
    Series A (II) No.32 → 1964 →   S. 1823 b

Report of the Kihara Institute for Biological Research. Kyoto.
    Rep.Kihara Inst.biol.Res. 1942-1963.   S. 1981 b

Report. King's School Canterbury, Natural History Society
    and Field Club.
    Rep.King's Sch.Canterb.nat.Hist.Soc. 1945-1946.
    Continued as Report. Natural History Society.
    King's School. Canterbury.   S. 24 b

Report of the Kumaun Government Gardens. Allahabad.
    Rep.Kumaun Govt Gdns 1909-1916.   B.S 1633

Report. Laboratory and Museum of Comparative Pathology of the
    Zoological Society of Philadelphia.
    Rep.Lab.Mus.comp.Path.zool.Soc.Philad. 1924-1933.
    Previous Reports Contained in Report of the Zoological
    Society of Philadelphia.
    Continued as Report of the Penrose Research Laboratory,
    Zoological Society of Philadelphia.   Z.S 2500

Report. Laboratory of Vertebrate Biology. University of Michigan.
    Ann Arbor, Michigan.
    Rep.Lab.vertebr.Biol.Univ.Mich. 1945-1946 - 1949-1950.   Z.S 2425 E

Report. Lake & Pond Survey Unit. State Board of Fisheries
    & Game. Hartford, Conn.
    See Report. Connecticut State Board of Fisheries
    & Game, Lake & Pond Survey Unit. Hartford.   Z. 75C o C

Report of the Lake Sevan Limnological Station.
    See Trudy Sevanskoi Ozernoi Stantsii.   Z.S 1905

Report Lambeth Field Club & Scientific Society.
    Rep.Lambeth Fld Club scient.Soc. 1881-1883.   S. 188

Report of the Lancashire & Cheshire Entomological Society, Liverpool.
    Rep.Lancs.Chesh.ent.Soc. 1881-1901. (imp.)
    Continued as Report and Proceedings of the Lancashire & Cheshire
    Entomological Society. Liverpool.   E.S 3

Report. Lancashire & Cheshire Fauna Committee. Liverpool.
    Rep.Lancs.Chesh.Fauna Comm. 1914-1919 (imp.), 1921-1965.
    Continued as Report. Lancashire and Cheshire Fauna Society.   Z.S 301

Report. Lancashire and Cheshire Fauna Society. Liverpool.
    Rep.Lancs.Chesh.Fauna Soc. 1966-1967.
    Formerly Report. Lancashire and Cheshire Fauna Committee.
    Continued as Publications. Lancashire and Cheshire
    Fauna Society.   Z.S 301

TITLE	SERIAL No.

Report. Lancaster Astronomical and Scientific Association.
    Rep.Lancaster astr.scient.Ass. 1903-1966.
    Continued as Report. Lancaster Lecture Association.   S. 143

Report. Lancaster Lecture Association.
    Rep.Lancaster Lect.Ass. 1967-1968.
    Formerly Report. Lancaster Astronomical &
    Scientific Association.   S. 143

Report of the Lancaster Literary, Scientific & Natural
History Society.
    Rep.Lancaster Lit.scient.nat.Hist.Soc. 1836-1837.   S. 141

Report of the Land Utilisation Survey of Britain. London.
    See Land of Britain. Report of the Land Utilisation
    Survey of Britain. London.   72.A.q.L

Report on the Lands and Mines Department, British Guiana.
Georgetown.
    Rep.Lds Mines Dep.Br.Guiana 1923-1944 (imp.)   M.S 2506

Report. Lands, Mines and Surveys. Annual Report. Kenya.
    Rep.Lds Mines Surv.a.Rep.Kenya 1945-1946, 1948.
    Formerly Report. Mining and Geological Department,
    Kenya Colony and Protectorate. Annual Report.
    Continued as Report. Mines and Geological Department,
    Kenya.   M.S 2109 A

Report. Lands, Mines and Surveys Department, Mines Division.
Kenya. Nairobi.
    Rep.Lds Mines Surv.Dep.Kenya No.14. 1948.
    Formerly Report. Mining and Geological Department.Kenya
    Colony and Protectorate. Nairobi.
    Continued as Report. Geological Survey of Kenya. Nairobi.   M.S 2109 A

Report. Laura Spelman Rockefeller Memorial. New York.
    Rep.L.Spelman Rockefeller Meml 1924 →   Z.S 2470 D

Report. Leeds City Museums. Leeds.
    Rep.Leeds Cy Mus. 1962-1966.   S. 145 a

Report. Leeds Naturalists' Club & Scientific Association.
    Rep.Leeds Nat.Club scient.Ass. 6-8. 1875-1878.   S. 145 C

Report of the Leeds Philosophical & Literary Society.
    Rep.Leeds phil.lit.Soc. 1824-1935.   S. 146 A

Report of the Leeds University.
    Rep.Leeds Univ. 3 → 1905 → 1954; 1963-1964 →   S. 148

Report. Leicester Literary & Philosophical Society.
    See Report of the Council of the Leicester Literary
    & Philosophical Society.   S. 151 A

Report of the (City of) Leicester (Town) Museum & Art Gallery.
    Rep.Leicester Mus.Art Gall. 1872 →   S. 150 A

| TITLE | SERIAL No. |

Report. Leicestershire and Rutland Ornithological Society.
    See Report on the Wild Birds of Leicestershire and
    Rutland.   T.B.S 260

Report. Leicestershire and Rutland Trust for Nature Conservation
    Limited. Leicester.
    Rep.Leics.Rutland Trust Nat.Conserv. 1972/1973 →   S. 150 b A

Report. Leland Stanford Junior University. Palo Alto.
    Rep.Leland Stanf.jr Univ. 1915-1918.   S. 2407 C

Report. Liberal Arts Faculty, Shizuoka University. Shizuoka.
    See Report. Liberal Arts and Science Faculty, Shizuoka
    University. Shizuoka.   S. 1995 e B

Report. Liberal Arts and Science Faculty, Shizuoka University.
    Shizuoka.
    Rep.lib.Arts Sci.Fac.Shizuoka Univ.
    Ser.B. Natural Science. 1950-1965.
    Continued as Report of Faculty of Science, Shizuoka
    University. Shizuoka.   S. 1995 e B

Report of the Librarian U.S.Department of Agriculture, Washington.
    Rep.Libr.U.S.Dep.Agric. 1931-1953 (imp.).   E.S 2446 a

Report. Libraries, Museums & Arts Committee, City of Liverpool.
    Rep.Mus.Lpool 1954 →
    Formerly Report of the Free Public Museum, Liverpool.   S. 160 B

Report of the Library. University College of Wales. Aberystwyth.
    Rep.Libr.Univ.Coll.Wales Aber. 1974 →   S. 7 C

Report of the Lichen Exchange Club of the British Isles. Leicester.
    Rep.Lichen Exch.Club 1908-1911.   B.L.S 1

Report of the Lincolnshire Naturalists' Trust Ltd. Alford.
    Rep.Lincs.Nat.Trust 1949 →   S. 153

Report of the Literary & Philosophical Society at
    Kingston-upon-Hull.
    Rep.lit.Phil.Soc.Kingston-upon-Hull 1825-1852 (imp.)
    Continued as Report (Proceedings & Transactions)
    Hull Literary & Philosophical Society.   S. 125 A

Report of the Literary & Philosophical Society of
    Newcastle-upon-Tyne.
    Rep.lit.phil.Soc.Newcastle 1904-1941. (imp.)   S. 284 A

Report (of the Council) of the Liverpool Geographical Society.
    Rep.Lpool geogr.Soc. 2-4. 1893-1895.
    Continued as Transactions and Annual Report Liverpool
    Geographical Society.   S. 162 A

Report. Liverpool Geological Association.
    See Proceedings. (Annual Report) Liverpool Geological
    Association.   P.S 132

| TITLE | SERIAL No. |

Report of the Liverpool Marine Biological Station on Puffin
    Island. Liverpool.
    Rep.Lpool mar.biol.Stn Puffin Isl. 1888-1893.   Z.S 313

Report of the Liverpool Marine Biology Committee.
    Rep.Lpool mar.Biol.Comm. 1893-1919 (imp.)   Z.S 270
                                  1903-1904.   T.R.S 133
    Continued as Report of the Oceanic Department of
    the University of Liverpool.   Z.S 270

Report of the Liverpool Microscopical Society.
    Rep.Lpool microsc.Soc. 18-53, 1887-1923, (imp.) & 67. 1934. S. 161 A

Report of the Liverpool Naturalists' Field Club.
    Rep.Lpool Nat.Fld Club 1860-1868.
    Continued as Proceedings of the Liverpool Naturalists'
    Field Club.   S. 158 A & T.R.S 135

Report. Livingstone Museum. Livingstone.
    Rep.Livingstone Mus. 1965-1966.
    Formerly Report. Rhodes-Livingstone Museum. The National
    Museum of Northern Rhodesia. Livingstone.
    Continued in Report. National Museums Board of Zambia.
    Lusaka.   S. 2075 A

Report of the Llandudno & District Field Club.
    Rep.Llandudno Distr.Fld Club 3 & 6. 1908-1909, 1911-1912. S. 152 A

Report. London Natural History Society.
    Rep.Lond.nat.Hist.Soc. 1914.
    Continued in Transactions of the London Natural
    History Society.   S. 174 A

Report. London School of Hygiene and Tropical Medicine.
    Rep.Lond.Sch.Hyg. 1934 →   S. 208 B

Report. Loughborough Naturalists' Club.
    Rep.Loughborough Nats' Cl. 1966 →   S. 252 B

Report of the Louisiana State Museum. New Orleans.
    Rep.La St.Mus. 1906.
    Continued as Biennial Report of the Louisiana State Museum. S. 2365 a

Report to the Louisiana Wildlife and Fisheries Commission.
    See Report. Biological Study of Lake Pontchartrain.
    Zoology Department, Tulane University.   Z.S 2415 A

Report. Louth Antiquarian and Naturalists' Society. Louth.
    Rep.Louth antiq.Nat.Soc. 1895-1939 (imp.)   S. 250

Report. Lowestoft and North Suffolk Field Naturalists' Club.
    Rep.Lowestoft N.Suff.Fld Nat.Club 1946 →   S. 251

Report of the Ludlow Natural History Society.
    Rep.Ludlow nat.Hist.Soc. 2nd.1836.(photostat)   S. 253

TITLE	SERIAL No.

Report. Lundy Field Society. Exeter.
    Rep.Lundy Fld Soc. 1947 →   S. 76

Report from the McLean Foraminiferal Laboratory. Alexandria.
    Rep.McLean foram.Lab. No.2, 1955.
    Continued as Report from the McLean Paleontological
    Laboratory.   P.S 826

Report from the McLean Paleontological Laboratory. Alexandria.
    Rep.McLean paleont.Lab. No.4 → 1960 →
    Formerly Report from the McLean Foraminiferal Laboratory.   P.S 826

Report. Maidenhead Naturalists' Field Club & Thames Valley
    Antiquarian Society.
    Rep.Maidenhead Nat.Fld Club 1884-1891.   S. 254

Report. Maidstone Museum, Public Library and Bentlif Art Gallery.
    Rep.Maidstone Mus.publ.Libr. 1902-1910
      (Wanting 1905-1906.)   S. 255 a

Report. Maine Aububon Society. Portland, Maine.
    Rep.Maine Audubon Soc. 1966-1968.
    Continued as Newsletter. Maine Audubon Society.   T.B.S 5205

Report to the Malaria Committee. Royal Society London.
    Rep.Malar.Comm.R.Soc. 1899-1903.   S. 3 H2

Report of the Malton Field Naturalists' & Scientific Society.
    Rep.Malton Fld Nat.scient.Soc. 1884-1887.   S. 258 A

Report. Malvern College Natural History Society.
    Rep.Malvern Coll.nat.Hist.Soc. 1935.
    Formerly Year Book. Malvern College Natural History Society.   S. 257

Report. Manawatu Philosophical Society.
    Rep.Manawatu phil.Soc. 1923.   S. 2162

Report. Manchester Field Naturalists' and Archaeologists' Society.
    See Report & Proceedings of the Manchester Field Naturalists'
    and Archaeologists' Society.   S. 260

Report. Manchester Microscopical Society.
    Rep.Manchr microsc.Soc. 1883-1884.
    Continued as Transactions & Annual Report of the
    Manchester Microscopical Society.   S. 262

Report. Manchester Museum.
    Rep.Manchr Mus. 1889-1901.
    Continued as Report of the Museum Committee. University
    of Manchester.   S. 263 C & T.R.S 170 B

Report of the Manchester Scientific Students Association.
    Rep.Manchr scient.Students' Ass. 1862-1876 (imp.)
    Continued as Report & Proceedings of the Manchester
    Scientific Students' Association.   S. 264

TITLE	SERIAL No.

Report of the Manchuria Research Institute. Harbin.
    Rep.Manchuria Res.Inst. 1936.   S. 1986 G

Report. Manitoba Historical & Scientific Society. Winnipeg.
    Rep.Manitoba hist.& scient.Soc. 1884-1885.
    Formerly Transactions. Manitoba Historical & Scientific
    Society. Winnipeg.   S. 2681

Report. The Manx Museum & Ancient Monument Trustees.
    Douglas, Isle of Man.
    Rep.Manx Mus. 1909-1939.   S. 38 A

Report. Marine Biochemistry Unit. C.S.I.R.O. Sydney.
    Rep.mar.Biochem.Unit C.S.I.R.O. 1971/1972 →   S. 2113 P

Report. Marine Biological Association of China. Amoy.
    Rep.mar.biol.Ass.China Nos. 2-3. 1933-1934.   S. 1945

Report. Marine Biological Association of the West of Scotland. Glasgow.
    Rep.mar.biol.Ass.W.Scotl. 1900-1913.
    Continued as Report. Scottish Marine Biological Association.Glasgow.
    Formerly Report. Millport Marine Biological Association.
    Glasgow.   S. 98

Report of the Marine Biological Station at Port Erin,
    Isle of Man. Liverpool.
    Rep.mar.Biol.Stn Port Erin No.47 → 1934 →
    Formerly Report of the Oceanic Department of the University
    of Liverpool.   Z.S 270

Report of the Marine Biologist. Department of Agriculture,
    Cape of Good Hope. Cape Town.
    Rep.mar.Biol.Cape Town 1896-1900.
    Continued as Report of the Government Biologist, Department
    of Agriculture, Cape of Good Hope & Cape Town.   Z.S 2030 B

Report of the Marine Department, New Zealand. Wellington.
    Rep.mar.Dep.N.Z. 1915-1916, 1919-1927.   Z. 77D f N

Report on Marine and Freshwater Investigations. Department
    of Zoology, University College of Wales. Aberystwyth.
    New Series.
    Rep.mar.Freshwat.Invest.Aberyst. N.S.I-II, 1923-1927.   Z.S 370

Report of Marine Geology and Geophysics. Seoul, Korea.
    Rep.mar.Geol.Geophys.Seoul 1970 →   P.S 1794 E

Report of the Marlborough College Natural History Society.
    Rep.Marlboro.Coll.nat.Hist.Soc. 1865 →   S. 266

| TITLE | SERIAL No. |

Report of the Massachusetts Agricultural Experiment Station. Amherst.
    Rep.Mass.agric.Exp.Stn 1906-1911.
    Formerly Report of the Hatch Agricultural Experiment Station.
    Amherst, Mass.　　　　　　　　　　　　　　　　　　　E.S 2479 a

Report of the Mauritius Institute. Port Louis.
    Rep.Maurit.Inst. 1901-1913: 1934 →　　　　　　　　S. 2092 B

Report. Mauritius Sugar Industry Research Institute. Port Louis.
    Rep.Maurit.Sug.Ind.Res.Inst. 1966 →　　　　　　　　B.S 2299 a

Report of the Medical Research Committee. London.
    See National Health Insurance. Annual Report of the Medical
    Research Committee. London.　　　　　　　　　　　　S. 191 A

Report. Medical Research Council. London.
    Rep.med.Res.Coun. 1966 →
    Formerly National Health Insurance. Annual Report of the
    Medical Research Committee. London.　　　　　　　　S. 191 A

Report of the Meetings. Caribbean Geological Conference.
    Rep.Meet.Caribb.geol.Conf. 1955 →　　　　　　　　　P.S 964

Report to Members. Wildfowlers' Association of Great Britain
    and Ireland. Liverpool.
    Rep.Memb.Wildfowl.Ass.Gt Br.Ir. 1953/1954.
    Continued as Report. Wildfowlers' Association of
    Great Britain and Ireland.　　　　　　　　　　　　　T.B.S 195

Report of the Metropolitan Water Board. London.
    Rep.metrop.Wat.Bd No.19-32, 1922-1937.　　　　　　Z.S 263

Report of the Michigan Academy of Science. Lansing.
    Rep.Mich.Acad.Sci. 1894-1968.　　　　　　　　　　　S. 2314 A

Report of the Microbiological Laboratory, Department of
    Public Health, New South Wales, Sydney.
    Rep.microbiol.Lab.Dep.publ.Hlth N.S.W. 1913-1921.
    (Extract from Report of the Director General of
    of Public Health, N.S.W.)
    Formerly Report of the Government Bureau of Microbiology,
    New South Wales, Sydney.　　　　　　　　　　　　　S. 2130 a

Report of the Microscopist, United States Department of
    Agriculture. Washington.
    Rep.Microscop.U.S.Dep.Agric. 1891-1892. (1892-1893).　　B.M.S 95 a

Report of the Middle-Thames Natural History Society.
    See Middle Thames Naturalist.　　　　　　　　　　　S. 363

Report Mid-Somerset Naturalist Society.
    See Report and Reference Book. Mid-Somerset Naturalist Society. S. 88

TITLE	SERIAL No.

Report. Millport Marine Biological Association. Glasgow.
    Rep.Millport mar.biol.Ass. 1896-1899.
    Continued as Report. Marine Biological Association of the
    West of Scotland.   S. 98

Report on the Mineral Industries of Canada. Ottawa.
    Rep.Miner.Inds Can. 1904-1905.
    Continued as Report on the Mineral Production
    of Canada. Ottawa.   M.S 2702

Report on Mineral Industry Operations in Ontario. Toronto.
    See Report. Ontario Department of Mines.   M.S 2708

Report (Annual General) upon the Mineral Industry of the
    United Kingdom, London.
    Rep.Miner.Indust. U.K. 1894-1896
    Continued as Mines and Quarries
    General Report and Statistics. London.   M.S 135

Report on the Mineral Production of Canada. Ottawa.
    Rep.Miner.Prod.Can. 1906-1918 (imp.)
    Formerly Report on the Mineral Industries of Canada. Ottawa.   M.S 2702

Report on the Mineral Resources of Alberta. Edmonton.
    Rep.Miner.Resour.Alberta 1919-1920.
    Continued in Report of the Scientific and Industrial
    Research Council of Alberta.   S. 2611

Report. Mineral Resources of the United States. Washington.
    See Mineral Resources of the United States. Washington.   M.S 2618

Report. Mines Branch, Department of Mines. Canada, Ottawa.
    Rep.Mines Brch Can. 1905-1935.
    Continued as Report. Mines & Geology Branch. Department of
    Mines & Resources, Canada. Ottawa.   M.S 2705

Report of the Mines Department, Ghana. Accra.
    Rep.Mines Dep.Ghana 1955 →
    Formerly Report of the Mines Department, Gold Coast.Accra.   M.S 2105 A

Report of the Mines Department, Gold Coast. Accra.
    Rep.Mines Dep.Gold Cst   1952-1955
    Continued as Report of the Mines Department. Ghana. Accra.   M.S 2105 A

Report of the Mines Department, Victoria. Melbourne.
    Rep.Min.Dep.Victoria 1968 →
    Formerly Report of the Secretary for Mines (and Water Supply).
    Victoria, Melbourne.   M.S 2424

Report of Mines, Forests and Scientific Services Branch.
    Department of Mines and Resources, Canada. Ottawa.
    Rep.Mines, For.scient.Serv.Brch Can. 1946-1949.
    Formerly Report. Department of Mines and Resources,
    Canada, Ottawa.
    Continued as Report. Department of Mines and
    Technical Surveys. Ottawa.   M.S 2703

Report. Mines and Geological Department, Kenya. Nairobi.
    Rep.Mines geol.Dep.Kenya 1949 →
    Formerly Report. Lands, Mines and Surveys. Annual
    Report. Kenya.   M.S 2109 A

TITLE	SERIAL No.

Report. Mines and Geology Branch, Canada. Department of
    Mines and Resources. Ottawa.
        Rep.Mines Brch Can.   1939-1946.   P.S 1082
                                1938-1944.   M.S 2705
    Formerly Report. Mines Branch. Department of Mines,
    Canada. Ottawa.

Report. Mining and Geological Department. Kenya Colony
    and Protectorate. Nairobi.
        Rep.Min.geol.Dep.Kenya Nos.1-13. 1933-1947.
    Continued as Report. Lands, Mines and Surveys Department
    Mines Division, Kenya. Nairobi.   M.S 2109 A

Report. Mining and Geological Department, Kenya Colony and
    Protectorate, Nairobi. Annual Report.
        Rep.Min.geol.Dep.a.Rep.Kenya 1933-1940 (imp.)
    Continued as Report. Lands, Mines and Surveys,
    Annual Report. Kenya.   M.S 2109 A

Report on the Mining Industry, New Zealand, Wellington.
        Rep.Min.Ind.N.Z. 1889-1891.
    Formerly Mining Industry, New Zealand.
    Continued as Mines Statement. Mines Department,
    New Zealand, Wellington.   M.S 2441

Report on the Mining and Metallurgical Industries of
    Canada. Ottawa.
        Rep.Min.Metall.Ind.Canada   1907-08   M.S 2702

Report of the Mining Registrars, Victoria Melbourne.
        Rep.Min.Reg.Vict. 1884-1889.
    Formerly Report of the Mining Surveyors and
    Registrars. Victoria, Melbourne.   M.S 2422

Report of the Mining Surveyors and Registrars, Victoria. Melbourne.
        Rep.Min.Surv.Reg.Vict. 1866-1883.
    Continued as Report of the Mining Registrars,
    Victoria. Melbourne.   M.S 2422

Report of the Minister for Agriculture, Eire. Dublin.
        Rep.Minister Agric.Eire   1945 →
    Formerly General Report of the Department of Agriculture
    and Technical Instruction for Ireland.   E.S 82

Report of the Minister of Mines, British Columbia. Victoria, B.C.
        Rep.Minister Mines B.C.   1899-1959 (imp.)
    Continued as Report of the Minister of Mines and
    Petroleum Resources. Province of British Columbia.   M.S 2713

Report of the Minister of Mines and Petroleum Resources,
    Province of British Columbia. Victoria, B.C.
        Rep.Minister Mines petrol.Res.B.C.   1960 →
    Formerly Report of the Minister of Mines, British Columbia. M.S 2713

| TITLE | SERIAL No. |

Report of the Ministry of African Agriculture, Northern Rhodesia.
   Lusaka.
     Rep.Minist.afric.Agric.Nth.Rhod. 1961 →   E.S 2177 a

Report. Ministry of Agriculture and Lands, Jamaica. Kingston.
     Rep.Minist.Agric.Lds Jamaica   1957-1961.
     Formerly Report of the Department of Science and
     Agriculture. Jamaica.   E.S 2375

Report of the Ministry of Agriculture and Natural Resources,
   Mauritius. Port Louis.
     Rep.Minist.Agric.nat.Resour.Mauritius 1969 →
     Formerly Report of the Department of Agriculture,
     Mauritius.   E.S 2183 a

Report of the Ministry of Fisheries, Irish Free State.
     See Report on the Sea and Inland Fisheries of Ireland.   Z.S 400 A

Report of the Minnesota Agricultural Experiment Station. St. Paul.
     Rep.Minn.agric.Exp.Stn 1935 →   E.S 2482 a

Report of the Minnesota Geological and Natural History Survey.
   Botanical Series. Minneapolis.
     Rep.Minn.geol.nat.Hist.Surv. 1892-1910.   B.S 4260

Report of the Minnesota State Entomologist. St. Paul.
     Rep.Minn.St.Ent. 1904-1922.
     Formerly Report of the Entomologist. Minnesota University
     Experiment Station. Minneapolis.   E.S 2480

Report. Missouri Botanical Gardens, St.Louis.
     Rep.Mo.bot.Gdn 1890-1912.   B.S 4270
                  1893-1905.   T.R.S 5153
     Continued as Annals of Missouri Botanical Garden, St.Louis.

Report. Missouri Bureau of Geology and Mines. Jefferson City.
     Rep.Mo.Bur.geol.Mines 1900-1930.
     Formerly Report. Geological Survey of Missouri.
     Continued as Report. Missouri Geological Survey and
     Mineral Resources. Rolla.   P.S 1952

Report. Missouri Geological Survey.
     See Report. Geological Survey of Missouri.   P.S 1950 & 1952

Report. Missouri Geological Survey and Mineral Resources. Rolla.
     Rep.Mo.geol.Surv.Wat.Resour. 1938 →
     Formerly Report. Missouri Bureau of Geology and Mines.
     Jefferson City.   P.S 1952

Report. Monmouthshire Naturalists' Trust Ltd. Newport.
     Rep.Monmouth.Nat.Trust 1970 →   S. 289 a

Report. Montana Agricultural Experiment Station. Bozeman.
     Rep.Mont.agric.Exp.Stn 1907-1915.   E.S 2485 a

Report of the Montrose Natural History & Antiquarian Society.
     Rep.Montrose nat.Hist.Soc. 1850-1902 (imp.)   S. 274

| TITLE | SERIAL No. |

Report. Moss Exchange Club. Stroud.
    Rep.Moss Exch.Club 1896-1922.
    Continued as Report British Bryological Society.   B.B.S 4

Report of the Museum and Art Gallery Committee. Birmingham.
    Rep.Mus.Art Gallery Comm. 1896-1948.   O.72.Aa.o.B

Report. Museum and Art Gallery of Western Australia. Perth.
    Rep.Mus.Art Gall.W.Aust. 1956-1959.
    Formerly Report. Public Library, Museum and Art Gallery
    of Western Australia. Perth.
    Continued as Report. Western Australian Museum.   S. 2156 B

Report of the Museum Committee to the City Council. Norwich.
    Rep.Mus.Comm.Norwich 1929 →
    Formerly Report of the Castle Museum Committee to the
    (Town) Council. Norwich.   S. 298 A

Report of the Museum Committee. University of Manchester.
    Rep.Mus.Comm.Univ.Manchr. 1901 →   S. 263 C
        1901-1917.   T.R.S 170 B
    Formerly Report. Manchester Museum.

Report... of the Museum of Comparative Zoology at Harvard
College. Cambridge, Mass.
    Rep.Mus.comp.Zool.Harv. 1863 → (imp.)   Z.S 2375
        1875-1885.   E.S 2479 d
        1937 →   T.R.S 5119 B

Report. Museum of History and Technology. U.S. National Museum.
    Washington.
    Published in Smithsonian Year. Washington.   S. 2426 A

Report of the Museum & Lecture Room Syndicate of the
University of Cambridge.
    Rep.Mus.& Lect.R.Synd.Camb.Univ. 1866-1912.   S. 35

Report. Museum of Natural History and Anthropology, Province of
    British Columbia.
    See Report of the Provincial Museum (of Natural History
    and Anthropology) Province of British Columbia. Victoria.   S. 2658 A

Report. Museum of Natural History, U.S. National Museum. Washington.
    Published in Smithsonian Year. Washington.   S. 2426 A

Report. Museum of Natural History, Wisconsin State University.
    Stevens Point.
    Rep.Mus.nat.Hist.Wis.St.Univ. 1969 →   S. 2537

Report. Museum and Science Center. Rochester, New York.
    Rep.Mus.Sci.Cent.Rochester 1968/1969 →   S. 2362

Report of the Museum of Technology and Applied Science. Sydney.
    See Report of the Trustees of the Museum of Applied Arts
    and Sciences. Sydney.   S. 2135 C

Report. Museum Texas Tech University. Lubbock.
    Rep.Mus.Texas Tech.Univ. 1970 →   S. 2564 D

| TITLE | SERIAL No. |

Report. Museums and Art Galleries. Corporation of Glasgow.
    Rep.Mus.Art Gall.Glasgow 1896-1915.
    Formerly Report on the Kelvingrove Museum and the Corporation
Galleries of Art, Glasgow.   S. 95

Report. Museums Association (Annual Report). London.
    See Report of the Council. Museums Association. London.   S. 2 K

Report. Museums of the Brooklyn Institute of Arts and Sciences.
    Brooklyn.
    Rep.Museums Brooklyn Inst. 1904-1937.   S. 2324 E

Report on Museums, Colleges and Institutions under the
    administration of the Board of Education. London.
    Rep.Mus.Bd Educ.,Lond. 1901-1919.   O. 72 Aa. o. G

Report on the Museums Department. Federated Malay States.
    Rep.Mus.Dep.F.M.S. 1911-1931 (imp.)   S. 1925 B

Report. Museums Trustees of Kenya and of the Coryndon Memorial
    Museum. Nairobi.
    Rep.Mus.Kenya Coryndon Mus. 1942 →   S. 2024 a A

Report. Mysore Department of Mines and Geology (and Explosives).
    Bangalore.
    Rep.Mysore Dep.Mines Geol. 1914-1923.
    Formerly Report of the Chief Inspector of Mines.
Mysore Geological Department. Madras.   M.S 1915

Report. Mysore Government Museum. Bangalore.
    Rep.Mysore Govt Mus. 1905-1914.   S. 1911 c

Report. Natal Botanic Gardens and Colonial Herbarium. Durban.
    Rep.Natal Bot.Gard.Col.Herb. 1885-1910 (imp.)   B.S 2303

Report of the Natal Fisheries Department. Pietermaritzburg.
    Rep.Natal Fish.Dep. 1918-1934. (imp.).   Z.S 2029

Report of the Natal (Government) Museum. Pietermaritzburg.
    Rep.Natal Mus. 1904-1905; 1950 →   S. 2061 B

Report of the Natal Parks, Game and Fish Preservation Board.
    Rep.Natal Pks-Game-Fish Preserv.Bd No.2 → 1950 →   Z.S 2084 A

Report of the National Academy of Sciences. National Research
    Council. Washington.
    See Report of the National Research Council. Washington.   S. 2420 E

Report. National Art Gallery and Dominion Museum. Wellington, N.Z.
    Rep.Dom.Mus.N.Z. 1915-1939; 1946 →   S. 2165 G

Report. National Association of Fishery Boards. Lancaster.
    Rep.natn.Ass.Fishery Bds Nos. 3-16, 1922-1935. (imp.)   Z.S 328

Report. National Botanic Gardens, Lucknow.
    Rep.natn.bot.Gdns Lucknow 1966 →   B.S 1624 a

Report of the National Botanic Gardens. South Africa.
    Rep.natn.bot.Gdns S.Afr. 1913 →   B.S 2281

Report. National Central Library. London.
    Rep.natn.Cent.Lib. 1958/1959 - 1972/1973.   S. 497 b

TITLE	SERIAL No.

Report. National Council of Alberta. Edmonton.
   See Report of the Scientific and Industrial Research
Council of Alberta. Edmonton.   S. 2611

Report. National Federation of Abstracting and Indexing Services.
   Philadelphia, Pa.
     Rep.natn.Fed.Abstr.Index.Serv. No.5 → 1973 →
     Formerly Report. National Federation of Science Abstracting
and Indexing Services. Philadelphia, Pa.   S. 2563 B

Report. National Federation of Science Abstracting and Indexing
   Services. Philadelphia, Pa.
     Rep.natn.Fed.Sci.Abstr.Index.Serv. No.3-4, 1971-1972.
     Continued as Report. National Federation of Abstracting
and Indexing Services. Philadelphia, Pa.   S. 2563 B

Report. National Institute of Genetics, Japan. Misima.
     Rep.natn.Inst.Genet.Misima 1951 →   S. 1998 d

Report of the National Institute of Oceanography.
   Cambridge and Wormley.
     Rep.natn.Inst.Oceanogr. 1949-1950, 1965-1968.
     Published as Report of the National Oceanographic Council.
Cambridge, 1950-1965.   Z.S 15, M.S 381 & S. 180

Report. National Institute of Oceanography, India. New Delhi.
     Rep.natn.Inst.Oceanogr.India 1965-1966 →   S. 1903 b C

Report. National Library of Wales. Aberystwyth.
     Rep.natn.Libr.Wales 1909 →
     (1909-1963 at Tring.)   S. 6 e

Report. National Maritime Museum. London.
     Rep.natn.marit.Mus. 1970 →   S. 173 b

Report of the National Museum Bloemfontein.
     See Jaarverslag van die Nasionale Museum Bloemfontein.   S. 2010.B

Report. National Museum of Botswana. Gaberones.
     Rep.natn.Mus.Botswana 1967 →   S. 2024 b

Report. National Museum, Department of Education. Republic
   of the Philippines. Manila.
     Rep.natn.Mus.Manila 1966 →   S. 1975 B

Report on the National Museum of Ireland. Dublin.
     Rep.natn.Mus.Ire. 1927-1936.
     Formerly Report on the National Museum of Science and
Art. Dublin.   S. 48 a A

Report on the National Museum of Science and Art. Dublin.
     Rep.natn.Mus.Sci.Art Dubl. 1908-1915, 1920-1921.
     Formerly Report. Dublin Institutions of Science and Art.
     Continued as Report on the National Museum of
Ireland. Dublin.   S. 48 a A

| TITLE | SERIAL No. |

Report of the National Museum of Southern Rhodesia. Bulawayo.
    Rep.natn.Mus.Sth.Rhod. 1936-1964.
    Formerly Report. Rhodesia Museum. Bulawayo.
    Continued as Report of the Trustees and Directors of the
    National Museum of Rhodesia. Bulawayo.   S. 2019 A

Report. National Museum of Tanzania. Dar es Salaam.
    Rep.natn.Mus.Tanzania 1969/1970 →   S. 2027 b

Report. National Museum of Wales. Cardiff.
    Rep.natn.Mus.Wales 1907 →   S. 39 A

Report. National Museums Board of Zambia. Lusaka.
    Rep.natn.Mus.Bd Zambia 1967 →   S. 2075 A

Report. National Museums of Zambia. The Livingstone Museum.
    Rep.natn.Mus.Zambia, Livingstone Mus. 1966 →
    Formerly Report. Rhodes Livingstone Museum.
    The National Museum of Northern Rhodesia.   S. 2075 A

Report of the National Oceanographic Council. Cambridge.
    Rep.natn.oceanogr.Coun. 1950-1965.
    Formerly and Continued as Report of the National Institute
    of Oceanography. Cambridge.   Z.S 15, M.S 381 & S. 180

Report of the National Parks Advisory Board and Director of
    National Parks and Wild Life Management. Ministry of Lands
    and Mines, Rhodesia. Salisbury.
    Rep.natn.Pks advis.Bd Rhod. 1964 →
    Formerly Report of the Federal National Parks Board,
    Rhodesia & Nyasaland.   Z.S 2065 A

Report. National Parks Board of Trustees. Pretoria.
    Rep.natn.Parks Bd Trustees 32nd. → 1957-1958 →   Z.O 74E f N

Report of the National Parks Committee (Commission). London.
    Rep.natn.Parks Comm. 1931-1968.
    Continued as Report of the Countryside Commission.   S. 225 a

Report of the National Physical Laboratory. London.
    Rep.natn.phys.Lab. 1923-1951. (imp.)   S. 205 N

Report. National Reprographic Centre for Documentation. Hatfield.
    Rep.natn.reprogr.Cent.Documn 1970 →   S. 497 g B

Report of the National Research Council. Washington.
    Rep.natn.Res.Coun.,Wash. 1930-1954.
    (From 1943-1954 styled: Report National Academy of Science.
    National Research Council.)   S. 2420 E

Report of the National Research Council of Canada. Ottawa.
    Rep.natn.Res.Coun.Can. No.16-29. 1926-1936.
    Formerly Report. Advisory Council of Scientific and
    Industrial Research. Ottawa.   S. 2602 D

Report of the National Research Council of Canada. Ottawa.
    Annual Report.
    Rep.natn.Res.Coun.Can.a.Rep. 1924 →
    Formerly Report of the Advisory Council for Scientific and
    Industrial Research of Canada. Ottawa. Annual Report.   S. 2602 B

TITLE	SERIAL No.
Report of the National Trust for Places of Historic Interest and Natural Beauty. London. Rep.nat.Trust,Lond. 1914 →	S. 430 A
Report. National Zoological Gardens. Pretoria. Rep.natn zool.Gdns Pretoria 1939 → (imp.).	Z.O 74E f P
Report. Natural Environment Research Council. London. Rep.nat.Envir.Res.Coun. 1965 →	S. 224 a
Report of the Natural History Club of Philadelphia. Rep.nat.Hist.Club Philad. No.3-5. 1871-1873.	S. 2370
Report of the Natural History Museum of Stanford University. Palo Alto. Rep.nat.Hist.Mus.Stanford Univ. 1937-1948.	S. 2407 E
Report. Natural History Museum, University of Khartoum. Rep.nat.Hist.Mus.Univ.Khartoum 1956 →	S. 2047 B
Report of the Natural History Section: Prince of Wales Museum of Western India. Bombay. Rep.nat.Hist.Sect.Pr.Wales Mus.W.India 1923-1925. Continued as Report. Prince of Wales Museum of Western India. Bombay.	S. 1934 A
Report. Natural History Section. Wiltshire Archaeological and Natural History Society. Devizes. Rep.nat.Hist.Sect.Wilts.archaeol.nat.Hist.Soc. 1947-1968. Formerly Wiltshire Bird (and Plant) Notes. Devizes.	S. 31 B
Report. Natural History Society Bradfield College. See Report. Bradfield College Natural History Society.	S. 32 a
Report. Natural History Society. King's School. Canterbury. Rep.nat.Hist.Soc.King's Sch.Canterbury 1958 → Formerly Report. King's School Canterbury, Natural History Society.	S. 24 b
Report of the Natural History Society of Maryland. Baltimore. Rep.nat.Hist.Soc.Md No.2-10, 1930-1939 & 1947.	S. 2317 a A
Report of the Natural History Society of Northumberland, Durham & Newcastle-upon-Tyne. Rep.nat.Hist.Soc.Northumb. 1921 → (imp.)	S. 282 C
Report Natural Monuments Investigation, Plants. Tokyo. Rep.Nat.Monuments Invest.Plants Nos.10-18, 1930-1937	B.S 1999
Report. Natural Resources Board of Rhodesia. Salisbury. Rep.nat.Resour.Bd Rhod. 1969 →	S. 2018 b
Report. Natural Resources Section, General Headquarters. Supreme Commander for the Allied Powers. Tokyo. Rep.Nat.Resour.Sect. H.Q.S.C.A.P. No. 44-182; 1946-51 (imp.) No. 117-146, 1948-1951 (imp.)	M.S 1946 Z. 73I q U

TITLE	SERIAL No.

Report of the Nature Conservancy. London.
   Rep.Nat.Conserv. 1949-1964.   S. 225

Report. Nature Conservation Branch, Transvaal Provincial
   Administration. Pretoria.
   Rep.Nat.Conserv.Brch Transvaal 1965-1966.
   Continued as Report. Transvaal Nature Conservation
   Division.   Z.S 2036 & S. 2070 B

Report. Nature Conservation, Provincial Administration of
   the Orange Free State. Bloemfontein.
   Rep.Nat.Conserv.Orange Free St. 1971 →   S. 2011 b

Report. Netherlands Foundation for the Advancement of Tropical
   Research. The Hague.
   Rep.Neth.Fdn Advmt trop.Res. 1964/65 →   S. 618

Report. Netherlands Geological Survey. Haarlem.
   See Jaarverslag. Rijks Geologische Dienst. Haarlem.   P.S 1289

Report of the New Brunswick Museum. St.John.
   Rep.New Brunsw.Mus. 1940 →   S. 2641 a A

Report of the New Cross Microscopical & Natural History Society.
   Rep.New Cross microsc.nat.Hist.Soc. 9 & 12. 1881-1884.   S. 198

Report. New England Museum of Natural History. Boston.
   Rep.New.Engl.Mus.nat.Hist. 1936-1938.   S. 2321 K

Report of the New Hampshire Agricultural Experiment Station. Durham
   Rep.New Hamps.agric.Exp.Stn 1908.   E.S 2486

Report. New Jersey (afterwards State) Agricultural Experiment
   Station. Paterson, N.J.
   Rep.New Jers.agric.Exp.Stn 1893-1931.   E.S 2487

Report of the New Jersey Department of Conservation and Development.
   Union Hill.
   Rep.New Jers.Conserv.Dev. 1916-1921.   S. 2412 a

Report. New Jersey Geological Survey. Trenton.
   Rep.New Jers.geol.Surv. 1859-1909.
   Continued in Bulletin of the Geological Survey of New Jersey. P.S 1957

Report of the New Jersey State Museum. Trenton, N.J.
   Rep.New Jers.St.Mus. 1905-1912 & 1914; 1966 →   S. 2412 A

Report of the New Mexico Agricultural Experiment Station.
   College of Agriculture and Mechanic Arts. College Station.
   Rep.New Mex.agric.Exp.Stn 1900-1904.   E.S 2487 d

Report. New Mexico Bureau of Mines and Mineral Resources.
   Rep.New Mex.Bur.Mines Miner.Resour. 1964-1965 →   P.S 1956 E

| TITLE | SERIAL No. |

Report. New South Wales Department of Agriculture. Sydney.
    Rep.N.S.W.Dep.Agric. 1892-93, 1900.
    for 1894 see Report. New South Wales Department of
    Agriculture and Forests.   E.S 2248

Report. New South Wales Department of Agriculture and Forests.
    Sydney.
    Rep.N.S.W.Dep.Agric.For. 1894.
    see also Report. New South Wales Department of Agriculture. E.S 2248 a

Report. New South Wales State Fisheries. Sydney.
    Rep.N.S.W.St.Fish. 1936-1937.
    Continued as Report on the Fisheries of New South Wales.   Z.S 2106

Report. New York Botanical Garden. New York.
    Rep.N.Y.bot.Gdn 1966-1967 →   B.S 4316 f

Report of the New York State College of Agriculture at Cornell
    University Agricultural Experiment Station. Albany.
    Rep.N.Y.St.Coll.Agric.Cornell 1914-1919.   E.S 2470 b

Report. New York State Conservation (Commission) Department.
    Rep.N.Y.St.Conserv.Dep. 1911-1939. (imp.)   S. 2364

Report. New York State Geological Survey. Albany.
    Rep.N.Y.St.geol.Surv. 1837-1841.
    See also Report of the State Geologist New York State.
    Albany.   P.S 1960

Report on the New York State Museum of Natural History.
    Albany, N.Y.
    Rep.N.Y.St.Mus.nat.Hist. 1871-1918.
    Formerly Report of the Regents of the University
    (of the State of New York) on the Condition of the State
    Cabinet of Natural History. Albany, N.Y.   S. 2313 A

Report of the New York Zoological Society.
    Rep.N.Y.zool.Soc. 1897 →   T.R.S 5134 A & Z.S 2465 B

Report. New Zealand Ecological Society. Wellington.
    See Report of Annual Meeting. New Zealand Ecological
    Society. Wellington.   S. 2166

Report. New Zealand Geological Survey Department. Wellington.
    Rep.N.Z.geol.Surv.Dep. New Series, 1906-1912.
    Continued as Report. Geological Survey Branch.
    New Zealand. Wellington.   P.S 1163

Report. New Zealand Ornithological Society. Christchurch.
    Rep.N.Z.orn.Soc. 1939-1940.   T.B.S 7301 B

TITLE	SERIAL No.

Report. New Zealand Science Congress.
    Rep.N.Z.Sci.Congr. 7th 1951.
    (For earlier reports See New Zealand Journal of Science
    & Technology, Transactions New Zealand Institute &
    Transactions Royal Society of New Zealand.   S. 2161 D

Report. Newark Museum Association.
    Rep.Newark Mus.Ass. 1909-1934.   S. 2355 A

Report of the Newfoundland Fishery Research Commission.
    St. John's.
    Rep.Newfoundld Fishery Res.Commn 1931-1935.
    Continued as Report of the Fishery Research Laboratory.
    The Series was split up into (Annual) Report, Research
    Bulletins, Service Bulletins, Economic Bulletins, and
    Contributions.   Z.S 2670

Report. Newfoundland Geological Survey.
    See Report. Geological Survey, Province of Newfoundland.
    St. John's.   P.S 1096 A

Report. Newport Free (Public) Libraries, Museum and Art Gallery.
    Rep.Newport Free (Public) Lbr., Mus.Art Gall.
    No.43-50, 1912-1920.   S. 289

Report. Nigerian Institute for Trypanosomiasis Research. Kaduna.
    Rep.niger.Inst.Trypan.Res. 1964-1966.
    Formerly Report. West African Institute for Trypanosomiasis
    Research. Kaduna.   E.S 2173 c

Report. Norfolk Naturalists Trust.
    Rep.Norfolk Nat.Trust 1966 →   S. 295

Report. Norsk Institutt for Tang- og Tareforskning. Oslo.
    See Rapport. Norsk Instituut for Tang- og Tareforskning.
    Oslo.   B.A.S 22

Report North Carolina Geological & Economic Survey.
    Chapel Hill, N.C.
    Rep.N.Carol.geol.econ.Surv. 1905-1924.   S. 2327 a
                    Vol.1 & 5, 1905 & 1923.   P.S 1964 A

Report of the North Central States Entomologists' Conference.
    Rep.N.cent.St.Ent.Conf. 1930.
    Continued as Proceedings. North Central States
    Entomologists.   E.S 2421 c

Report. North Gloucestershire Naturalists' Society. Cheltenham.
    Rep.N.Gloucs.Nat.Soc. 1957 →
    Formerly Report. Cheltenham and District Naturalists'
    Society.   S. 40 a B

Report North London Natural History Society.
    Rep.N.Lond.nat.Hist.Soc. 1904-1913.   S. 175

| TITLE | SERIAL No. |

Report. North Staffordshire Naturalists' Field Club &
    Archaeological Society.
      Rep.N.Staffs.Fld Club 1883-1886.
      Continued as Report and Transactions. North Staffordshire
(Naturalists') Field Club.   S. 371 B

Report. North Wales Naturalists' Trust Ltd. Bangor.
      Rep.N.Wales Nat.Trust 1967/1968 →   S. 25 b

Report. Northamptonshire Naturalists' Trust. Northampton.
      Rep.Northamps.Nat.Trust 1971 →   S. 279 B

Report. Northumberland Sea Fisheries Committee.
      See Report on the Scientific Investigations.
Northumberland Sea Fisheries Committee. Newcastle-upon-Tyne.   Z.S 300

Report from the Norwegian Fisheries Research Laboratory.
      See Fiskeridirektoratets Skrifter. Bergen.   Z.S 530

Report on Norwegian Fishery and Marine Investigations. Kristiana.
      Rep.Norw.Fishery mar.Invest. 1900-1930.
      Continued as Fiskeridirektoratets Skrifter.   Z.S 530

Report of the Norwich Geological Society.
      Rep.Norwich geol.Soc. 1867-1877.   P.S 149

Report. Noto Marine Laboratory of the Faculty of Science.
    University of Kanazawa. Ogi, Noto.
      Rep.Noto Mar.Lab.Fac.Sci.Univ.Kanazawa Vol.3 → 1963 →   S. 1999 b.B

Report. Nottinghamshire Trust for Nature Conservation Ltd.
    Nottingham.
      Rep.Notts.Trust Nat.Conserv. 1969 →   S. 302 B

Report. Noxious & Beneficial Insects in the State of Missouri.
      Rep.Noxious Benefic.Ins.Missouri 1869-1881.   E.S 2484 b

Report of Observations. Radley College Natural History Society. Oxford.
      See Report. Radley College Natural History Society. Oxford.   S. 309

Report on the Occurrence of Insect and Fungus Pests on Plants
    and Crops, Board of Agriculture and Fisheries. London.
      Rep.Occur.Insect Fungus Pests 1917-1918.
      Published as Miscellaneous Publications, Board of Agriculture
and Fisheries.
      Continued as Insect Pests on Crops. London.   E.S 55

Report of the Oceanic Department of the University of Liverpool.
      Rep.oceanic Dep.Univ.Lpool 1920-1933.
      Formerly Report of the Liverpool Marine Biology Committee.
      Continued as Report of the Marine Biological Station
at Port Erin.   Z.S 270

Report of Oceanographic Cruise. United States Coast Guard
    Cutter "Chelan." Bering Sea and Bering Strait. 1934.
      Rep.oceanogr.Cruise "Chelan" 1936.   75 A.q.U

| TITLE | SERIAL No. |

Report of the Oceanographical Institute. Moscow.
   See Doklady Gosudarstvennogo Okeanograficheskogo
   Instituta. Moskva.   S. 1845 C

Report. Oceanographical Investigation. Tokai Regional Fisheries
   Research Laboratory. Tokyo.
   Rep.oceanogr.Invest.Tokai reg.fish.res.Lab.   No.73 → 1967 →   Z.S 1971 B

Report of the Officer i/c Fisheries. Hong Kong.
   Rep.Fish.Hong Kong 1948.
   Continued as Report of the Director of Fisheries.
   Hong Kong.   Z.S 1946 A

Report of the Officers and Committees. Geological Society
   of America. New York.
   Rep.Offrs Comm.geol.Soc.Am.   1960.   P.S 865 c

Report of the Ohara Institute for Agricultural Biology.
   See Bericht des Ohara Instituts für Landwirtschaftliche
   Biologie. Kuraschiki.   E.S 1926

Report of the Ohara Institute for Agricultural Research.
   See Bericht des Ohara Instituts für Landwirtschaftliche
   Forschungen, Okayama Universitat. Kuraschiki.   E.S 1926

Report. Ohio Department of Natural Resources. Columbus.
   See Report of the Directors of the Ohio Department of Natural
   Resources. Columbus.   S. 2332 c A

Report. Ohio Geological Survey.
   See Report of Investigations. Division of Geological
   Survey, Ohio. Columbus.   P.S 1967 A

Report. Ohio Herpetological Society. Columbus, Ohio.
   See Trimonthly Report. Ohio Herpetological Society. Z. Reptile Section

Report of the Ohio State Academy of Science. Columbus.
   Rep.Ohio St.Acad.Sci. 1892-1902.
   Continued in Proceedings of the Ohio State Academy of
   Science. Columbus.   S. 2366

Report. Ontario Bureau of Mines. Toronto.
   Rep.Ont.Bur.Mines   1892-1919 (imp.)
   Continued as Report. Ontario Department of Mines. Toronto.   M.S 2708

Report. Ontario Department of Mines, Toronto.
   Rep.Ont.Dep.Mines   1920-1970.
   Formerly Report. Ontario Bureau of Mines, Toronto.   M.S 2708

Report. Ornamental Pheasant Trust and Norfolk Wildlife Park.
   Norwich.
   Rep.ornament.Pheasant Trust Norfolk Wildl.Pk   1964-1968.
   Continued as Report. Pheasant Trust and Norfolk
   Wildlife Park.   T.B.S 253

Report for Ornithology. Harrogate and District Naturalist
   and Scientific Society. Harrogate.
   Rep.Orn.Harrogate   1948 →   T.B.S 278

TITLE	SERIAL No.

Report of the Osaka Museum of Natural History. Osaka.
    Rep.Osaka Mus.nat.Hist. 1964 →   S. 1992 c D

Report. Oswestry & Welshpool Naturalists' Field Club &
    Archaeological Society.
    Rep.Oswestry Nat.Fld Club 1865.   S. 307

Report of the Otago Museum. Dunedin.
    Rep.Otago Mus. 1937 → imp.
    Formerly Report of the Otago University Museum. Dunedin.   S. 2192

Report of the Otago University Museum. Dunedin.
    Rep.Otago Univ.Mus. 1902-1936.
    Continued as Report of the Otago Museum. Dunedin.   S. 2192

Report of the Oundle School Natural History Society.
    Rep.Oundle Sch.nat.Hist.Soc. 1932 →   S. 305

Report of the Overseas Geological Surveys. London.
    Rep.Overseas geol.Survs 1957-1964.
    Continued in Report. Institute of Geological Sciences.
    London.   P.S 1026

Report of the Oxford Ornithological Society on the Birds
    of Oxfordshire, Berkshire and Buckinghamshire.
    Rep.Oxf.orn.Soc. 1950 →   T.B.S 264

Report. Oxfordshire Natural History Society & Field Club.
    Rep.Oxfsh.nat.Hist.Soc. 1900.
    Continued as Report. Ashmolean Natural History Society
    of Oxfordshire.   S. 313 A

Report. Pacific Division, American Malacological Union.
    See Report. American Malacological Union,
    Pacific Division.   Z. Mollusca Section

Report of the Pacific Marine Fisheries Commission. Portland.
    Oregon.
    Rep.Pacif.mar.Fish.Commn 1948 →   Z.S 2476 A

Report. Pacific Science Board. Washington.
    Rep.Pacif.Sci.Bd 1947-1958.   S. 2419 G

Report of the Paisley Philosophical Institution.
    Rep.Paisley phil.Instn 1884-1938. (imp.)   S. 318 A

Report. Pakistan Association for the Advancement of Science.
    Lahore.
    Rep.Pakist.Ass.Advmt Sci. 1949 →
        (Wanting Report for 1956.)   S. 1940 D

Report on Paleontology. Geological Survey of New Jersey. Trenton.
    Rep.Paleont.geol.Surv.New Jers. 1886-1903.   P.S 1957 A

TITLE	SERIAL No.

Report for Papua.
    See Papua. Annual Report. Melbourne.   S. 2143 B

Report. Papua New Guinea. Canberra.
    Rep.Papua New Guinea 1970/1971 →   S. 2146 b
    Replaces Report of the Territory of Papua. Canberra.   S. 2143 B
    & Report to the General Assembly of the United Nations
    on the Administration of the Territory of New Guinea. Canberra. S. 2146

Report on the Parasite Service. Imperial Agricultural Bureau. London.
    Rep.Parasite Serv.imp.agric.Bur. 1940-1943.   E.S 65

Report. Parks, Museum and Libraries Committee. Stockport.
    Rep.Parks Mus.Libr.Comm.Stockport 1911-1913.   S. 373 b

Report. Pasteur Institute, Viet-Nam, Saigon.
    See Rapport Annuel sur le Fonctionnement Technique
    de l'Institut Pasteur du Việt-Nam.   S. 1922 C

Report of the Peabody Academy of Science. Salem, Mass.
    Rep.Peabody Acad.Sci. 1868-1870.   Z.S 2376

Report. Peabody Museum. New Haven.
    See Report of the Directors of the Peabody Museum.
    New Haven.   S. 2352 Q

Report of the Peabody Museum of American Archaeology and
    Ethnology. Cambridge, Mass.
    Rep.Peabody Mus. 1868-1896.   Z.S 2375

Report. Pembrokeshire Bird Protection Society.
    Rep.Pembr.Bird Prot.Soc. 1938-1945.
    Continued as Report. West Wales Field Society. Tenby.   S. 378

Report of the Penrose Research Laboratory, Zoological Society
    of Philadelphia.
    Rep.Penrose Res.Lab.zool.Soc.Philad. 1937-1942.
    Formerly Report. Laboratory & Museum of Comparative
    Pathology of the Zoological Society of Philadelphia.   Z.S 2500

Report. Penzance Natural History & Antiquarian Society.
    Rep.Penzance nat.Hist.Soc. 1893-1899.
    For previous Reports See Transactions of the Natural History
    and Antiquarian Society of Penzance.   S. 321 B

Report on the Perak Museum.
    Rep.Perak Mus. 1901-1902, 1904-1908.
    Continued in Report on the Museums Department, Federated
    Malay States. S. 1925 B.   S. 1926 a

Report. Percy Fitzpatrick Institute of African Ornithology.
    Rondebosch, Cape.
    Rep.P.Fitzpatrick Inst.Afr.Orn 1965 →   Z. 74G f P

TITLE	SERIAL No.

Report. Perthshire Natural History Museum. Perth.
    Rep.Perthshire nat.Hist.Mus. 1902-1913.　　　　　　　　　　S. 324 a

Report of the Pest Infestation Laboratory, Ministry of Agriculture,
    Springförbi, Denmark.
    See Arsberetning Skogsinsektlaboratorium. Springförbi.　　　E.S 506

Report of the Pest Infestation Research Board. D.S.I.R. London.
    See Pest Infestation Research. Slough.　　　　　　　　　　　E.S 60

Report of the Peterborough Natural History, Scientific &
    Archaeological Society. Peterborough.
    Rep.Peterboro.nat.Hist.scient.archaeol.Soc. 1882-1936 (imp.)　S. 325

Report. Pheasant Trust and Norfolk Wildlife Park. Norwich.
    Rep.Pheasant Trust Norfolk Wildl.Pk 1969 →
    Formerly Report. Ornamental Pheasant Trust and Norfolk
    Wildlife Park.　　　　　　　　　　　　　　　　　　　　　　T.B.S 253

Report on the Phenological Observations in the British Isles
    London.
    Rep.Phenol.Obs.Brit.Is. 1914-1916; 1924-1928.
    Continued as Phenological Report. Royal Meteorological
    Society.　　　　　　　　　　　　　　　　　　　　　　　　　S. 210 A

Report of the Philadelphia Museums.
    Rep.Philad.Mus. 1912-1934. (imp.)　　　　　　　　　　　　　S. 2368 A

Report of the Plant Pathology Laboratory, Harpenden.
    (Ministry of Agriculture & Fisheries.)
    Rep.Pl.Path.Lab.Harpenden 1930-1942　　　　　　　　　　　　E.S 59

Report. Plymouth Municipal Museum and Art Gallery.
    Rep.Plymouth munic.Mus.Art Gall. 1899-1922.　　　　　　　　S. 326 a

Report. Port Elizabeth Museum. Port Elizabeth.
    Rep.Port Eliz.Mus. 1967 →　　　　　　　　　　　　　　　　　S. 2071

Report. Port Erin Marine Biological Station.
    See Report of the Marine Biological Station, Port Erin.　　Z.S 270

Report. Porto Rico Federal Agricultural Experiment Station,
    Mayaguez. Washington.
    Rep.P.Rico fed.agric.Exp.Stn 1914-1929.　　　　　　　　　　E.S 2392

Report. Porto Rico Insular Agricultural Experiment Station.
    Rio Piedras.
    Rep.P.Rico insul.agric.Exp.Stn 1924-1940. (imp.)　　　　　E.S 2393

Report. Postal Microscopical Society. Bath.
    Rep.Postal microsc.Soc. 1881.　　　　　　　　　　　　　　　S. 411 A

Report of the Premier (Transvaal) Diamond Mining Co. Johannesburg.
    Rep.Prem.Diam.Min.Co. 1904-23　　　　　　　　　　　　　　　M.S 2103

TITLE	SERIAL No.

Report of the Prince of Wales Museum of Western India. Bombay.
    Rep.Prince Wales Mus.W.India 1927 → (imp.)
    Formerly Report of the Natural History Section: Prince of
Wales Museum of Western India. Bombay.   S. 1934 A

Report of the Princeton University Expeditions to Patagonia
    1896-1899. Princeton, N.J.
    Rep.Princeton Univ.Exped.Patagonia 1901-1932.   76 F.q.P

Report of the Principal to the Governing Body. Wye College,
    University of London. Ashford.
    Rep.Princ.gov.Body Wye Coll.Ashford 1964-1965 →   B.S 1 A

Report on the Proceedings under the Acts relating to the Sea
    Fisheries, England and Wales. London.
    Rep.Proc.Acts Sea Fish., Lond. 1903-1911.
    Formerly Report of the Inspectors. Sea Fisheries
England & Wales.
Continued as Report on Sea Fisheries. London.   Z.S 460 F

Report of Proceedings. Annual General Meetings.
    Museums Association.
    Rep.Proc.Mus.Ass. 1890-1900.
    (Continued in Museums Journal.)   S. 2 A

Report of Proceedings. Association of Special Libraries &
    Information Bureau. London.
    Rep.Proc.Aslib, Lond. 1924-1947.   S. 231 A

Report and Proceedings. Barrow Naturalists' Field Club.
    Rep.Proc.Barrow Nat.Fld Club 1876-1912: 1928-1939.
        (Wanting Vol.7, 1891-1892.)
    Continued as Proceedings of the Barrow Naturalists'
Field Club.   S. 9

Report and Proceedings of the Belfast Natural History and
    Philosophical Society.
    Rep.Proc.Belf.nat.Hist.phil.Soc. 1882-1920.   S. 14 A
                                    1882-1904.   T.R.S 513
    Formerly Proceedings of the Belfast Natural History
& Philosophical Society.
Continued as Proceedings & Report of the Belfast
Natural History & Philosophical Society.   S. 14 A

Report & Proceedings of the Belfast Naturalists' Field Club.
    Rep.Proc.Belf.Nat.Fld Club N.S. 1873-1922.
    Formerly Report of the Belfast Naturalists' Field Club.
    Continued as Proceedings & Report of the Belfast
Naturalists' Field Club.   S. 15

Report and Proceedings Botanical Society Edinburgh.
    Rep.Proc.bot.Soc.Edinb. 1836-1844.   B.S 6a

Report of Proceedings. British Commonwealth Scientific
    Conference. London.
    Rep.Proc.Br.Commonw.scient.Conf. 1936-1952.   7.o.C

| TITLE | SERIAL No. |

Report and Proceedings. Challenger Society. London.
    Rep.Proc.Challenger Soc. 1967-1968.
    Formerly Report. Challenger Society for the Promotion
    of the Study of Oceanography.
    Continued as Proceedings of the Challenger Society.   S. 167 A

Report and Proceedings of the Chester Society of Natural Science,
    Literature and Art.
    Rep.Proc.Chester Soc.nat.Sci. No.28-50, 1898-1930 & 1947.
    Formerly Report of the Chester Society of Natural Science
    Literature and Art.
    Continued as Proceedings of the Chester Society of
    Natural Science Literature and Art.   S. 27 A

Report of Proceedings of Entomological Meetings at Pusa. Calcutta.
    Rep.Proc.ent.Meet Pusa 1917-1923.   E.S 1995

Report of the Proceeding of the Felsted School Natural
    Science Society. Cambridge &c.
    Rep.Felsted Sch.nat.sci.Soc. 1877-1885.
    Continued as Report of the Felsted School Natural
    History Society.   S. 83

Report of the Proceedings of the Fifth Pan-African
    Veterinary Conference. Nairobi.
    Rep.Proc.5th Pan-Afr.vet.Conf. 1923 (1924).   74 D.q.P

Report of Proceedings. International Congress for Microbiology.
    See International Congress of Microbiology.   S. 2713

Report and Proceedings of the Lancashire & Cheshire Entomological
    Society. Liverpool.
    Rep.Proc.Lancs.Chesh.ent.Soc. 1905-1910, 1912-1953, 1955 →
    Formerly Report of the Lancashire & Cheshire Entomological
    Society. Liverpool.   E.S 3

Report of the Proceedings of the Literary & Philosophical
    Society of Liverpool.
    Rep.Proc.lit.phil.Soc.Lpool 1844-1845.
    Continued as Proceedings of the Literary & Philosophical
    Society of Liverpool.   S. 156 B & T.R.S 136

Report and Proceedings. Liverpool Science Students' Association.
    Rep.Lpool Sci.Stud.Ass. 9-10, 26. 1889-1892, 1907.   S. 159

Report & Proceedings of the Manchester Field Naturalists'
    & Archaeologists' Society.
    Rep.Proc.Manchr Fld Nat.Archaeol.Soc. 1860-1914 & 1932.   S. 260

Report & Proceedings of the Manchester Scientific Students'
    Association.
    Rep.Manchr scient.Students' Ass. 1878-1887. (imp.)
    Formerly Report of the Manchester Scientific Students'
    Association.   S. 264

Report & Proceedings. Merseyside Aquarium Society. Wallasey.
    Rep.Proc.Mersey.Aquar.Soc. 1926-1931.   S. 389

TITLE	SERIAL No.

Report and Proceedings. Miner's Association of
    Cornwall and Devon. Truro.
    Rep.Proc.Miners' Assoc.Truro. 1865, 1872-81.   M.S 159 A

Report of the Proceedings of the Natural History Society,
    Bishop's Stortford College.
    Rep.Proc.nat.Hist.Soc.Bishop's Stortford Coll. 1925, 1928-1930.
    Continued as Coturnix. Report, Bishop's Stortford College
    Natural History Society.   S. 21

Report of Proceedings. Natural Science and Archaeological
    Society. Littlehampton.
    Rep.Proc.nat.Sci.archaeol.Soc.Littlehampton 1928-1945.
    Formerly Reports of Proceedings. Nature & Archaeological
    Circle, Littlehampton.   S. 151 a

Report of Proceedings. Nature and Archaeological Circle,
    Littlehampton.
    Rep.Proc.Nat.Archaeol.Circle Littlehampton 1924-1927.
    Continued as Report and Proceedings. Natural Science and
    Archaeological Society. Littlehampton.   S. 151 a

Report of Proceedings of the Norwich Museum Association.
    Rep.Proc.Norwich Mus.Ass. 1907-1912.   S. 298 B

Report of Proceedings. Norwich Science Gossip Club.
    Rep.Proc.Norwich Sci.Gossip Club 1903-1914.
    Formerly Report of Proceedings. Science Gossip Club.   S. 297

Report & Proceedings. Reading Literary & Scientific Society. Reading.
    Rep.Proc.Reading lit.scient.Soc. 1887-1920, 1922-1924.   S. 338 A

Report of Proceedings under the Salmon and Freshwater
    Fisheries Acts. London.
    Rep.Proc.Salm.Freshwat.Fish.Acts   1904-1914.
      (Wanting 1912.)
    Continued as Report on Salmon and Freshwater Fisheries
    Ministry of Agriculture and Fisheries. London.   Z.S 460 E

Report of Proceedings Science Gossip Club.
    Rep.Proc.Sci.Gossip Club   1874-1903.
    Continued as Report of Proceedings. Norwich Science
    Gossip Club.   S. 297

Report & Proceedings. South Western Naturalists' Union. Bristol.
    Rep.S.W.Nat.Un. 1924-1927.
    Continued as Proceedings of the South Western
    Naturalists' Union.   S. 370

Report of the Proceedings of the Teign Naturalists' Field
    Club. Exeter.
    Rep.Proc.Teign.Nat.Fld Club 1865-1917.
    Continued as Proceedings of the Teign Naturalists'
    Field Club.   S. 75

TITLE	SERIAL No.

Report, Proceedings and Transactions. Manchester Entomological Society.
   Rep.Proc.Trans.Manchr ent.Soc. 1961 →
   Formerly Report & Transactions of the Manchester Entomological Society.   E.S 38

Report on the Progress of Agriculture in India. Calcutta.
   Rep.Prog.Agric.India 1907-1919.
   Formerly Report of the Imperial Department of Agriculture, India.
   Continued as Review of Agricultural Operations in India. Calcutta.   E.S 1997

Report of Progress. Alabama State Geological Survey.
   Rep.Prog.Ala St.geol.Surv. 1874-1882; 1914 →   P.S 1870

Report on the Progress of Chemistry. London.
   Rep.Prog.Chem. 1904-1966: Sect.A, 1967 →   M.S 112

Report of the Progress of the "Discovery" Committee's Investigations. London.
   Rep.Prog.Discovery Comm.Invest. 1937.
   Formerly Report. "Discovery" Investigation. London.   S. 172 B

Report of Progress. Geological Survey of Alabama. Montgomery.
   See Report of Progress. Alabama State Geological Survey.   P.S 1870

Report of Progress. Geological Survey, Canada, Toronto.
   Rep.Prog.geol.Surv.Canada 1853-1856   M.S 2705 B

Report of Progress. Geological Survey of Japan.
   Rep.Prog.geol.Surv.Japan 1878-1879.   P.S 1750

Report.of Progress. Geological Survey of Ohio. Columbus.
   Rep.Prog.geol.Surv.Ohio 1869-1870.   P.S 1966

Report of Progress. Geological Survey of Pennsylvania. Harrisburg.
   See Report. Geological Survey of Pennsylvania. Harrisburg.   P.S 1970

Report of Progress. Geological Survey of Victoria. Melbourne.
   Rep.Prog.geol.Surv.Vict. 1874-1899.   P.S 1130
                               No.2-7 & 11, 1875-1899.   M.S 2415
   Continued as Monthly Progress Report. Geological Survey of Victoria. Melbourne.

Report of Progress. Indiana Division of Geology. Indianapolis.
   Rep.Prog.Indiana Div.Geol. 1946-1949.
   Continued as Report of Progress. Indiana Geological Survey. Bloomington.   P.S 1909

Report of Progress. Indiana Geological Survey. Bloomington.
   Rep.Prog.Indiana geol.Surv. No.4 → 1952 →
   Formerly Report of Progress. Indiana Division of Geology. Indianapolis.   P.S 1909

TITLE	SERIAL No.

Report on Progress in Physics. London.
    Rep.Prog.Phys.   Vol.3-7, 9-10; 1937-1940, 1942-1945.   M.S 117

Report of Progress and Synopsis of the Field Work.
    Geological Survey of California.
    Rep.Progr.Synop.Fld Work,geol.Surv.Calif.   1860-64   M.S 2685

Report of the Provancher Society of Natural History of Canada.
    Quebec.
    Rep.Provancher Soc.nat.Hist.Can.   1938-1951.   S. 2636

Report of the Provincial Museum (of Natural History and
    Anthropology), Province of British Columbia. Victoria, B.C.
    Rep.prov.Mus.nat.Hist.Anthrop.Br.Columb.   1912-1968.
    Continued in   Report. British Columbia Department of
    Recreation and Conservation. Victoria, B.C.   S. 2658 A

Report. Public Gardens and Plantations, Kingston. Jamaica.
    Rep.publ.Gdns Plantns Jamaica   1880-1900 (imp.)
    Continued as   Report of the Board of Agriculture,
    and Department of Public Gardens & Plantations, Jamaica.
    Kingston.   B.S 4100 a

Report of the (Free) Public Library, Museum. Liverpool.
    Rep.Mus.Lpool   1853-1913.
    Continued as   Report of the Free Public Museum, Liverpool.   S. 160 B

Report. Public Library, Museum and Art Gallery of Western
    Australia. Perth.
    Rep.publ.Libr.Mus.W.Aust.   1914-1955.
    Continued as   Report. Museum and Art Gallery of Western
    Australia. Perth.   S. 2156 B

Report of the Public Library and Museum of South Australia.
    Adelaide.
    Rep.publ.Libr.Mus.S.Aust.   1884-1940.
    Continued as   Report. South Australian Museum. Adelaide.   S. 2109 C

Report. Public Museum and Art Gallery. Cardiff.
    Rep.publ.Mus.Cardiff   1895.   S. 39 B

Report. Public Museum and Art Gallery of Papua and New Guinea.
    See   Report of the Trustees of the Museum & Art Gallery of
    Papua & New Guinea.   S. 2142 a

Report. Public Museum of the City of Milwaukee.
    Rep.publ.Mus.Milwaukee   1883-1911 & 1919.
    Replaced by   Yearbook of the Public Museum of the
    City of Milwaukee.   S. 2348 A

Report of the Public Museum Committee. Gloucester.
    Rep.publ.Mus.Comm.Gloucester   Nos.1-4 & 6, 1910-1920.   S. 104

TITLE	SERIAL No.

Report of the Public Museums. City of Sheffield.
    Rep.Publ.Mus. 1910-1914.
    Continued as Report of the Public Museums and Mappin
Art Gallery. City of Sheffield.   S. 358 a

Report of the Public Museums and Mappin Art Gallery. City
    of Sheffield.
    Rep.Publ.Mus.Mappin Art Gall. 1914-1932.
    Formerly Report of the Public Museum. City of Sheffield.
    Continued as Report. City Museums & Mappin Art Gallery.
Sheffield.   S. 358 a

Report. Puerto Rico Federal Agricultural Experiment Station.
    See Report. Porto Rico Federal Agricultural Experiment
Station, Mayaguez. Washington.   E.S 2392

Report. Puerto Rico University of Agricultural Experiment Station.
    See Report. Porto Rico Insular Agricultural Experiment
Station. Rio Piedras.   E.S 2393

Report on the Puffin Island Biological Station.
    See Report of the Liverpool Marine Biological Station
on Puffin Island.   Z.S 313

Report. Quebec Bureau of Mines. Quebec.
    Rep.Quebec.Bur.Min. 1929-30(imp.)   M.S 2710

Report. Queen Victoria Museum and Art Gallery. Launceston,
    Tasmania.
    Rep.Queen Vict.Mus.Launceston 1903 → (imp.)   S. 2155 A

Report of the Queensland Museum. Brisbane.
    See Report of the Trustees of the Queensland Museum.
Brisbane.   S. 2136 C

Report of the Quekett Microscopical Club. London.
    Rep.Quekett microsc.Club No.3-16. 1868-1881. (imp.)   S. 413 B

Report. Radley College Natural History and Biological Society.
    Abingdon.
    Rep.Radley Coll.nat.Hist.biol.Soc. 1965 →   S. 309 C

Report. Radley College Natural History Society. Oxford.
    Rep.Radley Coll.nat.Hist.Soc. 1944-1949.   S. 309

Report of the Raffles Museum and Library. Singapore.
    Rep.Raffles Mus. 1879, 1887-1956.   S. 1970 C

Report of the Raven Entomological and Natural History Society.
    Liverpool.
    Rep.Raven ent.nat.Hist.Soc. 1946-1951.   E.S 40

Report of the Ray Society. London.
    Rep.Ray Soc. 1855 → (imp.)   S. 203

TITLE	SERIAL No.

Report of the Reading Literary & Scientific Society. Reading.
    See Report & Proceedings. Reading Literary & Scientific
    Society. Reading.   S. 338 A

Report of the Reading Natural History Society. Reading.
    Rep.Reading nat.Hist.Soc. 1903-1904.   S. 339 A

Report. Reading Ornithological Club.
    See Reading Ornithological Club Reports.   T.B.S 262

Report and Records of the Colchester and District Natural
    History Society and Field Club.
    Rep.Rec.Colchester Distr.nat.Hist.Soc.Fld Club 1964-1965 → S. 37 a

Report. Red Deer Commission. Edinburgh.
    Rep.Red Deer Commn 1959 →   Z. 072Ab o G

Report. Red House Museum, Art Gallery and Gardens.
    Christchurch, Hants.
    Rep.Red House Mus. 1951-1970.   S. 28 a

Report & Reference Book. Mid-Somerset Naturalist Society. Bridgwater.
    Rep.Ref.Bk mid-Somers.Nat.Soc. 1949/1951 - 1965.   S. 88

Report on Reforms and Progress in Chosen.
    Rep.Progr.Chosen 1918-1921.
    Continued as Report on Administration of Chosen.   S. 1977 b

Report of Refractory Materials. London.
    See Special Report. Iron and Steel Institute. London.   M.S 146 B

Report of the Regents of the University (of the State of New York)
on the Condition of the State Cabinet of Natural History.
Albany, N.Y.
    Rep.Reg.Univ.N.Y. 1848-1870.   S. 2313 A
                  No.20, 1866.   P.S 1961
    Continued as Report on the New York State Museum of
    Natural History. Albany, N.Y.

Report. Research Council of Alberta. Edmonton.
    Rep.Res.Coun.Alberta 1930-1964.
    (A Report published 1966 → covering geological investigations
    is in the Palaeontological Library at P.S 1089 D. The Annual
    Report is also continuing.)
    Formerly Report. Scientific and Industrial Research
    Council of Alberta. Edmonton.   S. 2611

Report. Research Council of Alberta. Annual Report. Edmonton.
    Rep.Res.Coun.Alberta a.Rep. 1920 →
    (Nos.1-45 were issued within the Council's Report
    Series, No.3-86.)   S. 2611

Report (Geology). Research Council of Alberta. Edmonton.
    Rep.Res.Coun.Alberta 66-1, 1966 →
    Formerly Preliminary Report. Research Council of
    Alberta. Edmonton.   P.S 1089 D

TITLE	SERIAL No.

Report of the Research Council, D.S.I.R. London.
    Rep.Res.Coun., D.S.I.R. 1957-1964.
    Formerly Report of the Department of Scientific and
    Industrial Research. London.   S. 205 P

Report. Research Institute for Nature Management. Arnhem.
    See Jaarverslag. Rijksinstituut voor Natuurbeheer. Arnhem.   S. 620 A

Report on Research. Moss Landing Marine Laboratories. Moss Landing.
    Rep.Res.Moss Landing mar.Lab. 1970 →   S. 2390 a B

Report. Research Projects, Applied Scientific Research Corporation
    of Thailand. Bangkok.
    Rep.Res.Proj.appl.scient.Res.Corp.Thailand
    No.18/1 → 1966 → (imp.)   Z.S 1988 A

Report. Research School of Biological Sciences. Australian National
    University. Canberra.
    Rep.Res.Sch.biol.Sci.Aust.natn.Univ. 1969 →   S. 2148 a

Report and Review of Events. Smithsonian Institution Center
    for Short-lived Phenomena. Cambridge, Mass.
    Rep.Rev.Events Smithson.Inst.Cent.short-lived Phen. 1972 →
    Formerly Report. Smithsonian Institution Center for
    Short-lived Phenomena. Cambridge, Mass.   P. REF

Report. Rhodes-Livingstone Museum. The National Museum of Northern
    Rhodesia. Livingstone.
    Rep.Rhodes-Livingstone Mus. 1948-1964
    Continued as Report. Livingstone Museum. Livingstone.   S. 2075 A

Report. Rhodesia Museum. Bulawayo.
    Rep.Rhod.Mus. 1902-1935.
    Continued as Report of the National Museum of Southern
    Rhodesia. Bulawayo.   S. 2019 A

Report of the Rhodesia Scientific Association. Bulawayo.
    Rep.Rhod.Scient.Ass. 1904-1946. (imp.)   S. 2018 A

Report. Robert H. Lowie Museum of Anthropology. Berkeley.
    Rep.R.H.Lowie Mus.Anthrop. 1963 →   P.A.S 776 E

Report of the Rochdale Literary & Scientific Society.
    Rep.Rochdale lit.scient.Soc. 1885-1904 (imp.)   S. 342 B

Report of the Rockefeller Foundation. New York.
    Rep.Rockefeller Fdn   1918-1935.   Z.S 2470 B

Report of the Rockefeller Sanitary Commission for the
    Eradication of Hookworm Disease. Washington, D.C.
    Rep.Rockefeller sanit.Commn 1910-1911 - 1914.   Z.S 2520

Report. Rondevlei Bird Sanctuary. Cape Town.
    See Report of the Warden of the Ron Devlei Bird Sanctuary.   T.B.S 4005

TITLE	SERIAL No.

Report. Rothamsted Experimental Station. Harpenden.
    Rep.Rothamst.exp.Stn 1918-1933 (imp.)   B.S 14
    1964 →   E.S 449

Report. Royal Botanic Garden, Edinburgh.
    Rep.R.Bot.Gdn Edinb. 1876-1879.   B.S 4 a

Report. Royal Botanic Gardens and National Herbarium. Sydney.
    Rep.R.bot.Gdn natn.Herb. 1968 →   B.S 2440 a

Report of the Royal Botanic Gardens, Peradeniya. Colombo.
    Rep.R.bot.Gdns Peradeniya 1883-1910/1911 (imp.)
    Continued in Report of the Department of Agriculture,
Ceylon. Colombo.   B.S 1720 b

Report. Royal Botanic Gardens, Trinidad. Port of Spain.
    Rep.R.bot.Gdns Trin. 1888-1890, 1901-1902, 1906-1907.   B.S 4125 a

Report of the Royal Botanical Gardens Calcutta.
    Rep.R.bot.Gdns, Calcutta 1898-1949 (imp.)
    Continued as Report of the Indian Botanic Garden and
the Gardens in Calcutta, (Parks & Gardens in Cooch Behar)
and the Lloyd Botanic Garden, Darjeeling. Alipore.   B.S 1602 a

Report. Royal Commission on Historical Manuscripts. London.
    Rep.R.Commn Hist.MSS 1968/1969 →   S. 213 a

Report of the Royal Cornwall Polytechnic Society. Falmouth.
    Rep.R.Cornwall polytech.Soc. 1833 →   S. 81

Report. Royal Dublin Society.
    Rep.R.Dubl.Soc. 1965 →   S. 46 L

Report of the Royal Institute of Science Bombay.
    Rep.R.Inst.Sci.Bombay 1925-1934.   S. 1903 a

Report. Royal Institution of Cornwall. Truro.
    Rep.R.Instn Cornwall No.20-22, 25, 27-29, 32,
38-54, 1839-1872.
    (Subsequent Reports published in Journal of the Royal
Institution of Cornwall in General Library S. 381.)   Z. 5 o (Tracts)

Report. Royal Institution of South Wales. Swansea.
    Rep.R.Instn S.Wales Nos.16-112. 1850-1947.   S. 375

Report. Royal Naval Bird Watching Society.
    Rep.R.nav.Bird Watch.Soc. No.1, 1947.
    Continued as Sea Swallow.   T.B.S 106

Report. Royal Ontario Museum. Toronto.
    Rep.R.Ont.Mus. 1956 →
    Formerly Report. Royal Ontario Museum of Zoology and
Palaeontology. Toronto.   S. 2606 A

Report. Royal Ontario Museum of Zoology and Palaeontology. Toronto.
    Rep.R.Ont.Mus.Zool.Palaeont. 1949-1955.
    Continued as Report Royal Ontario Museum. Toronto.   S. 2606 A

TITLE	SERIAL No.

Report of the Royal Scottish Museum. Edinburgh.
    Rep.R.Scott.Mus. 1901-1915: 1919-1922: 1963 →   S. 60 a

Report. Royal Society for the Protection of Birds. London.
    Rep.R.Soc.Prot.Birds Nos.3-65, 70 → 1892-1893 → (imp.)
    Nos.66-69 published in Bird Notes.   T.B.S 105 A

Report of the Royal Zoological Society of Ireland. Dublin.
    Rep.R.zool.Soc.Ire. (No.11) - No.78, 1840-1909 (imp.)
    No.126 → 1958 →   Z.S 450

Report. Royal Zoological Society of South Australia. Adelaide.
    Rep.R.zool.Soc.S.Aust. 1961/1962 →   Z.S 2123

Report of the Rugby School Natural History Society.
    Rep.Rugby Sch.nat.Hist.Soc. 1867 →   S. 345

Report. Rye Meads Ringing Group.
    Rep.Rye Meads Ringing Gr. 1961 →   T.B.S 259

Report of the Sado Marine Biological Station, Niigata University.
    Rep.Sado mar.biol.Stn 1971 →   S. 1988 c B

Report. St.Agnes Bird Observatory, Isles of Scilly. London.
    Rep.St.Agnes Bird Observ. 1957 →   T.B.S 276

Report of the Salisbury, South Wilts and Blackmore Museum. Salisbury.
    Rep.Salisbury S.Wilts Blackmore Mus. 1904-1905; 1914-1958.   S. 351

Report on Salmon Fisheries, Scotland.
    See Report of the Fishery Board for Scotland. Edinburgh.   Z.S 380

Report on Salmon and Freshwater Fisheries. Ministry of
    Agriculture and Fisheries. London.
    Rep.Salm.Freshwat.Fish., Lond. 1925-1937.
    Formerly Report of Proceedings under the Salmon
    and Freshwater Fisheries Acts.   Z.S 460 F

Report of the San Diego Society of Natural History.
    Rep.S.Diego Soc.nat.Hist. 1923-1926.   S. 2399 A

Report. Santa Barbara Botanic Garden.
    Rep.Santa Barbara bot.Gdn 1966 →   B.S 4172 b

Report. Santa Barbara Museum of Natural History.
    Santa Barbara, California.
    Rep.Santa Barbara Mus.nat.Hist. 1932 →   S. 2402 B

Report on the Sarawak Museum.
    Rep.Sarawak Mus. 1900-1939.   S. 1974

| TITLE | SERIAL No. |

Report. Saudi Arabian Natural History Society. Jeddah.
   Rep.Saudi Arab.nat.Hist.Soc. 1971.
     Continued as Journal. Saudi Arabian Natural History Society.
   Jeddah.   S. 1922 d

Report of the Scarborough Philosophical & Archaeological Society.
   Rep.Scarboro.phil.Soc. 1834-1926 (imp.)   S. 357

Report. School House Natural History Society. Dauntsey's
   School West Lavington.
     See School House Natural History Society. Dauntsey's
   School West Lavington.   S. 393 a B

Report. School of Mines and Industries, South Australia, Adelaide.
     See Report. South Australian School of Mines and
   Industries. Adelaide.   M.S 2411

Report. School of Public Health and Tropical Medicine. University
   of Sydney.
   Rep.Sch.publ.Hlth trop.Med. 1968/69 →   S. 2133 C

Report of the Science Council of Japan. Tokyo.
   Rep.Sci.Coun.Japan 1949-1950 →   S. 1994 B

Report of Science Masters' Association.
   Rep.Sci.Masters' Ass. 1919.
     Formerly Report of the Association of Public School
   Science Masters.   S. 2 a

Report on the Science Museum. London.
   Rep.Sci.Mus.,Lond. 1920-1929.
     Formerly Report on the Science Museum and on the
   Geological Survey and Museum and Museum of Practical
   Geology, London.
     Continued as Report of the Advisory Council of the
   Science Museum. London.   S. 241 E

Report on the Science Museum and on the Geological Survey and
   Museum of Practical Geology. London.
   Rep.Sci.Mus.Lond. 1913-1919.
     Formerly Report on the Geological Survey and Museum,
   the Science Museum, and the work of the Solar Physics
   Committee. London.
     Continued as Report on the Science Museum, London.   O. 72Aa o G

Report. Science Museum of Minnesota. St.Paul.
   Rep.Sci.Mus.Minn. 1972/1973 →   S. 2470 D

Report. Science Research Council. London.
   Rep.Sci.res.Coun. 1965 →   S. 205 a

Report on Scientific Activities. Institute of Ecology, etc.
   Polish Academy of Sciences. Warsaw.
   Rep.scient.Activ.Inst.Ecol.Polish Acad.Sci. 1963 →   S. 1870 a K

| TITLE | SERIAL No. |

Report of Scientific Activities. The Weizmann Institute
of Science. Rehovoth.
Rep.scient.Activ.Weizmann Inst.Sci. 1960-1961.
Formerly Report. The Weizmann Institute of Science. Rehovoth.
Continued as Scientific Activities. The Weizmann Institute
of Science. Rehovoth.   S. 1928 c

Report from the Scientific Expedition to the North-Western
Provinces of China under the Leadership of Dr.Sven Hedin.
Stockholm.
Rep.scient.Exped.N.-W.Prov.China 1937 →   73 H.q.H

Report of the Scientific and Industrial Research Council of
Alberta. Edmonton.
Rep.scient.ind.Res.Coun.Alberta 1921-1929.
Formerly Report of the Advisory Council of Scientific
and Industrial Research of Alberta. Edmonton.
Continued as Report. Research Council of Alberta.
Edmonton.   S. 2611

Report on the Scientific Investigations. British Antarctic
Expedition 1907-9. London.
Rep.Brit.antarct.Exped. 1910-1930.   80.q.S

Report. Scientific Investigations in Micronesia. Washington, D.C.
Rep.scient.Invest.Micronesia No.4 → 1950 → (imp.).   Z. 77F q N

Report on the Scientific Investigations. Northumberland Sea
Fisheries Committee. Newcastle-on-Tyne.
Rep.scient.Invest.Northumb.Sea Fish.Comm. 1900-1911.
Formerly Report on the Trawling Excursions &c.
Continued as Report. Dove Marine Laboratory.   Z.S 300

Report on the Scientific Results. Bathymetrical Survey of the
Scottish Fresh-water Lochs. Edinburgh.
Rep.scient.Results Bathymetr.Surv.Scot.Freshw.Lochs 1910.   72 Ab.o.M

Report on the Scientific Results of the "Michael Sars"
North Atlantic Deep-Sea Expedition 1910. Bergen.
Rep.scient.Results Michael Sars N.Atlant.deep Sea Exped.
  1913 →   78 q.B

Report of the Scientific Results of the Norwegian Expedition to
Novaya Zemlya 1921. Kristiania.
Rep.scient.Results Norw.Exped.Nova Zemlya 1922-1930.   71 q.C

Report on Scientific Results. Research Institute of African
Geology. University of Leeds.
Rep.scient.Results res.Inst.Afr.Geol. 1960 →   P.S 140 a

Report on the Scientific Results. Scottish National Antarctic
Expedition. S.Y. "Scotia" 1902-1904. Edinburgh.
Rep.scient.Results Scott.natn.antarct.Exped. 1907-1920.   80 q.S

| TITLE | SERIAL No. |

Report of the Scientific Results of the Voyage of H.M.S.
  Challenger 1873-76. London.
  <u>Rep.scient.Results Voy.Challenger</u> 1880-1895.   70 q.G & P.S 1065
                                    T.R.S 1 G, Z. 70 q G & Sections

Report. The Scientific Society. University College of Wales.
  Aberystwyth.
  <u>Rep.scient.Soc.Univ.Coll.Wales</u> 1892-1893.   S. 7 C

Report of the Scientific Station of Fisheries of the Azov
  and Black Seas.
  <u>See</u> Trudȳ Azovsko-Chernomorskogo Nauchno-Issledovatel'skogo
  Instituta Rȳbnogo Khozyaistva i Okeanografii. Simferopol.   Z.S 1838

Report of the Scientific Station of Fisheries in Kertch.
  <u>See</u> Trudȳ Kerchenskoi Nauchnoi Rȳbokhozyaistvennoi Stantsii.   Z.S 1838

Report Scottish Field Studies Association. Glasgow.
  <u>Rep.Scott.Fld Stud.Ass</u>. 1962 →   S. 96

Report. Scottish Marine Biological Association. Glasgow.
  <u>Rep.Scott.mar.biol.Ass</u>. 1914 →
  <u>Formerly</u> Report. Marine Biological Association of the
  West of Scotland.   S. 98

Report by the Scottish National Parks Committee and the Scottish
  Wild Life Conservation Committee. Edinburgh.
  <u>Rep.Scott.natn.Parks Comm</u>. 1945, 1947, 1949.   S. 57 a

Report. Scottish Plant Breeding Station. Pentlandfield, Roslin.
  <u>Rep.Scott.Pl.Breed.Stn</u> 1955-1961; 1965/1966 →
  <u>Between 1962 and 1965 entitled</u> Record. Scottish Plant
  Breeding Station.   B.S 131

Report Scottish Seaweed Research Association Edinburgh.
  <u>Rep.Scott.Seaweed Res.Ass</u>. 1945-1950.
  <u>Continued as</u> Report Institute of Seaweed Research.   B.A.S 1

Report. Scripps Institution for Biological Research Berkeley.
  <u>Rep.Scripps Instn biol.Res</u>. 1923-1926.   S. 2319 N

Report. Scripps Institution of Oceanography. Berkeley,
  California.
  <u>Rep.Scripps Instn Oceanogr</u>. 1930-1934, 1939, 1967 →   S. 2319 Ma

Report on Sea Fisheries. London.
  <u>Rep.Sea Fish., Lond</u>. 1912-1914, 1926-1937. (imp.)
  <u>From 1915-1918 styled</u> Fisheries in the Great War.
  <u>Formerly</u> Report on the Proceedings under the Acts
  relating to the Sea Fisheries.
  <u>Replaced by</u> Fisheries in War Time. London.   Z.S 460 F

TITLE	SERIAL No.

Report of the Sea Fisheries Institute in Gdynia.
    See Prace Morskiego Instytutu Rybackiego w Gdyni.   Z.S 1804

Report on the Sea and Inland Fisheries of Ireland. Dublin.
    Rep.Sea Inld Fish.Ire. 1900 →
    Part II - Scientific Investigations issued separately
    after 1904.   Z.S 400 A

Report of the Second Norwegian Arctic Expedition in the
    "Fram" 1898-1902. Kristiania.
    Rep.2nd Norw.Arctic Exped.Fram 1907-1930.   71.q.C

Report of the Secretary of Agriculture for Northern Rhodesia.
    See Report. Department of Agriculture, Northern Rhodesia.
    Lusaka.   E.S 2177 a

Report. Secretary of Agriculture, United States Department
    of Agriculture.
    See Report. United States Department of Agriculture.
    Washington.   E.S 2450

Report of the Secretary of the Maine Board of Agriculture. Augusta.
    Rep.Secr.Maine Bd Agri. No. 6-7 1861-1862   P.S 1927

Report of the Secretary of the Massachusetts Board of
    Agriculture. Boston.
    Rep.Mass.Bd Agric. No.19-21. 1871-73.   75 C.o.M

Report. Secretary for Mines, Tasmania. Hobart.
    Rep.Secr.Mines Tasm. 1900-1939.
    Continued as Report. Director for Mines, Tasmania, Hobart.   M.S 2444

Report of the Secretary for Mines (and Water Supply).
    Victoria, Melbourne.
    Rep.Secr.Min.,Vict. 1884-1917.
    Continued as Report of the Mines Department, Victoria.   M.S 2424

Report. Section of Mines, Geological Survey, Canada. Ottawa.
    See Report on the Mineral Industries of Canada. Ottawa.   M.S 2702

Report. Selborne Society. Whitgift School. (Croydon.)
    Rep.Selborne Soc.Whitgift Sch. 1951 →   S. 397 a

Report. Sericultural Institute. Chungking.
    Rep.seric.Inst.,Chungking 1945.   E.S 1946

Report of the Severn Wildfowl Trust. London.
    Rep.Severn Wildfowl Trust 1948 - 1951-1952.
    Continued as Report of the Wildfowl Trust.   T.B.S 193

TITLE	SERIAL No.

Report. Sheffield City Museums.
    Rep.Sheffield City Mus. 1965/1966 - 1966/1967.
    Formerly Report. City Museums & Mappin Art Gallery.
    Sheffield.   S. 358 a

Report. Sheffield Naturalists' Club.
    Rep.Sheffield Nat.Club 1914-1915.   S. 358 B

Report of the Sherborne School Field Society.
    Rep.Sherborne Sch.Fld Soc. 1878.   S. 359

Report and Short Papers. Institute of Freshwater Research, Drottningholm.
    Rep.short Pap.Inst.Freshwat.Res.Drottningholm 1949-1954.
    Formerly Meddelanden fran Statens Undersöknings-och Försoksänstalt för Sötvattensfisket. Stockholm.
    Continued as Report. Institute of Freshwater Research. Drottningholm.   Z.S 510

Report of the Shropshire & North Wales Natural History & Antiquarian Society. Shrewsbury.
    Rep.Shropsh.nat.Hist.Soc. 1835-1836, 1859, 1863.   S. 362

Report. Sidcup Natural History Society.
    Rep.Sidcup nat.Hist.Soc. 1950-1963.   S. 354

Report. Skokholm Bird Observatory. Haverfordwest.
    Rep.Skokholm Bird Obs. 1937 →   T.B.S 461

Report of the Sleeping Sickness Commission. Royal Society. London.
    Rep.sleep.Sickn.Commn R.Soc. 1903-1915.   E.S 94

Report of the Slough Natural History Society.
    Rep.Slough nat.Hist.Soc. 1947.
    Continued as Middle Thames Naturalist.   S. 363

Report. Smithsonian Institution. Washington.
    See Report of the Board of Regents of the Smithsonian Institution. Washington.   S. 2426 A

Report. Smithsonian Institution Center for Short-lived Phenomena. Cambridge, Mass.
    Rep.Smithson.Inst.Cent.short-lived Phen. 1969-1971.
    Continued as Report and Review of Events. Smithsonian Institution Center for Short-lived Phenomena. Cambridge, Mass.   P. REF

Report. Société Guernésiase.
    See Report and Transactions. Société Guernésiaise. Guernsey.   S. 113

Report of the Society for the Acclimatisation of Animals, Birds, Insects & Vegetables London.
    Rep.Soc.Acclim.Lond. 1, 3-5. 1861-1865.   S. 164

TITLE	SERIAL No.

Report. Society for the Promotion of Nature Reserves. London.
   <u>See</u> Handbook (and Annual Report) of the Society for
   the Promotion of Nature Reserves. London.   S. 429 A

Report. Soil Conservation Service, United States Department
   of Agriculture. Washington.
   <u>Rep.Soil Conserv.Serv.U.S.Dep.Agric.</u> 1936 →   E.S 2458 h

Report on Somerset Birds. Taunton.
   <u>Rep.Somerset Birds</u> No.11 → 1924 →   T.B.S 254

Report. Somerset Trust for Nature Conservation. Taunton.
   <u>Rep.Somerset Trust Nat.Conserv.</u> No.7 → 1971 →   S. 376 a B

Report. Sorby Natural History Society. Sheffield.
   <u>Rep.Sorby nat.Hist.Soc.</u> 1942-1950 (imp.)   S. 358 D

Report of the South African Association for the Advancement of
   Science. Cape Town.
   <u>Rep.S.Afr.Ass.Advmt Sci.</u> 1903-1908.
   <u>Continued as</u> South African Journal of Science.   S. 2001 A

Report of the South African Central Locust Bureau.
   <u>See</u> Report of Committee of Control S.African Central
   Locust Bureau.   E.S 2168 c

Report. South African Institute for Medical Research. Johannesburg.
   <u>Rep.S.Afr.Inst.med.Res.</u> 1954 →   E.S 2159

Report of the South African Museum. Cape Town.
   <u>Rep.S.Afr.Mus.</u> 1855 → (imp.)   S. 2011 B

Report. South Australian Museum. Adelaide.
   <u>Rep.S.Aust.Mus.</u> 1940 →
   <u>Formerly</u> Report of the Public Library and Museum of
   South Australia. Adelaide.   S. 2109 C

Report. South Australian School of Mines & Industries. Adelaide.
   <u>Rep.S.Aust.Sch.Mines Ind.</u> 1918-1919   M.S 2411

Report of the South Eastern Agricultural College, Wye, Kent.
   London and Ashford.
   <u>Rep.S.East.agric.Coll.Wye</u> 1895-1939.   E.S 70

Report of the South London Entomological & Natural History Society.
   <u>Rep.S.Lond.ent.nat.Hist.Soc.</u> 1879-1884.
   <u>Continued as</u> Abstracts of the Proceedings of the South
   London Entomological & Natural History Society.   S. 216

Report of the South London Microscopical & Natural History Club.
   <u>Rep.S.Lond.microsc.nat.Hist.Club</u> 1872-1892.   S. 217

Report of the South West African Scientific Society. Windhoek.
   <u>Rep.S.W.afr.Scient.Soc.</u> 1974 →   S. 2077 E

| TITLE | SERIAL No. |

Report. Southport Scientific Society.
    Rep.Southport scient.Soc. 1933-1946.
    Formerly Report of the Southport Society of Natural History.   S. 368

Report of the Southport Society of Natural Science.
    Rep.Southport Soc.nat.Sci. 1890-1932.   S. 368
                                  1890-1904.   T.R.S 149
    Continued as Report. Southport Scientific Society.

Report of the Southwest Museum. Los Angeles. California.
    Rep.Sthwest Mus.,Los Angeles 1920-1926.   S. 2404 a A

Report. Staffordshire Nature Conservation Trust Ltd. Stoke-on-Trent.
    Rep.Staffs.Nat.Conserv.Trust 1970 →   S. 372

Report. Stanford University.
    See Report. Leland Stanford Junior University.   S. 2407 C

Report. State Biological Survey of Kansas. Lawrence.
    Rep.St.biol.Surv.Kans. 1964/1965.   S. 2409 a

Report. State Board of Fisheries and Game, Lake & Pond
    Survey Unit. Hartford, Conn.
    See Report. Connecticut State Board of Fisheries
    & Game, Lake & Pond Survey Unit. Hartford.   Z. 75C o C

Report of the State Board of Geological Survey of Michigan. Lansing.
    Rep.St.Bd geol.Surv.Mich. 1899-1908
    Continued in Publications of the Michigan Geological
    & Biological Survey. Lansing.   P.S 1936

Report of the State Entomologist, New York State Museum. Albany.
    Rep.St.Ent.N.Y. 1882-1916.   E.S 2487 k

Report of the State Geologist, Madras. Travancore.
    Rep.State Geol.Madras For the year 1089-1091 & 1093.
    (1915-1919.)   P.S 1116

Report of the State Geologist. Maine Geological Survey. Augusta.
    Rep.St.Geol.Maine geol.Surv. 1953/1954 →   P.S 1927 B

Report of the State Geologist on the Mineral Industries and
    Geology of (Certain Areas of) Vermont. Burlington.
    Rep.St.Geol.Miner.Ind.Geol.Vt 1901-1910, 1913-1946.
    Continued as Report of the State Geologist and Vermont
    Geological Survey.   P.S 1980

Report of the State Geologist of New Jersey.
    See Report. New Jersey Geological Survey. Trenton.   P.S 1957

Report of the State Geologist. New York State. Albany.
    Rep.St.Geol.N.Y. 1881-1903.
    See also Report. New York State Geological Survey.
    Albany.   P.S 1960

| TITLE | SERIAL No. |

Report of the State Geologist and Vermont Geological Survey.
    Rep.St.Geol.St.Vt geol.Surv. 1946-1948, 1952-1958.
    (From 1946-48 published in "Biennial Report. Vermont
    Development Commission".)
    Formerly Report of the State Geologist on the Mineral
    Industries & Geology of (Certain areas of) Vermont.   P.S 1980

Report of the State Mineralogist. California Division
    of Mines, San Francisco.
    Rep.St.Miner.Calif. No.55-58, 1958-1966.
    Formerly California Journal of Mines and Geology.
    San Francisco.   M.S 2634

Report of the State Mineralogist, California State
    Mining Bureau, San Francisco. etc.
    Rep.St.Miner.Calif. 1880-1896   M.S 2683

Report of the State Museum, North Carolina. Raleigh, N.C.
    Rep.St.Mus.N.Carolina 1966-1968 →   S. 2478

Report. Station Agronomique, Mauritius. Port Lewis.
    Rep.Stat.Agron., Mauritius 1897   B.S 2298

Report and Statistics, of the Mining Department, Victoria
    Melbourne.
    Rep.Statist.Min.Dep.Vict. 1890-91   M.S 2421

Report. Stichting voor Wetenschappelijk Onderzoek van de Tropen.
    The Hague.
    See Report. Netherlands Foundation for the Advancement of
    Tropical Research. The Hague.   S. 618

Report. Stockport Parks, Museums and Libraries Committee.
    See Report. Parks, Museum and Libraries Committee. Stockport. S. 373 a

Report of the Stoneham Museum. Kitale.
    Rep.Stoneham Mus. 1958 →
    Formerly published within Bulletin of the Stoneham Museum.   S. 2027 D

Report. Strangeways Research Laboratory. Cambridge.
    Rep.Strangeways Res.Lab. 1955-1959.   S. 35 a

Report of Sub-Committee on British Intelligence Objectives. London.
    Rep.Sub-comm.Brit.Intell.Obj. 1945-1951.   E.S 67

Report of the Subcommittee on the Ecology of Marine Organisms
    Washington, D.C.
    Rep.Subcomm.Ecol.mar.Org.Wash. 1940-1941.   S. 2419 E
                                          1941.   P.S 884 A
    Formerly Report on the Committee on Paleoecology.
    Washington, D.C.
    Continued as Report of the Committee on Marine Ecology
    as related to Paleontology. Washington, D.C.

TITLE	SERIAL No.

Report Sub-Department of Quaternary Research. University
of Cambridge.
Rep.Sub-Dep.Quaternary Res.Univ.Camb. 1957 →   P.S 122

Report to Subscribers. Cousin Island Nature Reserve. (Oxford.)
Rep.Subs.Cousin Isl.Nat.Reserve 1968 →   S. 2757

Report of the Sudan Geological Survey. Khartoum.
See Report. Geological Survey of the Anglo-Egyptian
Sudan. Khartoum.   P.S 1183

Report. Suffolk Trust for Nature Conservation. Ipswich.
Rep.Suffolk Trust Nat.Conserv. 1969/1970 →   S. 299 a

Report of Sugar Cane Experiments. Department of Agriculture,
Barbados. Bridgetown.
Rep.Sug.Cane Exps Barbados 1913-1926.   E.S 2384 a

Report. Sugar Cane Research Station, Mauritius.
Rep.Sug.Cane Res.Stn Maurit. 1936-1952.   E.S 2183 b

Report and Summary of Proceedings. Conference on Cotton Growing
Problems. Empire Cotton Growing Corporation.
See Conference on Cotton Growing Problems.   E.S 76

Report of the Superintendent of Government Laboratories
in the Philippine Islands. Manila.
Rep.Govt.Labs.Phil.Is. Nos. 2-4. 1902-1905.
Continued as Report of the Bureau of Science, Philippine
Islands. Manila.   S. 1976 C

Report. Surrey Naturalists' Trust. Croydon.
See Surrey Naturalist.   S. 174 a

Report of the Survey Branch of the Department of Lands. Province of
British Columbia. Victoria, B.C.
Rep.Surv.Brch Dep.Lands Br.Columb. 1913 & 1916.   S. 2658 a

Report. Sussex Naturalists' Trust Ltd. Henfield.
Rep.Sussex Nat.Trust 1961-1970.
Continued as Report. Sussex Trust for Nature Conservation.
Henfield.   S. 277

Report. Sussex Trust for Nature Conservation. Henfield.
Rep.Sussex Trust Nat.Conserv. 1971 →
Formerly Report. Sussex Naturalists' Trust Ltd. Henfield.   S. 277

Report of the Svalbard Commissioner. Copenhagen.
Rep.Svalb.Comm. 1927.   S. 539 A

Report. Swansea Philosophical & Literary Institution.
Rep.Swansea phil.Instn 1835-1836.   S. 375

TITLE	SERIAL No.

Report of the Swedish Deep-Sea Expedition 1947-1948. Göteborg.
    Rep.Swed.deep Sea Exped. 1951 →   70.q.G

Report. Taihoku Botanic Garden. Taihoku Imperial University, Formosa.
    Rep.Taihoku bot.Gdn   1931-1933.   B.S 1970 b

Report. Tasmanian Museum and Art Gallery. Hobart.
    Rep.Tasm.Mus.Art Gall. 1972/73 →
    Formerly   Report. Tasmanian Museum and Botanical Gardens.   S. 2152 A

Report. Tasmanian Museum and Botanical Gardens. Hobart.
    Rep.Tasm.Mus.bot.Gdn   1914/15 - 1922/23.
    Continued as   Report. Tasmanian Museum and Art Gallery.   S. 2152 A

Report. Technological Museum. Sydney.
    Rep.technol.Mus.Sydney   1906-1945.
    Continued as   Report of the Trustees of the Museum of
    Technology and Applied Science. Sydney.   S. 2135 C

Report on Technological Research concerning Norwegian Fish Industry.
    See Fiskeridirektoratets Skrifter, Serie Teknologiske
    Undersøkelser.   Z.S 530 A

Report. Teign Naturalists' Field Club.
    See   Report of the Proceedings of the Teign Naturalists'
    Field Club. Exeter.   S. 75

Report. Territory of New Guinea. Canberra.
    See   Report to the General Assembly of the United Nations on the
    Administration of the Territory of New Guinea. Canberra.   S. 2146

Report. Territory of Papua. Anthropology. Fort Moresby.
    See   Anthropology Reports. Territory of Papua. Fort Moresby.   P.A.S 861

Report of the Territory of Papua. Canberra.
    Rep.Terr.Papua 1967/1968 - 1969/1970.
    Formerly   Papua. Annual Report. Melbourne.   S. 2143 B
    Replaced by   Report. Papua New Guinea. Canberra.   S. 2146 b

Report. Texas Agricultural Experiment Station. College Station.
    Rep.Tex.agric.Exp.Stn   1934-1935.   E.S 2490 c

Report. Thomas Burke Memorial Washington State Museum. University of Washington. Seattle.
    Rep.Thos.Burke meml Wash.St.Mus.   1969/1970 →   S. 2405 G

Report on the II. Thule-Expedition to Greenland, 1916-1918. København.
    Rep.II.Thule Exped.   1922-1928.   71.o.D

Report of the Tokyo University of Fisheries. Tokyo.
    Rep.Tokyo Univ.Fish.   1966 →   Z.S 1965 A

| TITLE | SERIAL No. |

Report of the Topographical Surveys Branch Canada. Ottawa.
    Rep.topogr.Survs Brch Can. 1912-1937.   S. 2601 B

Report. Torry Research Station on the Handling and Preservation
    of Fish and Fish Products. Edinburgh.
    See Torry Research on the Handling and Preservation of
    Fish and Fish Products.   Z.S 482

Report and Transactions. Birmingham Natural History and
    Microscopical Society.
    Rep.Trans.Bgham nat.Hist & microsc.Soc. 1872-1886 (imp.)
    Formerly Proceedings of the Birmingham Natural History
    and Microscopical Society.   S. 17

Report & Transactions. Cardiff Naturalists' Society.
    Rep.Trans.Cardiff Nat.Soc. 1867-1902.
    Continued as Transactions. Cardiff Naturalists' Society.
     S. 25 & T.R.S 463

Report and Transactions of the Devonshire Association for the
    Advancement of Science. Plymouth.
    Rep.Trans.Devon.Ass.Advmt Sci. 1862 →   S. 36

Report and Transactions. East Kent Scientific and Natural
    History Society. Canterbury.
    Rep.Trans.E.Kent scient.nat.Hist.Soc. 1901-1913.
    Formerly Report of the East Kent Natural History Society.
    Canterbury.   S. 24 A

Report and Transactions of the Glasgow Society of Field
    Naturalists. Glasgow.
    Rep.Trans.Glasgow Soc.Fld Nat. 1872-1878.   B.L.S 2

Report and Transactions. Guernsey Society of Natural Science
    (& Local Research)
    Rep.Trans.Guernsey Soc.nat.Sci. 1882-1921.   S. 113
                                   1890-1921.   T.R.S 582
    Continued as Report and Transactions. Société Guernésiaise.

Report and Transactions of the Manchester Entomological
    Society. Manchester.
    Rep.Trans Manchr ent.Soc. 1903-1951 (imp.)
    Continued as Report, Proceedings and Transactions.
    Manchester Entomological Society.   E.S 38

Report and Transactions of the Manchester Microscopical Society.
    Rep.Trans.Manchr microsc.Soc. 1901-1930.
    Formerly Transactions and Annual Report of the
    Manchester Microscopical Society.   S. 262

TITLE	SERIAL No.

Report and Transactions. North Staffordshire (Naturalists')
Field Club. Stafford.
Rep.Trans.N.Staffs Fld Club 1887-1915.
Formerly Report. North Staffordshire Naturalists' Field Club
& Archaeological Society.
Continued as Transactions and Annual Report. North
Staffordshire Field Club.   S. 371 B

Report and Transactions. Nottingham Naturalists' Society.
Rep.Trans.Notts Nat.Soc. 1877-1918.   S. 301

Report & Transactions of the Penzance Natural History &
Antiquarian Society.
See Transactions of the Penzance Natural History &
Antiquarian Society.   S. 321 A

Report & Transactions of the Plymouth Institution & Devon &
Cornwall Natural History Society.
Rep.Trans.Plymouth Instn 1855-1961.
Continued as Proceedings of the Plymouth Athenaeum.   S. 326 B

Report and Transactions of the Société Guernésiaise. Guernsey.
Rep.Soc.quernés. 1922 →
    1922-1936.   S. 113
    T.R.S 582
Formerly Report and Transactions. Guernsey Society of
Natural Science (& Local Research).

Report & Transactions of the South-Eastern Union of
Scientific Societies.
Rep.S.-E.Un.scient.Socs. 1898-1899.
Continued as South Eastern Naturalist.
Formerly Transactions of the South-Eastern Union of
Scientific Societies.   S. 218 A

Report and Transactions. Yorkshire Philosopgical Society.
See Report of the Yorkshire Philosophical Society.   S. 403 A

Report of the Transvaal Department of Agriculture. Pretoria.
See Report. Department of Agriculture, Transvaal. Pretoria.   Z.S 2081

Report. Transvaal Museum. Pretoria.
Rep.Transv.Mus. 1906-1908, 1951-1970.   S. 2065 C

Report. Transvaal Nature Conservation Division. Pretoria.
Rep.Transv.Nat.Conserv.Div. 1966 →
Formerly Report. Nature Conservation Branch, Transvaal
Provincial Administration.   Z.S 2036 & S. 2070 B

Report on the Trawling Excursions... and on the Marine
Laboratory at Cullercoats. Newcastle-on-Tyne.
Rep.Trawl.mar.Lab.Cullercoats 1897-1899.
Continued as Report on the Scientific Investigations.
Northumberland sea Fisheries Committee.   Z.S 300

TITLE	SERIAL No.

Report on the Trivandrum Museum and Public Gardens.
    Rep.Trivandrum Mus.publ.Gdns 1899-1952 imp.   S. 1943 b

Report on the Tropical Diseases Library. London School of Hygiene
    & Tropical Medicine.
    Rep.trop.Dis.Libr.Lond.Sch.Hyg. 1921-1939.   S. 208 A
    Continued in Report. London School of Hygiene and
    Tropical Medicine.   S. 208 B

Report. Tropical Fish Culture Research Institute Batu Berendam,
    Malacca. Kuala Lumpur.
    Rep.trop.Fish Cult.Res.Inst.Malacca 1959-1960 →   Z.S 1984

Report of the Trustees of the Australian Museum. Sydney.
    See Report of the Australian Museum. Sydney.   S. 2126 F

Report of the Trustees. British Museum. London.
    Rep.Trust.Br.Mus. 1966 →
    (The 1966 Report includes a Survey of Activities from 1939 →)
    Formerly Report. British Museum & British Museum
    (Natural History). London.   B.M.o

Report of the Trustees and Directors of the National Museum of
    Rhodesia. Bulawayo.
    Rep.Trustees & Directors natn.Mus.Rhod. 1965 →
    Formerly Report of the National Museum of Southern
    Rhodesia. Bulawayo.   S. 2019 A

Report of the Trustees of the Museum of Applied Arts and
    Sciences. Sydney.
    Rep.Trust.Mus.appl.Art Sci.Sydney 1950 →
    Formerly Report of the Trustees of the Museum of Technology
    and Applied Science. Sydney.   S. 2135 C

Report of the Trustees of the Museum of Technology and Applied
    Science. Sydney.
    Rep.Trust.Mus.Technol.Appl.Sci.Sydney 1946-1949.
    Formerly Report Technological Museum. Sydney.
    Continued as Report of the Trustees of the Museum of
    Applied Arts and Sciences. Sydney.   S. 2135 C

Report of the Trustees of the National Museum of Victoria.
    Melbourne.
    Rep.Trust.Nat.Mus.Vict. 1945-1953. (imp.)
    Formerly Report of the Trustees of the Public Library,
    Museum and National Gallery of Victoria. Melbourne.   S. 2112 B

Report of the Trustees. National Portrait Gallery. London.
    Rep.Trust.natn.Portr.Gall. 1962-1963 →   S. 236

Report of the Trustees of the Peabody Academy of Science.
    Salem, Mass.
    See Report of the Peabody Academy of Science. Salem, Mass.   Z.S 2376

TITLE	SERIAL No.

Report of the Trustees of the Peabody Museum of American
    Archaeology & Ethnology. Cambridge, Mass.
    See Report of the Peabody Museum of American Archaeology
    & Ethnology. Cambridge, Mass.   Z.S 2375

Report of the Trustees of the Public Library, Museum and
    National Gallery of Victoria. Melbourne.
    Rep.Trust.publ.Libr.Mus.Vict. 1870-1944. (imp.)
    Continued as Report of the Trustees of the National
    Museum of Victoria. Melbourne.   S. 2112 B

Report of the Trustees of the Public Museum and Art Gallery of
    Papua and New Guinea. Port Moresby.
    Rep.Trust.publ.Mus.Art Gall.Papua & New Guinea 1966→   S. 2142 a

Report of the Trustees of the Queensland Museum. Brisbane.
    Rep.Trust.Qd Mus. 1876-1893 & 1898.   S. 2136 C

Report. Tunbridge Wells Natural History & Philosophical Society.
    Rep.Tunbridge Wells nat.Hist.phil.Soc. 1902-1907. (imp.)   S. 385

Report. Tyneside Geographical Society. Newcastle-upon-Tyne.
    Rep.Tyneside geogr.Soc. 1888.   S. 285 B

Report from the U.K. Scientific Mission (North America).
    Washington, D.C.
    Rep.U.K.scient.Miss.(N.Am.) 1970 → (imp.)   S. 2431 a

Report. The Uganda Museum. Kampala.
    Rep.Uganda Mus. 1962/1963.   S. 2025 b

Report of the Uganda Society. Kampala.
    Rep.Uganda Soc. 1934-1935: 1947 →   S. 2025 a B

Report. Ulster Museum. Belfast.
    Rep.Ulster Mus. 1966 →   S. 14 a

Report. Ulster University Bird Club. Coleraine.
    Rep.Ulster Univ.Bird Club 1970 →   T.B.S 503

Report. Underwater Association. Carshalton, Surrey.
    Rep.underwat.Ass. 1965 →
    (The first Report was entitled "Malta" '65.)   S. 272

Report of the Union County Mosquito Extermination Commission.
    East Cranford, N.J.
    Rep.Un.Cty Mosq.Exterm.Commn 1914-1948 (imp.)   E.S 2487 d

Report. United Fruit Company. Medical Department. New York.
    Rep.un.Fruit Co.med.Dep.N.Y. Nos.12-20, 1923-1931.   Z.O 75C q U

TITLE	SERIAL No.

Report of the United States Bureau of Animal Industry. Washington.
    Rep.U.S.Bur.Anim.Ind. Nos.3-38, 1887-1913 (imp.)   Z.S 2523 A

Report of the United States Bureau of Entomology and Plant
    Quarantine. Washington.
    Rep.U.S.Bur.Ent.Pl.Quarant. 1929-1942.
    Formerly Report of the Entomologist. United States
    Department of Agriculture. Washington.   E.S 2458 1

Report of the United States Bureau of Fisheries.
    See Report of the United States Commissioner of (Fish and)
    Fisheries. Washington.   Z.S 2510 B

Report of the United States Bureau of Mines. Washington.
    Rep.U.S.Bur.Mines 1911-1915.   M.S 2620

Report. United States Coast and Geodetic Survey. Washington.
    Rep.U.S.Cst geod.Surv. 1851-1943; 1946-1947.   S. 2428 A

Report United States Commissioner of (Fish and) Fisheries.
    Washington.
    Rep.U.S.Commnr Fish.   1871 →   Z.S 2510 B

Report United States Commissioner of Patents (Agriculture)
    Washington.
    Rep.U.S.Commnr Patents 1849-1858.   E.S 2450

Report. United States Department of Agriculture. Washington.
    Rep.U.S.Dep.Agric. 1862 → (imp.)   E.S 2450

Report. United States Entomological Commission. Washington.
    Rep.U.S.Ent.Commn 1877-1890.   E.S 2441 b

Report of the United States Geological Survey. Washington.
    Rep.U.S.geol.Surv. 1879-1949   P.S 1860
                        1890-1900 imp., 1915.   M.S 2616

Report of the United States Geological (and Geographical) Survey of
    the Territories. Washington.
    Rep.U.S. geol.geogr.Surv.Territ. 1867-1878.
    Vol.1-3 and 5-12. (1872-83)   P.S 1825

Report on the United States and Mexican Boundary Survey.
    Washington, D.C.
    Rep.U.S.Mex.Bound.Surv. 1856-1859.   75.q.U

Report of the United States National Museum. Washington.
    Rep.U.S.natn.Mus. 1884-1964.   S. 2427 D
                      1910-1964.   T.R.S 5145 C
    Continued in Smithsonian Year. Washington.

Report of the United States Section. Caribbean Commission.
    See Report. Caribbean Commission. United States Section.   75 F.o.C

| TITLE | SERIAL No. |

Report (of the President) of the University of California.
    Berkeley.
        Rep.Univ.Calif. 1920-1934.   S. 2319 K

Report of the University of California Archaeological Survey.
    Rep.Univ.Calif.archaeol.Surv. 1949 →
    Formerly Report of the California Archaeological Survey.   P.A.S 776 C

Report. University College of Wales. Aberystwyth.
    Rep.Univ.Coll.Wales 1916-1950.   S. 7 B

Report of the University of Kansas Museum of Natural History.
    Lawrence.
        Rep.Univ.Kans.Mus.nat.Hist. 1968 →   S. 2344 F

Report of the University Museum. Oxford.
    See Report of the Delegates of the University Museum. Oxford. S. 310 F

Report. University Museum, University of Tokyo. Tokyo.
    Rep.Univ.Mus.Tokyo 1973/74 →   S. 1990 H

Report of the University Museums. Ann Arbor, Michigan.
    Rep.Univ.Mus.Mich. 1937-1954. (imp.)   S. 2316 I

Report of the Uppingham School Field Club.
    Rep.Uppingham Sch.Fld Club No.2-23, 1947-1968.   S. 384

Report of the Usa Marine Biological Station. Kochi University.
    Rep.Usa mar.biol.Stn Kochi Univ. 1954 →   S. 1998 c A

Report of the Valetta Museum.
    See Report of the Working of the Museum Department. Malta.   S. 1135 B

Report of the Vermont Agricultural Experiment Station. Burlington.
    Rep.Vt agric.Exp.Stn 7-20, 1894-1907 (imp.).   E.S 2491 g

Report of the Veterinary and Agricultural Department,
    British Somaliland. Burao.
        Rep.vet.agric.Dep.Somalild 1936-1938.   Z. 74C q S

Report. Veterinary Department, Colony and Protectorate of
    Kenya. Nairobi.
    Rep.vet.Dep.Kenya 1937-1938.
    Continued as Report. Department of Veterinary
    Services, Kenya.   Z. Mammal Section

Report of the Veterinary Department, Gold Coast. Accra.
    Rep.vet.Dep.,Gold Cst 1925-1926 - 1929-1930.
    Continued as Report of the Department of Animal Health,
    Gold Coast.   Z. 74M f G

Report of the Veterinary Department, Nigeria. Lagos.
    Rep.vet.Dep.Nigeria 1933-1953-1954.
    Continued as Report of the Department of Veterinary
    Research of the Federation of Nigeria.   Z. 74M f N

| TITLE | SERIAL No. |

Report. Veterinary Department, Northern Rhodesia. Livingstone.
    Rep.vet.Dep.Nth.Rhod. 1928; 1935-1948.
    Between 1929-1933, entitled Report. Department of Animal
    Health, Northern Rhodesia.
    Continued as Report. Department of Veterinary Services,
    Northern Rhodesia. Lusaka.   Z. 74D f R

Report of the Veterinary Department, Nyasaland. Zomba.
    Rep.vet.Dep.Nyasald 1931-1948.
    Continued as Report of the Department of Veterinary
    Services & Animal Industry, Nyasaland. Zomba.   Z. 74D f N

Report of the Veterinary Department, Sierra Leone. Freetown.
    Rep.vet.Dep.Sierra Leone 1949-1961. (imp.)
    Continued as Report of the Veterinary Division of the
    Ministry of Natural Resources, Sierra Leone.   Z. 74N o S

Report of the Veterinary Department, Tanganyika. Dar-es-Salaam.
    Rep.vet.Dep.Tanganyika 1954-1962.
    Formerly Report of the Department of Veterinary Science
    and Animal Husbandry. Tanganyika.
    Continued as Report of the Veterinary Division,
    Tanzania.   Z.O 74Db f T

Report of the Veterinary Department, Uganda Protectorate.
    Entebbe.
    Rep.vet.Dep.,Uganda 1930-1952.
    Continued as Report of the Department of Veterinary
    Services and Animal Industry. Uganda.   Z. 74Da f U

Report on the Veterinary Departments, Malaya. Kuala Lumpur.
    Rep.vet.Deps Malaya 1937-1939.
    Formerly Report of the Veterinary Departments of the
    Straits Settlements and the Federated Malay States.   Z. 73G o M

Report of the Veterinary Departments, Straits Settlements
    and Federated Malay States. Singapore.
    Rep.vet.Deps Str.Settl.F.M.S. 1936.
    Continued as Report on the Veterinary Departments. Malaya.   Z. 73G o M

Report of the Veterinary Division of the Ministry of Natural
    Resources, Sierra Leone. Freetown.
    Rep.vet.Div.Minist.nat.Resour.Sierra Leone 1962 →
    Formerly Report of the Veterinary Department, Sierra Leone. Z. 74N o S

Report of the Veterinary Division, Tanzania. Dar-es-Salaam.
    Rep.vet.Div.Tanzania 1963 →
    Formerly Report of the Veterinary Department,
    Tanganyika.   Z.O 74Db f T

Report. Veterinary Laboratory. Chester Zoological Gardens. Chester.
    Rep.vet.Lab.Chester zool.Gdn 1971 →   Z.S 67

| TITLE | SERIAL No. |

Report on Veterinary Research. Department of Agriculture,
   Union of South Africa. Pretoria.
    Rep.vet.Res.Un.S.Afr. Nos.2-18, 1913-1932.
    Formerly Report of the Government Veterinary
   Bacteriologist, Union of South Africa. Pretoria.
    Continued as Onderstepoort Journal of Veterinary
   Science and Animal Industry.   Z.S 2081

Report. Veterinary Research Organization, East Africa.
   See Report. East African Veterinary Research
   Organization.   Z. Mammal Section

Report. Veterinary Service, Egypt. Cairo.
   Rep.vet.Serv.Egypt 1935-1936 - 1936-1937.   Z. 74B o E

Report. Veterinary Services, Northern Rhodesia.
   See Report. Department of Veterinary and Tsetse Control
   Services, Northern Rhodesia & Report. Veterinary
   Department of Northern Rhodesia.

Report. Veterinary Services, Uganda.
   See Report of the Department of Veterinary Services
   and Animal Industry, Uganda.   Z. 74Da f U

Report. Victoria Institute of Trinidad and Tobago. Port of Spain.
   Rep.Vict.Inst.Trin. 1905-1908.
   Formerly Proceedings of the Victoria Institute of Trinidad.
   Port of Spain.   S. 2299

Report. The Viking Fund, Inc. New York.
   Rep.Viking Fund 1941-1951.
   Continued as Report Wenner-Gren Foundation for
   Anthropological Research, Incorporated. New York.   P.A.S 806 A

Report. Virgin Islands Agricultural Experiment Station. Washington.
   Rep.Virg.Isl.agric.Exp.Stn Wash. 1920-1925.   E.S 2381 g

Report. Walthamstow Natural History & Microscopical Society.
   Rep.Walthamstow nat.Hist.microsc.Soc. 1-3. 1882-1883.   S. 235

Report. Wanganui Public Museum.
   Rep.Wanganui publ.Mus. No.14 & 17. 1908-1912.   S. 2198

Report of the Warden of the Ron Devlei Bird Sanctuary. Capetown.
   Rep.Ward.Ron Devlei Bird Sanct. 1952 →   T.B.S 4005

Report of Warren Spring Laboratory. London.
   Rep.Warren Spring Lab. 1961-1968.
   Continued as Review. Warren Spring Laboratory. Stevenage. M.S 143

Report. Warwick Natural History Society.
   Rep.Warwick nat.Hist.Soc. 1955 →   S. 386 a

Report. Warwickshire Natural History & Archaeological Society.
   Rep.Warwicksh.nat.Hist.archaeol.Soc. 1837-1892.   S. 386

TITLE	SERIAL No.

Report. Washington Geological Survey. Olympia.
    Rep.Wash.geol.Surv. 1901-1902.   P.S 1988

Report of the Water Pollution Research Board. D.S.I.R. London.
    Rep.Wat.Pollut.Res.Bd., Lond. 1927-1956. (imp.)
    Continued as Water Pollution Research, Water Pollution
    Research Board. D.S.I.R. London.   S. 205 W

Report of the Watson Botanical Exchange Club.
    Rep.Watson bot.Exch.Club 1884-1934.   B.H.S 104

Report Wayne State University. Detroit.
    Rep.Wayne St.Univ. 1961 →   P.A.S 820

Report. Weizmann Institute of Science. Rehovoth.
    Rep.Weizmann Inst.Sci. 1953-1959.
    Continued as Report of Scientific Activities. The Weizmann
    Institute of Science. Rehovoth.   S. 1928 c

Report of the Wellcome Tropical Research Laboratories
at the Gordon Memorial College. Khartoum.
    Rep.Wellcome trop.Res.Labs 1904-1911.   Z.S 2080
                                1904-1906.   E.S 2153 a

Report of the Wellington College Natural History Society.
    Rep.Wellington Coll.nat.Hist.Soc. 1868-1872.
    Continued as Report of the Wellington College Natural
    Science Society.   S. 393

Report of the Wellington College Natural Science Society.
    Rep.Wellington Coll.nat.Sci.Soc. 1872-1912.
    Formerly Report of the Wellington College Natural
    History Society.   S. 393

Report of the Welsh Plant Breeding Station. Aberystwyth.
    Rep.Welsh Pl.Breed.Stn 1965 →   B.S 145

Report Wenner-Gren Foundation for Anthropological Research,
Incorporated. New York.
    Rep.Wenner-Gren Fdn anthrop.Res. 1952 →
    Formerly Report. The Viking Fund, Inc. New York.   P.A.S 806 A

Report. West African Cacao Research Institute. Tafo.
    Rep.W.Afr.Cacao Res.Inst. 1944-1954.   B.S 2315
                            1949-1954.   E.S 2174 A c
    Continued as Report West African Cocoa Research
    Institute. Tafo.

Report of the West African Cocoa Research Institute (Nigeria). Ibadan.
    Rep.W.Afr.Cocoa Res.Inst.(Nigeria) 1963/1964.
    Continued as Report of the Cocoa Research Institute of
    Nigeria. Ibadan.   B.S 2316

Report. West African Cocoa Research Institute. Tafo.
    Rep.W.Afr.Cocoa Res.Inst. 1955-1957.   E.S 2174 A c &
    Formerly Report. West African Cacao Research Institute. Tafo. B.S 2315

| TITLE | SERIAL No. |

Report. West African Institute for Trypanosomiasis Research. Kaduna.
    Rep.W.Afr.Inst.Trypan.Res. 1952-1963.
    Continued as Report. Nigerian Institute for Trypanosomiasis Research. Kaduna.   E.S 2173 c

Report. West African Rice Research Station. Rokupr.
    Rep.W.Afr.Rice Res.Stn 1954 →   B.S 2314

Report of the West African Timber Borer Research Unit, Kumasi. Princes Risborough.
    Rep.W.Afr.Timb.Borer Res.Unit 1957 →   E.S 2173 a

Report. West Indian Conference. Caribbean Commission.
    See Report. Caribbean Commission. West Indian Conference.   75 F.o.C

Report. West Virginia Agricultural Experiment Station. Morgantown.
    Rep.W.Va agric.Exp.Stn 1936-1944.   E.S 2491 j

Report. West Wales Field Society. Tenby.
    Rep.W.Wales Fld Soc. 1945 →
    Formerly Report. Pembrokeshire Bird Protection Society.   S. 378

Report. West Wales Naturalists' Trust Ltd. Haverfordwest.
    Rep.W.Wales Nat.Trust 1968 →   S. 120 a

Report. Western Australian Museum.
    Rep.W.Aust.Mus. 1959 →
    Formerly Report. Museum and Art Gallery of Western Australia. Perth.   S. 2156 B

Report. Western Society of Malacologists. Pomona (Ca.)
    Rep.west.Soc.Malac. Vol.7 → 1974 →   Z. Mollusca Section
    Formerly Echo. San Diego.   (S. 2320)

Report. Weston-Super-Mare Public Library and Museum.
    Rep.Weston-Super-Mare publ.Libr.Mus. 1901-1914.
      (Wanting 1904.)   S. 393 b

Report of the Whitby Literary & Philosophical Society.
    Rep.Whitby lit.phil.Soc. 1824-1932.   S. 394

Report of the White Sea Biological Station of the State University of Moscow.
    See Trudy Belomorskoi Biologicheskoi Stantsii, M.G.U.   S. 1844 E

Report on the Wild Birds of Leicestershire and Rutland. Leicester.
    Rep.wild Birds Leics.Rutl. 1946-1947.
    Continued as Report. Birds of Leicestershire and Rutland.   T.B.S 260

Report of the Wildfowl Trust. London.
    Rep.Wildfowl Trust No.6, 1952/1953 - 1965/1966.
    Formerly Report of the Severn Wildfowl Trust.
    Continued as Wildfowl. London.   T.B.S 193

TITLE	SERIAL No.

Report. Wildfowlers' Association of Great Britain and Ireland.
    Liverpool.
    Rep.Wildfowl.Ass.Gt Br.Ir. 1958-1964.
    Formerly Report to Members. Wildfowlers' Association of
    Great Britain and Ireland.
    Continued as Report and Yearbook. Wildfowlers' Association
    of Great Britain and Ireland.	T.B.S 195

Report of the Wildlife Ecology Section. Department of Agriculture,
    Stock and Fisheries, Territory of Papua and New Guinea.
    Port Moresby.
    Rep.Wildl.Ecol.Sect.Terr.Papua New Guinea 1968-1969 →	Z.S 1985 A

Report. Wildlife Survey Section, C.S.I.R.O., Australia. Melbourne.
    Rep.Wildl.Surv.Sect.C.S.I.R.O.Aust. 1958-1961.
    Continued as Report. Division of Wildlife Research,
    C.S.I.R.O., Australia.	Z.S 2112 H

Report. Wiltshire Trust for Nature Conservation. Trowbridge.
    Rep.Wilts.Trust Nat.Conserv. 1964 →	S. 31 a

Report. Wincanton Field Club.
    Rep.Wincanton Fld Club 1889-1901.	S. 394 a

Report of the Winchester College Natural History Society.
    Rep.Winchester Coll.nat.Hist.Soc. 1871-1882, 1906-1957.	S. 395

Report of the Wisconsin Geological Survey. Madison.
    Rep.Wisc.geol.Surv. 1854, 1856, 1878-1879.	P.S 1990

Report of the Woburn Experimental Fruit Farm. London.
    Rep.Woburn exp.Fruit Fm 1-5, 8, 10-11, 14, 16. 1897-1917.	B.S 78

Report. Woods Hole Oceanographic Institution.
    Rep.Woods Hole oceanogr.Instn 1930 →	S. 2487 A

Report. Worcester Natural History Society.
    See Report & Addresses. Worcester Natural History Society.	S. 400

Report on the Work of Agricultural Research Institutes in the
    United Kingdom. London.
    Rep.Wk agric.Res.Insts U.K. 1930-1933.	E.S 57

Report on the Work of the Biological Board of Canada. Ottawa.
    See Report. Biological Board of Canada.	Z.S 2628 F

Report on the Work of the Entomological Section, Ministry of
    Agriculture, Egypt. Cairo.
    Rep.Wk ent.Sect.Minist.Agric.Egypt 1911/1923 - 1923/1924.	E.S 2156

Report on the Work of the Horn Scientific Expedition to
    Central Australia. London.
    Rep.Wk Horn Exped.C.Aust. 1896.	77 Cb.q.H

TITLE	SERIAL No.

Report on the Work of the London School of Hygiene and
 Tropical Medicine.
  <u>See</u> Report. London School of Hygiene & Tropical Medicine.   S. 208 B

Report of the Work. Saito Ho-On Kai. Sendai.
  <u>Rep.Wk Saito Ho-On Kai</u> No.1-2, 1926-1927.   S. 1988 c A

Report on the Work of the Survey Department. Egypt. Cairo.
  <u>Rep Wk Surv.Dep.Egypt</u> 1905-12.   P.S 1180

Report on the Work of the Szechenyi Scientific Societies. Budapest.
  <u>Rep.Wk Szechenyi scient.Soc.</u> 1934-1936.   S. 1717

Report on the Working of the Lucknow Provincial Museum. Allahabad.
  <u>Rep.Wkg Lucknow prov.Mus</u>. 1898-1925.
  <u>Continued as</u> Report on the Working of the United Provinces
  Provincial Museum. Lucknow.   S. 1927 a B

Report on the Working of the Museum Department. Malta.
  <u>Rep.Wkg Mus.Dep.Malta</u> 1920-1950.   S. 1135 B

Report on the Working of the United Provinces Provincial Museum,
 Lucknow. Allahabad.
  <u>Rep.Wkg United Prov.prov.Mus.Lucknow</u> 1925-1939 imp.
  <u>Formerly</u> Report on the Working of the Lucknow
  Provincial Museum. Allahabad.   S. 1927 a B

Report of the World Wildlife Fund. Morges.
  <u>Rep.Wld Wildl.Fund</u> 1961-1967.
  <u>Continued as</u> Yearbook. World Wildlife Fund.   Z.S 2703

Report. Worthing Archaeological Society.
  <u>Rep.Worthing Arch.Soc.</u> 1923-1925.   P.S 145

Report of the Wyoming Agricultural Experiment Station. Laramie.
  <u>Rep.Wyo.agric.Exp.Stn</u> 1893-1917 (imp.).   E.S 2496 c

Report. Yale Peabody Museum of Natural History. New Haven, Conn.
  <u>Rep.Yale Peabody Mus.nat.Hist.</u> 1960-1964.   S. 2352 S

Report and Yearbook. Wildfowlers' Association of Great Britain
 and Ireland. Liverpool.
  <u>Rep.Yb.Wildfowl.Ass.Gt Br.Ir.</u> 1964/1965 →
  <u>Formerly</u> Report. Wildfowlers' Association of Great Britain
  and Ireland.   T.B.S 195

Report. Yorkshire Natural Science Association. York.
  <u>Rep.Yorks.nat.Sci.Ass.</u> 1918-1928.   S. 404

Report of the Yorkshire Naturalists' Trust Ltd. York.
  <u>Rep.Yorks.Nat.Trust</u> 1970 →   S. 406 a

TITLE	SERIAL No.

Report. Yorkshire Naturalists' Union. Leeds.
    Rep.Yorks.Nat.Un. 1887-1910.
    Continued in Naturalist. Hull, London. — S. 474

Report of the Yorkshire Philosophical Society.
    Rep.Yorks.phil.Soc. 1823 → — S. 403 A

Report. Yorkshire Underground Research Team. Pateley Bridge.
    Rep.Yorks.Undergr.Res.Team 1968 → — S. 382

Report. Zambia Department of Game and Fisheries. Lusaka.
    Rep.Zambia Dep.Game Fish. 1964, 1966.
    Continued as Report. Zambia Department of Wildlife, Fisheries and National Parks. Lusaka. — Z.S 2067

Report. Zambia Department of Wildlife, Fisheries and National Parks. Lusaka.
    Rep.Zambia Dep.Wild.Fish.natn.Parks 1971 →
    Formerly Report. Zambia Department of Game and Fisheries. Lusaka. — Z.S 2067

Report of the Zoological and Acclimatisation Society of Victoria. Melbourne.
    Rep.zool.acclim.Soc.Vict. Nos. 29,33,39,40-42, 1893-1906. — Z.S 2135 A

Report of the Zoological Garden of Prague.
    Rep.zool.Gdn Prague 1973 → — Z.S 1763

Report. Zoological Gardens, Gizeh.
    Rep.zool.Gdns Gizeh 1899-1911.
    Continued in Report of the Zoological Service, Ministry of Public Works, Egypt. Cairo. — Z. 074B o C

Report. Zoological Institute, University of Tokyo.
    Rep.zool.Inst.Univ.Tokyo 1972 → — Z.S 1964 A

Report on the Zoological Service, Ministry of Public Works, Egypt. Cairo.
    Rep.zool.Serv.Egypt 1912-1923.
    Some Parts issued as Publications. — Z.O 74B o C

Report of the Zoological Society of London.
    Rep.zool.Soc.Lond. 1833 → — Z. 1 D

Report of the Zoological Society of Philadelphia.
    Rep.zool.Soc.Philad. Nos.32-64, 1904-1936. — Z.S 2500

Report of the Zoological Society of Scotland. Edinburgh.
    Rep.zool.Soc.Scotl. 1926-1929, 1931, 1934-1936. — Z.S 95 A

Report on the Zoological Survey of India. Calcutta.
    Rep.zool.Surv.India 1916-1917 → — Z.S 1912

Reports.
    See Report.

TITLE	SERIAL No.

Reprint and Circular Series of the National Research Council. Washington, D.C.
    Repr.Circ.Ser.natn.Res.Coun.Wash. 1919-1947 (imp.)   S. 2419 B

Reprint of Geological Papers on Burma. Calcutta.
    Repr.Geol.Pap.Burma 1833-1881 (1882).   P.S 1110 A

Reprint Series Geological Survey of Alabama. Alabama.
    Repr.Ser.geol.Surv.Ala 1963 →   P.S 1870 D

Reprint Series. Geological Survey of Ohio. Columbus.
    Repr.Ser.geol.Surv.Ohio 1939 →   P.S 1967 B

Reprints of Papers from the Science Laboratory of the University of Sydney.
    Repr.Pap.Sci.Lab.Univ.Sydney
       Anatomy, Biology, (Botany), Geology, (Pathology), Physiology, (Veterinary Science & Zoology). 1894-1920.
       Mathematics, Physics, Chemistry & Engineering, 1916-1920.   S. 2130

Reprints from the Public Health Reports. Washington.
    Repr.publ.Hlth Rep.,Wash. 1920-1940 (imp.).   E.S 2460 a

Reprints. Wye Agricultural College. Wye.
    Repr.Wye agric.Coll. 1947-1952.
    Continued as Wye College Reprints. Ashford.   E.S 70

Reprographics Quarterly. Hatfield.
    Reprogr.Q. 1974 →
    Formerly NRCd Bulletin.   S. 497 g

Res Biologicae. Torino.
    Res biol. Vol.1 Nos.1-4, 1926-1927.   Z.S 1151

Research. Liverpool.
    Research Lpool 1888-1890.   S. 484

Research. London.
    Research Lond. 1947-1961.   S. 436 a

Research Briefs. Fish Commission of Oregon. Portland.
    Res.Briefs Fish Commn Ore. 1955-1967.
    Formerly Fish Commission Research Briefs.   Z.S 2477

Research Branch Report. Entomology Research Institute. Ottawa.
    Res.Brch Rep.Ent.Res.Inst.Ottawa 1971 →
    Formerly Research Report. Entomology Research Institute. Ottawa.   E.S 2536

TITLE	SERIAL No.

Research Bulletin. Department of Agriculture, Stock & Fisheries. Port Moresby.
    Res.Bull.Dep.Agric.Stock Fish., Port Moresby 1963 →   S. 2143 a

Research Bulletin. Division of Fishery Research, Department of Natural Resources, Newfoundland. St. John's.
    Res.Bull.Div.Fish.Res.Newfoundld 3–16, 1936–1945.
    Formerly Research Bulletin. Newfoundland Fishery Research Commission.
    Continued as Bulletin of the Newfoundland Government Laboratory.   Z.S 2670 B

Research Bulletin of the East Panjab University. Hoshiarpur.
    Res.Bull.E.Panjab Univ. 1950–1953, (Nat.Hist.only.)
    Continued as Research Bulletin of the Panjab University of Science. Hoshiarpur.   S. 1938 a

Research Bulletin of the Faculty of Education, Oita University. Oita.
    Res.Bull.Fac.Educ.Oita Univ. Natural Science. 1952 → (imp.)   S. 1988 g

Research Bulletin of the Faculty of Liberal Arts, Oita University. Oita.
    See Research Bulletin of the Faculty of Education, Oita University. Oita.   S. 1988 g

Research Bulletin of the Geological and Mineralogical Institute, Tokyo University of Education.
    Res.Bull.geol.miner.Inst.Tokyo Univ.Educ. No.4, 1955.
    Formerly Studies from the Geological and Mineralogical Institute, Tokyo University of Education.   P.S 1773

Research Bulletin. Hiroshima Institute of Technology. Hiroshima.
    Res.Bull.Hirosh.Inst.Technol. 1966 →   S. 1985 a

Research Bulletin. Imperial Tokyo Sericultural College. Tokyo.
    Res.Bull.imp.Tokyo seric.Coll. 1936–1940.   E.S 1931

Research Bulletin. International Commission for the Northwest Atlantic Fisheries. Dartmouth, N.S.
    Res.Bull.int.Commn NW.Atlant.Fish. 1964 →   Z.S 2716 A

Research Bulletin. Iowa Agricultural Experiment Station. Ames.
    Res.Bull.Iowa agric.Exp.Stn 1911–1937 (imp.)   E.S 2471 f

Research Bulletin of the Meguro Parasitological Museum. Tokyo.
    Res.Bull.Meguro parasit.Mus. 1967 →   S. 1991 c

Research Bulletin. Missouri Agricultural Experiment Station. Columbia.
    Res.Bull.Mo.agric.Exp.Stn 1936–1939.   E.S 2484 a

| TITLE | SERIAL No. |

Research Bulletin. Newfoundland Fishery Research
    Commission. St. John's.
    Res.Bull.Newfoundld Fish.Res.Commn Nos.1-2, 1932-1933.
    Originally issued as Reports of the Newfoundland Fishery
    Research Commission, and later considered to be
    Research Bulletins Nos. 1-2.
    Continued as Research Bulletin. Division of Fishery Research.
    Department of Natural Resources, Newfoundland.   Z.S 2670

Research Bulletin. New South Wales State Fisheries.
    See Research Bulletin. State Fisheries, New South
    Wales. Sydney.   Z.S 2105

Research Bulletin. Obihiro Zootechnical University. Obihiro.
    Res.Bull Obihiro zootech.Univ. Series 1.
    Vol.1, No.4 → 1954 → (imp.)   S. 1993 b

Research Bulletin of the Panjab University. Hoshiarpur.
    Res.Bull.Panjab Univ. 1954-1958. (Nat.Hist.only.)
    New Series Science, 1959 →
    Formerly Research Bulletin of the East Panjab University.
    Hoshiarpur.   S. 1938 a

Research Bulletin. Porto Rico University Agricultural Experiment
    Station. Rio Piedras.
    Res.Bull.P.Rico Univ.agric.Exp.Stn 1941-1945.   E.S 2391 a

Research Bulletin. Saito Ho-on Kai Museum. Sendai.
    Res.Bull.Saito Ho-on Kai Mus. 1934 →   S. 1988 a

Research Bulletin on State Fisheries, New South Wales. Sydney.
    Res.Bull.St.Fish.N.S.W. 1938 No.1.   Z.S 2105

Research Division Bulletin. Virginia Polytechnic Institute.
    Blacksburg.
    Res.Div.Bull.Va polytech.Inst. No. 48 → 1969 → (imp.)   E. 75C o B

Research Division Monographs. Virginia Polytechnic Institute.
    Blacksburg.
    Res.Div.Monogr.Va Polytech.Inst. 1969-1971.
    Continued as Research Monographs. Virginia Polytechnic
    Institute and State University. Blacksburg.   S. 2320 a

Research in Fisheries. University of Washington College
    (formerly School) of Fisheries. Seattle.
    Res.Fish.Univ.Wash.Coll.(Sch.)Fish. 1962 →
    Forms part of the Contributions. College of Fisheries.
    University of Washington.   Z.S 2384 A

Research Journal. Directorate General of Higher Education. Djakarta.
    Res.J.Dir.gen.Educ.Djakarta Series B. Vol.2 → 1968 →   S. 2178

Research Journal of the Hindi Science Academy.
    Res.J.Hindi Sci.Acad. 1958 →   S. 1911 b

TITLE	SERIAL No.

Research Memoirs of the London School of Tropical Medicine.
London.
Res.Mem.Lond.Sch.trop.Med. 1912-1924
Continued as Memoir Series. London School of Hygiene
and Tropical Medicine.   Z.S 48

Research Monographs. Virginia Polytechnic Institute and State
University. Blacksburg.
Res.Monogr.Va Polytech.Inst.St.Univ. 1972 →
Formerly Research Division Monographs. Virginia Polytechnic
Institute. Blacksburg.   S. 2320 a

Research News. National Research Council, Canada. Ottawa.
See N.R.C. Research News. National Research Council, Canada.
Ottawa.   S. 2602 E

Research Notes. Department of Forest Entomology, Royal College
of Forestry. Stockholm.
See Rapport och Uppsatser. Institutionen för Skogsentomologi.
Stockholm.   E.S 511

Research Notes. Division of Forest Research, Zambia. Kitwe.
Res.Notes Div.For.Res.Zambia 1968 →   B.S 2313 c

Research Notes Series. Entomology Division, Department of
Agriculture, Canada. Ottawa.
Res.Notes Ser.Ent.Div.Can. 1948 →   E.S 2523 c

Research Pamphlet. Division of Forest Research. Zambia.
Res.Pamph.Div.for.Res. No.15 → 1968 → (imp.)   B.S 2313 b

Research Pamphlet. Faculty of Agriculture, University of Aleppo.
Res.Pamph.Fac.Agric.Univ.Aleppo 1971 →   E.S 2042

Research Papers. Federal Department of Forest Research, Nigeria.
Ibadan.
Res.Pap.Fed.Dep.For.Res.Nigeria
Forest Series no.13 → 1973 →
Savanna Series no.26 → 1973 →   B.S 2294 c

Research Papers. Pacific Northwest Forest and Range Experiment
Station. Portland, Oregon.
Res.Pap.Pacif.NW For.Range Exp.Stn No.80 → 1969 →   B.S 4363

Research on the Past Climate and Continental Drift. Taipei.
Res.past Clim.contin.Drift.,Taipei 1943 →   P.S 14 q.M

Research Progress Report. Canadian Wildlife Service. Ottawa.
Res.Prog.Rep.Can.Wildl.Serv. 1961.   Z.S 2635 A

Research & Progress. Review of German Science. Berlin.
Res.Prog. 1935-1939.   S. 1663

Research Publications. University of Hawaii. Honolulu.
Res.Publs Univ.Hawaii No.11 1935.   M.R.GAZ.77 E.o.C
                               No.21, 1944.   B. 581.144/145 ENG.
                               No.23, 1946.   Z. 40C o A

| TITLE | SERIAL No. |

Research Report. Applied Geology. University of Strathclyde. Glasgow.
　　See Applied Geology Research Report. University of Strathclyde.
　Glasgow.   P.S 168 a

Research Report. Division of Nutritional Biochemistry.
　　Commonwealth Scientific and Industrial Research Organization,
　　Australia. Adelaide.
　　　Res.Rep.Div.Nutr.Bioch. 1966 →
　　　Formerly in Report. Animal Research Laboratories.
　　C.S.I.R.O. Australia.   Z.S 2112 C a

Research Report. Entomology Research Institute. Ottawa.
　　　Res.Rep.Ent.Res.Inst.Ottawa 1965/1966, 1970.
　　　Continued as Research Branch Report. Entomology Research
　　Institute. Ottawa.   E.S 2536

Research Report. Hebrew Iniversity of Jerusalem.
　　Science, Agriculture.
　　　Res.Rep.hebrew Univ.Sci.agric. 1963-1964 →   S. 1919 d

Research Report. Institute of Animal Genetics, Edinburgh
　　University.
　　　Res.Rep.Inst.Anim.Genet. 1955-1957 →   Z.S 55

Research Report of the Kanagawa Prefectural Museum. Yokohama.
　　　Res.Rep.Kanagawa Prefect.Mus. Nat. Hist. 1970 →   S. 1998 f B

Research Report. Marine Research Laboratory. South West Africa.
　　Windhoek.
　　　Res.Rep.mar.Res.Lab.S.W.Afr. No.4, (1962).
　　　Formerly & Continued as Investigational Report.
　　Marine Research Laboratory.   Z.S 2070

Research Report. New Jersey State Museum. Trenton, N.J.
　　　Res.Rep.New Jers.St.Mus. No.2 → 1970 →
　　　(No.1 not on Natural History.)   S. 2412 D

Research Report. Plant Research Institute. Ottawa.
　　　Res.Rep.Pl.Res.Inst.Ottawa 1963/1966 →   B.S 4519

Research Report. Royal College of Science. London.
　　　Res.Rep.R.Coll.Sci.Lond. 1965-1968 →   S. 233 B

Research Report. Tay Estuary Research Centre. Dundee.
　　　Res.Rep.Tay Estuary Res.Centre 1971 →   S. 52 c

Research Report. Thomas Burke Memorial Washington State Museum.
　　Washington.
　　　Res.Rep.Thos.Burke Mem.Wash.St.Mus. 1968 →   S. 2405 I

Research Report. United States Fish & Wildlife Service,
　　Washington D.C.
　　　Res.Rep.U.S.Fish Wildl.Serv. 1941 →
　　　Formerly Investigational Report. United States
　　Bureau of Fisheries.   Z.S 2510 D

TITLE	SERIAL No.

Research Reports of the Fish Commission of Oregon. Portland.
   Res.Rep.Fish Commn Ore. 1969 →   Z.S 2477

Research Reports of the Forest Experiment Station, Office of
   Rural Development, Ministry of Agriculture and Forestry,
   Seoul, Korea.
   Res.Rep.Forest exp.Stn No.10-14, 1965-1968.
   Formerly Bulletin of the Forest Experimental Station.
   Seoul, Korea.
   Continued as Research Reports of the Forest Research Institute.
   Seoul, Korea.   B.S 1940

Research Reports of the Forest Research Institute. Seoul, Korea.
   Res.Rep.forest Res.Inst. No.15 → 1968 →
   Formerly Research Reports of the Forest Experiment Station.
   Seoul, Korea.   B.S 1940

Research Reports. Kasetsart University. Bangkok.
   Res.Rep.Kasetsart Univ. 1971/1972 →
   Formerly Kasetsart University Research Activities.   S. 1913 c B

Research Reports of the Kôchi University.
   Res.Rep.Kôchi Univ. Vol.1, No.3 → 1952 → (imp.)   S. 1998 c B

Research Reports. National Geographic Society. Washington, D.C.
   Res.Rep.Natn.geogr.Soc. 1963 →   S. 2422 C

Research Reports. Smithsonian Institution. Washington, D.C.
   Res.Rep.Smithson.Inst. 1972 →   S. 2426 M

Research Review of the Commonwealth Scientific & Industrial
   Research Organization. Commonwealth of Australia. Melbourne.
   Res.Rev.C.S.I.R.O.Aust. 1959-1960.   S. 2113 K

Research in Review. Massachusetts Agricultural Experiment
   Station. Amherst.
   Res.Rev.Mass.agric.Exp.Stn 1952-1956.   E.S 2479 c

Research Series. Geological Center Lansdale, Pa.
   See Geological Center Research Series.   P.S 838

Research Series. Museum of Systematic Biology. University of
   California, Irvine.
   Res.Ser.Mus.syst.Biol. 1968 →   S. 2319 b

Research Series. Royal Geographical Society.
   See R.G.S. Research Series.   S. 211 L

Research Studies. State College of Washington. Pullman.
   See Research Studies. Washington State University. Pullman.

Research Studies. Washington State University. Pullman.
   Res.Stud.Wash.St.Univ. 1929 →   B.S 4406 & S. 2377 D

TITLE	SERIAL No.

Researches on Crustacea. The Carcinological Society of Japan.
    Researches Crust. 1963 →   Z. Malacostraca Section

Researches of the National Museum. Bloemfontein.
    See Navorsinge van die Nasionale Museum. Bloemfontein
    and Memoir. Natural Museum. Bloemfontein.   S. 2010 A

Researches on Population Ecology. Entomological Laboratory,
    Kyoto University.
    Researches Popul.Ecol.Kyoto Univ.   Vol.2 → 1953 →   E.S 1904 a

Reseñas Cientificas de la Real Sociedad Espanola de Historia
    Natural. Madrid.
    Reseñ.cient.R.Soc.esp.Hist.nat.   1931-1936.
    Formerly Conferencias y Reseñas Cientificas de la Real
    Sociedad Española de Historia Natural. Madrid.   S. 1003 F

Resource Publications. Bureau of Sport Fisheries and Wildlife.
    Washington.
    Resour.Publs Bur.Sport Fish.Wildl.   1965 →   Z.S 2510 L

Resources of Tennessee. Geological Survey of Tennessee. Nashville.
    Resour.Tenn.   1911-1919. (imp.)   P.S 1975

Respiration Physiology. Amsterdam.
    Respiration Physiol.   1966 →   S. 692

Restaurator. Copenhagen.
    Restaurator 1969 →   S. 536 a

Resultados de las Campanas Realizadas por Acuerdos Internacionales,
    Instituto Español de Oceanografia. Madrid.
    Resultados Camp.int.Inst.esp.Oceanogr.   No.2-4, 1926-1927.   Z.S 1001 D

Resultados. Expediciones Científicas del Buque Oceanografico
    "Cornide de Saavedra". Madrid.
    Resultados Exped.cient.Buque oceanogr.Cornide
    Saavedra 1972 →   Z.S 1006 C

Resultate der Wissenschaftlichen Erforschung des Balatonsees. Wien.
    Resultate Wiss.Erforsch.Balatonsees   1897-1920.   72 M.q.B

Resultater av de Norske Statsunderstøttede Spitzbergenekspeditioner.
    Oslo.
    Resultater Norske Spitzbergeneksped.   1922-1929.
    Continued as Skrifter om Svalbard og Ishavet
    (formerly og Nordishavet.) Oslo.   S. 539 C

Resultati Scientifici della Missione alla Oasi di Giarabub.
    (1926-1927). Roma.
    Resultati scient.Miss.Oasi Giarabub   1928-1931.   74 Ad.o.R

TITLE	SERIAL No.

Resultati Scientifici della Missione Stefanini-Paoli
   nella Somalia Italiana.
   <u>See</u> Pubblicazioni del R. Istituto di Studi Superiori
   Pratici e di Perfezionamento in Firenze.   S. 1122

Résultats des Campagnes Scientifiques accomplies par le
   Prince Albert I. Monaco.
   <u>Résult.Camp.scient.Prince Albert I</u> 1889-1950.   70 f.A

Résultats des Recherches Scientifiques Entreprises au Parc
   National Suisse.
   <u>See</u> Ergebnisse der Wissenschaftlichen Untersuchungen des
   Schweizerischen Nationalparks.   S. 1201 H

Résultats Scientifiques de la Campagne du N.R.P. "FAIAL" dans
   les Eaux Cotières du Portugal (1957). Lisboa.
   <u>Résult.scient.Camp.N.R.P."FAIAL"</u> 1959-1961.   72 H.q.L

Résultats Scientifiques. Expédition Antarctique Belge 1957-1958.
   Bruxelles.
   <u>Résult.scient.Expéd.antarct.belge</u> 1961 →   80.q.B

Résultats Scientifiques. Expédition Océanographique Belge
   dans les Eaux Côtières Africaines de l'Atlantique Sud
   (1948-1949). Bruxelles.
   <u>Result.scient.Expéd.océanogr.belge Eaux côt.afr.Atlant.Sud</u>
   1951-1965.   74 q B

Résultats Scientifiques. Exploration Hydrobiologique du Bassin
   du Lac Bangweulu et du Luapula. Bruxelles.
   <u>Résult.scient.Explor.hydrobiol.Bassin Lac</u>
     <u>Bangweulu & Luapula</u> 1964 →   74 Dd q B

Résultats Scientifiques. Exploration Hydrobiologique du Lac
   Tanganika (1946-1947). Bruxelles.
   <u>Résult.scient.Explor.hydrobiol.Lac Tanganika</u> 1949-1958.   74 Db.q.B

Resultats Scientifiques. Exploration Hydrobiologique des Lacs
   Kivu, Edouard et Albert (1952-1954). Bruxelles.
   <u>Résult.scient.Explor.hydrobiol.Lacs Kivu,</u>
     <u>Edouard et Albert</u> 1957 →   74 K q B

Résultats Scientifiques. Mission Robert Ph. Dollfus en Egypte. Paris.
   <u>Résult.scient.Miss.R.P.Dollfus en Egypte</u>
     Pt.3. 1959.   74 B.q.D

Résultats Scientifiques. Voyage de M. le Baron Maurice de
   Rothschild en Ethiopie et en Afrique Orientale Anglaise
   (1904-1905). Paris.
   <u>Résult.scient.Voy.Baron M.de Rothschild Ethiopie</u> 1922.   74.f.P

TITLE	SERIAL No.

Resultats Scientifiques des Voyages en Afrique (1880-1897)
d'Edouard Foà. Paris.
Result.scient.Voy.Afr.E.Foà 1908.   74.q.P

Résultats du Voyage du S.Y. Belgica en 1897-1899. Anvers.
Résult.Voyage S.Y. Belgica 1901-1949.   80 f.B

Results of the Expedition to Tristan da Cunha, 1937-1938. Oslo.
See Results of the Norwegian Scientific Expedition to
Tristan da Cunha, 1937-1938. Oslo.

Results of Fisheries Oceanographical Observation. (Tokyo.)
Res.Fish.oceanogr.Obs. 1952 →   Z.S 1971 C

Results of the Harriman Alaska Expedition.
See Harriman Alaska Series. Smithsonian Institution.
Washington.   75 A.q.W

Results of the Kyoto University Scientific Expedition to the
Karakoram and Hindukush, 1955. Kyoto.
Results Kyoto Univ.scient.Exped.Karakoram Hindukush 1963-1965. 73 F.o.K
Vol.2, 1960.   B. See Bot.Lbry Catalogue

Results of the Norwegian Scientific Expedition to Tristan
da Cunha 1937-1938. Oslo.
Results Norw.scient.Exped.Tristan da Cunha 1946 →   78.o.C

Results and Problems in Cell Differentiation. Berlin.
Results Probl.Cell.Differ. 1968 →   10 C o R

Résumé des Communications présentées au Congrès de l'Association
Française pour l'Avancement des Sciences. Paris.
See Compte Rendu de l'Association Française pour
l'Avancement des Sciences. Paris.   S. 801 A

Résumé du Compte-Rendu Annuel de la Société Royale des
Sciences de Bohôme.
See Výroční Zpráva Královské Ceské Společnosti Nauk. v   S. 1703 B
Praze.

Resumenes. Congreso Argentino de Paleontologia y Biostratigrafia.
Tucuman.
Resum.Congr.argent.Paleont.Biostratigr. 1974 →   P.S 2007

Resumenes de Investigacion. Centro de Investigaciones Pesqueras.
La Habana.
Resum.Invest.Cent.Invest.Pesq. 1974 →   Z.S 2252 C

Résumés des Communicationes Congrès International des Sciences
Anthropologiques et Ethnologiques.
Résumés Com.Congr.Int.Sci.anthrop.ethnol. Sess.6. 1960.   P.A.S 906

Resumptio Genetica. s'Gravenhage.
Resumptio genet. 1924-1953.   S. 688 C

| TITLE | SERIAL No. |

Retrospective Geological Bibliography of Poland.
    See Retrospektywna Bibliografia Geologiczna Polski.   P.S 571 A

Retrospektivnaya Geologicheskaya Bibliografiya Pol'shi.
    See Retrospektywna Bibliografia Geologiczna Polski.   P.S 571 A

Retrospektywna Bibliografia Geologiczna Polski. Warszawa.
    Retrosp.biblfia geol.Pol. 1900–1950 → (1957 →)   P.S 571 A

Return. British Museum. London.
    Return Br.Mus. 1811–1921.
    (From 1882 includes Return. British Museum (Natural History.)
    Previous to 1843 Styled Account of Receipts & Income
    & Expenditure.
    Continued as Report. British Museum & British Museum
    (Natural History.) London.   B.M.c

Return of Mineral Production in India. Calcutta.
    Return Miner.Prod.India 1891–1892.
    Formerly Statement Showing Quantities and Values of Minerals
    and Gems. India.
    Continued as Review of Mineral Production in India.
    Calcutta.   M.S 1916

Reunión Nacional de la Sociedad Argentina de Ciencias
    Naturales. Buenos Aires.
    Reun.nac.Soc.argent.Cienc.nat. 1916 (1918–1919.)   S. 2332 B

Reunión sobre Productividad y Pesquerias. Barcelona.
    Reun.Product.Pesq. 1954 → (imp.)   Z.S 1006 A

Reunión de la Sociedad Argentina de Patologia Regional del
    Norte. Buenos Aires.
    Reun.Soc.argent.Patol.reg.N. 1935 (1936–1939.)   S. 2242 a

Review. Academy of Natural Sciences of Philadelphia.
    Rev.Acad.nat.Sci.Philad. 1932–1934.
    Formerly Yearbook. Academy of Natural Sciences of
    Philadelphia.
    Continued as Annual Review. Academy of Natural Sciences
    of Philadelphia.   S. 2305 E

Review of Activities. National Research Council. Ottawa.
    Rev.Activ.natn.Res.Coun.Can. 1939–1941.
    Continued as National Research Council Review. Ottawa.   S. 2602 **F**

Review of Agricultural Operations in India. Calcutta.
    Rev.agric.Ops India 1919–1933.
    Formerly Report on the Progress of Agriculture in India.
    Continued as Agriculture and Animal Husbandry in India.
    Calcutta.   E.S 1997

Review of Applied Entomology, London.
    Rev.appl.Ent. 1913 →
    (a) Agriculture (b) Medical & Veterinary.   E.S 33

| TITLE | SERIAL No. |

Review of Applied Mycology. London.
    Rev.appl.Mycol. 1922-1969.
    Continued as Review of Plant Pathology.   B.M.S 5 a

Review of the Bulgarian Geological Society.
    See Spisanie na Bŭlgarskoto Geologichesko Druzhestvo.   P.S 490

Review. Fisheries Research Board of Canada. Ottawa.
    Rev.Fish.Res.Bd Can. 1964 →
    Formerly contained in Report. Fisheries Research
    Board of Canada.   Z.S 2628 G

Review. International Council of Scientific Unions (I.C.S.U.)
    Amsterdam.
    See ICSU Review. Amsterdam.   S. 2701 A

Review of the Manchuria Research Society. Harbin.
    See Otdel'noe Izdanie. Obshchestvo Izucheniya
    Man'chzhurskago Kraya. Ser.C.   S. 1986 D

Review of Medical & Veterinary Mycology. Kew.
    Rev.med.vet.Mycol. 1951 →
    Formerly Annotated Bibliography of Medical Mycology. Kew.   B.M.S 5 d

Review of Mineral Production in India. Calcutta.
    Rev.Miner.Prod.India 1893-1897.
    Formerly Return of Mineral Production in India. Calcutta.   M.S 1916

Review of Mining Operations in the State of South
    Australia. Adelaide.
    Rev.Min.Ops S.Aust. 1904-1917.
    Formerly Short Review of Mining Operations in the State
    of South Australia. Adeladie.
    Continued as Mining Review. Department of Mines,
    South Australia. Adelaide.   M.S 2410

Review. National Research Council, Canada. Ottawa.
    Rev.natn.Res.Coun.Can. 1957-1968.
    Formerly National Research Council Review. Ottawa.
    Continued as NRCL. Review of the Activities of the
    Laboratories. Ottawa.   S. 2602 F

Review of the Oilseed, Oil and Oil Cake Markets. London.
    Rev.Oilseed, Oil, Oil Cake Markets 1943-1946.
    Continued as Annual Review of Oilseeds, Oils, Oilcakes,
    and other Commodities. London.   S. 155 a

Review of the Ontario Department of Mines. Ottawa.
    Rev.Ont.Dep.Mines 1966-1969.
    Continued as Review. Ontario Department of Mines and
    Northern Affairs. Ottawa.   M.S 2708 B

Review. Ontario Department of Mines and Northern Affairs. Ottawa.
    Rev.Ont.Dep.Mines nth.Affairs 1970-1971.
    Formerly Review of the Ontario Department of Mines. Ottawa.
    Continued as Annual Review Mineral Industry. Ontario
    Ministry of Natural Resources.   M.S 2708 B

TITLE	SERIAL No.
Review of Palaeobotany and Palynology. Amsterdam. Rev.Palaeobot.Palynol. 1967 →	P.S 992
Review of Plant Pathology. London. Rev.Pl.Path. 1970 → Formerly Review of Applied Mycology.	B.M.S 5 a
Review of Plant Protection Research. Nishigahara. Rev.Pl.Prot.Res. 1968 →	E.S 1930 B
Review. Rockefeller Foundation. New York. Rev.Rockefeller Fdn 1917 → (imp.)	Z.S 2470 C
Review of the Society for Scientific Investigation of Manchuria. See Otdel'noe Izdanie. Obshchestvo Izucheniya Man'chzhurskago Kraya. Ser.C.	S. 1986 D
Review. Warren Spring Laboratory. Stevenage. Rev.Warren Spring Lab. 1969/1970 → Formerly Report of Warren Spring Laboratory. London.	M.S 143
Reviews in Engineering Geology. New York. Rev.Engng Geol. 1962 →	P.S 865 D
Revista de la Academia de Ciencias Exactas, Fisico-Químicas y Naturales de Zaragoza. Revta Acad.Cienc.exact.fis.quím.nat.Zaragoza 1916-1937.	S. 1021 A
Revista de la Academia Colombiana de Ciencias Exactas, Fisicas y Naturales. Bogotá. Revta Acad.colomb.Cienc.exact.fis.nat. 1936 → 1936-1947. Formerly Revista de la Sociedad Colombiana de Ciencias Naturales. Bogotá.	S. 2247 a T.R.S 6505 S. 2247 a
Revista Agrícola. Guatemala. Revta agríc.,Guatem. 1945-1946.	E.S 2395
Revista Agricola. Rio de Janeiro. Revta agric.Rio de J. 1869-1872.	E.S 2365
Revista Agronomica. Lisboa. Revta agron.,Lisb. 1903-1908.	B.S 612
Revista Argentina de Agronomia. Buenos Aires. Revta argent.Agron. 1934-1962.	B.S 3001
Revista Argentina de Botanica. La Plata. Revta argent.Bot. 1925-1926.	B.S 3009
Revista Argentina de Entomologia. Buenos Aires. Revta argent.Ent. 1935-1944.	E.S 2354 a
Revista Argentina de Historia Natural. Buenos Aires. Revta argent.Hist.nat. 1891.	S. 2233

TITLE	SERIAL No.

Revista Argentina de Paleontologia y Antropologia Ameghinia.
   Buenos Aires.
   <u>Revta argent.Paleont.Antrop.Ameghinia</u> 1935.

Revista do Arquivo Publico Mineiro. Belo Horizonte.
   <u>Revta Arquiv.Publ.Mineiro.</u> No.24; 1933   M.S 2511 C

Revista de la Asociacion Geologica Argentina. Buenos Aires.
   <u>Revta Asoc.geol.argent.</u> 1948 →
   <u>Formerly</u> Revista de la Sociedad Geologica Argentina.
   Buenos Aires.   P.S 950

Revista de la Asociacion Kraglieviana del Uruguay. Montevideo.
   <u>See</u> Kraglieviana. Revista de la Asociacion Kraglieviana
   del Uruguay.   P.S 937

Revista de la Asociación Paleontologica Argentina. Buenos Aires.
   <u>See</u> Ameghiniana   P.S 959

Revista de Biologia. Lisboa.
   <u>Revta Biol.Lisb.</u> 1956 →   S. 1064

Revista de Biologia Forestal y Limnologia. Madrid.
   <u>Revta Biol.for.Limnol.</u> 1929-1930.   Z.S 1005

Revista de Biologia e Hygiene. São Paulo.
   <u>Revta Biol.Hyg.</u> 1927-1939.   S. 2221 b

Revista de Biologia Marina. Valparaiso.
   <u>Revta Biol.mar.</u> 1948-1960, 1965 →
   <u>From 1961-1964 styled</u> Montemar.   S. 2237 D

Revista de Biologia Tropical. San José, Costa Rica.
   <u>Revta Biol.trop.</u> 1953 →   S. 2276

Revista de Biologia del Uruguay. Montevideo.
   <u>Revta Biol.Uruguay</u> 1973 →   S. 2222 b

Revista Brasileira. Rio de Janeiro.
   <u>Revta bras.Rio de J.</u> 1857-1861.   S. 2216

Revista Brasileira de Biologia. Rio de Janeiro.
   <u>Revta bras.Biol.</u> 1941 →   S. 2214 a A & T.R.S 6518

Revista Brasileira de Entomologia. São Paulo.
   <u>Revta bras.Ent.</u> 1954 →   E.S 2351

Revista Brasileira de Geociências. São Paulo.
   <u>Revta bras.Geociênc.</u> 1971 →   P.S 911 C

TITLE	SERIAL No.

Revista Brasileira de Malariologia e Doenças Tropicais.
　Rio de Janeiro.
　　Revta bras.Malar.Doenç.trop. 1949 →　　　　　　E.S 2353 a

Revista Brasileira de Pesquisas Médicas e Biologicas. São Paulo.
　　Revta bras.Pesq.méd.biol. 1968 →　　　　　　　S. 2220 b

Revista de la Catedra de Microbiologia y Parasitologia. Buenos Aires.
　　Revta Cat.Microbiol.Parasit.B.Aires No. 89, 1960.
　　Formerly Publicaciones. Misión de Estudios de Patologia
　　Regional Argentina, Jujuy. Universidad de Buenos Aires.　　S. 2242 F

Revista. Centro de Cultura Scientifica. Pelotas.
　　Revta Cent.Cult.scient.Pelotas 1918-1919.　　　　S. 2210

Revista del Centro de Estudiantes del Doctorado en Ciencias
　Naturales. Buenos Aires.
　　Revta Cent.Estud.Doct.Cienc.nat.B.Aires 1935-1941.
　　Continued as Holmbergia. Revista del Centro de Estudiantes
　　del Doctorado en Ciencias Naturales. Buenos Aires.　　S. 2226 a

Revista do Centro de Estudos de Cabo Verde. Praia.
　　Revta Cent.Estud.Cabo Verde
　　　Serie de Ciencias Biologicas. 1972 →　　　　　　S. 20·7

Revista Ceres. Vicosa.
　　Revta Ceres Vol.6 (No.31) → 1944 → (imp.)
　　Formerly Ceres. Vicosa.　　　　　　　　　　　　S. 2248 a A

Revista Chilena de Entomologia. Santiago.
　　Revta chil.Ent. 1951 →　　　　　　　　　　　　E.S 2368

Revista Chilena de Historia Natural. Santiago de Chili.
　　Revta chil.Hist.nat. 1899-1948 (imp.)　　　　　　S. 2239 a

Revista de Ciencias. Lima.
　　Revta Cienc.Lima Num.428-490. 1939-1954 (imp.)　　S. 2207 B

Revista de Ciencias Agronomicas. Lourenco Marques.
　　Revta Cienc.agron. Serie B Vol.3 No.2 → 1970 →　　E.S 2192

Revista de Ciências Biológicas. Belém.
　　Revta Cienc.biol.Belém 1963-1964.　　　　　　　S. 2209 d

Revista de Ciências Biológicas. Lourenco Marques.
　　Revta Ciênc.biol.Lourenco Marques Série A 1968 →
　　　　　　　　　　　　　　　　　　Série B 1970 →
　　Formerly Revista dos Estudos Gerais Universitários de Mocambique.
　　Serie II. Ciencias Biológicas e Agronómicas.　　　S. 2087 a B

Revista de Ciências Geológicas. Lourenco Marques.
　　Revta Cienc.geol. 1968 →
　　Formerly Revista dos Estudos Gerais Universitários de
　　Mocambique. Serie VI. Ciências Geológicas.　　　　P.S 1692

| TITLE | SERIAL No. |

Revista de Ciências Veterinárias. Lourenco Marques.
   Revta Ciênc.vet. 1968 →
   Formerly Revista dos Estudos Gerais Universitarios de
   Mocambique. Ser.IV Ciências Veterinárias.   Z.S 2099

Revista Cientifica de Investigaciones del Museo de Historia
   Natural de San Rafael. Mendoza.
   Revta cient.Invest.Mus.Hist.nat.S.Rafael 1956 →   S. 2231 b

Revista Cientifica (Mensual) de la Universidad Central de
   Venezuela. Caracas.
   Revta cient.Univ.cent.Venez. 1887-1891.   S. 2201 A

Revista del Colegio Nacional Vicente Rocafuerte, Guayaquil.
   Revta Col.nac.Vicente Rocafuerte 1919-1939, 1944.   S. 2203

Revista de Colonizacion y Agricultura. La Paz.
   Revta Colon.Agric.,La Paz Nos.8-14, 1937.
   Formerly Colonizacion y Agricultura. La Paz.   E.S 2367

Revista Cubana de Ciencias Veterinarias. La Habana.
   Revta cub.Cienc.vet. 1970 →   Z.S 2255

Revista Ecuatoriana de Entomologia y Parasitologia. Quito.
   Revta ecuat.Ent.Parasit. 1953 →   E.S 2370

Revista de Entomologia. Rio de Janeiro.
   Revta Ent.,Rio de J. 1931-1951.   E.S 2352

Revista de Entomologia de Moçambique. Lourenço Marques.
   Revta Ent.Moçamb. 1958 →   E.S 2190

Revista da Escola Superior Colonial. Lisboa.
   Revta Esc.sup.colon.Lisb. Vol.3. 1953.   S. 1057

Revista Espanola de Biologia. Madrid.
   Revta esp.Biol. 1932-1936.   T.R.S 1106 & S. 1003 G

Revista Española de Micropaleontologia. Madrid.
   Revta esp.Micropaleont. 1969 →   P.S 661

Revista dos Estudantes da Universidade de Lourenço Marques.
   Revta Estud.Univ.Lourenço Marques 1970 →   S. 2087 a C

Revista dos Estudos Gerais Universitários de Moçambique.
   Lourenço Marques.
   Revta Estud.ger.Univ.Moçambique
   Serie II. Ciências Biológicas e Agronómicas. 1964.
   Continued as Revista de Ciencias Biologicas.
   Serie IV. Ciências Veterinárias. 1964-1966.
   Continued as Revista de Ciencias Veteinárias.
   Serie VI. Ciências Geológicas. Vol.2-4, 1965-1967.
   Continued as Revista de Ciencias Geológicas.   S. 2087 a

TITLE	SERIAL No.
Revista da Faculdade de Ciencias. Universidade de Coímbra. Revta Fac.Cienc.Univ.Coímbra 1931 →	S. 1016 D
Revista da Faculdade de Ciências. Universidade de Lisboa. Série C. Ciências Naturais. Revta Fac.Ciênc.Univ.Lisb. 1950 →	S. 1050
Revista da Faculdade de Medicina Veterinaria. Universidade de São Paulo. Revta Fac.Med.vet.Univ.S.Paulo 1938-1971. Continued as Revista da Faculdade de Medicina Veterinaria e Zootecnia. Universidade de São Paulo.	S. 2221 c A
Revista da Faculdade de Medicina Veterinaria e Zootecnia. Universidade de São Paulo. Revta Fac.Med.vet.Zootec.Univ.S.Paulo 1972 → Formerly Revista da Faculdade de Medicina Veterinaria. Universidade de São Paulo.	S. 2221 c A
Revista de la Facultad de Agricultura, Universidad Central de Venezuela. Maracay. Revta Fac.Agric.Univ.cent.Venez. 1953-1955. Formerly Revista de la Facultad de Ingeniera Agronomica, Universidad Central de Venezuela, Maracay. Continued as Revista de la Facultad de Agronomia, Universidad Central de Venezuela, Maracay.	E.S 2355 B
Revista de la Facultad de Agronomia, Universidad Central de Venezuela. Maracay. Revta Fac.Agron.Univ.cent.Venez. 1956 → Alcance. 1956 → Formerly Revista de la Facultad de Agricultura, Universidad Central de Venezuela.	E.S 2355 B E.S 2355 B b E.S 2355 B
Revista de la Facultad de Agronomia y Veterinaria. Universidad de Buenos Aires. Revta Fac.Agron.Vet.Univ.Buenos Aires Tomo 17 → 1967/1968 →	S. 2242 N
Revista de la Facultad de Ciencias Agrarias, Universidad Nacional de Cuyo. Mendoza. Revta Fac.Cienc.Agrar.Univ.nac.Cuyo 1949 → imp.	B.S 3016
Revista de la Facultad de Ciencias Biologicas. Universidad Nacional de Trujillo. Peru. Revta Fac.Cienc.biol.Univ.nac.Trujillo 1964.	S. 2207 d
Revista de la Facultad de Ciencias Naturales de Salta. Universidad Nacional de Tucuman. Salta. Revta Fac.Cienc.nat.Salta 1959-1962.	S. 2234 a D
Revista de la Facultad de Ingeniera Agronomica, Universidad Central de Venezuela. Maracay. Revta Fac.Ing.agron.Univ.cent.Venez. 1952. Continued as Revista de la Facultad de Agricultura, Universidad Central de Venezuela, Maracay.	E.S 2355 B

| TITLE | SERIAL No. |

Revista de la Facultad de Medicina, Universidad del Zulia.
Maracaibo.
  Revta Fac.Med.Univ.Zulia 1968 →   S. 2201 b C

Revista de la Facultad Nacional de Agronomia. Medellín.
  Revta Fac.nac.Agron.,Medellín 1946 →   E.S 2320

Revista de Fitopatologia. Madrid.
  Revta Fitopatol. 1923-1938.   E.S 1004

Revista de Fomento, Caracas.
  Revta Fom.,Caracas Año 3, no.19. 1939.   76 B.o.V

Revista Forestal del Peru. Lima.
  Revta for.Peru 1967 →   B.S 4036

Revista Forestal Venezolana. Merida.
  Revta for.venez. 1958 →   B.S 4052

Revista Geologica de Chile. Santiago.
  Rev.geol.Chile 1974 →   P.S 2037

Revista Ibérica de Parasitologia. Granada.
  Revta ibér.Parasit. 1941 →   S. 1024

Revista de Industria Animal. São Paulo.
  Revta Ind.anim. N.S. Vol.1-3. 1938-1940.
  Continued as Boletim de Industria Animal. São Paulo.   S. 2217 F

Revista del Instituto Bacteriológico. Buenos Aires.
  Revta Inst.bact.B.Aires 1917-1941.
    (Vols.1-3 incomplete.)
  Continued as Revista del Instituto Bacteriológico
  'Dr. Carlos G. Malbrán'. Buenos Aires.   S. 2244 A

Revista del Instituto Bacteriológico 'Dr. Carlos G. Malbrán'.
  Buenos Aires.
  Revta Inst.bact.Dr.Carlos G. Malbrán
  Tom.10, No.2 - Tom.14, 1941-1949.
  Formerly Revista del Instituto Bacteriologico Buenos Aires.
  Continued as Revista del Instituto Malbrán. Buenos Aires.   S. 2244 A

Revista Instituto de Biologia e Pesquisas Tecnologicas. Curitiba.
  Revta Inst.Biol.Pesq.tecnol. No.10 → 1958 →   S. 2210 b D

Revista del Instituto Malbrán. Buenos Aires.
  Revta Inst.Malbrán Tom.15-16, 1950-1954.
  Formerly Revista del Instituto Bacteriologico
  'Dr. Carlos G. Malbrán'. Buenos Aires.
  Continued as Anales del Instituto Nacional de Microbiologia.
  Buenos Aires.   S. 2244 A

Revista do Instituto de Medicina Tropical de São Paulo.
  Revta Inst.Med.trop.S.Paulo 1959 →   S. 2217 a

Revista del Instituto Municipal de Botanica, Buenos Aires.
  Revta Inst.munic.Bot. 1 → 1961 →   B.S 3008

| TITLE | SERIAL No. |

Revista. Instituto Nacional de Geologia y Mineria. Buenos Aires.
   Revta Inst.nac.Geol.Min.B.Aires   1965 →   P.S 2008 A

Revista del Instituto Nacional de Investigación de las Ciencias
   Naturales y Museo Argentino de Ciencias Naturales
   "Bernardino Rivadavia". Buenos Aires.
   Revta Inst.nac.Invest.Cienc.nat.Mus.argent.Cienc.nat.
   Bernardino Rivadavia   Ciencias Botanicas 1948-1951.   B.S 3007 a
                            Ciencias Geologicas 1949-1957.   P.S 2005 A
   Ciencias Zoologicas Vol.1, No.2 - Vol.3, No.2. 1948-1954.
                                    Z.S 2228 A & T.R.S 6509 C
   Formerly Revista del Museo Argentino de Ciencias
   Naturales "Bernardino Rivadavia".
   Continued as Revista del Museo Argentino de Ciencias
   Naturales "Bernardino Rivadavia" e Instituto Nacional de
   Investigacion de la Ciencias Naturales. Buenos Aires.

Revista del Instituto de Salubridad y Enfermedades Tropicales.
   México.
   Revta Inst.Salubr.Enferm.trop.Méx. 1939-1965.   S. 2258 a A
                                    1950-1965.   E.S 2398
   Continued as Revista de Investigacion en Salud Publica. Mexico.

Revista. International de Botanica Experimentalis. Buenos Aires.
   See Phyton. International Journal of Experimental
   Botany. Buenos Aires.   B.S 4048

Revista de Investigacion en Salud Publica. Mexico.
   Revta Invest.Salud.Publ. Vol.26 → 1966 →
   Formerly Revista del Instituto de Salubridad y   S. 2258 a A
   Enfermedades Tropicales. Mexico.   E.S 2398

Revista de Investigaciones Agricolas. Buenos Aires.
   Revta Invest.agric.,B.Aires 1947-1959.   E.S 2348
                           1947-1959. (imp.)   B.S 3005 c
   Subsequently merged with Revista de Investigaciones
   Ganaderas to form Revista de Investigaciones Agropecuarias.

Revista de Investigaciones Agropecuarias. Buenos Aires.
   Revta Investnes Agropec. 1964 →
   (published in series.)   E.S 2348

Revista del Jardin Botanico y Museo de Historia Natural del
   Paraguay. Asunción.
   Revta Jard.bot.Mus.Hist.nat.Parag. 1921-1930   B.S 4030

Revista del Jardin Zoologico. Buenos Aires.
   Revta Jard.zool.B.Aires   1893-1918.   Z.S 2224

Revista Latinoamericana de Microbiologia. Mexico, D.F.
   Revta lat.-am.Microbiol. 1958-1964, 1970 →
   Issued as Revista Latinoamericana de Microbiologia y
   Parasitologia 1966-1969.   S. 2260

TITLE	SERIAL No.

Revista Latinoamericana de Microbiologia y Parasitologia.
   Mexico, D.F.
     Revta lat.-am.Microbiol.Parasit. 1966-1969.
     Formerly and Continued as Revista Latinoamericana
   de Microbiologia.   S. 2260

Revista de Lepidopterologia. SHILAP, Madrid.
     Revta Lepid.   1973 →   E.S 1008

Revista de Medicina Tropical y Parasitologia, Bacteriologia,
   Clinica y Laboratorio. Habana.
     Revta Med.trop.Parasit.Habana 1937-1946.
     Formerly Revista de Parasitologia, Clinica y Laboratorio.
   Habana.   S. 2288

Revista Mensual de Entomologia Economica y Galliniculture.
     See Entomologisto Brasileiro.   E.S 2361

Revista Minera de Bolivia. Oruro.
     Revta min.Bolivia   1926-1928   M.S 2503

Revista Minera. Sociedad Argentina de Mineria y Geologia.
   Buenos Aires.
     Revta min.,B.Aires   1929-1938   M.S 2504

Revista del Ministerio de Colonias y Agricultura. La Paz.
     Revta Minist.Colon.Agric, La Paz 1905-1907.   S. 2245 A

Revista del Ministerio de Obras Publicas y Fomento. Bogotá.
     Revta Minist.Obr.publ.Fom., Bogotá   Tom 3-4. 1908-1909.   S. 2247

Revista del Museo Argentino de Ciencias Naturales
   "Bernardino Rivadavia", Buenos Aires.
     Revta Mus.argent.Cienc.nat.Bernardino Rivadavia
     Serie Ciencias Zoologicas, Vol.1 No.1 1948.
     Continued as Revista del Instituto Nacional de
   Investigacion de las Ciencias Naturales y Museo Argentino
   de Ciencias Naturales "Bernardino Rivadavia". T.R.S 6509 C & Z.S 2228 A

Revista del Museo Argentino de Ciencias Naturales "Bernardino
   Rivadavia" e Instituto Nacional de Investigacion de las
   Ciencias Naturales. Buenos Aires.
     Revta Mus.argent.Cienc.nat.Bernardino Rivadavia
     Inst.nac.Invest.Cienc.nat.

Ecologia 1963 →	S. 2225 K
Hidrologia 1963 →	S. 2225 K
Paleontologia 1963 →	P.S 960
Parasitologia 1968 →	S. 2225 L
Ciencias Entomologica 1964 →	E.S 2354 b
Ciencias Botanicas 1961 →	B.S 3007 a
Ciencias Geológicas 1957 →	P.S 2005 A
Ciencias Paleontologia 1964 →	P.S 2005 B
Ciencias Zoologicas Vol.3 No.3 → 1957 →	Z.S 2228 A
	& T.R.S 6509 C

     Formerly Revista del Instituto Nacional de Investigacion
   de las Ciencias Naturales y Museo Argentino de Ciencias
   Naturales "Bernardino Rivadavia". Buenos Aires.

TITLE	SERIAL No.

Revista del Museo de Historia Natural de Mendoza. R.A.
    Revta Mus.Hist.nat.Mendoza   Vol.1, No.1, Vol.2, No.1→ 1947→ S. 2231 a

Revista del Museo de La Plata.
    Revta Mus.La Plata 1890-1934:
      Sección Oficial 1935-1945.   S. 2231 A
      Sección Anthropologia N.S. 1936 →   P.A.S 702
      Sección Botánica N.S. 1936 →   B.S 3004
      Sección Geologia 1936 →   P.S 956
      Sección Paleont. N.S. Vol.1 → 1936 →   P.S 956
      Sección Zoologia N.S. 1937 →   Z.S 2221

Revista del Museo Municipal de Ciencias Naturales y Tradicional
   de Mar del Plata.
    Revta Mus.munic.Cien.nat.Mar del Plata 1952 →   S. 2229 b

Revista do Museo Municipal de Iguape. S.Paulo.
    Revta Mus.mun.Iguape 1907.   S. 2220

Revista do Museu Nacional. Rio de Janeiro.
    Revta Mus.nac.Rio de J. 1944-1945.   S. 2213 E

Revista do Museu Paulista. São Paulo.
    Revta Mus.paul. 1895-1938: 1947 →   S. 2219 A
                   1895-1938.   T.R.S 6511

Revista Muzeelor. Bucuresti.
    Revta Muzeelor 1969 →   S. 1894 d

Revista Muzeului Geologic-Mineralogic al Universitătii din Cluj.
    Revta Muz.geol.-miner.Univ.Cluj 1924-1937.
    Formerly Múzeumi Füzetek. Az Erdélyi Nemzeti Múzeum
    Asványtárának Ertesitöje.   P.S 470

Revista Paraguaya de Microbiologia. Facultad de Ciencias
   Medicas de la Universidad Nacional. Asuncion.
    Revta parag.Microbiol. 1966 →   S. 2204 a

Revista de Parasitologia, Clinica y Laboratorio. Habana.
    Revta Parasit.clin.Lab. 1935-1936.
    Continued as Revista de Medicina Tropical y Parasitologia,
    Bacteriologia, Clinica y Laboratorio. Habana.   S. 2288

Revista Peruana de Entomologia. Lima.
    Revta peru.Ent. 1964 →
    Formerly Revista Peruana de Entomologia Agrícola.   E.S 2346

Revista Peruana de Entomologia Agrícola. Lima.
    Revta peru.Ent.agríc. Vol.2-6, 1959-1963.
    Continued as Revista Peruana de Entomologia.   E.S 2346

Revista Portuguesa de Zoologia e Biologia Geral. Lisbon.
    Revta port.Zool.Biol.ger. 1957-1963.   T.R.S 1163 &
    Formerly and Continued as Archivos do Museu Bocage. Z.S 1070

TITLE   SERIAL No.

Revista de los Progresos de Ciencias Exactas Físicas y
    Naturales. Madrid.
    Revta Prog.Cienc.Madrid   1850-1899.   S. 1011

Revista de la R.Academia de Ciencias Exactas Fisicas y
    Naturales de Madrid.
    Revta R.Acad.Cienc.exact.fis.nat.Madr.   1904 →   S. 1001 B

Revista de Saúde Pública. Universidade de São Paulo.
    Revta Saúde públ.   1967 →   E.S 2358 a

Revista de la Sociedad Científica del Paraguay. Asunción.
    Revta Soc.cient.Parag.   1921-1947. (imp.)   S. 2204 A

Revista de la Sociedad Colombiana de Ciencias Naturales. Bogotá.
    Revta Soc.colomb.Cienc.nat.   1929-1931.
    Formerly Boletin de la Sociedad Colombiana de Ciencias
    Naturales. Bogotá.   T.R.S 6507
    Continued as Revista de la Académia Colombiana de
    Ciencias Exactas, Fisicas y Naturales Bogotá.   S. 2247 a

Revista de la Sociedad Colombiana de Orquideología.
    Medellin, Colombia.
    Revta Soc.Colomb.Orquideol.   1966 →   B.S 4011

Revista de la Sociedad Cubana de Botánica. La Habana.
    Revta Soc.cub.Bot.   1944 →   B.S 4090

Revista de la Sociedad Entomológica Argentina. Buenos Aires.
    Revta Soc.ent.argent.   1926 →   E.S 2353

Revista de la Sociedad Geológica Argentina. Buenos Aires.
    Revta Soc.geol.argent.   1946-1948.
    Continued as Revista de la Asociación Geologica Argentina.
    Buenos Aires.   P.S 950

Revista de la Sociedad Malacológica "Carlos de la Torre". Habana.
    Revta Soc.malac.Carlos de la Torre   1943 →   Z. Mollusca Section

Revista de la Sociedad Mexicana de Entomologia, Mexico.
    Revta Soc.mex.Ent.   1955.   E.S 2399

Revista de la Sociedad Mexicana de Historia Natural. Mexico.
    Revta Soc.mex.Hist.nat.   1939 →   S. 2252 B

Revista. Sociedad del Mismo Nombre en Las Palmas de G. Canaria.
    See Museo Canario. Las Palmas.   S. 1037

Revista de la Sociedad Ornitológica de la Plata.   T.R.S 6501 &
    See Hornero. Revista de Sociedad Ornitológica de la Plata. Z.S 2229

TITLE	SERIAL No.

Revista de la Sociedad Uruguaya de Entomologia. Montevideo.
    Revta Soc.urug.Ent. 1956 →   E.S 2372

Revista Sudamericana de Botanica. Montevideo.
    Revta sudam.Bot. 1934 →   B.S 4045

Revista de la Universidad de Buenos Aires.
    Revta Univ.B.Aires 1924-1931. (imp.)   S. 2242 A

Revista de la Universidad de Madrid.
    Revta Univ.Madr. Vol.2 No.5, Vol.8, Nos.29-31. 1942, 1959.   S. 1032

Revista da Universidade de Coímbra.
    Revta Univ.Coímbra 1912-1914.   S. 1016 A

Revista Universitaria. Universidad Católica de Chile.
Santiago de Chile.
    Revta univ.Santiago 1915-1968. (imp.)
    (Includes: Anales de la Academia Chilena de Ciencias
    Naturales.)   S. 2237 a

Revista Universității "Al I. Cuza" și a Institutului
Politehnic din Iasi.
    Revta Univ."Al I. Cuza" Inst.politeh.Iasi 1954-1955.   S. 1883 B

Revista Universității "C.I.Parhon" si a Politehnicii. Bucurešti.
    Revta Univ.C.I.Parhon Politeh.Bucuresti 1952-1955.
    Continued as Analele Universității C.I. Parhon. Bucuresti.
Series Stiintelor Naturii.   S. 1892 a

Revue Agricole de l'Ile Maurice. Port Louis.
    Revue agric.Ile Maurice 1946-1949.   E.S 2182

Revue Algologique. Paris.
    Revue algol. 1924-1941, 1954 →
    For 1942 See Travaux Algologiques. Paris.   B.A.S 14

Revue d'Anthropologie. Paris.
    Revue anthrop. 1872-1888.   P.A.S 224

Revue de Biologie. Académie de la République Populaire
Roumaine. Bucarest.
    Revue Biol.Buc. Tom 1-8. 1954-1963.
    Continued as Revue Roumaine de Biologie.
Academie la République Populaire Roumaine. Bucarest.   S. 1893 K

Revue Biologique du Nord de la France. Lille.
    Revue biol.N.Fr. 1888-1895.   S. 989

Revue de Botanique. Courrensan.
    Revue Bot.,Courrensan 1882-1895.   B.S 393

TITLE	SERIAL No.

Revue Botanique. Paris.
    Revue Bot.Paris   1845-1847.   B.S 421

Revue de Botanique Systématique et de Géographie Botanique. Asnières.
    Revue Bot.syst.Géogr.bot.   1903-1905.   B.S 388

Revue Bretonne de Botanique Pure et Appliquée. Rennes.
    Rev.Bret.Bot.Pure Appl. 1906-1907, 1923-1930 (imp.)   B.S 458

Revue Bryologique. Paris.
    Revue bryol.   1874-1931.
    Continued as   Revue Bryologique et Lichenologique, Paris.   B.B.S 12

Revue Bryologique et Lichenologique. Paris.
    Revue bryol.lichen.   1932 →
    Formerly   Revue Bryologique. Paris.   B.B.S 12

Revue Canadienne de Biologie. Montréal.
    Revue can.Biol.   1942 →   S. 2663

Revue Coléoptérologique, Bruxelles.
    Revue coléopt.Bruxelles   1882.   E.S 705

Revue du Comportement Animal. Paris.
    Revue Comporte.Anim.   Tome 4 → 1970 →   Z.S 821

Revue des Cours Scientifiques de la France et de l'Etranger. Paris.
    Revue Cours scient.Paris   1863-1870.
    Continued as   Revue Scientifique de la France et de
    l'Etranger. Paris.   S. 998

Revue Critique de Paléozoologie. Paris.
    Revue crit.Paléozool.   1897-1924.   P.S 204

Revue des Cultures Coloniales. Paris.
    Revue Cult.colon.   1897-1904.   B.S 415

Revue de Cytologie et de Biologie Végétales. Paris.
    Revue Cytol.Biol.vég.   Vol.12 → 1951 →   B.S 472

Revue d'Ecologie et de Biologie du Sol. Paris.
    Revue Ecol.Biol.Sol.   1964 →   S. 951 a

Revue Encyclopédique. Paris.
    Revue encycl.   1820-1832.
    Formerly   Annales Encyclopédiques. Paris.   S. 1000

Revue d'Entomologie, Caen.
    Revue Ent.   1882-1910.   E.S 806

Revue d'Entomologie de l'URSS.
    See   Entomologicheskoe Obozrenie. Moskva.   E.S 1804

TITLE	SERIAL No.

Revue Entomologique. (Silbermann).
    Revue Ent.(Silbermann) 1833-1840.   E.S 807

Revue Entomologique Internationale. Paris.
    See Miscellanea Entomologica.   E.S 813

Revue de la Faculté des Sciences Forestières de l'Université d'Istanbul.
    See Istanbul Universitesi Orman Fakültesi Dergisi.   B.S 1545 a

Revue de la Faculté des Sciences de l'Université d'Istanbul.
    See Istanbul Universitesi Fen Fakültesi Mecmuasi. Istanbul.   S. 1887 A

Revue de la Faculté Vétérinaire, Université de Teheran.
    Revue Fac.vét.Univ.Teheran Tome 26 → 1970 →   Z.S 1995

Revue de la Fédération Francaise des Sociétés des Sciences Naturelles. 3rd Series. Paris.
    Revue Féd.fr.Soc.Sci.nat. 1962 →
    Formerly Bulletin Trimestriel. Fédération Française des Sociétés des Sciences Naturelles. Versailles.   S. 978 a

Revue der Fortschritte der Naturwissenschaften in Theoretischer und Praktischer Beziehung. Cöln und Leipzig.
    Revue Fortschr.Naturw.Cöln 1873-1889.   S. 1628

Revue Française d'Entomologie. Paris.
    Revue fr.Ent. 1934-1965.
    Merged (1965) with Annales de la Société Entomologique de France. Paris.   E.S 812

Revue Française de Lépidoptérologie. Paris.
    Revue fr.Lépidopt. 1938-1958.
    Formerly Amateur de Papillons.   E.S 802

Revue Française de Mammalogie. Paris.
    Revue fr.Mammal. Nos.1-2, 1927-1928.   Z.S 960 A

Revue Francaise d'Ornithologie Scientifique et Pratique. Paris.
    Revue fr.Orn.Scient.prat. 1909-1928.
    Continued in Oiseau et la Revue Francaise d'Ornithologie, Paris.   T.R.S 801 & Z.S 813

Revue Générale de Botanique. Paris.
    Revue gén.Bot. 1889 →   B.S 420

Revue de Geographie Physique et de Geologie Dynamique. Paris.
    Revue Géogr.phys.Géol.dyn. Vol.8 → 1966 →   P.S 193

Revue de Géologie. Paris.
    Revue géol.Paris 1860-1878   P.S 200

| TITLE | SERIAL No. |

Revue de Géologie et de Géographie. Bucarest.
   Revue Géol.Géogr.Bucr. 1957-1963.
    Continued as Revue Roumaine de Géologie Géophysique
et Géographie. Bucarest.   P.S 464

Revue de Géologie et des Sciences Connexes. Liége.
   Revue Géol.   1920-1940 (imp.)   P.S 250

Revue Géologique Polonaise.
   See Wiadomości Muzeum Ziemi.   P.S 571

Revue d'Histoire Naturelle Appliquée. Paris.
   Revue Hist.nat.appl. (1920-1930).
    Première partie: not held.
    Deuxième partie: L'Oiseau, Paris. 1920-1930.
    Continued as Oiseau et la Revue Francaise
d'Ornithologie. Paris.   T.R.S 802 & Z.S 814

Revue d'Histoire des Sciences. Paris.
   Revue Hist.Sci. Tom.21 → 1968 →   S. 992 a

Revue Hongroise de Géologie et Paléontologie.
   See Földtani Szemle. Budapest.   P.S 367

Revue Horticole. Paris.
   Revue hort. 1832-1904.   B.S 414

Revue d'Hydrologie (Hydrographie, Hydrobiologie, Pisciculture).
   Aarau.
   Revue Hydrol. 1920-1939.
   Continued as Zeitschrift für Hydrologie (Hydrographie,
Hydrobiologie, Fischereiwissenschaft.) Aarau.   S. 1201 F

Revue Illustrée d'Entomologie.
   See Insecta. Rennes.   E.S 804

Revue de l'Institut Français du Pétrole et Annales des
   Combustibles Liquides. Paris.
   Revue Inst.fr.Pétrole 19 → 1964 →
   Certain earlier parts are catalogued by authors and are
in presses 25 q; 30 q; & 48 q.   P.S 214

Revue Internationale des Archives des Bibliothèques et des
   Musées. Paris.
   Revue int.Arch.Bibl.Mus. 1895-1896.   S. 982

Revue Internationale de Botanique Expérimentale. Buenos Aires.
   See Phyton. International Journal of Experimental Botany.
Buenos Aires.   B.S 4048

Revue Internationale de Législation pour la Protection de la
   Nature. Bruxelles.
   Revue int.Législ.Prot.Nat. 1916-1934.   S. 2728 A

TITLE	SERIAL No.

Revue Linnéenne, Lyon.
    See Echange.   S. 992

Revue et Magasin de Zoologie Pure et Appliquée. Paris.
    Revue Mag.Zool.   2e série: Tomes 1-23,   1849-1872.   T.R.S 808
                    (Tweeddale: T. 1-17,   1849-1865).   Z.S 940
                    3e série: Tomes 24-30,   1873-1879.   Z.S 940
                                             & T.R.S 808
                                         1849-1866.   E.S 805
    Formerly Magasin de Zoologie & Revue Zoologique.

Revue Mensuelle d'Entomologie. St. Petersburg.
    Revue mens.ent.St.Petersburg 1883-1884.   E.S 1803

Revue Mensuelle des Musées et des Collections de la Ville
de Genève.
    Revue mens.Mus.Genève   1960 →   S. 1231 B

Revue Mensuelle de la Société Entomologique Namuroise. Namur.
    Revue mens.Soc.ent.namur. 1901-1925.
    Formerly Revue de la Société Entomologique Namuroise. Namur.
    Continued as   Lambillionea. Revue Mensuelle de l'Union
des Entomologistes Belges. Bruxelles.   E.S 703

Revue Méthodique et Critique des Collections. Muséum
d'Histoire Naturelle des Pays-Bas. Leide.
    Revue méth.Mus.Hist.nat.Pays-Bas 1862-1880.
    Continued as Catalogue Systématique. Muséum d'Histoire   Z.S 601
Naturelle des Pays-Bas.   & Tweeddale

Revue de Micropaléontologie. Paris.
    Revue Micropaléont. 1958 →   P.S 231

Revue du Musée Géologique-Minéralogique de l'Université de Cluj.
    See Revista Muzeului Geologic-Mineralogic al Universităţii
din Cluj.   P.S 470

Revue de Mycologie. Paris.
    Revue Mycol. 1936 →
    Formerly Annales de Cryptogamie Exotique. Paris.   B.M.S 15

Revue Mycologique. Toulouse.
    Revue mycol. 1879-1906.   B.M.S 18

Revue Nationale de la Chasse (et La Sauvagine.) Paris.
    Revue Natn.Chasse   No.62-196, 1952-1963.   Z.S 816

Revue de Pathologie Végétale et d'Entomologie Agricole de
France. Paris.
    Revue Path.vég.Ent.agric.Fr. 1923-1964.
    Formerly Bulletin de la Société de Pathologie
Végétale de France.
    Merged (1965) with Annales de la Société Entomologique
de France. Paris.   E.S 942

TITLE	SERIAL No.

Revue Portugaise de Zoologie et de Biologie Générale.
   See Revista Portuguesa de Zoologia e Biologia Geral.   Z.S 1070

Revue des Questions Scientifiques, Louvain.
   Revue Quest.scient. 1877-1910; 1960 →   S. 708 B

Revue Rosé. Paris.
   See Revue Scientifique. Revue Rosé. Paris.   S. 998

Revue Roumaine de Biologie. Académie la République Populaire Roumaine. Bucarest.
   Revue roum.Biol. Série Botanique, Tom.9-18,1964-1973.
                    Série de Zoologie, Tom.9-18,1964-1973.
                        These series re-united, Tom.19 → 1974 →
   Formerly Revue de Biologie. Académie la République
Populaire Roumaine. Bucarest.   S. 1893 K

Revue Roumaine d'Embryologie et de Cytologie. Jassy.
   Sér. de Cytologie.
   Revue roum.Embryol.Cytol. 1964-1973.
   Continued in Revue Roumaine de Morphologie et
d'Embryologie.   S. 1893 Ub

Revue Roumaine de Géologie Géophysique et Géographie. Bucarest.
   Revue roum.Geol.geophys.Geogr. Série de Géologie, Vol.8 → 1964 →
   Formerly Revue de Géologie et de Géographie. Bucurest.   P.S 464

Revue Roumaine de Morphologie et d'Embryologie. Bucarest.
   Revue roum.Morph.Embryol. Tome 19. 1974.
   Formerly Revue Roumaine d'Embryologie et de Cytologie.
   Continued as Revue Roumaine de Morphologie et de Physiologie.
Bucarest.   S. 1893 Ub

Revue Roumaine de Morphologie et de Physiologie. Bucarest.
   Revue roum.Morph.Physiol. Tom.20 → 1974 →
   Formerly Revue Roumaine de Morphologie et d'Embryologie.
Bucarest.   S. 1893 U b

Revue Russe d'Entomologie.
   See Entomologicheskoe Obozrenie.   E.S 1804

Revue des Sciences Naturelles. Montpellier.
   Revue Sci.nat.Montpellier 1872-1885   S. 984

Revue des Sciences Naturelles. St.Pétersbourg.
   See Vestnik Estestvoznaniya. S.-Peterburg.   S. 1855 C

Revue des Sciences Naturelles Appliquées. Paris.
   Revue Sci.nat.appl. 1889-1890.   Z.S 830
                    1890-1895.   T.R.S 828
   Formerly & Continued as Bulletin de la Société Nationale
d'Acclimatation de France.

Revue des Sciences Naturelles d'Auvergne. Clermont-Ferrant.
   Revue Sci.nat.Auvergne 1935 →
   Formerly Bulletin de la Société d'Histoire Naturelle
d'Auvergne. Clermont-Ferrant.   S. 853

TITLE	SERIAL No.

Revue des Sciences Naturelles de l'Ouest, Paris.
   Revue Sci.nat.Ouest 1891-1897.   S. 990

Revue Scientifique du Bourbonnais et du Centre de la France. Moulins.
   Revue scient.Bourbon.Cent.Fr. 1888 → (suspended 1915-1921)   S. 993
                              1888-1914.   T.R.S 812

Revue Scientifique de la France et de l'Etranger. Paris.
   Revue scient.Paris 1871-1884.
   Formerly Revue des Cours Scientifiques de la France et de l'Etranger. Paris.
   Continued as Revue Scientifique. Revue Rosé.   S. 998

Revue Scientifique du Limousin, Limoges.
   Revue scient.Limousin 1893-1914. (imp.)   S. 989 a

Revue Scientifique. Revue Rosé. Paris.
   Revue scient.Paris 1884-1924.
   Formerly Revue Scientifique de la France et de l'Etranger. Paris.   S. 998

Revue de Sériculture Comparée. Paris.
   Revue Séricult.comp. 1863-1867.   Z.S 950

Revue de la Société Entomologique Namuroise. Namur.
   Revue Soc.ent.namur. 1896-1900.
   Continued as Revue Mensuelle de la Société Entomologique Namuroise.   E.S 703

Révue de la Société Géologique Bulgare.
   See Spisanie na Bŭlgarskoto Geologichesko Druzhestvo.   P.S 490

Revue des Sociétés Savantes de Haute Normandie. Rouen.
   Revue Socs sav.haute Normandie 1956 →   S. 962
                                    No.2 → 1956 →   P.A.S 233

Revue Suisse d'Hydrologie.
   See Schweizerische Zeitschrift für Hydrologie. Basel.   S. 1201 F

Revue Suisse de Zoologie. Genève.
   Revue suisse Zool. 1893 →   T.R.S 1209 & Z.S 1240

Revue des Travaux de l'Institut des Pêches Maritimes. Paris.
   Revue Trav.Inst.Pêch.marit. 1955 →
   Formerly Revue des Travaux de l'Institut Scientifique et Technique des Pêches Maritimes.   Z.S 980

Revue des Travaux de l'Institut Scientifique et Technique des Pêches Maritimes. Paris.
   Revue Trav.Inst. scient.tech. Pêch.marit. 1953-1954.
   Formerly Revue des Travaux de l'Office Scientifique Technique des Pêches Maritimes.
   Continued as Revue des Travaux de l'Institut des Pêches Maritimes.   Z.S 980

| TITLE | SERIAL No. |

Revue des Travaux de l'Office des Pêches Maritimes. Paris.
    Revue Trav.Off.Pêch.marit. 1928-1938.
    Formerly Mémoires. Office Scientifique et Technique
    des Pêches Maritimes. Paris.
    Continued as Revue des Travaux de l'Office Scientifique
    et Technique des Pêches Maritimes.   Z.S 980

Revue des Travaux de l'Office Scientifique et Technique des
    Pêches Maritimes. Paris.
    Revue Trav.Off. scient.tech. Pêch.marit. 1939-1952.
    Formerly Revue des Travaux de l'Office des Pêches Maritimes.
    Continued as Revue des Travaux de l'Institut Scientifique
    et Technique des Pêches Maritimes.   Z.S 980

Revue des Travaux Scientifiques. Paris.
    Revue Trav.scient. 1881-1898.   S. 929 D

Revue Trimestrielle Consacrée à la Protection des Plantes de Pologne.
    See Choroby i Szkodniki Roślin. Warszawa.   B.S 1301

Revue Trimestrielle des Sociétes des Amis des Arbres et du
    Reboisement à Nice et d'Histoire Naturelle de Haute-Provence
    à Digne.
    Revue trim.Soc.Amis Arbres Rebois.Soc.Hist.nat.Haute-Prov.
      1966-1967.
    Formerly Bulletin Trimestriel de la Société des Amis des
    Arbres et du Reboisement et Société d'Histoire Naturelle
    de Haute Provence. Nice.
    Continued as Revue Trimestrielle de la Société d'Histoire
    Naturelle de Haute-Provence. Digne.   S. 923 a

Revue Trimestrielle de la Société d'Histoire Naturelle de
    Haute-Provence. Digne.
    Revue trim.Soc.Hist.nat.Haute-Prov. 1968 →
    Formerly Revue Trimestrielle des Societés des Amis des
    Arbres et du Reboisement à Nice et d'Histoire Naturelle
    de Haute-Provence à Digne.   S. 923 a

Revue Trimestrielle. Terres Australes et Antarctiques
    Françaises. Paris.
    See Terres Australes et Antarctiques Françaises. Paris.   S. 947 a

Revue Universelle des Mines, de la Metallurgie des
    Sciences et des Arts. Liége, etc.
    Revue univlle Mines 1857-1898   M.S 909

Revue de l'Université Officielle de Bujumbura. Burundi.
    Revue Univ.off.Bujumbura 1967-1968.   S. 2053 a

| TITLE | SERIAL No. |

Revue de l'Université Officielle du Congo à Lubumbashi.
    Revue Univ.off.Congo
      Ser.B, Sciences. 1970 →
    Formerly Publications de l'Université Officielle du Congo
à Lubumbashi.   S. 2058 A

Revue du Ver à Soie. Arles.
    Revue Ver Soie 1949 →   E.S 818

Revue Verviétoise d'Histoire Naturelle. Verviers.
    Revue verviét.Hist.nat. 1949 →
    Formerly Naturaliste Amateur.   S. 756

Revue de Zoologie Africaine. Tervuren.
    Revue Zool.afr. 1974 →
    Formerly Revue de Zoologie et de Botanique Africaine.   Z.S 770

Revue de Zoologie et de Botanique Africaines. Bruxelles.
    Revue Zool.Bot.afr. 1928-1973.
    Formerly Revue Zoologique Africaine.
    Continued as Revue de Zoologie Africaine.   T.R.S 706 & Z.S 770

Revue Zoologique. Paris.
    Revue zool. 1838-1848.   E.S 805
    Continued as Revue et Magasin de Zoologie.   Z.S 940 & Tweeddale

Revue Zoologique Africaine. Bruxelles & Paris.
    Revue zool.afr. 1911-1927.   T.R.S 706 & Z.S 770
      Suppl.Botanique. 1919-1927.   B.S 355
    Continued as Revue de Zoologie et de Botanique
Africaines. Bruxelles.

Revue Zoologique Russe.
    See Russkiĭ Zoologicheskiĭ Zhurnal.   Z.S 1850

Rhea. Zeitschrift für die Gesammte Ornithologie.
    Herausgegeben von F.A.L. Thienemann. Leipzig.   Z. 18 o T
    Rhea 1846-1849.   & Tweeddale

Rheinisches Magazin zur Erweiterung der Naturkunde. Giessen.
    Rhein.Mag.Naturk. 1793.   S. 1629 a

Rhizocrinus. Oslo.
    Rhizocrinus 1969 →   Z.S 544

Rhodesia Agricultural Journal. Salisbury.
    Rhodesia agric.J. 1932 →   E.S 2178

Rhodesia Geological Survey Bulletin. Salisbury.
    See Bulletin. Geological Survey. Rhodesia. Salisbury.   P.S 1171

Rhodesia Science News. Salisbury.
    Rhodesia Sci.News 1967 →   S. 2018 a

Rhodesia, Zambia and Malawi Journal of Agricultural Research.
    Salisbury.
    Rhodesia Zambia Malawi J.agric.Res. 1965-1967.
    Formerly and continued as Rhodesian Journal of Agricultural
Research. Salisbury.   E.S 2178 a

TITLE	SERIAL No.

Rhodesian Journal of Agricultural Research. Salisbury, Southern Rhodesia.
    Rhod.J.agric.Res. 1963-1965, 1968 →
    (1966-1967 entitled Rhodesia, Zambia and Malawi Journal of Agricultural Research. Salisbury.)   E.S 2178 a

Rhododendron. Olinda, Victoria.
    Rhododendron 1962 →   B.S 2421

Rhododendron. Puyallup, Wash.
    Rhododendron, Puyallup Vol.5 → 1955 →   B.S 4404

Rhododendron and Camellia Year Book. London.
    Rhodod.Camellia Yb. 1954-1971.
    Formerly Rhododendron Year Book. London.   B.S 50 f

Rhododendron und Immergrüne Laubgehölze Jahrbuch. Bremen.
    Rhodod.immergrüne Laubehölze 1955 →
    Formerly Jahrbuch der Rhododendron Gesellschaft. Bremen.   B.S 956

Rhododendron Society Notes. Ipswich.
    Rhodod.Soc.Notes 1916-1931.   B.S 137

Rhododendron Year Book. Royal Horticultural Society. London.
    Rhodod.Yb., Lond. 1946-1953.
    Continued as Rhododendron and Camellia Year Book. London.   B.S 50 f

Rhodora. Journal of the New England Botanical Club. Boston.
    Rhodora 1899 →   B.S 4236

Ricerca Scientifica. Consiglio Nazionale delle Ricerche. Roma. Supplemento.
    Ricerca scient.Suppl. Vol.1, No.5 → 1962 → (imp.)   P.S 1602 a

Ricerche di Biologia della Selvaggina. Bologna.
    Ric.Biol.Selvag. 1971 →
    Formerly Ricerche di Zoologia Applicata alla Caccia. Bologna.   Z.S 1199

Ricerche e Lavori eseguiti nell'Istituto Botanico della R. Università di Pisa.
    Ric.Lav.Ist.bot.R.Univ.Pisa 1882-1885.   S. 1172 B

Richerche di Zoologia Applicata alla Caccia. Bologna.
    Ric.Zool.appl.Caccia 1930-1971.
        Supplemento. 1939 →
    Continued as Ricerche di Biologia della Selvaggina. Bologna.   Z.S 1199

Rickia. Cryptogamic Series of the Arquivos de Botânica do Estado de São Paulo. São Paulo.
    Rickia 1962 →
    Supplemento 1963 →   B.C.S 55 & 55 a

TITLE	SERIAL No.

Riistatieteellisiä Julkaisuja. Helsinki.
    Riistat.Julk. 1948 →   Z.S 1864

Rika Kwai Siu. Tokyo.
    See Memoirs of the Science Department, University of Tokyo.   S. 1990 B

Rimba Indonesia. Bogor.
    Rimba Indonesia Vol.1, No.2 → 1952 → (imp.)   B.S 1805

Ring. Croydon (later Wrocław).
    Ring 1954 → (From No.42, 1965 styled Series A.)
    Series B, 1963-1964.   T.B.S 8003 A

Risultati delle Osservazioni e Studi. Ruwenzori. Milano.
    Risult.Osserv.Stud.Ruwenzori Parte Scientifica 1909.   74 Da.o.L

Rit Fiskideildar. Reykjavik.
    Rit.Fiskideild. 1940 →   Z.S 550

Rit Visindafjelags Islendinga. Akureyri.
    Rit.Visindafj.isl. 1923 →   S. 502 A

Riviera Nature Notes. London.
    Riviera Nat.Notes 2nd edn. 1903.   72.o.C

Riviera Scientifique. Nice.
    Riviera scient. 1914 →   S. 924

Rivista di Agricoltura Subtropicale e Tropicale. Firenze.
    Riv.Agric.subtrop.trop. Vol.64 → 1970 →   E.S 1119

Rivista di Biologia. Roma.
    Riv.biol. 1919 →   S. 1194 a

Rivista di Biologia Coloniale. Roma.
    Riv.Biol.colon. 1938-1958.   S. 1194 b

Rivista di Biologia Generale. Turino.
    Riv.Biol.gen. 1901.
    Formerly Rivista di Scienze Biologiche. Turino.   S. 1183

Rivista di Coleotterologia. Genova.
    Riv.Coleott. 1923-1924.   E.S 1112

Rivista Coleotterologica Italiana. Parma.
    Riv.coleott.ital. 1903-1915.   E.S 1111

Rivista di Idrobiologia. Istituto di Biologia Generale
    dell'Università di Perugia.
    Riv.Idrobiol. 1960 →   S. 1198 a

| TITLE | SERIAL No. |

Rivista Internazionale di Cecidologia. Avellino.
   See Marcellia.   E.S 1104

Rivista Italiana di Ornitologia. Bologna (later Milano).
   Riv.ital.Orn. 1911 →   T.R.S 1252 & Z.S 1182

Rivista Italiana di Paleontologia. Bologna, &c.
   Riv.ital.Paleont. 1895-1946.
   Continued as Rivista Italiana di Paleontologia e (di)
   Stratigrafia.   P.S 610

Rivista Italiana de Paleontologia. Memoria. Milano.
   Riv.ital.Paleont.Mem. 4, 1942.
   (Two copies in Pal.Lib. One bound with Rivista Vol. 48, the other separately.)
   Formerly Rivista Italiana di Paleontologia. Supplementi.
   Continued as Rivista Italiana di Paleontologia e
   Stratigrafia. Memoria.   P.S 610 & P.S 611

Rivista Italiana di Paleontologia. Supplementi. Pavia.
   Riv.ital.Paleont.Supp. 1935-1938.
   (1-3 bound with Vol. 40 of Rivista. Duplicate copies of 2-3 bound separately)
   Continued as Rivista Italiana de Paleontologia.
   Memoria.   P.S 610 & P.S 611

Rivista Italiana di Paleontologia e Stratigrafia. Milano.
   Riv.ital.Paleont.Stratigr. Anno 53 → 1947 →
   Formerly Rivista Italiana di Paleontologia.   P.S 610

Rivista Italiana di Paleontologia e Stratigrafia. Memoria. Milano.
   Riv.ital.Paleont.Stratigr.Mem. 5 → 1952 →
   Formerly Rivista Italiana di Paleontologia. Memoria.   P.S 611

Rivista Italiana di Scienze Naturali. Napoli.
   Riv.ital.Sci.nat.Napoli 1885-1886.   S. 1152 A

Rivista Italiana di Scienze Naturali e Bollettino del Naturalista. Siena.
   Riv.ital.Sci.nat. 1889-1910.
   Formerly Bollettino del Naturalista.   S. 1195

Rivista Ligure. Giornale di Lettere, Scienze ed Arti. Genova.
   Riv.ligure 1843.   S. 1189

Rivista de Mineralogia e Cristallografia Italiana. Padova.
   Riv.Miner.Cristall.ital. 1887-1918.   M.S 1151

Rivista Mineraria Siciliana. Palermo.
   Riv.Miner.sicil. Anno 15, num.85 → 1964 →   M.S 1105

Rivista de Parassitologia. Roma.
   Riv.Parassit. 1937 →   Z.S 1135

TITLE	SERIAL No.

Rivista di Patologia Vegetale. Padova.
    Riv.Patol.veg., Padova 1892-1904.   B.S.S 3
                            1895-1900.   E.S 1117

Rivista Periodica dei Lavori della I.R. Accademia di Scienze,
    Lettere ed Arti di Padova.
    Riv.Period.Accad.Padova 1851-1884.   S. 1154 A

Rivista di Scienza. Bologna.
    Riv.Sci. 1907-1909.
    Continued as Scientia, Bologna.   S. 1179

Rivista di Scienze Biologiche. Turino.
    Riv.Sci.biol. 1899-1900.
    Continued as Rivista di Biologia Generale. Turino.   S. 1183

Riz et Riziculture. Paris.
    Riz Rizic. 1925-1932.   B.S 408

Riz et Riziculture et Cultures Vivrières Tropicales. Supplément
    à l'Agronomie Tropicale. Nogent-sur-Marne.
    Riz Rizic.Cult.vivr.trop. N.S. Année 3 → 1957 →   B.S 406 b

Roboti Sektsii Zoologiï Bezhrebtovikh Kharkivs' koï
    Naukovo-Doslidochoï Katedri Zoologii. Kharkiv.
    See Trudy Khar'kovskogo Tovaristva Doslidnikiv Prirodi.
    Kharkiv.   S. 1833 A

Ročenka Československé Společnosti Entomologické. Praha.
    Roč.čsl.Spol.ent. 1953-1956.
    Formerly and Re-continued as Časopis Ceskoslovenské Spolecnosti
    Entomologické. Praha.   E.S 1720

Ročenka Museí Olomouckého Kraje. Olomouc.
    Roč.mus.olomouc.Kr. 1955.   S. 1748 A

Ročenka Severočeského Přírodovědeckého Klubu v Liberci.
    Roč.severočesk.přír.Klubu Liberci 1947.   S. 1769

Rochester Naturalist.
    Rochester Nat. 1883-1932, 1967 →   S. 343
                    Vol.1, 1883-1890.   T.R.S 148

Rocks and Minerals. Peekskill, N.Y.
    Rocks Miner. 1926 →   M.S 2606

Rocznik. Arboretum Kórnicke. Poznań.
    See Arboretum Kórnickie. Poznań.   B.S 1303

Rocznik Muzeum w Częstochowie. Częstochowa.
    Roczn.Muz.Czestochowie 1965 →   S. 1727 a

| TITLE | SERIAL No. |

Rocznik. Muzeum Górnoślaskie w Bytomiu. Przyroda. Bytom.
  Roczn.Muz.Górnoślaskie 1962 →   S. 1730

Rocznik. Polska Akademia Nauk. Oddział w Krakowie.
  Roczn.polsk.Akad.Nauk 1959 →
  Formerly Rocznik Polskiej Akademji Umiejetnosci. Krakow.   S. 1725 I

Rocznik Polskiego Towarzystwa Geologicznego (w Krakowie.) Kraków.
  Roczn.pol.Tow.geol. 1923 → (imp.)   P.S 585

Rocznik Polskiej Akademji Umiejetności. Krakow.
  Roczn.pol.Akad.Umiejet. 1928-1947.
  Continued as Rocznik. Polska Akademia Nauk.
  Oddział w Krakowie.   S. 1725 I

Rocznik Sekcji Dendrologicznej Polskiego Towarzystwa
  Botanicznego. Warszawa.
  Roczn.Sekc.dendrol.pol.Tow.bot. Vol.7, 1951.   B.S 1300

Rocznik Towarzystwa Naukowego Krakowskiego z Uniwersytetem
  Jagiellónskim. Kraków.
  Roczn.Tow.Nauk.Kraków 1841-1843.   S. 1725 L

Rodriguesia. Revista do Instituto de Biologia Vegetal. Rio de Janeiro.
  Rodriquesia 1935 →   B.S 3061 a

Roebuck. Journal of the Northumberland Wildlife Trust. Newcastle.
  Roebuck 1973 →   S. 282 a B

Roemeriana. Clausthal-Zellerfeld.
  Roemeriana 1954-1964.
  Replaced by Clausthaler Geologische Abhandlungen.   P.S 357

Rohstoffe des Tierreichs. Von F.A. Pax & W. Arndt. Berlin.
  Rohstoffe Tierreichs 1928 →   Z. 11 o P

Romanes Lectures. Oxford.
  Romanes Lect. 1898-1928.   S. 310 G

Romanian Scientific Abstracts. Bucharest.
  Rom.scient.Abstr. Vol.5 → 1968 →
  (Wanting Vol.7, No.10 & 11.)   REF.ABS.100

Roosevelt Wild Life·Annals of the Roosevelt Wild Life
  Forest Experiment Station of the New York State
  College of Forestry at Syracuse University.
  Roosevelt wild Life Ann. 1926-1936, imp.   Z.S 2475 A
                            1926-1936.   T.R.S 5140 A

Roosevelt Wild Life·Bulletin of the Roosevelt Wild Life
  Forest Experiment Station of the New York State
  College of Forestry at Syracuse University.
  Roosevelt wild Life Bull. 1921-1936.   Z.S 2475
                             1923-1948 (imp.)   T.R.S 5140 B

TITLE	SERIAL No.

Rose Annual. London.
See National Rose Society's Rose Annual. Croydon. — B.S 38

Rose Bulletin of the Royal National Rose Society. St.Albans.
Rose Bull.R.natn.Rose Soc. 1970 → — B.S 38 a

Rostlinná Výroba. Praha.
Rostl.Vyr. Roč.9 (36), Cis.12; Roč.10 (37), Cis.12;
Roč.11 (38) → 1963 →
Formerly Sbornik Ceskoslovenské Akademie Zemědělských
Věd. Praha. — B.S 1239 c

Rostria. Sakuragaoka, Japan.
Rostria 1962 → — E.S 1920 C

Rotifer News. Hanover (N.H.)
Rotifer News 1973 → — Z. Parasitic Worms Section

Rotunda, the Bulletin of the Royal Ontario Museum.
Rotunda 1968 → — S. 2606 C

Roux Archiv fur Entwicklungsmechanik der Organismen.
See Wilhelm Roux Archiv für Entwicklungsmechanik
der Organismen. — Z.S 1440

Rovartani Közlemenyek, Budapest.
See Folia Entomologica Hungarica. Budapest. — E.S 1731

Rovartani Lapok. Budapest.
Rovart.Lap. 1884-1926.
(Suspended 1887-96.) — E.S 1730

Royal Geographical Society. Research Series.
See R.G.S. Research Series. — S. 211 L

Royal Institute of Chemistry Reviews.
See RIC Reviews.

Royal Society Catalogue of Scientific Papers (1800-1900.) London.
See Catalogue of Scientific Papers. — REF.

Rozpravy České Akademie Cisaře Františka Josefa pro Vědy, Slovesnost
a Uměni. Praha.
Rozpr.česke Akad. 1891-1916
Continued as Rozpravy Ceské Akademie věd a Uměni. Praha. — S. 1760 A

Rozpravy Ceské Akademie Věd a Uměni. Praha.
Rozpr.česke Akad.Ved Uměni 1917-1954.
Formerly Rozpravy Ceské Akademie Cisaře Františka Josefa
pro Vědy, Sovesnost a Uměni. Praha.
Continued as Rozpravy Ceskoslovenské Akademie Ved. Praha. — S. 1760 A

Rozpravy Ceskoslovenské Akademie Ved. Praha. Rada MPV
(Matematických a Přírodnich Věd).
Rozpr.česl.Akad.Ved 1953 →
Formerly Rozpravy Ceské Akademie Věd a Uměni. Praha. — S. 1760 A

| TITLE | SERIAL No. |

Rozpravy Státního Geologického Ustavu Ceskoslovenské Republiky. Praha.
  Rozpr.st.geol.Ust. 1926-1949
  Continued as Rozpravy Ustředního Ustavu Geologického. Praha.   P.S 1355

Rozpravy Třídy Matematicko-Přírodovědecké Královské Ceské
  Společnosti Náuk. Praha.
  Rozpr.math.-přir.K.ceské Spol.Náuk 1885-1892.
  Formerly Abhandlungen der Königlichen Böhmischen
  Gesellschaft der Wissenschaften. Prag.   S. 1703 C

Rozpravy Ustředního Ustavu Geologického. Praha.
  Rozpr.ústred.Ust.geol. Vol.13 → 1950 →
  Formerly Rozpravy Státniho Geologického Ustavu
  Ceskoslovenské Republiky. Praha.   P.S 1355

Rozpravy Vědecké Společnosti Badatelské pří Ruské Svobodné
  Université v Praze.
  See Zapiski Nauchno-Izsledovatel'skoe Ob'edineniya
  Risskii Svobodnyĭ Uníversitet v Prage.   S. 1762 a

Rozpravy Vědecké Společnosti při Ruské Svobodné Univesitě v Praze.
  See Zapiski. Nauchno-Issledovatel'skoe Ob'edinenie, Russkiĭ
  Svobodnyĭ Universitet v Prage.   S. 1762 a

Rozprawy Akademia Umiejetnosci. w Krakowie. Wydzial
  Matematyczno-Przyrodniczy.
  Rozpr.Akad.Umiejet.   1874-1902.
  Continued as Rozprawy Wydzialu Matematyczno-Przyrodniczego
  Polskiej Akademii Umiejetności. Krakow.   S. 1725 B

Rozprawy i Sprawozdania. Instytut Badawczy Lasów Państwowych
  w Warszawie.
  Rozpr.Spraw.Inst.badaw.Las.państ. 1934-1939.
  Formerly Razprawy i Sprawozdania. Zakład Doświadczalny
  Lasów Państwowych w Warszawie.
  Continued as Rozprawy i Sprawozdania Instytut Badawczy
  Lésnictwa Kraków.   S. 1874 A

Rozprawy i Sprawozdania. Instytut Badawczy Lésnictwa. Kraków.
  Rozpr.Spraw.Inst.badaw.Lésn. 1946-1949.
  Formerly Rozprawy i Sprawozdania. Instytut Badawczy Lasów
  Państwowych w Warszawie.   S. 1874 A

Rozprawy i Sprawozdania z Posiedzen Wydziału Matematyczno-
  Przyrodniczego Akademii Umiejetnosci. w Krakowie.
  See Rozprawy Akademia Umiejetnosci. w Krakowie.   S. 1725 B

Rozprawy i Sprawozdania. Zakład Doświadczalny Lasów
  Państwowych w Warszawie.
  Rozpr.Spraw.Zakł.doświad.Las.państ.Warsz. Ser.A. 1933-1934.
  Continued as Rozprawy i Sprawozdania. Instytut Badawczy
  Lasów Państwowych w Warszawie.   S. 1874 A

| TITLE | SERIAL No. |

Rozprawy i Wiadomości z Muzeum im. Dzieduszyckich Lwow.
    Rozpr.Wiad.Muz.Dziedusz. 1915-1924.   S. 1791 B

Rozprawy Wydziału Matematyczno-Przyrodniczego Polskiej Akademii
    Umiejetnosci. Krakow.
    Rozpr.Wydz.mat.-przyr.pol.Akad.Umiejet. 1902-1938.
    Formerly Rozprawy Akademia Umiejetnosci.w Krakowie. Wydzial
    Matematyczno-Przyrodniczy.   S. 1725 B

Runa. Minnesblåd från Nordiska Museet. Stockholm.
    Runa. 1888.   S. 583 D

Russian Biological Journal.
    See Biologicheskii Zhurnal. Moskva.   S. 1845 a

Russian Review of Biology. Translated from Uspekhi Sovremennoĭ
    Biologii. London & Edinburgh.
    Russ.Rev.Biol. vol.48-50, No.3. 1959-1961.   S. 1802 a B

Russian Technical Literature. Organisation for Economic
    Co-operation and Development. Paris.
    Russ.tech.Lit. No.6-12. 1962-1963.
    Continued as Science East to West. Paris.   S. 2713 a

Russische Hydrobiologische Zeitschrift.
    See Russkiĭ Gidrobiologicheskiĭ Zhurnal.   Z.S 1840 A

Russkaya Bibliografiya po Estestvoznaniyu i Matematikê.
    Sanktpeterburg.
    Russk.Biblfiya Estest.Mat. 1901-1902 (1904-1906.)   R. 72 Q.o.S

Russkaya Geologicheskaya Biblioteka. S.-Peterburg.
    See Izvestiya Geologicheskago Komiteta.   P.S 1506

Russkiĭ Arkhiv Protistologii. Moskva.
    Russk.Arkh.Protist. 1925-1928.
    Formerly Arkhiv Russkago Protistologicheskago
    Obshchestva. Moskva.   S. 1846 a

Russkii Botanicheskii Zhurnal. S.-Peterburg.
    Russk.bot.Zh. 1909-1915.   B.S 1404

Russkiĭ Gidrobiologicheskiĭ Zhurnal Izdavaemȳi pri Volzhskoĭ
    Biologicheskoi Stantsii. Saratov.
    Russk.gidrobiol.Zh. 1921-1930.
    Continued as Gidrobiologicheskii Zhurnal S.S.S.R.   Z.S 1840 A

Russkiĭ Zoologicheskiĭ Zhurnal. Moskva.
    Russk.zool.Zh. 1916-1930.
    Continued as Zoologicheskiĭ Zhurnal.   T.R.S 2030 & Z.S 1850

TITLE	SERIAL No.

Russkoe Entomologicheskoe Obozrênīe. S.-Peterburg.
    <u>Russk.ent.Obozr.</u> 1901-1930. (Suspended 1918-1921).
    <u>Continued as</u> Entomologicheskoe Obozrênīe.   E.S 1804

Rüsta- ja Kalatalouden Tutkimuslaitoksen Kalantutkimusosaston
    Tiedonantoja. Helsinki.
    <u>See</u> Tiedonantoja. Rüsta- ja Kalatalouden Tutkimuslaitos
    Kalantutkimusosasto.   Z.S 1866

Ruwenzori Expedition 1934-35. British Museum (Natural History).
    London.
    <u>Ruwenzori Exped. 1934-35</u>. 1939-1957.   E. 74 o BM & BM.Ae.o

Ruwenzori Expedition 1952. British Museum (Natural History).
    London.
    <u>Ruwenzori Exped. 1952.</u> 1958 →   E. 74 o BM & BM.Ae.o

Rȳbnoe Khozyaĭstvo Karelii (Sbornik Stateĭ). Leningrad.
    <u>Rȳb.Khoz.Karel</u>. 1932.   Z.S 1857

Rybnoe Khozyaĭstvo. Ministerstvo Rybnogo Khozyaĭstva SSSR. Kiev.
    <u>Ryb.Khoz.Kiev</u> Vyp. 2 → 1965 → (imp.)   Z.S 1816 a A

Rybnoe Khozyaĭstvo Volgo-Kaspiĭskogo Raĭona. Astrakhan.
    <u>Ryb.Khoz.volgo-kasp.Raĭona</u> No.1-12, 1931.   Z.S 1830 A

S.A.G. A Magazine of Outdoor Interests. Cape Town.
    <u>S.A.G.</u> vol.27-30 1936-1940
    <u>Formerly</u> South African Country Life. Cape Town.   B.S 2266

SAMAB
    <u>See</u> Bulletin. South African Museums Association.   S. 2013 a

SCAR Bulletin. International Council of Scientific Unions.
    <u>SCAR Bull</u>. 1959 →
        (Reprinted from the Polar Record: S.6 b.)   S. 2701 B

SHILAP. Revista de Lepidopterologia. Madrid.
    <u>See</u> Revista de Lepidopterologia.   E.S 1008

SWANEWS. Southwestern Association of Naturalists.
    (University of Tulsa) Oklahoma.
    <u>SWANEWS</u> 1966 →   S. 2463 B

S.W.A.N.S. Department of Fisheries and Fauna. Perth, W.A.
    <u>S.W.A.N.S.</u> 1970 →   S. 2157 a

S.W.G. Tabellenserie. Hoogwoud.
    <u>S.W.G.Tabellenser</u>. No.21, 1966.
    <u>Continued as</u> Tabellenserie van de Strandwerkgemeenschap.
    Hoogwoud.   S. 700 H

TITLE	SERIAL No.

Sabah Forest Record. Sandakan.
    Sabah Forest Rec. 1964 →
    Formerly North Borneo Forest Records. Sandakan.   B.S 1827

Sabah Society Journal. Jesselton.
    Sabah Soc.J. 1961 →   S. 1972 b

Sabouraudia. Edinburgh & London.
    Sabouraudia 1 → 1961 →   B.M.S 105

Saccardoa. Pavia.
    Saccardoa 1960 →   B.S.M 41

Saggi Scientifici e Litterarj dell'Accademia di Padova.
    Saggi sci.Accad.Padova 1786-1794.
    Continued as Memorie dell'Accademia di Scienze, Lettere
    ed Arti di Padova.   S. 1154 B

Saint-Hubert, La. Revue Mensuelle du Saint-Hubert Club
    de France. Paris.
    St.Hubert 1950-1967 (imp.)   Z.S 815

St. Paul's Review of Arts & Sciences. Department of General
    Education. St. Paul's University. Tokyo.
    St.Paul's Rev.Arts Sci.
    Natural Science. (Vol.1) No.1, 7-10, 1956-1961.
    Continued as St. Paul's Review of Science. College of
    General Education, St. Paul's University. Tokyo.   S. 1992 d

St. Paul's Review of Science. College of General Education,
    St. Paul's University. Tokyo.
    St.Paul's Rev.Sci. 1962 →
    Formerly St. Paul's Review of Arts & Sciences. Department of
    General Education. St. Paul's University. Tokyo.   S. 1992 d

Salamandra. Zeitschrift für Herpetologie und Terrarienkunde.
    Deutsche Gesellschaft für Herpetologie und Terrarienkunde.
    Frankfurt am Main.
    Salamandra 1965 →   Z. Reptile Section

Salmon Fisheries. Fisheries Board for Scotland. Edinburgh.
    Salm.Fish.Edinb. 1910-1915, 1919-1939, 1948, No.1.   Z.S 380 C

Salmon and Trout Magazine. Journal of the Salmon and Trout
    Association. London.
    Salm.Trout Mag. 1911 →
    Formerly Journal of the Salmon & Trout Association.   Z.S 323

Salt Water Aquarium. Miami.
    Salt Wat.Aquar. 1965 →   Z.S 2554

Samaru Agricultural Newsletter. Samaru.
    Samaru agric.Newsl. Vol.10 → 1968 →   E.S 2174 F

| TITLE | SERIAL No. |

Samaru Miscellaneous Papers. Institute for Agricultural Research.
Ahmadu Bello University. Samaru, Zaria.
    Samaru misc.Pap. No.3 → 1964 → (imp.)   S. 2035 C

Samaru Research Bulletin. Kaduna.
    Samaru Res.Bull. 1960 →   S. 2035

Samling af Afhandlinger E Museo Lundii. Kjøbenhavn.
    Saml.Afh.Mus.Lundii 1888-1915.   S. 511 A

Samling af Rön och Uptäkter, Götheborg.
    Saml.Rön Uptäkter 1781.   S. 545

Sammlung der Deutschen Abhandlungen welche in der Königlichen
Akademie der Wissenschaften zu Berlin vorgelesen worden in
den Jahren 1788-1803.
    Samml.Abh.Akad.Wiss.Berlin 1788-1806.   S. 1305 D

Sammlung Gemeinnütziger Vorträge. Deutscher Verein zur
Verbreitung Gemeinnütziger Kenntnisse. Prag.
    Samml.gemeinnütz.Vortr.Prag 1921-1925 (imp.)   S. 1759

Sammlung des Geologischen Reichsmuseums in Leiden.
    Samml.geol.Reichsmus.Leiden 1881-1923.
    Continued as Leidsche Geologische Mededelingen.   P.S 290 & P.S 291

Sammlung Geologischer Führer. Berlin.
    Samml.geol.Führ. 1897 →   P.S 344 a

Sammlung des Kaukasischen Museums.
    See Kollektsii Kavkazkago Muzeya. Tiflis.   S. 1858 C

Sammlung Naturhistorischer und Physikalischer Aufsäze.
F. von Paula von Schrank. Nürnberg.
    Samml.naturh.u.phys.Aufsäze 1796   7.o.S

Sammlung Palaeontologischer Abhandlungen. Cassel.
    Samml.palaeont.Abh.Cassel 1883.   P.S 331

Sammlung Physikalisch-Okonomischer Aufsätze. F.W. Schmidt. Prag.
    Samml.phys.-ökon.Aufsätze 1795   B. 58 SCH-

Sammlung Verschiedener Schriften. Halle.
    See D. Daniel Gottfried Schrebers Sammlung Verschiedener
Schriften. Halle.   7.o.S

Sammlung Zoologischer Feldführer. Frankfurt-am-Main.
    Samml.zool.Feldführ. 1951-1952.   Z. 72 o D

Sammlungen des Kaukasischen Museums, Tiflis.
    See Kollektsii Kavkazskago Muzeya. Tiflis.   S. 1858 F

San Carlos Publications. University of San Carlos, Cebu City.
    San Carlos Publs Series B, 1964 →   S. 1974 a A

TITLE	SERIAL No.

Sandoz Bulletin. Basle.
    Sandoz Bull. 1965 → (imp.)   P.S 704

Sands, Clays and Minerals. Chatteris.
    Sands Clays Miner. 1932-1939   M.S 131

Sanitary Injurious Insects. (Tokyo)
    Sanit.injur.Insects 1955 →   E.S 1925 a

Sanitation Supplements of the Tropical Diseases Bulletin. London.
    Sanitn Suppl.trop.Dis.Bull. 1921-1925.   Z.S 49 C

Sapporo Bulletin of the Botanic Garden. Hokkaido University. Sapporo.
    Sapporo Bull.bot.Gdn Hokkaido 1963 →   B.S 1960 a

Sarawak Museum Journal. Sarawak.
    Sarawak Mus.J. 1911 →   S. 1974 A
                   1911 → (imp.)   T.R.S 7502

Sarawak Papers. Scientific Results of the Oxford University Expedition to Sarawak (Borneo) in 1932. London.
    Sarawak Pap. 1952.   77 Ad.o.O

Sargentia. Jamaica Plain, Mass.
    Sargentia 1942-1949.
    Formerly Contributions to the Arnold Arboretum of Harvard University. Cambridge, Mass.   B.S 4238 a

Sargetia. Deva.
    Sargetia Series Scientia Naturae Vol.8 → 1971 →   S. 1899 c

Sarracenia. Montreal.
    Sarracenia 1959 →   B.S 4510 a

Sarsia. Universitetet i Bergen.
    Sarsia 1961 →   S. 552 F
           No.51 → 1972 →   Z.S 542

Saugar University Journal.
    Saugar Univ.J. 1951-1956.
    Continued as Journal of the University of Saugar.   S. 1943 a

Säugetierkundliche Mitteilungen. Stuttgart.
    Säugetierk.Mitt. 1953 →   Z. Mammal Section

Säugetierschutz. Zeitschrift für Theriophylaxe. Hohenbüchen.
    Säugetierschutz 1970 →   Z. Mammal Section

Saussurea. Genève.
    Saussurea 1970 →
    Formerly Travaux de la Société Botanique de Genève.   B.S 822

Sauvagine, La. (Les Oiseaux Migrateurs.) Noyelles-sur-Mer.
    Sauvagine No.54-64, 1950-1952.
    Merged in Revue Nationale de la Chasse (et La Sauvagine). Paris.   Z.S 817

TITLE	SERIAL No.

Savanna Research Series. McGill University Savanna Research
Project. Montreal.
Savanna Res.Ser. 1964 →   S. 2667

Savaria. Savaria Muzeum. Szombathely.
Savaria 1963 →   S. 1715 B

Savon Luonto. Kuopio.
Savon Luonto 1974 →
Formerly Savonia. Kuopio.   S. 1829 a D

Savonia. Kuopio.
Savonia 1972.
Formerly Kuopion Luonnon Ystäväin Yhdistyksen Julkaisuja.
Kuopio.
Continued as Savon Luonto. Kuopio.   S. 1829 a D

Sbírka Prací Výzkumných Ustavú. Prague.
Sbír.Prací Výzkumn. Ust. 1954-56   M.S 1730

Sbornik Biologicheskogo Fakul'teta. Odesskiĭ Gosudarstvennyi
Universitet im. I.I. Mechnikova. Kiev.
Sb.biol.Fak.odessk.gos.Univ. Tom 6, 1953.   S. 1847 b

Sbornik Ceskoslovenské Akademie Zemědělských Věd. Praha.
Sb.čsl.Akad.zeměd.Věd. Rostlinná Výroba, Roč.7 (34),
Cis.7, 1961.
Continued as Rostlinná Výroba. Praha.   B.S 1239 c

Sbornik Dokladov Ornitologicheskoĭ Konferentsii. Riga.
Sb.Dokl.Orn.Konf. 1951 (1953).
Continued as Trudȳ Pribaltiĭskoĭ Ornitologicheskoĭ
Konferentsii.   Z. 72Q q C

Sborník Entomologického Oddělení Národního Musea v Praze.
Sb.ent.Odd.nár.Mus.Praze 1923-1965.
Continued as Acta Entomologica Musei Nationalis Pragae.   E.S 1723

Sborník Fakulty Hospodarské Vysoké Zemědelské. Brno.
See Sbornik Vysoké Skoly Zemědelské v Brné. Brno.
Fakulta Hospodarská.   S. 1710 a A

Sborník Fakulty Lesnické Vysoké Skoly Zemědelské. Brno. Sign. D.
See Sbornik Vysoké Skoly Zemědelské v Brné. Brno.
Fakulta Lesnická.   S. 1710 a D

Sborník Faunistických Prací Entomologického Oddeleni Národního
Musea v Praze. Praha.
Sb.faun.Prací ent.Odd.nár.Mus.Praze 1956 →   E.S 1723 a
  Supplementum 1959 →   E.S 1723 c

Sborník Geologickýck Věd. Praha.
Sb.geol.Věd.Praha Rada A. Antropozoikum 1963 →   P.A.S 77
Formerly Anthropozoikum. Praha.
Rada G. Geologie, Rada P. Paleontologie 1963 →   P.S 1351
Formerly Sborník Ustredního Ustavu Geologického. Praha.

| TITLE | SERIAL No. |

Sborník Geologických Vied. Rad.Zk. Západné Karpaty Praha.
  Sb.geol.Vied Praha 1964 →
  Formerly Geologické Práce Bratislava.   P.S 1351 A

Sbornik. Institut Botaniki. Akademiya Nauk Gruzinskoĭ SSR. Tbilisi.
  Sb.Inst.Bot.Tbilisi   1965 →   B.S 1430 d

Sbornik. Institut Zoologii. Akademiya Nauk Gruzinskoĭ SSR.
  Tbilisi.
  Sb.Inst.zool.Akad.Nauk Gruz. SSR 1966 →   Z.S 1837 A

Sbornik Izdan. Imperatorskim S. Peterburgskim Mineralogicheskim
  Obshchestvom. St. Petersburg.
  Sb.Izdan.Imper.S.Peterburg.Miner.Obshch.   1867.   M.S 1818 A

Sborník Jihočeského Muzea v Ceských Budějovicích.
  Sb.Jihočesk.Muz.Ceských Budějovicích Přír.Vědy 1964-1972.
  (Wanting roč.5, čís.2; roč.6, čís.1; roč.7, čís.1 & 2)
  Formerly Sborník Krajského Vlastivědného Musea v
  Ceských Budějovicích.
  Continued as Přírodovědecký Casopis Jihočeský.
  České Budějovice.   S. 1745

Sborník Klubu Přírodovédeckého v Brně.
  Sb.Klubu přír.Brně   1913-1961.   S. 1714 B

Sborník Klubu Přírodovedéckého v Praze.
  Sb.Klubu přír.Praze 1911-1924.   S. 1706

Sborník Krajského Múzea v Trnave. Trnava.
  Sb.Krajsk.Muz.Trnave 1955-1956.
  Continued as Prírodovedné Práce Slovenských Múzeí. Trnava. S. 1754 a A

Sborník Krajského Vlastivědného Musea v Ceských Budějovicích.
  Sb.Krajsk.Vlast.Mus.Ceských Budějovicích Přír, Vědy 1958-1961.
  Continued as Sbornik Jihočeského Muzea v Ceských
  Budějovicích.   S. 1745

Sborník Krajského Vlastivědného Musea v Olomouci.
  Oddíl A, Přírodní Vědy.
  Sb.Krajsk.Vlast.Mus.A.Přír.Vědy Vol.3 1955.
  Formerly Sborník SLUKO. Olomouc.A-Přír.Vědy.
  Continued as Sborník Vlastivědného Ustavu v Olomouci.
  Oddíl A, Přír. Vědy.   S. 1748 B

Sbornik. Krymskoe Obshchestvo Estestvoispytateleĭ i
  Lyubiteleĭ Prirody. Simferopol.
  Sb.Krymsk.Obshch.Estest.Simferopol   1914-1918.   S. 1813 B

Sbornik. Národního Musea v Praze. Praha. Rada B. Prírodni Vědy.
  Sb.nar.mus.Praze 1938 →   S. 1761 a A & T.R.S 1852

| TITLE | SERIAL No. |

Sborník Nauchno-Issledovatel'skogo Instituta Zoologii Moskovskogo
Gosudarstvennogo Universiteta im. M.N. Pokrovskogo.
Moskva, Leningrad.
  Sb.nauchno-issled.Inst.Zool.mosk.gos.Univ. No.3. 1936.
  Formerly Byulleten'Nauchno-Issledovatel'skogo Instituta
  Zoologii Moskovskogo Gosudarstvennogo Universiteta.   Z.S 1828 B

Sbornik Nauchnykh Rabot. Nauchno-Issledovatel'skiĭ Sektor,
  Kievskiĭ Universitet. Kiev.
  Sb.nauch.Rabot.nauchno-issled.Sektor kiev.Univ. No.6 → 1970 → P.S 531 GA

Sbornik Nauchnȳkh Rabot Studentov Saratovskogo Gosudarstvennogo
  Universiteta. Saratov.
  Sb.Nauch.Rabot Student.Saratov 1938.   S. 1879 a B

Sbornik Nauchnȳkh Statei. Vilnius.
  See Straipsniu Rinkinys. Vilnius.   B.S 1516

Sbornik Nauchnȳkh Trudov. Gosudarstvennȳi Ordena Trudovogo
  Krasnogo Znameni Nikitskiĭ Botanicheskiĭ Sad.
  Sb.nauch.Trud.gos.nikit.bot.Sad Tom.37, 1964.
  Formerly Trudȳ Gosudarstvennogo Nikitskogo Botanicheskogo
  Sada. Yalta.
  Continued as Nauchnȳe Trudȳ. Gosudarstvennȳĭ Ordena
  Trudovogo Krasnogo Znameni Nikitskiĭ Botanicheskiĭ
  Sad. Moskva.   B.S 1445

Sbornik Nauchnȳkh Trudov. Institut Biologii. Akademiya
  Nauk Belorusskoĭ SSR. Minsk.
  Sb.nauch.Trud.Inst.Biol.Minsk 1950-1952.   S. 1814 a A

Sbornik Pedagogického Inštitútu v Trnave. Bratislava.
  Sb.pedag.Inst.Trnave
    Prírodné Vedy. 1963.
  Continued as Zborník Pedagogickej Fakulty Univerzity
  Komenského v Bratislave. Prírodné Vedy.   S. 1740 B

Sborník Prác Prirodoveckej Fakulty Slovenskej Univerzity v
  Bratislave. Prace Zoologického Ustavu.
  Sb.Prać prir.Fak.slov.Univ.Bratisl. 1-6, 1942-1944.   Z.S 1886

Sborník Prací Pedagogické Fakulty v Ostravě.
  Sb.Prací pedagog.Fak.Ostrave
    Přírodní Vědy (Series E) 1972 →   S. 1758 c

Sborník Prací Přírodovědecké Fakulty Palackého University
  v Olomouci.
  Sb.Prací přír.Fak.palack.Univ.Olomouci 1960 →   S. 1744

Sborník Přírodovědecké Společnosti v Mor. Ostravě. Mor. Ostrava.
  Sb.přír.Spol.Mor.Ostravé 1921-1935.
  Continued as Přírodovědecký Sborník Ostravského
  Kraje. Opava.   S. 1714 a

| TITLE | SERIAL No. |

Sborník Přírodovědeckého Klubu v Třebíči.
    Sb.přír.Klubu Třebíči 1936-1954.   S. 1714 b

Sborník Přírodovědecký. Ceské Akademie v Praze.
    Sb.přír.čes.Akad.Praze 1925-1930.   S. 1760 H

Sbornik Rabot po Ikhtiologii i Gidrobiologii. Akademiya Nauk
    Kazakhskoĭ SSR. Alma-Ata.
    Sb.Rab.Ikhtiol.Gidrobiol.Alma-Ata 1956-1961.   Z.S 1813

Sborník Severočeského Musea, Liberec. Přírodní Vědy.
    Historia Naturalis.
    Sb.severočesk.Mus. Přírodní vědy. 1958 →   S. 1769 b

Sborník Slovenského Národného Múzea. Prírodné Vedy. Bratislavá.
    Sb.slov.narod.Muz. Prírodné Vedy 1962-1967.
    Formerly Prírodovedný Sborník Slovenského Múzea.
    Bratislavá.
    Continued as Zborník Slovenského Narodného Múzea.
    Bratislava.   S. 1754 a A

Sborník SLUKO. Olomouc. A-Přírodní Vědy.
    Sb.SLUKO 1951-1954.
    Continued as Sbornik Krajského Vlastivědného Musea
    v Olomouci Oddíl A, Přírodni Vědy.   S. 1748 B

Sbornik Stateĭ. Biologicheskie Resursy Dal'nego Vostoka. Moskva.
    Sborn.Stat.Biol.Res.Dal.Vost. 1959.   S. 73 A.o.L

Sbornik Statei po Geologii Arktiki. Leningrad.
    Published in Trudy Nauchno-Issledovatel'skogo Instituta Geologii
    Arktiki. Leningrad.   P.S 1553

Sbornik Stateĭ. Karel'skaya Nauchno-Issledovatel'skaya
    Rybokhozyaĭstvennaya Stantsiya.
    See Rybnoe Khozyaĭstvo Karelii.   Z.S 1857

Sborník Statei po Paleontologii i Biostratigrafii. Leningrad.
    Sb.Stat.Paleont.Biostratigr. Vȳp. 4-17, 1957-1959.
    (Wanting Vȳp. 13 & 16)   P.S 1553 B

Sbornik Statisticheskikh Svedenii po Gornoi i Gornozavodskoi
    Promyshlennosti SSSR. Leningrad.
    Sb.Stat.Sved.Gorn.Gornozav.Promyshlen.SSSR. 1924-1928.   M.S 1813

Sborník Státneho Banského Muzea Dionýza Stúra. Banská Stiavnica.
    Sb.st.bansk.Muz.Dionýza Stúra 1927-1937.   M.S 1731

Sborník Státních Výzkumných Ustavu Lesnických CSR. V Brne.
    Sb.st.vyzk.Úst.les.Brně Svazek 1, 1948.   Z.S 1758

Sborník Státního Geologického Ustavu Ceskoslovenské Republiky.
    v Praze.
    Sb.st.geol.Ust.čsl.Repub. 1921-1950
    Continued as Sborník Ustredniho Ustavu Geologického.Praha.   P.S 1351

| TITLE | SERIAL No. |

Sbornik Trudov Chlenov Petrovskago Obshchestva Izsledovateleĭ
Astrakhanskago Kraya. Astrakhan.
Sb.Obshch.Izsled.Astrakhan 1892.   S. 1807

Sbornik Trudov Gosudarstvennogo Zoologicheskogo Muzeya
(pri MGU). Moskva.
Sb.Trud.gos.zool.Muz. 1934-1951.   Z.S 1828 A
                     1934-1936.   T.R.S 2005
Continued as Sbornik Trudov Zoologicheskogo Muzeya MGU.

Sbornik Trudov.Institut Geologii i Mineralogii. Akademiya
Nauk Gruzinskoĭ SSR. Tbilisi.
Sb.Trud.Inst.Geol.Min.Akad.Nauk Gruz.SSR 1951 →   P.S 557 A

Sbornik Trudove na Mladite Nauchni Rabotnitsi i Studenti. Plovdiv.
Sb.Trud.mlad.nauch.Rabot.Student. 1973 →   S. 1897 a C

Sbornik Trudov Professorov i Prepodavateleĭ
Gosudarstvennogo Irkutskogo Universiteta. Irkutsk.
Sb.Trud.Prof.Prepod.gos.irkutsk.Univ. 1921-1931.   S. 1831 A

Sbornik Trudov Zoologicheskogo Muzeya MGU. Moskva.
Sb.Trud.zool.Muz.MGU 1961 →
Formerly Sbornik Trudov Gosudarstvennogo Zoologicheskogo
Muzeya (pri MGU).   Z.S 1828 A

Sborník ÚVTI. Ústav Vedeckotechnických Informací. Praha.
Sb.ÚVTI Genetika a Šlechtění, Roc.4 (41)→1968 →   S. 1760 a

Sborník Ustredního Ustavu Geologického. Praha.
Sb.Ustred.Ust.Geol. 1951-1963.
Formerly Sbornik Státního Geologického Ustavu Československé
Republiky v Praze.
Continued as Sbornik Geologických Věd. Rada G.
Geologie.   P.S 1351

Sbornik. Vedeckeho Lesnickeho Ustavu Vysoke Skoly Zemedelske
v Praze. Prague.
Sb.ved.lesn.Ust.vys.Sk.zemed.Praze 17 → 1974 →   S. 1755 a

Sborník Vlastivědného Muzea v Olomouci. Oddíl A, Prírodní Vědy.
Sb.Vlast.Muz.Olomouc.A.Přír.Vědy. Vol.5. 1962.
Formerly Sborník Vlastivědného Ustavu v Olomouci.
A.Přír. Vědy.   S. 1748 B

Sborník Vlastivědného Ustavu v Olomouci. Oddíl A, Přírodní Vědy.
Sb.Vlast.Ust.Olomouc.A,Přír.Vědy. Vol.4. 1959.
Formerly Sborník Krajskeho Vlastivědného Musea v Olomouci.
A. Přír. Vědy.
Continued as Sborník Vlastivědného Muzea v Olomouci.
A.Přír. Vědy.   S. 1748 B

Sbornik. Vostochnosibirskiĭ Geologo-Razvedochnyĭ Trest. Irkutsk.
Sb.Vost.geol.-Razv.Trest 1933.   P.S 1531 A

| TITLE | SERIAL No. |

Sborník Východoslovenského Múzea. Košice.
   Sb.východoslov.Múz.Košice Séria A, 1960-1966.
                        Séria B, 1965.
   Continued as Zbornik Východoslovenského Múzea. Košice.   S. 1763 d A-B

Sborník Vysoké Skoly Chemicko-Technologické v Praze. Oddil
   Fakulty Technologie Paliv a Vody.
   Sb.vys.Sk.chem-technol.Praze 1957 →   S. 1759 a

Sborník Vysoké Skoly Zemědelské v Brne. Brno.
   Sb.vys.Sk.zeměd.Brné 1924-1949.
   Fakulta Lesnická & Fakulta Hospodářská.
   Continued as Sborník Vysoké Skoly Zemědelské a
   Lesnické Fakulty v Brne.   S. 1710 a A-C

Sborník Vysoké Skoly Zemědelské a Lesnické Fakulty v Brne. Brno.
   Sb.vys.Sk.zeměd.les.Fak.Brné 1951-1952.
   Then in series.
   A. Spisy Fakulty Agronomické a Zootechnické 1956-1959:
   Spisy Fytotech. Zootech a Ekon. 1960-1963:
   Spisy Fakulty Agronomické, 1964-1966.
   B. Spisy Fakulty Veterinarní. 1953-1966.
   C. Spisy Fakulty Lesnické. 1953-1966.
   Formerly Sborník Vysoké Skoly Zemedelská v Brné. Brno.
   Fakulta Hospodarská & Fakulta Lesnická.
   Continued as Acta Universitatis Agriculturae, Brno.   S. 1710 a D-F

Sbornik. Západočeské Muzeum v Plzni. Plzen.
   Sb.Zapadoceské Muz.Plzni   Priroda. No.2 → 1968 →   S. 1703 a

Sborník Zoologického Oddělení Národního Musea v Praze.
   Sb.zool.Odd.nár.Mus.Praze   1934.   Z.S 1790

Scanning Electron Microscopy. Chicago.
   Scanning Electron Microsc. 1968 →   E.M. UNIT

Scar Bulletin. Special Committee on Antarctic Research.
   See SCAR Bulletin. International Council of Scientific
   Unions.   S. 2701 B

Scenic Trips to the Geologic Past. Santa Fé, Socorro.
   Scenic Trips geol.Past 1955 →   P.S 1956 D

Schedae ad Floram Exsiccatam Austro-Hungaricam a Museo
   Botanico Universitatis Vindobonensis.   B.S 581.9(439.1):
   Sched.Flor.exsicc.Austro-Hung. 1881-1913.   579 KER

Schedae ad Floram Exsiccatam Carniolicam. Laibach.
   Sched.Flor.exsicc.Carniol. 1901 →   B.S 581.9(43):579 PAU

Schedae ad Floram Hungaricam Exsiccatam a Sectione Botanica
   Musei Nationalis Hungarici Editam. Budapest.
   Sched.Flor.hung.exsicc. 1912-1932.   B.S 581.9(439.1):579

| TITLE | SERIAL No. |

Schedae ad Floram Ibericam Selectam. Barcelona.
    Sched.Flor.Iber.Sel. 1934.   B.S 581.9(46):579

Schedae ad Floram Romaniae Exsiccatam a Museo Botanico
    Universtaties Clusiensis Editam. Cluj.
    Sched.Flor.Roman. 1921-1936.   B.S 1520

Schedae ad Floram Stiriacam Exsiccatam. Vienna.
    Sched.Flor.Stiriac.exsicc. 1904-1912.   B.S 581.9(436):579 HAY

Schedae ad Herbarium Florae Reipublicae Sowjeticae Ucrainicae. Kiev.
    See Spisok Roslin Gerbari Flori USRR.   B.S 581.9(477):579 I.B

Schedae ad Herbarium Florae Rossicae.
    See Spisok Rastenii Gerbariya Russkoĭ Flory.
    Petropoli.   B. Eur.(581.9(47):579 AN

Schedae ad Herbarium Florae URSS.
    See Spisok Rastenii Gerbariya Flory SSSR. Moskva.
           B. Eur.(581.9(47):579 AN

Schedae ad Herbarium 'Plantae Orientales Exsiccatae'. Tiflis.
    Sched.Herb.Pl.orient.exsicc. 1924-1928.   B.S 581.9(479.22):579 GRO

Schedae Plantae Finlandiae Exsiccatae. Helsingfors.
    Sched.Pl.Finland.exsicc. 1906-1916.   B.S 581.9(480):579 B.M

Schedulae Orchidianae. Boston, Mass.
    Schedul.orchid. 1922-1930.   B.S 4237

Schlern-Schriften. Herausgegeben von R. Klebelsberg. Innsbruck.
    Schlern-Schriften No. 188. 1958.   72 M.o.J

School House Natural History Society Report. Dauntsey's School,
    West Lavington.
    Sch.Hse nat.Hist.Soc.Rep.Dauntsey's Sch. 1943-1951. (imp.)
    Continued as Report,Dauntsey's School Natural
    History Society.   S. 393 a B

School of Mines Quarterly. Columbia University. New York.
    See Columbia University School of Mines Quarterly. New York.   M.S 2691

School Nature Study. London.
    Sch.Nat.Study 1913-1924 (imp.) 1954-1962.
    Continued as Natural Science in Schools. The Journal
    of the School Natural Science Society (School Nature Study
    Union). London.   S. 473 a A

School Service Series. Department of Education. American Museum
    of Natural History. New York.
    Sch.Serv.Ser.Am.Mus.nat.Hist. 1929-1937.   S. 2356 T

TITLE	SERIAL No.

Schriften des Arbeitskreises für Naturwissenschaftliche
   Heimatforschung in Wedel. (Holst.) im Verkehrs- und
   Heimatverein Wedel und Umgebung e.V.
     Schr.ArbKreis.naturw.Heimatforsch.Wedel 1965 →   S. 1597

Schriften der Berlinischen Gesellschaft Naturforschender
   Freunde. Berlin.
     Schr.berl.Ges.naturf.Fr.Berl. 1780-1794.   T.R.S 1336 A & S. 1326 A
                            1780-1789.   Z. Tweeddale
     Formerly Beschäftigungen der Berlinischen Gesellschaft
     Naturforschender Freunde.
     Continued as Neue Schriften. Gesellschaft Naturforschender
     Freunde zu Berlin.

Schriften der Elsass-Lothringischen Wissenschaftlichen
   Gesellschaft. Heidelberg. Reihe A. Alsatica & Lotharingica.
     Schr.els.-loth.wiss.Ges. 1927-1938.   S. 1455

Schriften aus dem Geologisch-Paläontologischen Institut der
   Universität Kiel. Kiel.
     Schr.geol.paläont.Inst.Univ.Kiel 1933-1937.   P.S 326

Schriften der Gesellschaft zur Beförderung der Gesamten
   Naturwissenschaften zu Marburg.
     Schr.Ges.Beförd.ges.Naturw.Marburg 1823-1935.   S. 1531

Schriften der Herzoglichen Societät für die Gesammte
   Mineralogie zu Jena.
     See Annalen der Herzoglichen Societät für Die Gesammte
     Mineralogie zu Jena.   M.S 1421

Schriften der Königsberger Gelehrten Gesellschaft.
   Naturwissenschaftliche Klasse.
   Schr.königsb.gelehrt.Ges.naturw.Kl.
     Jahr.4, Hft.8; Jahr 15, Hft.1;
     Jahr.19, Hft.1. 1927, 1939-1944.   S. 1486 a

Schriften der Naturforschenden Gesellschaft in Danzig.
     Schr.naturf.Ges.Danzig 1863-1938.
     Formerly Neueste Schriften der Naturforschenden Gesellschaft
     in Danzig.   S. 1384 B

Schriften der Naturforschenden Gesellschaft zu Kopenhagen.
     Schr.naturf.Ges.Kopenhagen 1793.   S. 521 B

Schriften der Naturforschenden Gesellschaft zu Leipzig.
     Schr.naturf.Ges.Leipzig 1822.   S. 1516 A

Schriften der Naturforscher-Gesellschaft bei der Universität
   Jurjeff (Dorpat, Tartu).
     Schr.NaturfGes.Univ.Jurjeff 1884-1925.   S. 1816 D

| TITLE | SERIAL No. |

Schriften des Naturwissenschaften Vereins des Harzes, Wernigerode.
    Schr.naturw.Ver.Harzes 1886-1895.   S. 1598 B

Schriften des Naturwissenschaftlichen Vereins für Schleswig-
    Holstein. Kiel.
    Schr.naturw.Ver.Schlesw.-Holst. 1873 →   S. 1471

Schriften der (Königlichen) Physikalisch-Okonomischen
    Gesellschaft zu Königsberg.
    Schr.phys.-ökon.Ges.Königsb. 1860-1939.   S. 1486 B

Schriften der Reichszentrale für Pelztier- und Rauchwaren-
    Forschung. Leipzig.
    Schr.Reichszent.Pelztier-u.Rauchwar.-Forsch. 1926-1927.   Z.S 1375 A

Schriften der in St. Petersburg Gestifteten Russisch-Kaiserlichen
    Gesellschaft für die Gesammte Mineralogie. St. Petersburg.
    Schr.Russ.-Kais.Ges.ges.Miner. 1842.   M.S 1819 & P.S 500

Schriften der Trondheimischen Gesellschaft.
    See Drontheimischen Gesellschaft Schriften aus dem
    Dänischen übersetzt. Kopenhagen.   S. 554 B

Schriften der Universität zu Kiel.
    Schr.Univ.Kiel 1854-1881.   S. 1473

Schriften des Vereines zur Verbreitung Naturwissenschaftlicher
    Kenntnisse in Wien.
    Schr.Ver.Verbreit.naturw.Kennt.Wien 1860-1872.   S. 1775

Schriften des Vereins für Geschichte des Bodensee's und seiner
    Umgebung. Lindau.
    Schr.Ver.Gesch.Bodensee 1869-1876, 1893-1902.   S. 1517 A - B

Schriften der Zoologischen Station in Büsum für Meereskunde. Berlin.
    Schr.Z.S.B.Meeresk. 1919-1921. (imp.).   Z.S 1350

Schriftenreihe. Institut für Naturschutz. Darmstadt.
    SchrReihe Inst.Natursch.Darmstadt
    Bd.5, Hft.2 → 1960 → (imp.)
    Beiheft. Hft.12 → 1960 → (imp.)
    Formerly Schriftenreihe der Naturschutzstelle.   S. 1399 A-B

Schriftenreihe der Naturschutzstelle. Darmstadt.
    SchrReihe NaturschSt.Darmstadt
    Bd.1, Hft.6 - Bd.5, Hft.1, 1953-1959 (imp.)
    Beiheft. Hft.4-10, 1956-1959. (imp.)
    Continued as Schriftenreihe. Institut für Naturschutz.   S. 1399 A-B

TITLE	SERIAL No.

Schriftenreihe für Vegetationskunde. Bundesanstalt für
    Vegetationskunde, Naturschutz und Landschaftspflege.
    Bad Godesberg.
    SchrReihe Vegetationskde 1966 →   B.S 890

Schriftenreihe des Verein Naturschutzpark e.V. Stuttgart.
    See Naturschutzparke.   S. 1593 A

Schrifter der Dresdener Gesellschaft für Mineralogie. Leipzig.
    Schr.Dresd.Gesell.Miner. 1818-19   M.S 1426

Schwalbe. Berichte des Komitee für Ornithologische
    Beobachtungs-Stationen in Osterreich.
    Schwalbe 1898-1913.
    Formerly Mitteilungen des Ornithologischen Vereines
    in Wien.   T.R.S 1803

Schwedische Annalen der Medisin und Naturgeschichte Berlin
    und Stralsund.
    Schwed.Ann.Med.u.Naturg. 1799.   S. 600

Schweizer Entomologischer Anzeiger. Zürich.
    Schweizer ent.Anz. 1922-1926.   E.S 1212

Schweizer Naturschutz. Basel.
    Schweizer NatSchutz 1949 →
    Formerly Schweizerische Blätter für Naturschutz. Basel.   S. 1217 A

Schweizer Strahler, Luzern.
    Schweiz.Strahler 1967 →   M.S 1203

Schweizerische Blätter für Naturschutz. Basel.
    Schweiz.Bl.NatSchutz 1926-1934.
    Continued as Schweizer Naturschutz. Basel.   S. 1217 A

Schweizerische Jugendbücherei für Naturschutz. Basel.
    Schweiz.JugBüch.Natursch. Nos.2, 6, 9 & 21. 1925-1926.   S. 1217 A

Schweizerische Jugendflugblätter für Naturschutz. Basel.
    Schweiz.JugBl.Natursch. (1925-1927)   S. 1217 B

Schweizerische Mineralogische und Petrographische
    Mitteilungen. Zürich.
    Schweiz.miner.petrogr.Mitt. 1921 →   M.S 1202

Schweizerische Paläontologische Abhandlungen. Basel.
    Schweiz.palaeont.Abh. Vol. 63 → 1940 →
    Formerly Abhandlungen der Schweizerischen Paläontologischen
    Gesellschaft.   P.S 700

TITLE	SERIAL No.

Schweizerische Zeitschrift für Hydrologie. Basel.
   Schweiz.Z.Hydrol. 1948 →
   Formerly Zeitschrift für Hydrologie (Hydrographie,
Hydrobiologie, Fischereiwissenschaft.) Aarau.   S. 1201 F

Schweizerische Zeitschrift für Pilzkunde. Bern.
   Schweiz.Z.Pilzk. Vol.30 → 1952 →   B.M.S 44

Schweizerisches Archiv für Ornithologie.
   See Archives Suisse d'Ornithologie.   T.B.S 1201 B

Science. New York, etc.
   Science, N.Y. 1883 →   S. 2481
               Vol.75-114, 1932-1951.   T.R.S 5142

Science. Washington, D.C.
   See Science. New York, etc.   S. 2481

Science Abstracts. (a) Physics Abstracts (b) Electrical
   Engineering Abstracts. London.
   Sci.Abstr. 1898-1912   M.S 308

Science Abstracts of China. Peking.
   Sci.Abstr.China Biological Sciences. 1963 →   R.R.ABS 103

Science Bulletin of the College of Agriculture, University of
   the Ryukyus. Okinawa.
   Sci.Bull.Coll.Agric.Univ.Ryukyus No.19 → 1972 →   E.S 1899

Science Bulletin. College of Arts & Sciences, Pahlavi University.
   Shiraz.
   Sci.Bull.Coll.Arts Sci.Pahlavi Univ. 1966.   R. 73 C o B

Science Bulletin. Department of Agriculture. Bangkok.
   Sci.Bull.Dep.Agric.Bangkok 1966 →   E.S 2027

Science Bulletin. Department of Agriculture and Forestry.
   Union of South Africa. Pretoria.
   Sci.Bull.Dep.Agric.For.Un.S.Afr. 1942 → (imp.)   E.S 2166

Science Bulletin. Department of Agriculture,
   New South Wales, Sydney.
   Sci.Bull.Dep.Agric.N.S.W. 1911-1948 (imp.).   E.S 2254

Science Bulletin. Department of Education and Science. London.
   Sci.Bull.Dep.Educ.Sci. 1969-1970.   S. 206 a B

Science Bulletin of the Faculty of Agriculture, Kyushu University.
   Fukuoka.
   Sci.Bull.Fac.Agric.Kyushu Univ. 1950 →
   Formerly Bulteno Scienca de la Fakultato Terkultura
Kjusu Imperia Universitato.   E.S 1932 b

| TITLE | SERIAL No. |

Science Bulletin of the Faculty of Education, Nagasaki University. Nagasaki.
   See Science Bulletin of the Faculty of Liberal Arts and Education, Nagasaki University. Nagasaki.   S. 1994 f B

Science Bulletin of the Faculty of Liberal Arts and Education, Nagasaki University. Nagasaki.
   Sci.Bull.Fac.lib.Arts Educ.Nagasaki Univ. No.6 → 1957 →
   Formerly Science Reports of the Faculty of Arts and Literature, Nagasaki University.   S. 1994 f B

Science Bulletin of Kansas University.
   See Kansas University Science Bulletin. Lawrence.   S. 2344 B

Science Bulletin. Lingnan University. Canton.
   Sci.Bull.Lingnan Univ. Nos.1-7 & 10. 1930-1944.   S. 1963 B

Science Bulletin. Museum of the Brooklyn Institute of Arts and Sciences. New York.
   Sci.Bull.Mus.Brooklyn Inst.Arts Sciences   1901-1930.   S. 2324 A
                                              1901-1916.   T.R.S 5117 A

Science Bulletin. Natural History Museum, Los Angeles County.
   Sci.Bull.nat.Hist.Mus.Los Ang.Cty 1972 →
   Formerly Bulletin of the Los Angeles County Museum (of Natural History). Science.   S. 2309 C

Science Bulletin. Ohio State Museum. Columbus.
   Sci.Bull.Ohio St.Mus. 1928-1946.   S. 2366 a

Science Bulletin of the Science Museum of the Saint Paul Institute. St. Paul.
   Sci.Bull.Sci.Mus.St.Paul Inst. 1935-1960.   S. 2470 A

Science Bulletin. University of Kansas.
   See Kansas University Science Bulletin.   S. 2344 B

Science and Culture. Calcutta.
   Sci.& Cult. 1935 → (vols.7 & 8 imp.)   S. 1937

Science Dimension. National Research Council of Canada. Ottawa.
   Sci.Dimension 1969 →   S. 2602 E

Science East to West. Paris.
   Sci.East West No.13-17,1964.
   Formerly Russian Technical Literature. Organisation for Economic Co-operation and Development. Paris.   S. 2713 a

Science Gossip. London.
   Sci.Gossip 1865-1902.   S. 481 & T.R.S 156

Science Gossip & Country Queries and Notes. London.
   Sci.Gossip & Ctry Queries 1909-1910.   S. 436 B

TITLE	SERIAL No.

Science Guide. American Museum of Natural History. New York.
    Sci.Guide Am.Mus.nat.Hist. 1944-1958 (imp.)
    Formerly Guide Leaflet. American Museum of Natural History.
New York.   S. 2356 E

Science Horizons. London.
    Sci.Horiz. No.115-135, 1970-1972.   S. 482 a

Science in Iceland. Reykjavik.
    Sci.Iceland 1968 →   S. 502 D

Science and Industry. Pakistan Council of Scientific and
    Industrial Research. Karachi.
    Sci.Ind. Vol.5, No.3 → 1967 →   S. 1940 a C

Science Journal. London.
    Sci.J. 1965-1971.   S. 494 b

Science Lectures for the People. Manchester.
    Sci.Lect.Manchester 1866-1880.   7.o.S

Science on the March. Buffalo, New York.
    Sci.on the March Vol.38, No.4 → 1958 →
    Formerly Hobbies.   S. 2326 B

Science Monthly Illustrated. London.
    Sci.Mon.Ill. 1883-1884.
    Continued as Illustrated Science Monthly.   S. 483

Science et Nature. Paris.
    Sci.Nat., Paris 1954-1973 (wanting No. 4, 8, 11, 14, 19,
    20, 21, 25, 29, 31 & 33)   S. 931 a

Science in New Guinea. Port Moresby.
    Sci.New Guin. 1972 →   S. 2142 b

Science News. Harmondsworth.
    Sci.News, Harmondsworth 1946-1960.   S. 489 c
    1946-1951.   E.S 7 c

Science News. Washington, D.C.
    Sci.News, Washington Vol.98 → 1970 →   S. 2510 a

Science Notes. New Jersey State Museum. Trenton, N.J.
    Sci.Notes New Jers.St.Mus. 1971 →   S. 2412 E

Science Papers. Haslemere Microscope & Natural History
    Society. Guidford.
    Sci.Pap.Haslemere microsc.nat.Hist.Soc. 1903-1904.
    Continued as Science Papers. Haslemere Natural
History Society.   S. 117 B

TITLE	SERIAL No.

Science Papers. Haslemere Natural History Society.
    Sci.Pap.Haslemere nat.Hist.Soc. No.4-11. 1909-1934.
    Formerly Science Papers. Haslemere Microscope &
Natural History Society.   S. 117 B

Science Policy.
    Sci.Policy 1972-1973.
    Continued as Science and Public Policy.   S. 214 a

Science Polonaise. Ses Besoins, son Organisation et ses
    Progrès. Varsovie.
    See Nauka Polska. Jej Potrzeby, Organizacja Rozwoj.
    Warszawa.   S. 1873

Science Progrès Découverte. Paris.
    Sci.Prog.Découv. 1969-1972.
    Formerly Science Progrès, La Nature. Paris.   S. 997
    Absorbed by Recherche. Paris.   S. 996 b

Science Progrès, La Nature. Paris.
    Sci.Prog., Paris 1963-1969.
    Formerly Nature. Paris.
    Continued as Science Progrès Découverte. Paris.   S. 997

Science Progress. London.
    Sci.Progr., Lond. 1894-1898: 1906 →   S. 459

Science Progress in the Twentieth Century. London.
    See Science Progress. London.   S. 459

Science and Public Policy. London.
    Sci.publ.Policy 1974 →
    Formerly Science Policy.   S. 214 a

Science Quarterly. National University of Peking.
    Sci.Q.natn.Univ.Peking 1929-1935.
    Continued as Science Reports of the National University
of Peking.   S. 1950 A

Science Record. Academia Sinica. Peking.
    Sci.Rec.Acad.sin. vol.3, nos.2-4. 1950.   S. 1985 A

Science Record. University of Otago, Dunedin.
    Sci.Rec.Univ.Otago Vol.13, 15, 16, 18 → 1963 →   S. 2173

Science Reports. College of General Education. Osaka University.
    Sci.Rep.Coll.Gen.Educ.Osaka Univ. vol.11, no.1 → 1962 →   S. 1993 a

| TITLE | SERIAL No. |

Science Reports of Faculty of Agriculture, Kobe University.
   Sci.Rep.Fac.Agric.Kobe Univ. Vol.9 → 1971 →
   Formerly Science Reports of the Hyogo University of
Agriculture (and Faculty of Agriculture, Kobe University.) S. 1988 b B

Science Reports of the Faculty of Arts and Literature, Nagasaki
   University. Nagasaki.
   Sci.Rep.Fac.Arts Lit.Nagasaki Univ. 1951-1956.
   Continued as Science Bulletin of the Faculty of
Liberal Arts and Education, Nagasaki University. S. 1994 f B

Science Reports of the Faculty of Education, Gifu University.
   Sci.Rep.Fac.Educ.Gifu Univ. Vol.4 → 1967 →
   Formerly Science Report of the Faculty of Liberal Arts
and Education, Gifu University. S. 1997 e

Science Reports of the Faculty of Education, Gunma University.
   Maebashi.
   Sci.Rep.Fac.Educ.Gunma Univ. Vol.15 → 1966 →
   Formerly Science Reports of the Gunma University, Maebashi. S. 1984 f

Science Reports of the Faculty of Liberal Arts and Education,
   Gifu University.
   Sci.Rep.Fac.Lib.Arts Educ.Gifu Univ. 1953-1965.
   Continued as Science Report of the Faculty of Education,
Gifu University. S. 1997 e

Science Reports of the Faculty of Science, Kyushu University. Fukuoka.
   Sci.Rep.Fac.Sci.Kyushu Univ.Geol. 1950 → P.S 1768 A

Science Reports of the Gunma University. Maebashi.
   Sci.Rep.Gunma Univ. Vol.3 (imp), 13-14, 1953, 1965-1966.
   Continued as Science Reports of the Faculty of Education,
Gunma University. Maebashi. S. 1984 f

Science Reports of the Hirosaki University. Hirosaki.
   Sci.Rep.Hirosaki Univ. Vol.15 → 1968 → S. 1998 g

Science Reports of the Hyogo University of Agriculture (and
   Faculty of Agriculture, Kobe University). Sasayama, Hyogo.
   Sci.Rep.Hyogo Univ.Agric.
   Series: Agricultural Chemistry. Vol.7, No.1, 1965.
   Series: Agricultural Technology. Vol.7, No.1, 1965.
   Series: Agriculture and Horticulture. Vol.7, No.1, 1965.
   Series: Natural Science, Vol.1-7, No.1-2, 1953-1966.
   Series: Plant Protection. Vol.7, No.1-2, 1965-1966.
   Series: Zootechnical Science, No1.7, No.2, 1966.
   Unified in one series, Vol.8, 1967-1968.
   Continued as Science Reports of Faculty of Agriculture,
Kobe University. S. 1988 b B

| TITLE | SERIAL No. |

Science Reports of Kagoshima University.
    Sci.Rep.Kagoshima Univ. 1952-1967.
        Continued as Report of the Faculty of Science, Kagoshima
    University.   S. 1990 b

Science Reports of the Kanazawa University. Kanazawa.
    Sci.Rep.Kanazawa Univ. 1951 →   S. 1999 b

Science Reports. National Central University. Nanking. Series B.
    Biological Sciences.
    Sci.Rep.natn.cent.Univ.Nanking Ser.B. 1930-1935.   S. 1969 a

Science Reports of the National Taiwan University. Taipei. First Series.
    See Acta Geologica Taiwanica. Taipei &
    Acta Botanica Taiwanica. Taipei.

Science Reports of National Tsing Hua University. Peiping.
    Series B. Biological and Psychological Sciences.
    Sci.Rep.natn.Tsing Hua Univ. Ser.B. Vol.1, Nos.4-6;
    Vol.2, Nos.1-3. 1932-1937.   S. 1954

Science Reports of the National University of Peking.
    Sci.Rep.natn.Univ.Peking 1936-1937.
    Formerly Science Quarterly. National University of Peking. S. 1950 B

Science Reports of National Wu-Han University. Biological
    Science. Wuchang.
    Sci.Rep.natn.Wuhan Univ. Biol.Ser. Nos.1-3. 1939.   S. 1947 B

Science Reports of Niigata University. Series D (Biology).
    Sci.Rep.Niigata Univ. 1964 →
    Formerly Journal of the Faculty of Science Niigata
    University. Ser.II. Biology, Geology & Mineralogy.   S. 1988 c

Science Reports of the Research Institutes, Tohôku
    University. Sendai.
    Series A. (Physics Chemistry and Metallurgy)
    Sci.Rep.Res.Insts Tôhoku Univ. 1949 →   M.S 1941

Science Reports of the Saitama University. Urawa, Japan.
    Series B, (Biology and Earth Sciences.)
    Sci.Rep.Saitama Univ. (B) 1952 →   S. 1997 a

Science Reports of Shima Marineland. Shima.
    Sci.Rep.Shima Marineld 1972 →   S. 1999 h

| TITLE | SERIAL No. |

Science Reports of the Tôhoku (Imperial) University. Sendai.
    Sci.Rep.Tohoku Univ. First Series, Mathematics, Physics,
Chemistry. Vol.1, 1911-1912: Vol.26-34, 1937-1950 (imp.)
    Physics, Chemistry, Astronomy. Vol.35 → 1951 →
    Third Series, Mineralogy, Petrology, Economic Geology. 1921 →   M.S 1950
    Second Series, Geology 1912 →
      Special Volume 1936 →   P.S 1770
    Fourth Series, Biology 1924 →   S. 1988 A
    Seventh Series, Geography. No.2 → 1953 →   S. 1988 A b

Science Reports of the Tokyo Bunrika Daigaku. Tokyo.
    Sci.Rep.Tokyo Bunrika Daig. Sect.B. 1932-1954.   S. 1989
                                  Sect.C. 1932-1955.   P.S 1773 A
    Continued as Science Report of the Tokyo Kyoiku Daigaku.

Science Reports of the Tokyo Kyoiku Daigaku. Tokyo.
    Sci.Rep.Tokyo Kyoiku Daig. Sect.B. 1955 →   . 1989
                                  Sect.C. 1955 →   P.S 1773 A
    Formerly Science Reports of the Tokyo Bunyika Daigaku.

Science Reports. University of Chekiang.
    Sci.Rep.Univ.Chekiang 1934-1936.   S. 1949

Science Reports of the Yamaguchi University. Yamaguchi.
    Sci.Rep.Yamaguchi Univ. 1959 →
    Formerly Yamaguchi Journal of Science. Yamaguchi.   S. 1985 e C

Science Reports of the Yokohama National University. Section II.
Biological and Geological Sciences. Kamakura, Japan.
    Sci.Rep.Yokohama natn.Univ. Sect.II 1952 →   S. 1991 a A

Science Reports of the Yokosuka City Museum. Kurihama, Yokosuka, Japan.
    Sci.Rep.Yokosuka Cy Mus. 1956 →   S. 1991 b

Science Series. Colorado State University Range Science Department.
Fort Collins (Co.)
    Sci.Ser.Co St.Univ.Range Sci.Dep. Nos.3, 12, 15 → 1969 →   S. 2334 a

Science Series. Los Angeles (County) Museum.
    Sci.Ser.Los Ang.Mus.
      Anthropology No.1, 1955   R.A.S 789
      Palaeontology No.1-22, 1930-1963.   P.S 839
      Zoology No.1-12, 1942-1970.   Z.S 2570
    Continued as Science Series. Natural History Museum of
Los Angeles County.   S. 2309 E

Science Series. Natural History Museum of Los Angeles County.
    Sci.Ser.nat.Hist.Mus.Los.Ang.Cty No. 26 → 1973 →
    Formerly Science Series Los Angeles (County) Museum.   S. 2309 E

Science Society Report. Dulwich College.
    Sci.Soc.Rep.Dulwich Coll. 1890-1891.
    Formerly Report of the Dulwich College Science Society.   S. 171

TITLE	SERIAL No.

Science of the South Seas. Journal of the Palao Tropical Biological
    Station. Tokyo.
    See Kagaku Nan'yo. Tokyo.   S. 1991 B

Science Studies. London.
    Sci.Stud. 1971 →   S. 460 a

Science Studies. Montana Agricultural College. Bozeman.
    Sci.Stud.Mont.agric.Coll. 1904-1905   B.S 4280

Science Studies. St.Bonaventure University. New York.
    Sci.Stud.St.Bonaventure Univ. 1953 →   S. 2525

Science Survey. London.
    Sci.Surv. 1960-1962.   S. 483 a

Science To-day, London.
    Sci.to-day 1946-1951.   S. 432

Science of the Total Environment. Amsterdam.
    Sci.tot.Envir. 1972 →   S. 2765

Sciences. Paris.
    Sciences Tome 1, No.3 → 1970 →   S. 801 B a

Sciences: Revue de l'Association Francaise pour l'Avancement des
    Sciences. Paris.
    Sciences 1936-1952.
    Formerly Bulletin de l'Association Francaise pour l'Avancement
    des Sciences.   S. 801 B

Sciences Géologiques. Annales du Musée Royal du Congo Belge.
    See Annales du Musée Royal du Congo Belge. Tervuren.

Sciences Géologiques. Institut de Géologie. Strasbourg.
    Sciences géol.Inst.Géol.Strasbourg
      Mémoires. 1971 →
      Formerly Mémoires du Service de la Carte Géologique
      d'Alsace et de Lorraine. Strasbourg.   P.S 1304
      Bulletin. 1972 →
      Formerly Bulletin du Service de la Carte Géologique
      d'Alsace et de Lorraine. Strasbourg.   P.S 1303

Sciences de l'Homme. Annales du Musée Royal du Congo Belge.
    See Annales du Musée Royal du Congo Belge. Tervuren.

TITLE	SERIAL No.

Sciences Naturelles. Bulletin Mensuel des Naturalistes. Paris.
    Sciences nat. Tom.1, num.2-4. 1939.   S. 990 a

Sciences de la Terre. Nancy.
    Sciences Terre 1953 →
    Replaces Géologie Appliquée et Prospection Minière.
Nancy. (In the Mineral Department Library.)   P.S 218

Sciences de la Terre. Mémoirs. Nancy.
    Sci.Terre Mém. 1962 →   M.S 910

Sciences Zoologiques. Annales du Musée Royal du Congo Belge.
    See Annales du Musée Royal du Congo Belge. Tervuren.

Scientia. Bologna.
    Scientia, Bologna 1910 →
    Formerly Rivista di Scienza. Bologna.   S. 1179

Scientia Genetica. Torino.
    Scientia genet. 1939-1954.   S. 1183 a

Scientia Geologica Sinica. Peking.
    Scientia Geol.sin. Vol.3 → 1965 →   P.S 1793 F

Scientia Islandica. Reykjavik.
    See Science in Iceland. Reykjavik.   S. 502 D

Scientia Silvae.
    Scientia silv. Vol.9 → No.2 → 1964 →   B.S 1896

Scientia Sinica. Academia Sinica. Peking.
    Scientia sin. 1952/1953 → (imp.)   S. 1985 C

Scientific Activities. The Weizmann Institute of Science. Rehovoth.
    Scient.Activ.Weizmann Inst.Sci. 1962 →
    Formerly Report of Scientific Activities. The Weizmann
Institute of Science. Rehovoth.   S. 1928 c

Scientific African. Cape Town.
    Scient.African 1895-1896.   S. 2014

Scientific American. New York.
    Scient.Am. Vol. 202 → 1960 →   S. 2504

Scientific Enquirer. London.
    Scient.Enquirer 1886-1888.   S. 463

Scientific Information Notes. New York.
    Scient.Inf.Notes 1969-1970.
    Continued as Information. New York.   S. 2555

TITLE	SERIAL No.

Scientific Investigations. Department of Fisheries.
    Irish Free State.
        See Scientific Investigations. Fisheries Branch. Department
    of Agriculture for Ireland.   Z.S 400 B

Scientific Investigations. Fisheries Branch, Department
    of Agriculture for Ireland. Dublin.
    Scient.Invest.Fish.Brch Ire. 1904-1926.
    (Publication Suspended 1922-1925.)
    Formerly in Report on the Sea and Inland Fisheries
    of Ireland.   Z.S 400 B

Scientific Investigations. Fisheries Divison. Scottish Home
    Department. Edinburgh.
    Scient.Invest.Fish.Div.Scott.Home Dep. 1949, No.1.
    Previously published in Report of the Fishery Board
    for Scotland & Report on the Fisheries of Scotland.
    Continued as Marine Research. Scottish Home Department.
    Edinburgh.   Z.S 380 A

Scientific Investigations. Fishery Board for Scotland. Edinburgh.
    Scient.Invest.Fishery Bd Scotl. 1909-1939.   Z.S 380 B

Scientific Investigations. Freshwater and Salmon Fisheries Research.
    Scottish Home Department, Edinburgh.
    Scient.Invest.Freshwat.Salm.Fish.Res.Scott.Home Dep. 1948-1949.
    Continued as Freshwater & Salmon Fisheries Research.
    Scottish Home Department, Edinburgh.   Z.S 385

Scientific Investigations. Imperial Bureau of Fisheries. Tokyo.
    See Report. Imperial Bureau of Fisheries Scientific
    Investigations. Tokyo.   Z.S 1966

Scientific Investigations in Micronesia 1949-1956.
    Washington.   77 F.q.N
    Scient.Invest.Micronesia 1950-1962.   & See Gen.Lbry.Catalogue

Scientific Journal of the Royal College of Science. London.
    Scient.Jl R.Coll.Sci. vols.1 & 5-27. 1930-1959.   S. 233

Scientific Magazine of Biology. Kharkov.
    See Naukovi Zapȳskȳ po Biolohiyi. Kharkov.   S. 1833 a

Scientific Memoirs, London.
    Scient.Mem. 1837-1853.   S. 460

Scientific Memoirs by Medical Officers of the Army of India.
    Calcutta.
    Scient.Mem.med.Offrs Army India 1884-1901.   S. 1918 A
                                    1884-1889.   E.S 2001
    Continued as Scientific Memoirs by Officers of the
    Medical and Sanitary Department of the Government of
    India. Calcutta.

TITLE	SERIAL No.

Scientific Memoirs of the Museum Darwinianum in Moscow.
   See Trudy Zoopsikhologicheskoĭ Laboratorii.
   Gosudarstvennyi Darvinovskii Muzei. Moskva.   S. 1805

Scientific Memoirs by Officers of the Medical and Sanitary
   Department of the Government of India. New Series. Calcutta.
   <u>Scient.Mem.Offrs med.sanit.Deps India</u>   1902-1913.   S. 1918 a
                                                                         Z.S 1925
      Nos 20-60, 1905-1913.   E.S 2002
      <u>Formerly</u> Scientific Memoirs by Medical Officers of the
      Army of India. Calcutta.

Scientific Memoirs of the University of Perm.
   See Uchenȳe Zapiski. Permskiĭ Gosudarstvennyĭ Universitet
   im M. Gor'kogo. Perm.   S. 1849 a

Scientific Monograph. Pakistan Association for the Advancement
   of Science. Lahore.
   <u>Scient.Monogr.Pakist.Ass.Advmt Sci</u>.   No.1. 1952.   B. 632:633/5(54)

Scientific Opinion. London.
   <u>Scient.Opinion</u>   1868-1870.   S. 480

Scientific Papers of the Applied Sections of the Tiflis
   Botanical Garden.
   See Zapiski Nauchno-Prikladnȳkh Otdelov Tiflisskogo
   Botanicheskogo Sada. Tiflis.   B.S 1430 b

Scientific Papers. Institute of Algological Research,
   Hokkaido Imperial University. Sapporo.
   <u>Scient.Pap.Inst.algol.Res.Hokkaido Univ</u>.   1935 →   B.A.S 27

Scientific Papers from Institute of Chemical Technology, Prague.
   Faculty of Technology of Fuel and Water.
   See Sborník Vysoké Školy Chemicko-Technologické v Praze.
   Oddíl Fakulty Technologie Paliv a Vody.   S. 1759 a

Scientific Papers of the Namib Desert Research Station. Pretoria.
   <u>Scient.Pap.Namib Des.Res.Stn</u>   1961-1969.   S.·2004 a
     <u>Continued in</u> Madoqua. Series 2.   S. 2079 B

Scientific Proceedings of the Royal Dublin Society.
   <u>Scient.Proc.R.Dubl.Soc</u>.   1877 →   S. 46 C

TITLE	SERIAL No.

Scientific Publications of the Cleveland Museum of Natural History.
   Cleveland.
     Scient.Publs Cleveland Mus.nat.Hist. 1928-1950; 1962-1965.   S. 2452 A
                                                1928-1950.   T.R.S 5122

Scientific Publications. Cranbrook Institute of Science.
   Bloomfield Hills. Mich.
     Scient.Publs Cranbrook Inst.Sci.   1932.   S. 2316 a B

Scientific Publications. Freshwater Biological Association. Ambleside.
     Scient.Publs Freshwat.biol.Ass.   1949 →
     Formerly Scientific Publications. Freshwater Biological
     Association of the British Empire. Ambleside.   S. 7 b B

Scientific Publications. Freshwater Biological Association of the
   British Empire. Ambleside.
     Scient.Publs Freshwat.biol.Ass.Br.Emp.   1939-1947.
     Continued as Scientific Publications. Freshwater Biological
     Association. Ambleside.   S. 7 b B

Scientific Publications. Reading Public Museum and Art Gallery.
   Reading, Pa.
     Scient.Publs Reading publ.Mus.   1941-1967.   S. 2374 a A

Scientific Publications of the Science Museum, Saint Paul, Minn.
     Scient.Publs Sci.Mus.St.Paul   New Series 1967 →   S. 2470 B

Scientific Records of the S.M. Kirov Kazakh State University.
   See Uchenye Zapiski Kazakhskogo Gosudarstvennogo Universiteta
   im. S.M. Kirov. Alma-Ata.   S. 1832 a

Scientific Reports of the Agricultural Research Institute,
   Pusa. Calcutta.
     Scient.Rep.agric.Res.Inst.Pusa   1916-1929.
     Formerly Report of the Agricultural Research Institute
     & College, Pusa. Calcutta.
     Continued as Scientific Reports of the Imperial Institute of
     Agricultural Research, Pusa. Calcutta.   E.S 1997

Scientific Reports. Australasian Antarctic Expedition
   1911-14. Sydney.
     Scient.Rep.Austalas.antarct.Exped.   1916-1948.   80 q S

Scientific Reports. British Antarctic Survey. London.
     Scient.Rep.Br.Antarct.Surv.   No. 36 → 1962 →   80 q G
                              No. 36 → (imp.) 1962 →   P.S 1119 A
     Formerly Scientific Reports. Falkland Island Dependencies
   Survey. London.

Scientific Reports. British Graham Land Expedition 1934-37. London.
     Scient.Rep.Br.Grahamld Exped.   1940-1941.   Z.O 72Aa q B & BM.A.q

| TITLE | SERIAL No. |

Scientific Reports. Ehime Agricultural College. Matsuyama.
   See Scientific Reports of the Matsuyama Agricultural
   College. Matsuyama.   E.S 1937

Scientific Reports of the Faculty of Agriculture, Naniwa
   University. Sakai.
   Scient.Rep.Fac.Agric.Naniwa Univ. 1950.
   Continued as Bulletin of the Naniwa University.   S. 1985 b

Scientific Reports. Falkland Island Dependencies Survey. London.
   Scient.Rep.Falkld Isl.Depend.Surv. 1953-1962.   80 q G
                              Nos.3-35, 1953-1962.   P.S 1119 A
   Continued as Scientific Reports. British Antarctic Survey.
   London.

Scientific Reports. Geological and Geographical Institute. Vilnius.
   See Moksliniai Pranešimai. Geologijos ir Geografijos
   Institutas.   P.S 575 A

Scientific Reports. Great Barrier Reef Expedition 1928-29. London.
   Scient.Rep.Gt.Barrier Reef Exped. 1930 →   Z.O 72Aa q B & BM.A.q

Scientific Reports of the Hokkaido Fish Hatchery. Sapporo.
   Scient.Rep.Hokkaido Fish Hatch. Vol.9 → 1954 →   Z.S 1951 D

Scientific Reports of the Hokkaido Fisheries Experimental
   Station. Yoichi.
   Scient.Rep.Hokkaido Fish.exp.Stn No.2 → 1964 →   Z.S 1951 F

Scientific Reports of the Hokkaido Salmon Hatchery. Sapporo.
   Scient.Rep.Hokkaido Salm.Hatch. No.12 → 1958 →   Z.S 1951 H

Scientific Reports of the Imperial Institute of Agricultural
   Research, Pusa. Calcutta.
   Scient.Rep.imp.Inst.agric.Res.Pusa 1930-1935.
   Formerly Scientific Reports of the Agricultural Research
   Institute, Pusa, Calcutta.
   Continued as Scientific Reports of the Indian Agricultural
   Research Institute. New Delhi, Calcutta.   E.S 1997

Scientific Reports of the Indian Agricultural Research Institute,
   New Delhi. Calcutta.
   Scient.Rep.Indian agric.Res.Inst. 1936-1937.
   Formerly Scientific Reports of the Imperial Institute
   Agricultural Research. Pusa, Calcutta.   E.S 1997

Scientific Reports. Japanese Antarctic Research Expedition
   1956-1962. Tokyo.
   Scient.Rep.Jap.Antarct.res.Exped.
     Ser.A-F. 1963 →
     Special Issue, 1967-1971.
     Continued as Memoirs of National Institute of Polar Research.
    Ser.E. Biology was formerly published as Biological Results
   of the Japanese Antarctic Research Expedition as part of the
   Special Publications from the Seto Marine Biological Laboratory. 80.o.J

| TITLE | SERIAL No. |

Scientific Reports. The John Murray Expedition 1933-34. London.
   Scient.Rep.John Murray Exped. 1935 →   Z.O 72Aa q B & BM.A.q
   (Selected numbers only) 1935 →   T.R.S 1 E&M.S 384

Scientific Reports of the Kyoto Prefectural University.
   Scient.Rep.Kyoto prefect.Univ.
   Natural Science & Living Science
     Vol.3, No.1 → 1959 →   S. 1980 b
     Agriculture. 1959 →   E.S 1933 a
   Formerly Scientific Reports of the Saikyo University. Kyoto.

Scientific Reports of the Laboratory for Amphibian Biology.
   Hiroshima University.
   Scient.Rep.Lab.amphib.Biol.Hiroshima Univ. 1972 →   Z.S 1956 B

Scientific Reports of the Matsuyama Agricultural College. Matsuyama.
   Scient.Rep.Matsuyama agric.Coll. 1948-1954.
   Continued as Memoirs of Ehime University. Matsuyama.   E.S 1937

Scientific Reports of Mukogawa Women's University. Nishinomiya.
   Scient.Rep.Mukogawa Wom.Univ. Science. No.1-10, 1953-1962.
   (Wanting No.3-5, 8)
   Continued as Bulletin of Mukogawa Women's University.
   Nishinomiya.   S. 1984 q

Scientific Reports of the Saikyo University. Kyoto.
   Scient.Rep.Saikyo Univ. Natural Science & Living Science
     Vol.1, nos.1-4. 1952-1954.
   Natural Science & Living Science. Ser.A. Mathematical
   & Natural Science. Vol.2 Nos.1-3. 1955-1956.
   Natural & Living Science. Vol.2. No.4-5. 1957-1958.   S. 1980 b
   Agriculture. 1951-1958.   E.S 1933 a
   Continued as Scientific Reports of the Kyoto
   Prefectural University.

Scientific Reports. Trans-Antarctic Expedition 1955-1958. London.
   Scient.Rep.transantarct.Exped. No.2 → 1960 →   80.q.T

Scientific Reports. Ukrainian Technical University. Augsburg.
   See Naukoviȳi Byuleten. Ukrayins'kyyi Tekhnichno-Hospodarskȳi
   Instȳtut.   S. 1315

Scientific Reports of the Whales Research Institute. Tokyo.
   Scient.Rep.Whales Res.Inst., Tokyo 1948 →   Z.S 1961

Scientific Reports. Zoological Society of London.
   Scient.Rep.zool.Soc.Lond. 1966 →   Z.S 1 D a

Scientific Researches. Dacca.
   Scient.Res.Dacca 1964 →   S. 1940 a

| TITLE | SERIAL No. |

Scientific Researches of the Ozegahara Moor. Tokyo.
    Scient.Res.Ozegahara Moor   1954.   73 I o O

Scientific Results of the "Brategg" Expedition, 1947-1948.
    Published in Publikationer. Kommander Chr.Christensens
    Hvalfangstmuseum i Sandefjord.   Z.S 533

Scientific Results. Chesapeake Zoological Laboratory.
    Baltimore, Md.
    Scient.Res.Chesapeake zool.Lab.   1878.   Z.S 2315 A

Scientific Results of Cruise VII of the Carnegie during 1928-1929.
    Washington, D.C.
    Scient.Results Cruise VII Carnegie   1942-1946.   70.q.C

Scientific Results of the Danish Expedition to the Coasts of
    Tropical West Africa, 1945-1946. Copenhagen.
    See Atlantide Report.   Z. 78 o B

Scientific Results of the Danish Rennell Expedition 1951 and the
    British Museum (Natural History) Expedition, 1953.
    See Natural History of Rennell Island, British Solomon
    Islands. London and Copenhagen.   77 Bb.o.C

Scientific Results. Hydrobiological Survey of the Lake Bangweulu
    and Luapula River Basin. Brussels.
    See Résultats Scientifiques. Exploration Hydrobiologique
    du Bassin du Lac Bangweulu et du Luapula.   74 Dd q B

Scientific Results. Icefield Ranges Research Project. New York.
    Scient.Results Icefield Ranges Res.Proj.   1969 →   S. 2358 D

Scientific Results of the Japanese Expeditions to Nepal Himalaya.
    Kyoto 1952-1953.
    Scient.Results Jap.Exp.Nepal   1955-1957.   73 F.o.K

Scientific Results of the Noona Dan Expedition (Rennell Section, 1962)
    and the Danish Rennell Expedition, 1965.
    See Natural History of Rennell Island, British Solomon Islands.
    London and Copenhagen.   77 Bb.o.C

Scientific Results of the Norwegian Antarctic Expeditions
    1927-1928 et sqq. Oslo.
    Scient.Results Norw.Antarct.Exped.   1929 →   80.q.C

Scientific Results. Norwegian-British-Swedish Antarctic
    Expedition 1949-52. Oslo.
    Scient.Results Norw.-Br.Swed.Antarct.Exped.   1956 →   80.q.N

TITLE	SERIAL No.

Scientific Results. Norwegian North Polar Expedition 1893-1896. London.
    Scient.Results Norw.N.polar Exped. 1893-1896 1900-1906.   71.q.N

Scientific Results. Norwegian North Polar Expedition with the "Maud" 1918-1925. Bergen.
    Scient.Results Norw.N.polar Exped.Maud   1928-1936.   71.q.B

Scientific Results of the Norwegian-Swedish Spitzbergen Expedition in 1934. Stockholm.
    Scient.Results Norw.Swed.Spitzbergen Exped. 1935-1936.   71.o.S

Scientific Results of the Swedish-Norwegian Arctic Expedition in the Summer of 1931. Stockholm.
    Scient.Results Swed.Norw.Arct.Exped. 1933-1936.   71.o.S

Scientific Results of the Works of Tadjiko-Pamirian Expedition. Moscow.
    See Nauchnye Itogi Rabot Tadzhiksko-Pamirskoĭ Ekspeditsii. Moskva.   73 A.o.S

Scientific Series. Department of Agriculture, Federation of Malaya. Kuala Lumpur.
    Scient.Ser.Dep.Agric.Fed.Malaya 1949-1952.
    Formerly Scientific Series. Department of Agriculture, Straits Settlements & Federated Malay States. Kuala Lumpur.   E.S 1970

Scientific Series. Department of Agriculture, Straits Settlements and Federated Malay States. Kuala Lumpur.
    Scient.Ser.Dep.Agric.Straits Settl.F.M.S. 1930-1939.
    Continued as Scientific Series. Department of Agriculture, Federation of Malaya. Kuala Lumpur.   E.S 1970

Scientific Survey of Porto Rico and the Virgin Islands. New York.
    Scient.Surv.P.Rico 1919 →   S. 2361 D

Scientific Transactions of the Royal Dublin Society.
    Scient.Trans.R.Dubl.Soc. Ser.II, 1-9, 1877-1909.   S. 46 E

Scientific Worker (and the B.A.C. Bulletin). London.
    Scient.Wkr 1920-1947. (imp.)
    (1932-1935 published as "Progress and the Scientific Worker".)   S. 431

Scientific Works. Higher Agricultural Institute "Vassil Kolarov". Plovdiv.
    See Nauchni Trudove. Vissh Selskostopanski Institut "Vassil Kolarov". Plovdiv.   B.S 1533

TITLE	SERIAL No.

Scientist. Harrow.
    Scientist, Harrow 1931.   S. 434 a

Scientist. Karachi.
    Scientist, Karachi 1953-1966 (imp.)   S. 1924 a

Scope. Journal of the Radley College Natural History Society. Radley.
    Scope, Radley Coll. 1963-1964.   S. 309 B

Scottish Agriculture. Edinburgh.
    Scott.Agric. 1946-1958.
    Formerly Scottish Journal of Agriculture. Edinburgh.   E.S 79

Scottish Birds. Edinburgh.
    Scott.Birds 1958 →   T.B.S 201

Scottish Botanical Review. Edinburgh.
    Scot.bot.Rev. 1912.   B.S 7

Scottish Fisheries Bulletin. Edinburgh.
    Scott.Fish.Bull. 1952 →   Z.S 385 B

Scottish Geographical Magazine. Edinburgh.
    Scott.geogr.Mag. 1885 →   S. 60

Scottish Journal of Agriculture. Edinburgh.
    Scott.J.Agric. 1918-1946.
    Continued as Scottish Agriculture. Edinburgh.   E.S 79

Scottish Journal of Geology. Published for the Geological Societies of Edinburgh & Glasgow. Edinburgh.
    Scott.J.Geol. 1965 →
    Amalgamating the Transactions of the Edinburgh Geological Society & Transactions of the Geological Society of Glasgow.   P.S 163

Scottish Journal of Science. Penicuik.
    Scott.J.Sci. 1965 →   S. 476 b

Scottish Mountaineering Club Guide. Edinburgh.
    Scott.Mountg Club Guide 1920 →   72 Ab.o.S

Scottish Naturalist. Perth & Edinburgh.
    Scott.Nat. 1871-1891: 1912-1964.   S. 476
                  1871-1891: 1912-1956.   T.R.S 222
    (From 1892-1911 incorporated in "Annals of Scottish Natural History".)

Scottish Rock Garden Club.
    See Journal of the Scottish Rock Garden Club. Edinburgh.   B.S 9

TITLE	SERIAL No.

Scottish Wildlife. Edinburgh.
    Scott.Wildl. 1974 →
    Formerly Newsletter. Scottish Wildlife Trust.   S. 576

Scottish Zoo & Wild Life. Edinburgh.
    Scott.Zoo wild Life   Vol.1 - Vol.3 No.5, 1948-1951.   Z.S 95

Scripta Botanica. Tartu.
    See Botaanilised Uurimused. Tartu.   B.S 1478

Scripta Botanica Horti Universitatis Imperialis Petropolitanae.
    See Botanicheskiya Zapiski. Izdanȳya pri Botanicheskom
    Sadê Imperatorskago S.-Peterburgskago Universiteta.   B.S 1403 a

Scripta Facultatis Scientiarum Naturalium Universitatis
    Purkynianae Brunensis. Brno.
    Scr.Fac.Sci.nat.Univ.purkyn.brun. 1971 →
    Replaces Spisy Přírodovědecké Fakulty University
    J.E. Purkyné v Brne.   S. 1709 B

Scripta Geobotanica. Göttingen.
    Scr.geobot. 1970 →   B.S 999

Scripta Geologica. Leiden.
    Scr.geol. 1971 →   P.S 291 A

Scripta Horti Botanici Tallinnensis. Tallinn.
    See Tallinna Botaanikaaia Uurimused.   B.S 1423

Scripta Mycologica. Tartu.
    Scr.mycol. 1970 →   B. 582.28 Z.B.I.

Sea Bird Bulletin. The Sea Bird Group. London.
    Sea Bird Bull. 1965 →   T.B.S 107

Sea Fisheries Statistical Tables. London.
    Sea Fish.statist.Tabl.,Lond. 1929-1947 (imp.). 1949 →   Z.S 460 G

Sea Frontiers. Bulletin of the International Oceanographic
    Foundation. Coral Gables, Fla.
    Sea Front. Vol.3 → 1957 →
    Formerly Bulletin of the International Oceanographic
    Foundation. Coral Gables, Fla.   S. 2538 B

Sea Secrets. The International Oceanographic Foundation. Miami.
    Sea Secrets 1956 →   S. 2538 C

Sea Swallow. Being the Annual Report of the Royal Naval
    Bird Watching Society.
    Sea Swallow Vol.2 → 1949 →
    Formerly Report. Royal Naval Bird Watching Society.   T.B.S 106

| TITLE | SERIAL No. |

Séance Générale. Société des Lettres, Sciences et Arts de Metz.
   <u>Séans.Soc.Sci.Metz</u> 1819-1827.
   (Ann.I is of the 2e éd. published 1859.)
   <u>Continued as</u> Mémoires de la Société des Lettres, Sciences
   et Arts et d'Agriculture de Metz.   S. 1539

Séance Publique. Académie Royale des Sciences, Belles-Lettres
   et Arts de Bordeaux.
   <u>Séanc.publ.Acad.R.Sci.Bordeaux</u> 1819-1837.
   <u>Continued as</u> Actes de l'Académie Royale (Nationale)
   des Sciences, Belles-Lettres et Arts de Bordeaux.   S. 829 B

Séance Publique de l'Académie des Sciences, Arts et
   Belles-Lettres de Dijon.
   <u>Séance publ.Acad.Sci.Dijon</u> 1808-1829.
   <u>Formerly</u> Analyse des Travaux de l'Académie. Dijon.
   <u>Continued as</u> Mémoires de l'Académie des Sciences,
   Arts et Belles-Lettres de Dijon.   S. 861 B

Séance Publique de la Société Linnéenne de Normandie.
   <u>Séan.publ.Soc.linn.Normandie</u> 1834-1837.   S. 841 A

Séances et Travaux de l'Académie de Reims.
   <u>Séanc.Trav.Acad.Reims</u> 1844-1851.
   <u>Formerly</u> Annales de l'Académie de Reims.
   <u>Continued as</u> Travaux de l'Académie de Reims.   S. 956

Search. Ithaca, N.Y.
   <u>Search, Ithaca</u>
      Agriculture. Vol. 1, No. 3 → 1970 → (imp.)   E.S 2469 c

Search. Sydney.
   <u>Search</u> 1970 →
   <u>Formerly</u> Australian Journal of Science. Sydney.   S. 2131 B

Searchlight. Bulletin of the University of the Philippines
   Instructors Association. Manila.
   <u>Searchlight</u> vol.1, no.1 1936.   S. 1976 a B

Seasonable Hints. Dominion Experimental Farms. Ottawa.
   <u>Seasonable Hints Dom.exp.Fms</u> 1923-1929.   E.S 2523 e

Seaweed Research and Utilisation. Mandapam.
   <u>Seaweed Res.Util.</u> 1971 →   B.A.S 29

Second Danish Pamir Expedition. 1898-99. Copenhagen.
   <u>Second Danish Pamir Exped.</u> 1912, 1920.   73 A.o.D

| TITLE | SERIAL No. |

Sedimenta. Miami.
    Sedimenta 1971 →   P.S 804

Sedimentary Geology. International Journal of Applied and
    Regional Sedimentology. Amsterdam, etc.
    Sedim.Geol. 1967 →   P.S 994 B

Sedimentologia e Pedologia. Instituto de Geografia, Universidade
    de São Paulo.
    Sedim.Pedol.   1971 →   P.S 2014 a D

Sedimentologija. Beograd.
    Sedimentologija 1961 →   P.S 1461 A

Sedimentology. Amsterdam and New York.
    Sedimentology 1962 →   P.S 994

Seed Exchange List. Botanical Gardens, Faculty of Science. Tokyo.
    Seed Exch.List, Tokyo 1967 →   B.S 1995 b

Seiken Zihô. Kyoto.
    See Report of the Kihara Institute for Biological
    Research. Kyoto.   S. 1981 b

Seiro-seitai. Kyoto.
    Seiro-seitai 1947 →   S. 1979 a

Seitai - Konchu.
    See Insect Ecology.   E.S 1915 b

Seiva. Vicosa.
    Seiva Ano 4 (No.15) → 1944 → (imp.)   S. 2248 a B

Selborne Magazine. Ealing, Middx.
    Selborne Mag. Vol.70, No.1 - Vol.72, No.10, 1958-1960.
    Vol.79, No.34 → 1967 →
    (From Vol.73, No.11 - Vol.78, No.36 (340), 1960-1966
    incorporated in "Birds & Country".)
    Formerly Selborne Magazine and Nature Notes. London.   S. 477

Selborne Magazine for Lovers and Students of Living Nature.
    London.
    Selborne Mag. 1888-1889.
    Continued as Nature Notes. The Selborne Society's
    Magazine. London.   T.R.S 152 & S. 477

Selborne Magazine and Nature Notes. London.
    Selborne Mag. Vol.20-29, 1909-1925.
    Formerly Nature Notes. The Selborne Society's Magazine.
    London.   T.R.S 152
    Continued as Selborne Magazine. Ealing, Middx.   S. 477

| TITLE | SERIAL No. |

Selbstbiographien von Naturforschern. Halle (Saale).
    <u>Selbstbiogr.Naturf</u>. 1944—1948.
    <u>Continued as</u> Lebensdarstellungen Deutscher Naturforscher.   S. 1301 J

Select List of Accessions to the Library of the Institute
    of Geological Sciences. London.
    <u>Select List Access.Lbry Inst.geol.Sci</u>.   No.8 → 1968 →   P. REF.

Select Transactions of the Honourable the Society of Improvers
    of the Knowledge of Agriculture in Scotland. Edinburgh.
    <u>Select Trans.Soc.Impr.Agr.Scot</u>.   1743.   S. 57

Selected Bibliography on Algae. Nova Scotia Research   B. 582.26: 016
    Foundation, Halifax. N.S.   N.S.R.F
    <u>Sel.Biblphy Algae</u>   1953 →   Q

Sellowia. Itajai.
    <u>Sellowia</u>   1954 →
    <u>Formerly</u> Anais Bôtanicos do Hérbario "Barbosa Rodrigues".
    Santa Catarina.   B.S 3050

Selysia. Department of Entomology, Purdue University, Lafayette,
    Indiana.
    <u>Selysia</u>   1963 →   E.S 2414

Seminar Papers. Geology Department, Carleton University.
    <u>Semin.Pap.geol.Dep.Carleton Univ</u>.   1965 →   P.S 892

Seminarios de Estratigrafia. Departamento de Estratigrafia,
    Universidad de Madrid.
    <u>Semin.Estratigr</u>.   1969 →   P.S 672

Sempervivum Society Journal. Bishops Stortford.
    <u>Sempervivum Soc.J</u>.   1970 →   B.S 151

Senckenberg-Bücher. Frankfurt a.M.
    <u>Senckenberg-Büch</u>.   1926 → (imp.)   S. 1411 H

Senckenbergiana. Frankfurt a.M.
    <u>Senckenbergiana</u>   1918-1954.
    <u>Continued as</u>   Senckenbergiana Biologica <u>&</u> Senckenbergiana   S. 1411 E
    Lethaea q.v.   & T.R.S 1337 B

Senckenbergiana Biologica. Frankfurt a.M.
    <u>Senckenberg.biol</u>.   1954 →
    <u>Formerly</u> Senckenbergiana. Frankfurt a.M.   S. 1411 E a
                                                              & T.R.S 1337 B
Senckenbergiana Lethaea. Frankfurt a.M.
    <u>Senckenberg.leth</u>.   1954 →
    <u>Formerly</u> Senckenbergiana. Frankfurt a.M.   S. 1411 E b

TITLE	SERIAL No.

Senckenbergiana Maritima. Frankfurt a.M.
    Senckenberg.marit. 1969 →   S. 1411 E c

Seri Bibliografi. Bibliotheca Bogoriensis. Bogor.
    Seri biblfi Biblthca Bogor. No.4 → 1967 → (imp.)   S. 1953

Serial Atlas of the Marine Environment. New York.
    Ser.Atlas mar.Envir. Folio 1 → 1962 →   S. 2358 B

Serie "B" (Didactica y Complementaria.) Asociacion Geologica Argentina. Buenos Aires.
    Serie B (Didact.Complement) Asoc.geol.argent. 1972 →   P.S 950 A

Serie Biologica. Academia de Ciencias de Cuba. La Habana.
    Serie biol.Acad.Cienc.Cuba 1967 →   S. 2287 F

Série Botânica e Fisiologia Vegetal. Instituto de Pesquisas e Experimentação Agropecuárias do Norte. Belém.
    Série Bot.Fisiol.veg.Inst.Pesq.Exp.agropec.Norte
    1970-1972.   S. 2213 c H

Série Ciencias Naturales Facultad de Ciencias Exactas Fisicas y Naturales. Córdoba.
    Série Cienc.nat.Córdoba 1954-1969.   S. 2228 a B

Serie Cientifica. Instituto Antarctico Chileno. Santiago.
    Serie cient.Inst.antart.chil. 1970 →   S. 2235 a C

Série Científica. Instituto de Investigação Agronómica de Angola. Luanda.
    Série cient.Inst.Invest.agron.Angola 1968 → (imp.)   E.S 2158 a A

Série Circular. Instituto de Biologia e Tecnologia Pesqueira. Ceara, Brasil.
    Série Circ.Inst.Biol.Tecn.Pesq. 1967 →   Z.S 2219

Série Conférences et Documents. Laboratoire d'Océanographie Biologique. Rennes.
    Série Conf.Docum.Lab.Oceanogr.biol.Rennes 1971-1972.
    Formerly Série Conférences. Laboratoire d'Océanographie Biologique. Rennes.   S. 958 a C

Série Conférences. Laboratoire d'Océanographie Biologique. Rennes.
    Série Conf.Lab.Oceanogr.biol.Rennes 1969-1970.
    Continued as Série Conférences et Documents. Laboratoire d'Océanographie Biologique. Rennes.   S. 958 a C

Série Culturas da Amazônia. Instituto de Pesquisas e Experimentação Agropecuárias do Norte. Belém.
    Série Cult.Amazônia Inst.Pesq.Exp.agropec.Norte
    1970-1971.   S. 2213 c B

Serie Didactica. Departamento de Geologia. Universidad de Chile. Santiago.
    Serie didact.Dep.Geol.Univ.Chile 1968 →   P.S 2035 C

TITLE   SERIAL No.

Serie Didactica. Instituto de Estudios Geographicos Universidad
  Nacional de Tucuman.
  See Publicaciones Especiales. Instituto de Estudios
  Geographicos. Tucuman.   S. 2234 a C

Série Didática. Instituto de Geologia. Universidade do Recife.
  Série Didát.Inst.Geol.Univ.Recife No.3 → 1964 →   P.S 909

Serie de Divulgacion Cientifica. Servicio de Pesqueria, Peru. Lima.
  Serie Divulg.cient.Peru Nos.6, 14, 17-24, 1962-1966.
  (Numbers 6 & 14 are second editions both dated 1966,
  first appeared in 1955 & 1960 respectively.)   Z.S 2249

Série Documentation. Département de Géologie. Université
  d'Abidjan. (Abidjan.)
  Série Doc.Dép.Géol.Univ.Abidjan 1971 →   P.S 191 K

Serie de Estudios sobre Trabajos de Investigación. Instituto
  Cubano de Investigacines Tecnológicas. La Habana.
  Serie Estud.Trab.Invest.Inst.cub.Invest.tecnol.
  No.5-21, 1958-1962 (imp.)   S. 2285 a

Série Estudos sôbre Bubalinos. Instituto de Pesquisas e
  Experimentacão Agropecuárias do Norte. Belém.
  Série Estud.Bubalinos Inst.Pesq.Exp.Agropec.Norte
  1970-1971.   S. 2213 c C

Serie Fauna. Instituto de Investigacion de los Recursos
  Naturales Renovables. Salta.   See Gen.Lbry
  Serie Fauna Inst.Invest.Recurs.nat.renov.,Salta 1972 →   Catalogue

Série Fertilidade de Solo. Instituto de Pesquisas e Experimentacão
  Agropecuárias do Norte. Belém.
  Série Fertil.Solo Inst.Pesq.Exp.agropec.Norte 1971.   S. 2213 c G

Serie Fitogeografica. Ministerio de Agricultura y Ganaderia.
  Buenos Aires.
  Serie fitogeogr.Minist.Agric.Ganad. 1951 →   B. 581.9(82)I.B.

Série Fitotecnia. Instituto de Pesquisas e Experimentacão
  Agropecuárias do Norte. Belém.
  Serie Fitotec.Inst.Pesq.Exp.agropec.Norte 1970-1971.   S. 2213 c D

Série Géologie. Centre de Recherches sur les Zones Arides. CNRS. Paris.
  See Publications du Centre de Recherches sur les Zones Arides.
  CNRS. Paris. Série Géologie.   P.S 1170

Serie Historica. Academia de Ciencias de Cuba. La Habana.
  Serie hist.Acad.Cienc.Cuba
    No.7 → 1969 → (imp.)   S. 2287 G

| TITLE | SERIAL No. |

Serie Investigación Pesquera. Instituto Nacional de Investigaciones
    Biológico Pesqueras. Mexico.
        Serie Invest.Pesq.Mexico   Estudios. 1970 →   Z.S 2219 a

Serie Memorandum. Instituto Nacional de Investigacion y
    Fomento Mineros. Lima.
        Serie Memo.Inst.nac.Invest.Fom.min.,Lima   1954-1962.   M.S 2501

Serie Notas y Comunicaciones. Departamento de Geologia.
    Universidad de Chile. Santiago.
        Serie Notas Comun.Dep.Geol.Univ.Chile   1967 →   P.S 2035 B

Serie Oceanologica. Instituto de Oceanologia, Academia
    de Ciencias de Cuba. La Habana.
        Serie oceanol.Inst.Oceanol.Acad.Cienc.Cuba   1968 →   S. 2289 C

Serie Poeyana. La Habana.
        See Poeyana. La Habana.   S. 2287 E

Série Quimica de Solos. Instituto de Pesquisas e Experimentacão
    Agropecuárias do Norte. Belém.
        Série Quim.Solos Inst.Pesq.Exp.agropec.Norte   1970.   S. 2213 c F

Série Solos da Amazônia. Instituto de Pesquisas e Experimentacão
    Agropecuárias do Norte. Belém.
        Série Solos Amazônia Inst.Pesq.Exp.agropec.Norte   1967-1971.   S. 2213 c

Série Técnica y Didáctica. Facultad de Ciencias Naturales y
    Museo de La Plata.
        Série téc.didáct.Fac.Cienc.nat.Mus.La Plata   1949 →   S. 2231 K

Série Técnica. Instituto de Investigacão Agronómica de Angola.
    Nova Lisboa.
        Série tec.Inst.Invest.agron.Angola   No.2 → 1968 →   E.S 2158 a A

Serie Tecnologia y Ciencias, Universidade de Oriente. Santiago
    de Cuba.
        Serie Tecnol.Cienc.Univ.Oriente Santiago de Cuba   1968 →   S. 2289 a

Série Tecnologia. Instituto de Pesquisas e Experimentacão
    Agropecuárias.do Norte. Belém.
        Série tecnol.Inst.Pesq.Exp.agropec.Norte   1970.   S. 2213 c E

Série Transformacion de la Naturaleza. Academia de Ciencias
    de Cuba. Habana.
        Série Transform.Nat.Acad.Sci.Cuba   No.5 → 1968 →   S. 2289 E

Series in the Biological Sciences. Western Illinois University.
    Macomb.   See General
        Series biol.Sci.west.Ill.Univ.   No.3 → 1963 → (imp.)   Lbry Catalogue

Series Entomologica. Den Haag.
        Series Ent.   1966 →   E.S 610

TITLE	SERIAL No.

Seriya Broshyur. Botanicheskiĭ Kabinet i Botanicheskiĭ Sad
    Imperatorskago Nikitskago Sada. Yalta.
    Ser.Broshyur bot.Kab.bot.Sad imp.nikitsk.Sad
    No.2-3, 6-7. 1916-1917.   B.S 1445 a

Seriya Monograficheskaya. Geologicheskiĭ Muzei im. A.P. Karpinskogo.
    Akademiya Nauk SSSR. Moskva.
    Ser.Monogr.geol.Muz.A.P.Karpinskogo   1954 →   P.S 516 A

Service Bulletin. Division of Fishery Research, Department of
    Natural Resources, Newfoundland. St. John's.
    Serv.Bull.Div.Fish.Res.,Newfoundld   1935-1940 (imp.)
    No.1 was orginally issued as report of the Newfoundland
    Fishery Research Commission   Vol.2 No.4   Z.S 2670 C

Service Publications. Department of Health. Australia, Melbourne.
    Serv.Publs Dep.Hlth Aust.   1922 → (1927).
    Formerly Service Publications. Quarantine Service.
    Australia, Melbourne.   S. 2122 A

Service Publications. Department of Health Australia.
    Tropical Division. Melbourne.
    Serv.Publs Dep.Hlth Aust.trop.Div.   1923-1926.   S. 2122 B

Service Publications. Quarantine Service, Australia. Melbourne.
    Serv.Publs Quarant.Serv.Aust.   1912-1920.
    Continued as Service Publications. Department of Health.
    Australia, Melbourne.   S. 2122 A

Service Publications. School of Public Health and Tropical
    Medicine, University of Sydney.
    Serv.Publs Sch.publ.Hlth trop.Med.Univ.Sydney
      1934-1946: 1958 → (Not issued 1947-1957.)   S. 2133

Service and Regulatory Announcements. Bureau of Animal Industry.
    Washington.
    Serv.regul.Announc.Bur.Anim.Ind.,Wash.   1914-1954.   E.S 2458 m

Service and Regulatory Announcements. Bureau of Entomology and
    Plant Quarantine. Washington.
    Serv.regul.Announc.Bur.Ent.Pl.Quarant.,Wash.   1933-1951.
    Formerly Service and Regulatory Announcements. Plant
    Quarantine & Control Administration. Washington.
    Continued as Service and Regulatory Announcements. Plant Pest
    Control Branch and Plant Quarantine Branch. Washington.   E.S 2458 l

Service and Regulatory Announcements. Plant Pest Control Branch
    and Plant Quarantine Branch. Washington.
    Serv.regul.Announc.Pl.Pest Control Brch Wash.   1954-1956.
    Formerly Service and Regulatory Announcements. Bureau of
    Entomology and Plant Quarantine. Washington.   E.S 2458 l

| TITLE | SERIAL No. |

Service and Regulatory Announcements. Plant Quarantine and
    Control Administration. Washington.
        Serv.regul.Announc.Pl.Quarant.Control Adm.,Wash. 1931-1932.
        Continued as Service and Regulatory Announcements.
    Bureau of Entomology and Plant Quarantine. Washington.   E.S 2458 1

Session Report. General Fisheries Council for the Mediterranean.
    Food and Agriculture Organization of the United Nations. Rome.
        Sess.Rep.gen.Fish.Coun.Mediterr. No.9, 1968.
        Formerly Proceedings and Technical Papers. General Fisheries
    Council for the Mediterranean. Rome.
    Continued as Report. General Fisheries Council for the
    Mediterranean.   Z.S 2713 D

Severnaya Mongoliya. Leningrad.
    Sev.Mongol. 1926-1928.   73 B.o.S

Seward Memorial Lectures. Birbal Sahni Institute of Palaeobotany.
    Lucknow.
    Seward meml Lect.Birbal Sahni Inst.Palaeobot. 1954 →   P.S 187 E B

Shchorichnȳk Ukrayins'kogo Botanichnogo Tovarystva. Kȳyiv.
    Shchorichnȳk ukr.bot.Tov. 1959 →   B.S 1373 a

Sheet Memoirs. Geological Survey of Great Britain and the Museum
    of Practical Geology. London.
    See Memoirs of the Geological Survey of the United Kingdom.   P.S 1005

Sheffield University Geological Society Journal.
    See Journal of the University of Sheffield Geological
    Society.   P.S 129

Shellfish Information Leaflet. Burnham-on-Crouch.
    Shellfish Inf.Leafl. 1965 →   Z.S 460 H

Shikoku Chuho. Entomological Society of Japan, Shikoku Branch.
    Shikoku Chuho 1957 →   E.S 1915 c

Shin Konchū. Tokyo.
    Shin Konchū 1950-1959.   E.S 1919

Shiroari. Tokyo.
    Shiroari 1962 →   E.S 1913 a

Shizenshi-Kenkyu. Occasional Papers from the Osaka Museum of
    Natural History. Osaka.
    Shizenshi-Kenkyu 1968 →   S. 1992 c B & E.S 1901 b

TITLE	SERIAL No.

Shokubutsu Gaku Zasshi.
   See Botanical Magazine. Tokyo.   B.S 1997

Shooting Times & Country Magazine. London.
   Shooting Times 1951 → (imp.)   Z.S 316

Short Papers. Department of Geology and Mineral Industries,
   State of Oregon. Portland.
   Short Pap.Dep.Geol.miner.Ind.Oregon
   No.2 → 1940 → (imp.)   P.S 1968 D

Short Papers. Geological Division. Department of Lands and Mines.
   Tanganyika Territory. Dar Es Salaam.
   Short Pap.geol.Div.Tanganyika 1936-1947.   P.S 1179 A
     No.13-22, (imp.) 1936-1939.   M.S 2111
   Formerly and Continued as Short Papers. Geological Survey.
   Tanganyika Territory.

Short Papers. Geological Survey. Tanganyika Territory.
   Dar Es Salaam.
   Short Pap.geol.Surv.Tanganyika 1928-1934. 1950 →   P.S 1179 A
                                No.8, 1931.   M.S 2111
   (From 1936-1945 Styled Short Paper. Geological Division.
   Department of Lands and Mines. Tanganyika Territory.)

Short Papers from the Institute of Geology and Paleontology,
   Tôhoku University, Sendai.
   Short Pap.Inst.Geol.Paleont.Tôhoku Univ. Nos.1-5, 1950-1954. P.S 1770 D

Short Report. Geological Survey. Southern Rhodesia. Salisbury.
   Short Rep.Geol.Surv.Sth.Rhod. 1919 →   P.S 1172 A

Short Review of Mining Operations in the State of South
   Australia. Adelaide
   Short Rev.Min.Ops S.Aust. 1903-1904.
   Continued as Review of Mining Operations in the
   State of South Australia. Adelaide.   M.S 2410

Shropshire County Museum Publications. Ludlow.
   Shrops.Cty Mus.Publs 1971 →
   Formerly Ludlow Museum Publications. Ludlow.   S. 361 a

Siam Science Bulletin. Bangkok.
   Siam Sci.Bull. 1937-1939, 1947-1948.
   Continued as Thai Science Bulletin, Bangkok.
     which title it had from 1939-1941.   S. 1913 a

Siboga-Expeditie. Leiden.
   Siboga Exped. 1901 →   77.f.W

Sida Contributions to Botany. Dallas.
   Sida Contr.Bot. 1962 →   B.S 4388

TITLE	SERIAL No.

Sieboldia. Kyushu University. Fukuoka.
    Sieboldia   1952 →   E.S 1932 a

Sierra Leone Studies. Freetown.
    Sierra Leone Stud. Nos.1-3 (Abridged edition)
    No.5-22. 1922-1939.   S. 2020

Silva Fennica. Suomen Metsätieteellinen Seura. Helsinki.
    Silva fenn. 1926 →   B.S 1486

Silvaecultura Tropica et Subtropica. Prague.
    Silvic.trop.subtrop. 1969 →   B.S 1246

Sind University Research Journal. Hyderabad.
    Sind Univ.Res.J. Science Series 1965 →   S. 1940 b

Sinensia. Contributions from the Metropolitan Museum of
    Natural History and the Institute of Zoology, Academy
    Sinica. Shanghai, Nanking.
    Sinensia, Shanghai 1929-1949.   S. 1969 A

Sinensia. Special Bulletin of the Metropolitan Museum of Natural
    History. Nanking.
    Sinensia, Nanking No.1, 1930.   B.S 1880

Singapore Naturalist. Singapore.
    Singapore Nat. 1922-1925.   S. 1929 & T.R.S 3201
    Continued as Malayan Naturalist. Singapore.

Siruna Seva. Berlin.
    Siruna Seva 1940-1953.   E.S 1365

Sistematicheskie Zametki po Materialam Gerbariya Tomskogo
    Universiteta.
    See Animadversiones Systematicae ex Herbario
    Universitatis Tomskensis.   B.S 1440

Sistematicheskie Zametki po Materialam Zoologicheskogo Muzeya,
    Biologicheskogo Instituta pri Tomskom Gosudarstvennom
    Universitete imeni V.V. Kuĭbȳshėva.
    See Animadversiones Systematicae ex Museo Zoologico
    Instituti Biologici Universitatis Tomskensis.   Z.S 1846

Sitzungsberichte und Abhandlungen der Naturforschenden
    Gesellschaft zu Rostock.
    Sber.Abh.naturf.Ges.Rostock 1909-1918, 1925-1939.
    Discontinued in 1918, re-started in 1925.
    (From 1886-1908 the Sitzungsberichte published in
    Archiv des Vereins der Freunde der Naturgeschichte
    in Mecklenburg.   S. 1576

TITLE	SERIAL No.

Sitzungsberichte und Abhandlungen der Naturwissenschaftlichen
    Gesellschaft Isis zu Bautzen.
    <u>Sber.Abh.naturw.Ges.Isis Bautzen</u> 1898-1905.
    <u>Continued as</u> Bericht über die Tätigkeit der
    Naturwissenschaftlichen Gesellschaft Isis zu Bautzen.   S. 1320 A

Sitzungsberichte und Abhandlungen der Naturwissenschaftlichen
    Gesellschaft Isis in Dresden.
    <u>Sber.Abh.naturw.Ges.Isis Dresd.</u> 1881-1939 (1940).   S. 1388 B
                                    1881-1912.   T.R.S 1363
    <u>Formerly</u> Sitzungsberichte der Naturwissenschaftlichen
    Gesellschaft Isis zu Dresden.

Sitzungsberichte der Akademie der Wissenschaften. Mathematisch-
    Naturwissenschaftliche Classe. Wien.
    <u>Sber.Akad.Wiss.Wien</u> 1848-1947.
    From 1861 in sections.
    Abteiling I. Mineralogie, Biologie, Erdkunde und
    Verwandte Wissenschaft (formerly Mineralogie, Kristallographie,
    Botanik, Zoologie, etc.) 1861-1947.
    Abteilung II. Mathematik, Physik, Chemie, Mechanik, etc.
    1861-1887.
    Abt. II. then subdivided.
    Abteiling II a. Mathematik, Astronomie, Physik, etc. 1888-1947.
    Abteilung II b. Chemie. 1888-1944.
    <u>Continued in</u> Monatshefte für Chemie und Verwandte Teile
    anderer Wissenschaften, Wien.
    Abteilung III. Physiologie, Anatomie, Theoretische
    Medizin. 1872-1923.
    (Incorporated in Abt.I.& II.)
    <u>Continued as</u> Sitzungsberichte der Österreichischen
    Akademie der Wissenschaften. Mathematisch-Naturwissenschaftliche
    Klasse.   S. 1702 A

Sitzungsberichte. Bayerische Akademie der Wissenschaften.
    Mathematisch-Naturwissenschaftliche Klasse. München.
    <u>Sber.bayer.Akad.Wiss.</u> 1955 →
    <u>Formerly</u> Sitzungsberichte der Mathematisch-
    Naturwissenschaftlichen (Abteilung) Klasse der Bayerischen
    Akademie der Wissenschaften zu München.   S. 1310 B

Sitzungsberichte der Deutschen Akademie der Wissenschaften
    zu Berlin.
    <u>Sber.dt.Akad.Wiss.Berl.</u>
      Physikalisch-Mathematische Klasse 1948-1949.
      Klasse für Mathematik und Allgemeine
      Naturwissenschaften 1950-1954.
      Klasse für Chemie, Geologie und Biologie 1956-1968.
      Klasse für Bergbau, Hüttenwesen und Montangeologie 1960-1967.
    <u>Formerly</u> Sitzungsberichte der Preussischen Akademie
der Wissenschaften zu Berlin.   S. 1305 A.-Aa-Ab

TITLE	SERIAL No.

Sitzungsberichte des Deutschen Naturwissenschaftlich-Medizinischen
    Vereins fur Böhmen "Lotos" in Prag.
        Sber.dt.naturw.-med.Ver.Böhm "Lotos" 1896-1906.
        Formerly & Continued as Lotos. Zeitschrift fur
Naturwissenschaftlichen. Prag.   S. 1762 A

Sitzungsberichte der Finnischen Akademie der Wissenschaften.
    Helsinki.
        Sber.finn.Akad.Wiss. 1950 →   S. 1827 B

Sitzungsberichte der Geologischen Landesanstalt. Berlin.
        Sber.geol.Landesanst. 1926-1930
        Continued as Sitzungsberichte der Preussischen Geologischen
Landesanstalt. Berlin.   P.S 1331

Sitzungsberichte der Gesellschaft zur Beförderung der
    Gesamten Naturwissenschaften zu Marburg.
        Sber.Ges.Beförd.ges.Naturw.Marburg 1866-1966.
        Continued as Sitzungsberichte der Wissenschaftlichen
Gesellschaft zu Marburg.   S. 1531 C

Sitzungsberichte der Gesellschaft für Morphologie und
    Physiologie. München.
        Sber.Ges.Morph.Physiol.Münch. 1885-1909.   S. 1555

Sitzungsberichte der Gesellschaft Naturforschender Freunde
    zu Berlin.
        Sber.Ges.naturf.Freunde Berl. 1860-1942, 1961 →   S. 1326 D
        1839-1859, 1893-1938.   T.R.S 1336 B

Sitzungsberichte der Heidelberger Akademie der Wissenschaften.
    Mathematisch-Naturwissenschaftliche Klasse.
        Sber.heidelb.Akad.Wiss. 1909-1910.
        Math-nat.Kl.B. Biologische Wissenschaften. 1911-1924.
        Math-nat.Kl. (nat.hist.) 1925-1944.   S. 1456 B

Sitzungsberichte der Kaiserlichen Akademie der Wissenschaften.
    Mathematisch-Naturwissenschaftliche Classe, Wien.
        See Sitzungsberichte der Akademie der Wissenschaften.
Mathematisch-Naturwissenschaftliche Classe. Wien.   S. 1702 A

Sitzungsberichte der Königl.Bayerischen Akademie der
    Wissenschaften zu München.
        Sber.bayer.Akad.Wiss. 1860-1870.
        Continued as Sitzungsberichte der Mathematisch-
Physikalischen Classe der Königlichen Bayerischen Akademie
der Wissenschaften.   S. 1310 B

| TITLE | SERIAL No. |

Sitzungsberichte der Königlichen Böhmischen Gesellschaft der
    Wissenschaften. Prag.
    Sber.K.böhm.Ges.Wiss.Math.-nat.Kl. 1859-1917.
    Continued as Mémoires de la Société Royale des (Lettres)
    et des Sciences de Bohême. Prague. Classes des Sciences.   S. 1703 A

Sitzungsberichte der Königlichen Preussischen Akademie der
    Wissenschaften zu Berlin.
    See Sitzungsberichte der Preussischen Akademie der
    Wissenschaften zu Berlin.   S. 1305 A

Sitzungsberichte der Mathematisch-Naturwissenschaftlichen
    (Abteilung) Klasse der Bayerischen Akademie der
    Wissenschaften zu München.
    Sber.bayer.Akad.Wiss. 1924-1954.
    Formerly Sitzungsberichte der Mathematisch-Physikalischen
    Classe der Bayerischen Akademie der Wissenschaften
    zu München.
    Continued as Sitzungsberichte. Bayerische Akademie der
    Wissenschaften. Mathematisch-Naturwissenschaftliche
    Klasse. München.   S. 1310 B

Sitzungsberichte der Mathematischen-Physikalischen Classe der
    Königlich Bayerischen Akademie der Wissenschaften zu München.
    Sber.bayer.Akad.Wiss. 1871-1923.
    Formerly Sitzungsberichte der Königl.Bayer. Akademie
    der Wissenschaften.
    Continued as Sitzungsberichte der Mathematisch-
    Naturwissenschaftlichen (Abteilung) Klasse der Bayerischen
    Akademie der Wissenschaften zu München.   S. 1310 B

Sitzungsberichte der Medizinisch-Naturwissenschaftlichen
    Sektion des Siebenbürgischen Museumvereins. Kolozsvart.
    See Ertesitö az Erdélyi Muzeum Egyesület Orvos-
    Termeszettudományi Szakosztályabol. Kolozsvar.   S. 1743

Sitzungsberichte der Naturforschenden Gesellschaft zu Leipzig.
    Sber.naturf.Ges.Lpz. 1874-1932.   S. 1516 B

Sitzungsberichte der Naturforschenden Gesellschaft zu Rostock.
    See Sitzungsberichte und Abhandlungen der Naturforschenden
    Gesellschaft zu Rostock.   S. 1576

Sitzungsberichte der Naturforscher Gesellschaft zu Dorpat.
    Sber.naturf.Ges.Dorpat 1853-1898.
    Continued as Protokoly Obshchestva Estestvoispytateleĭ
    pri Imperatorskom Yur'evskom Universitete.Yur'ev. (Dorpat).   S. 1816 A

| TITLE | SERIAL No. |

Sitzungsberichte der Naturforscher-Gesellschaft bei der
   Universität Tartu (Dorpat).
   See Protokoly Obshchestva Estestvoispytateleĭ Imperatorskom
   Yur'evskom Universitete. Yur'ev. (Dorpat).   S. 1816 A

Sitzungsberichte. Naturhistorischen Verein der Preussischen
   Rheinlande und Westfalens.
   Sber.naturh.Ver.preuss.Rheinl.Westf. 1906-1933.
   Formerly Sitzungsberichte der Niederrheinischen Gesellschaft
   für Natur-und Heilkunde. Bonn.
   Continued in Decheniana. Verhandlungen des
   Naturhistorischen Vereins der Rheinlande und Westfalens.
   Bonn. 1936 →   S. 1361 B

Sitzungsberichte der Naturwissenschaften Gesellschaft Isis
   zu Dresden.
   Sber.naturw.Ges.Isis Dresden 1861-1880.
   Continued as Sitzungsberichte und Abhandlungen der
   Naturwissenschaften Gesellschaft Isis in Dresden.
         S. 1388 B & T.R.S 1363

Sitzungsberichte der Niederrheinischen Gesellschaft für
   Natur-und Heilkunde. Bonn.
   Sber.niederrhein.Ges.Nat.u.Heilk. 1854-1905.
   (The volumes for 1854-1894 were published in S.1361
   A. Verhandlungen.)
   Continued as Sitzungsberichte. Naturhistorischer Verein
   der Preussischen Rheinlande und Westfalens Bonn.   S. 1361 B

Sitzungsberichte der Österreichischen Akademie der Wissenschaften.
   Mathematisch-Naturwissenschaftliche Klasse.
   Sber.öst.Akad.Wiss. 1947 →
   Abt.I, Biologie, Mineralogie, Erdkunde und Verwandte
   Wissenschaften.
   Abt.IIa, then II, Mathematik, Astronomie, Physik, Meteorologie
   und Technik.
   Formerly Sitzungsberichte der Akademie der Wissenschaften.
   Mathematisch-Naturwissenschaftliche Klasse.   S. 1702 A

Sitzungsberichte der Physikalisch-Medizinischen Gesellschaft
   zu Würzburg.
   Sber.phys.-med.Ges.Würzb. 1881-1923.
   (From 1869-1880 and from 1923-1954 forms part of the
   Verhandlungen und Bericht der Physikalisch-Medizinischen
   Gesellschaft zu Würzburg.)   S. 1611

Sitzungsberichte der Physikalisch-Medizinischen Sozietät
   zu Erlangen.
   Sber.phys.-med.Soz.Erlangen 1870 →
   Formerly Verhandlungen der physikalisch-medizinischen
   Societät zu Erlangen.   S. 1405 A

| TITLE | SERIAL No. |

Sitzungsberichte der Preussischen Akademie der Wissenschaften
  zu Berlin.
    Sber.preuss.Akad.Wiss. 1882-1938.
    Formerly Monatsberichte der Königlichen Preuss.
    Akademie der Wissenschaften zu Berlin.
    Continued as Sitzungsberichte der Deutschen Akademie der
    Wissenschaften zu Berlin. — S. 1305 A

Sitzungsberichte der Preussischen Geologischen Landesanstalt. Berlin.
    Sber.preuss.geol.Landesanst. 1931-1932.
    Formerly Sitzungsberichte der Geologischen Landesanstalt.
    Berlin. — P.S 1331

Sitzungsberichte der Sächsischen Akademie der Wissenschaften
  zu Leipzig.
    Sber.sächs.Akad.Wiss. vol. 105 → 1961 →
    Math.-naturw.Kl.
    Formerly Bericht über die Verhandlungen der (Kgl.) Sächsischen
    Akademie der Wissenschaften zu Leipzig. — S. 1503 A

Sitzungsberichte der Serbischen Geologischen Gesellschaft.
    See Zapisnici Srpskog Geološkog Društva. — P.S 450

Sitzungsberichte. Vereins für Vaterländische Naturkunde
  in Württemberg. Stuttgart.
    See Jahreshefte des Vereins für Vaterländische
    Naturkunde in Württemberg. Stuttgart. — S. 1591 A

Sitzungsberichte der Wissenschaftlichen Gesellschaft zu Marburg.
    Sber.wiss.Ges.Marburg 1967.
    Formerly Sitzungsberichte der Gesellschaft zur Beförderung
    der Gesamten Naturwissenschaften zur Marburg. — S. 1531 C

Skandia. Tidskrift för Vetenskap och Konst. Upsala.
    Skandia 1833. — S. 593

Skånes Natur. Lund.
    Skånes Nat. Arsskrift No.9 → 1921 → — S. 577 a A
               Kontaktblad Vol.52 → 1965 → — S. 577 a B
    Formerly Skånes Naturskyddsförenings Arsberättelse. Lund. — S. 577 a A

Skånes Naturskyddsförenings Arsberättelse. Lund.
    Skånes Nat.Arsb. No.5 & 6. 1913-1915.
    Formerly Meddelanden från Skånes Naturskyddsförening.
    Continued as Skånes Natur. Skånes Naturskyddsförenings
    Arsskrift. — S. 577 a

| TITLE | SERIAL No. |

Skrifter fra Danmarks Fiskeri- og Havundersøgelser. København.
　Skr.Danm.Fisk.-og Havunders. No.15-31, 1953-1971.
　Formerly Skrifter udgivne af Kommissionen for Danmarks
　Fiskeri-og Havundersøgelser.
　Continued as Fisk og Hav. Kobenhavn.　　　　　　　　　　　Z.S 570 D

Skrifter som udi det Kiøbenhavnske Selskab. Kjøbenhavn.
　Skr.Kiøbenh.Selsk. 1743-1770.
　Continued as Skrifter som udi det Kongelige
　Videnskabers Selskab. Kjøbenhavn.　　　　　　　　　　　　S. 510 B

Skrifter udgivne af Kommissionen for Danmarks Fiskeri-og
　Havundersøgelser. København.
　Skr.Kommn Danm.Fisk.-og Havunders. No.10-14. 1927-1952.
　Formerly Skrifter udgivne af Kommissionen for Havundersøgelser.
　Continued as Skrifter fra Danmarks Fiskeri-og
　Havundersøgelser. København.　　　　　　　　　　　　　　　Z.S 570 D

Skrifter udgivne af Kommissionen for Havundersøgelser. København.
　Skr.Kommn Havunders., Kbh. Nos.3-9, 1905-1919.
　Continued as Skrifter udgivne af Kommissionen for Danmarks
　Fiskeri- og Havundersøgelser.　　　　　　　　　　　　　　　Z.S 570 D

Skrifter. Kongelige Danske Videnskabernes Selskabs.
　See Kongelige Danske Videnskabernes Selskabs Skrifter.
　Kjøbenhavn.　　　　　　　　　　　　　　　　　　　　　　　　S. 510 B

Skrifter. Kongelige Norske Videnskabers Selskabs. Kiøbenhavn.
　Skr.K.norske Vidensk.Selsk. 1768-1774.
　Formerly Skrifter. Trondhiemske Selskabs. Kiøbenhavn.
　Continued as Skrifter.Nye Samling af det Kongelige Norske
　Videnskabers Selskabs. Kiøbenhavn.　　　　　　　　　　　　S. 554 A

Skrifter som udi det Kongelige Videnskabers Selskab. Kjøbenhavn.
　Skr.K.Vidensk.Selsk. Deel 11-12. 1777-1779.
　Formerly Skrifter som udi det Kiøbenhavnske Selskab.
　Kjøbenhaven.
　Continued as Nye Samling af det Kongelige Danske
　Videnskabers Selskabs Skrifter. Kiøbenhavn.　　　　　　　S. 510 B

Skrifter från Mineralogisk-och Paleontologisk-Geologiska
　Institutionerna, Lund.
　Skr.miner.-o.paleont.-geol.Inst.Lund 1949-1957
　Formerly Meddelanden från Lunds Geologisk-Mineralogiska
　Institution, Lund.
　Continued as Publications from the Institutes of Mineralogy,
　Paleontology, and Quaternary Geology. University of Lund.　P.S 423

Skrifter af Naturhistorie-Selskabet. Kiøbenhavn.
　Skr.Naturh.-Selsk.Kiøbenhavn 1790-1810.　　　　S. 521 A & T.R.S 663

| TITLE | SERIAL No. |

Skrifter i Naturskyddsärenden. Kungliga Svenska Vetenskapsakademiens.
   See Kungliga Svenska Vetenskapsakademiens Skrifter i
   Naturskyddsärenden. Stockholm.   S. 570 O

Skrifter. Norges Geologiske Undersøkelse. Oslo.
   See Norges Geologiske Undersögelse. Kristiania.   P.S 1450

Skrifter. Norsk Polarinstitutt. Oslo.
   Skr.norsk Polarinst. 1948 →
   Formerly Skrifter om Svalbard og Ishavet (formerly og
   Nordishavet.) Oslo.   S. 539 C

Skrifter utgitt av det Norske Videnskaps-Akademi i Oslo.
   Mat.-naturv.Kl.
   Skr.norske Vidensk-Akad.mat.-nat.Kl. 1925 →
   Formerly Skrifter udgivne af Videnskabsselskabet i
   Christiania. Mat.-Nat.Klasse.   S. 551 A

Skrifter. Nye Samling af det Kongelige Norske Videnskabers
   Selskabs. Kiøbenhavn.
   Skr.Nye Saml.norske Vidensk.Selsk. 1784-1788.
   Formerly Skrifter. Kongelige Norske Videnskabers Selskabs.
   Kiøbenhavn.
   Continued as Skrifter. Nyeste Samling af det Kongelige
   Norske Videnskabers Selskabs. Kjøbenhavn.   S. 554 A

Skrifter. Nyeste Samling af det Kongelige Norske
   Videnskabers Selskabs. Kjøbenhavn.
   Skr.Nyeste Saml.norske Vidensk.Selsk. 1798.
   Formerly Skrifter. Nye Samling af det Kongelige Norske
   Videnskabs Selskabs. Kiøbenhavn.
   Continued as Kongelige Norske Videnskabers Selskabs Skrifter.
   Kjøbenhavn & Trondheim.   S. 554 A

Skrifter utgivna av Södra Sveriges Fiskeriförening. Lund.
   Skr.söd.Sver.FiskFör. 1949 →   Z.S 518

Skrifter Statens Växtskyddsanstalt. Stockholm.
   Skr.St.Växtskyddsanst. 1933-1952.   E.S 507

Skrifter om Svalbard og Ishavet (formerly og Nordishavet.) Oslo.
   Skr.Svalbard Ishavet 1927-1947
   Formerly Resultater av de Norske Statsunderstøttede
   Spitzbergenekspeditioner. Oslo.
   Continued as Skrifter om Norsk Polarinstitut. Oslo.   S. 539 C

Skrifter utgivna av Svenska Linné Sällskapet. Uppsala & Stockholm.
   Skr.svenska Linné-Sallsk. 1919-1925.   S. 592 B

Skrifter utgifna af Svenska Litteratursällskapet i Finland.
   Helsingfors.
   Skr.svenska LittSällsk.Finland Nos.16, 17, 19, 21, 23,
   26, 32, 35, 37, 42, 50, 53, 66, 80, 93, 97, 104, 114,
   120, 166, 203, 210 & 419. 1890-1966.   S. 1824

TITLE	SERIAL No.

Skrifter. Trondhiemske Selskabs. Kiøbenhavn.
    Skr.Trondh.Selsk. 1761-1765.
    Continued as Skrifter. Kongelige Norske Videnskabers
    Selskabs. Kiøbenhavn.   S. 554 A

Skrifter udgivet af Universitetets Zoologiske Museum.
    See Spolia Zoologica Musei Hauniensis.   Z.S 575

Skrifter udgivne af Videnskabsselskabet i Christiania.
    I Math.-Nat. Klasse.
    Skr.VidenskSelsk.Christiania   1894-1924.
    Continued as Skrifter utgitt av det Norske
    Videnskaps-Akademi i Oslo. Mat.-Nat.Kl.   S. 551 A

Skýrsla. Islenzka Náttúrufraedisfélag. Reykjavik.
    Skýrs.isl.NáttFraedFél. 1893-1946.   S. 503

Skýrsla. Vísindefélag Islendinga. Reykjavik.
    Skýrs.Vísindefél.Isl. 1933-1934.   S. 502 C

Slovenská Archeologia. Bratislava.
    Slov.Archeol. 1953 →   P.A.S 79

Små Godbiter fra Samlingene. Universitetet i Bergen.
    Små Godb.Saml. 1951-1958.
    Continued as Godbiter fra Samlingene.Universitetet
    i Bergen.   S. 551 H

Smaaskrifter. Norges Geologiske Undersøkelse. Kristiania.
    Smaaskr.Norg.geol.Unders. 1922-1927.   P.S 1450 A

Småskrifter. Stavanger Museum. Series Zoologica.
    Småskr.Stavanger Mus. 1951-1952.
    Continued as Sterna. Stavanger Museum.   T.B.S 622

Smith College Fiftieth Anniversary Publications.
    Northampton, Mass.
    See Publications. Smith College Fiftieth Anniversary.
    Northampton, Mass.   S. 2387

Smithsonian. Washington, D.C.
    Smithsonian 1970 →   S. 2426 J

Smithsonian Contributions in Anthropology. Washington.
    Smithson.Contr.Anthrop. 1965 →   P.A.S 800 C

Smithsonian Contributions to Astrophysics. Washington.
    Smithson.Contr.Astrophys. 1956-1973.   M.S 2651 B

Smithsonian Contributions to Botany. Washington, D.C.
    Smithson.Contr.Bot. 1969 →   B.S 4408

TITLE	SERIAL No.

Smithsonian Contributions to the Earth Sciences. Washington.
    Smithson.Contr.Earth Sci. 1969 →   P.S 891

Smithsonian Contributions to Knowledge. Washington.
    Smithson.Contr.Knowl. 1848-1916.   S. 2426 C

Smithsonian Contributions to Paleobiology. Washington.
    Smithson.Contr.Paleobiol. 1969 →   P.S 891 B

Smithsonian Contributions to Zoology. Washington, D.C.
    Smithson.Contr.Zool. 1969 →   T.R.S 5145 B & Z.S 2514 C

Smithsonian Institution Information Systems Innovations.
    Washington, D.C.
    Smithson.Inst.Inf.Syst.Innov. 1969 →   S. 2426 H

Smithsonian Journal of History. An Illustrated Journal of
    General History published quarterly by the Smithsonian
    Institution. Washington, D.C.
    Smithson.J.Hist. 1966-1968/1969.   S. 2426 F

Smithsonian Miscellaneous Collections. Washington.
    Smithson.misc.Collns 1862 →   S. 2426 B

Smithsonian Year. Washington.
    Smithson.Year 1965 →
    Formerly Report of the Board of Regents of the Smithsonian
    Institution. Washington.   S. 2426 A & T.R.S 5145 A

Snake. Nittagun.
    Snake 1969 →   Z. Reptile Section

Snellius-Expedition in the Eastern Part of the Netherlands
    East-Indies 1929-1930. Leiden.
    Snellius Exped.east.Part Neth.E.-Indies 1934 →   77 A.q.A

Soap and Chemical Specialties. New York.
    Soap chem.Spec. 1954-1962
    Formerly Soap and Sanitary Chemicals. New York. E.S 2461

Soap and Sanitary Chemicals. New York.
    Soap sanit.Chem. 1939-1954.
    Continued as Soap and Chemical Specialties. New York. E.S 2461

Social Biology. Chicago.
    Social Biol. Vol.16 → 1969 →   P.A.S 773

Sociedad Cubana de Historia Natural "Felipe Poey", Havana.
    Soc.cub.Hist.nat.Felipe Poey 1960.   S. 2287 D

TITLE	SERIAL No.

Sociedad Mexicana de Hidrobiologia. Mexico.
   Soc.mex.Hidrobiol. No.1. 1952.   S. 2252 a

Societas Entomologica. Stuttgart.
   Societas ent. 1886-1930.
   Supplementary to & Continued as Entomologische Rundschau.   E.S 1326

Société Agricole, Scientifique et Littéraire des Pyrénées-Orientales. Perpignan.
   See Bulletin de la Société Agricole, Scientifique et Litteraire des Pyrénées-Orientales. Perpignan.   S. 952

Société d'Histoire Naturelle du Creusot. Bulletin.
   Soc.Hist.nat.Creusot 1932-1949, 1974 →   S. 856
   Between 1950 and 1973, incorporated in Revue de la Fédération Francaise des Sociétés des Sciences Naturelles (S. 978a)

Société des Sciences et Arts de Vitry-le-François.
   See Mémoires. Société des Sciences et Arts de Vitry-le-François.   S. 979

Société des Sciences Naturelles de Grand-Duché de Luxembourg.
   See Publications. Société des Sciences Naturelles de Grand-Duché de Luxembourg.   S. 651

Soil Biology. Paris.
   Soil Biol. 1964 →   S. 2742

Soil Biology and Biochemistry. Oxford.
   Soil Biol.Biochem. 1969 →   S. 487 a

Soil Bulletin. Geological Survey of China. Peiping.
   Soil Bull.Peiping 1930-1944   P.S 1789

Soil Conservation. Washington.
   Soil Conserv. 1935 →   E.S 2458 h

Soil Science. New Brunswick, N.J.
   Soil Sci. 1916 →
   (1916-1960 at Tring.)   S. 2475

Soil Survey Reports. United States Bureau of Soils. Washington.
   Soil Surv.Rep.U.S.Bur.Soils 1923-1940
   Formerly Field Operations of the Bureau of Soils. Washington. P.S 1865

Soils and Fertilizers. Harpenden.
   Soils Fertil., Harpenden Vol.11 → 1948 →   B.S 140 a

Soils Quarterly. National Geological Survey of China. Nanking.
   Soils Q.Nanking Vol.6, nos 2-4; Vol.7, no.1. 1947-48   P.S 1789 C

| TITLE | SERIAL No. |

Solanaceae Newsletter. Birmingham.
    Solanaceae Newsl. 1974 →   B.S 153

Solanus. Bulletin of the Sub-Committee on Slavonic and
    East European Materials of the Standing Committee on
    National and University Libraries (SCONUL).
    Solanus 1966 →   S. 209 a

Solenni Adunanze del R. Istituto Lombardo di Scienze e
    Lettere. Milano.
    Solenni Adun.Ist.lombardo 1864-1868.   S. 1104 D

Somerset Birds. Taunton.
    See Report on Somerset Birds. Taunton.   T.B.S 254

Sonderausgabe. Geologisches Institut. Beograd.
    See Posebna Izdanja, Geoloshki Institut.   P.S 456

Sonderheft. Mitteilungen der Österreichischen Mineralogischen
    Gesellschaft. Vienna.
    See Mitteilungen der Österreichischen Mineralogischen
    Gesellschaft. Vienna.   M.S 1310 A

Sonderheft. Münchener Beitrage zur Geschichte und Literature der
    Naturwissenschaften und Medizin. München.
    Sonderh.Münch.Beitr.Gesch.Lit.Naturw.Med. 1926-1928.   3.o.D

Sonderhefte der Carinthia II.
    See Carinthia II.   S. 1741 B

Sonderhefte zum Repertorium Novarum Specierum Regni
    Vegetabilis. Berlin.
    Sonderh.Repert.nov.Spec.Regn.veg. 1925 →
    Supplement to Repertorium Novarum Specierum Regni
    Vegetabilis. Berlin.   B.S 904 b

Sonderschriften. Bayerische Akademie der Wissenschaften. München.
    Sonderschr.bayer.Akad.Wiss. 1963.   S. 1310 G

Sonderveröffentlichungen des Geologischen Institutes der
    Universität Köln. Köln.
    Sonderveröff.geol.Inst.Univ.Köln 1956 →   P.S 351

Soobshcheniya Akademii Nauk Gruzinskoĭ SSR. Tbilisi.
    Soobshch.Akad.Nauk gruz.SSR Tom 1, No.5 → 1940 → (imp.)   S. 1866 a
    1941-1944 (Min.& petrogr.papers only).   M.S 1827 A

Soobshcheniya Gruzinskogo Filiala Akademii Nauk SSSR. Tbilisi.
    See Soobshcheniya Akademii Nauk Gruzinskoĭ SSR. Tbilisi.   S. 1866 a

Soobshcheniya Instituta Lesa. Moskva.
    Soobshch.Inst.Lesa 1953-1958.   B.S 1383

TITLE	SERIAL No.

Soobshcheniya Laboratorii Lesovedeniya. Moskva.
   Soobshch.Lab.Lesov. Vyp.1, 1959.   B.S 1389 b

Soobshcheniya Pribaltiĭskoĭ Komissii po Izucheniyu Migratsiĭ
   Ptits. Tartu.
   Soobshch.pribalt.Kom.Izuch.Migr.Ptits 1961 →   Z.S 1870

Soölogiese Navorsing van die Nasionale Museum. Bloemfontein.
   Soöl.Navors.nas.Mus.Bloemfontein 1935-1947.
   Continued in Navorsing van die Nasionale Museum.
   Bloemfontein.   Z.S 2083

Sorby Record. Sheffield.
   Sorby Rec. 1958 →   S. 358 E

Sous le Plancher. Organe du Spéléo-Club du Dijon.
   Sous Plancher 1956 →   S. 983 a

South African Animal Life. Results of the Lund University
   Expedition in 1950-1951. Edited by B.Hanström, P. Brinck.
   & G. Rudebeck. Stockholm.
   S.Afr.anim.Life 1955 →   Z. 74G q H

South African Archaeological Bulletin. Clarement-Cape.
   S.Afr.archaeol.Bull. 1945 →   P.A.S 501

South African Avifauna Series of the Percy Fitzpatrick
   Institute of African Ornithology. Cape Town.
   S.Afr.Avifauna Ser.P.Fitzpatrick Inst.afr.Orn. 1961 →   T.B.S 4006

South African Country Life. Cape Town.
   S.Afr.Ctry Life Vol.25-26 1935-1936
   Formerly South African Gardening. Johannesburg.
   Continued as S.A.G. A Magazine of Outdoor Interests.
   Cape Town.   B.S 2266

South African Forum Botanicum. Grahamstown.
   S.Afr.Forum Bot. Vol.6-7, 1968-1969.
   Continued as Forum Botanicum.   B.S 2309

South African Gardening & Country Life. Johannesburg.
   S.Afr.Gdng Vol.17-24. 1927-1934.
   Continued as South African Country Life. Cape Town.   B.S 2266

South African Horticultural Journal. Johannesburg.
   S.Afr.hort.J. 1938-1940.   B.S 2275

South African Insect Life. Pretoria.
   S.Afr.Insect Life 1945.   E.S 2160 a

South African Jeweller. Kimberley.
   See Diamond News and South African Jeweller. Kimberley.   M.S 2101

TITLE	SERIAL NO.

South African Journal of Agricultural Science. Pretoria.
    S.Afr.J.agric.Sci. 1958-1968.
    Replaced by Agricultural Science in South Africa. Pretoria.   E.S 2163

South African Journal of Industries (& Labour Gazette). Pretoria.
    S.Afr.J.Inds   1917-1925.   S. 2004

South African Journal of Natural History. Pretoria.
    S.Afr.J.nat.Hist. 1918-1930.   S. 2066 A & T.R.S 4014

South African Journal of Science. Cape Town.
    S.Afr.J.Sci. 1909 →
    Formerly Report of the South African Association
    for the Advancement of Science.   S. 2001 A

South African Quarterly Journal. Cape Town.
    S.Afr.Q.Jl 1829-1835.   S. 2006

South African Science. Johannesburg.
    S.Afr.Sci. 1947-1949.
    (Amalgamated with South African Journal of Science.)   S. 2001 B

South African Watchmaker and Jeweller. Kimberley.
    See Diamond News (and South African Watchmaker & Jeweller).
    Kimberley.   M.S 2101

South Australian Naturalist. Adelaide.
    S.Aust.Nat. 1919 →   S. 2102 C

South Australian Ornithologist. A Magazine of Ornithology.
    Adelaide.
    S.Aust.Orn. Vols.2-3, 1916-1918 (imp.)., Vol.6 → 1921 →   Z.S 2143
        1914 →   T.R.S 7201

South Dakota Geological Survey Guidebook. Vermillion.
    See Guidebook. South Dakota Geological Survey. Vermillion.   P.S 1887 D

South Devon Monthly Museum. Plymouth.
    S.Devon Mon.Mus. 1833-1836.   S. 329

South Eastern Naturalist. Canterbury.
    S.East.Nat.Canterbury   1890-1899.   S. 24 C

South Essex Naturalist.
    S.Essex Nat. 1951 →   S. 367 b B

South Tow Prospectus. Scripps Institution of Oceanography. La Jolla.
    S.Tow Prospectus 1971-1972.   S. 2319 T

| TITLE | SERIAL No. |

Southeast Asian Journal of Tropical Medicine and Public Health.
Bangkok.
SE.asian J.trop.Med.publ.Hlth 1970 →   S. 1913 e

South-Eastern Bird Report. Being an Account of Bird-Life
in Hampshire, Kent, Surrey and Sussex during 1934.
Edited by R. Whitlock. Salisbury.
SEast.Bird Rep. 1934-1936, 1943-1945.   Z.S 18 o W

Southeastern Geology. Duke University. North Carolina. Durham.
SEast.Geol. Vol.6, No.2 → 1965 →   P.S 807
Special Publ. 1968 →   P.S 807 A

South-Eastern Naturalist. London.
SEast.Nat. 1900-1927.
Formerly Report & Transactions of the South-Eastern
Union of Scientific Societies.
Continued as South-Eastern Naturalist & Antiquary.   S. 218 A

South-Eastern Naturalist & Antiquary. London.
SEast.Nat. 1928 →
Formerly South-Eastern Naturalist.   S. 218 A

Southern Methodist University Studies. Dallas, Texas.
Sth.Methodist Univ.Stud. 1946-1952.   S. 2462 C

Southern Rhodesia Geological Survey Bulletin. Bulawayo, etc.
See Bulletin. Geological Survey. Southern Rhodesia. Bulawayo, etc.   P.S 1171

Southern Science Record. Melbourne.
Sth.Sci.Rec. 1880-1883.   S. 2118

Southwest Museum Papers. Los Angeles.
SW.Mus.Pap. 1928 →   S. 2404 a D

Southwest Science Bulletin. Los Angeles, California.
SW.Sci.Bull. 1920.   S. 2404 a B

Southwestern Journal of Anthropology. Albuquerque.
SWest.J.Anthrop. Vol.13-28, 1957-1972.
Continued as Journal of Anthropological Research.   P.A.S 802

Southwestern Naturalist, The. Dallas, Tex.
SWest.Nat. 1956 →   S. 2463

Sovetskaya Antropologiya. Moskva Gosudarstvennyi Universitet
im M.V. Lomonsova.
Sov.Antrop. 1957-1960.
Continued as Voprosy Antropologii. Moskva.   P.A.S 350

Sovetskaya Botanika. Leningrad.
Sov.Bot. 1933-1946.
Formerly Izvestiya Glavnogo Botanicheskogo Sada SSSR.
(afterwards R.S.F.S.R.)   B.S 1403 c

TITLE	SERIAL No.

Sovetskaya Geologiya. Moskva.
    Sov.Geol. 1938-1940 & 1957 → (imp.)
    Formerly Problemy Sovetskoĭ Geologii. Moskva.
	P.S 1511

Soviet Botany. See Sovetskaya Botanika. Leningrad.	B.S 1403 c
Soviet Geology. Moscow. See Sovetskaya Geologiya. Moskva.	P.S 1511

Soviet Journal of Ecology. New York.
    Soviet J.Ecol. 1971 →
    Formerly Ecology. New York.
    (Translation of Ekologiya, Sverdlovsk.)
	S. 1804 d T

Soviet Physics. Crystallography, New York. Soviet Phys.Crystallogr. Vol.2 → 1957 →	M.S 1826 A

Soviet Plant Physiology. Washington.
    Soviet Pl.Physiol. vol. 8, No.3 → 1961 →
    Formerly Plant Physiology. Washington.
	B.S 1387 b

Soviet Science Review. Guildford, Surrey. Soviet Sci.Rev. 1970-1972.	S. 2744

Space Life Sciences. Dordrecht.
    Space Life Sci. 1968-1973.
    Continued as Origins of Life. Dordrecht.
	S. 691

Span. (Shell Public Health & Agricultural News). London. Span 1958 →	E.S 48

Specchio delle Scienze o Giornale Enciclopedico di Sicilia,
    (Rafinesque). Palermo.
    Specchio Sci.Palermo 1814.
        (Photostat reproduction).
	S. 1190 a

Special Bulletin. Baylor University Museum, Waco, Texas. Spec.Bull.Baylor Univ.Mus. 1927-1931.	S. 2513 C

Special Bulletin. Bureau of Entomology of Chekiang Province.
    Hangchow.
    Spec.Bull.Bur.Ent.Chekiang.Prov. 1933-34.
	E.S 1954

Special Bulletin. Department of Agriculture, Nigeria. Lagos. Spec.Bull.Dep.Agric.Nigeria 1924-1925.	E.S 2172 b

Special Bulletin of the Hatch Agricultural Experiment Station.
    Amherst, Mass.
    Spec.Bull.Hatch agric.Exp.Stn 1889-1900 (imp.)
	E.S 2479 b

TITLE	SERIAL No.

Special Bulletin of the Lepidopterological Society of Japan. Osaka.
    Spec.Bull.lepid.Soc.Japan 1965 →   E.S 1909 B

Special Bulletin of the Metropolitan Museum of Natural History.
    Nanking.
    See Sinensia. Special Bulletin of the Metropolitan Museum
    of Natural History. Nanking.   B.S 1880

Special Bulletin. Middle East Biological Scheme.
    Spec.Bull.Mid.East biol.Scheme 1945-1946. (imp.)   S. 1919 a B

Special Bulletin. Minnesota Agricultural Experiment Station.
    Extension Division. St. Paul.
    Spec.Bull.Minn.agric.Exp.Stn Ext.Div. 8-48, 1916-20 (imp.)   E.S 2482 b

Special Bulletin. United States National Museum, Washington.
    Spec.Bull.U.S.natn.Mus. 1892-1915.   S. 2427 C

Special Circular. Department of Agriculture, Trinidad and
    Tobago. Port of Spain.
    Spec.Circ.Dep.Agric.Trin. 1913-1914.   E.S 2378 a

Special Communications. Central Research Institute for
    Forestry. Bogor.
    See Pengumuman Istimewa. Lembaga Pusat Penjelidikan
    Kehutanan. Bogor.   B.S 1801 a

Special Contribution. National Museum of Canada. Ottawa.
    Spec.Contr.nat.Mus.Can. Nos.43, 45 & 46.
        1943-1946.   S. 2632 B

Special Distribution Publications. State Geological Survey of
    Kansas. Lawrence.
    Spec.Distrib.Publs St.geol.Surv.Kans. 1969 →   P.S 1920 B

Special Guide. British Museum (Natural History). London.
    Spec.Guide Br.Mus.nat.Hist. 1905-1932.   SBM.2

Special Guide. Peabody Museum of Natural History. Yale
    University, New Haven.
    Spec.Guide Peabody Mus. 1931.   S. 2352 R

Special Leaflet. Board of Agriculture and Fisheries. London.
    Spec.Leafl.Bd Agric.Fish. 1914-1917.   E.S 51 a

Special Memoirs. Essex Field Club.
    See Essex Field Club. Special Memoirs.   S. 22 D

Special Monograph. Sarawak Museum. 1966 →
    Published as a numbered sequence of the
    Sarawak Museum Journal.   S. 1974 A

TITLE	SERIAL No.

Special Occasional Publications. Department of Mollusks,
    Museum of Comparative Zoology, Harvard University.
    Cambridge, Mass.
        Spec.Occ.Publs Dep.Moll.Mus.Comp.Zool.Harv. 1973 → Z. Mollusca Section

Special Pamphlet, War-time Production Series. Department of
    Agriculture, Canada.
        Spec.Pamphl.War-time Prod.Ser.Dep.Agric.Can. 13-39, 1940.   E.S 2522 g

Special Papers. Centre for Precambrian Research, University
    of Adelaide.
        Spec.Pap.Cent.Precamb.Res.Univ.Adelaide 1972 →   P.S 1139 K

Special Papers. Directorate of Geology. Dehra Dun, India.
        Spec.Pap.Dir.Geol. 1972 →   P.S 1104 A

Special Papers. Geological Association of Canada. Toronto.
        Spec.Pap.geol.Ass.Can. 1956 →   P.S 898 A

Special Papers. Geological Society of America. New York.
        Spec.Pap.geol.Soc.Am. 1934 →   P.S 864 A

Special Papers. Geological Survey of Ireland. Dublin.
        Spec.Pap.geol.Surv.Ir. 1971 →   P.S 1020 C

Special Papers. Mineralogical Society of America. Washington.
        Spec.Pap.mineralog.Soc.Am. 1963 →   M.S 2604 B

Special Papers. Ohio State Academy of Science. Columbus.
    Spec.Pap.Ohio St.Acad.Sci. 1899-1903.
    Continued in Proceedings of the Ohio State Academy
    of Science. Columbus.   S. 2366

Special Papers. Palaeontological Society of Japan.
        Spec.Pap.palaeont.Soc.Japan 1951 →   P.S 752 A

Special Papers in Palaeontology. Palaeontological Association.
    London.
        Spec.Pap.Palaeont. 1967 →   P.S 173 A

Special Publication.
    See Special Publications.

Special Publication Series. Minnesota Geological Survey.
    Minneapolis.
        Spec.Publ.Ser.Minn.geol.Surv. 1964 →   P.S 1943 C

Special Publications. Academy of Natural Sciences of Philadelphia.
        Spec.Publs Acad.nat.Sci.Philad. 1922 →   S. 2305 F

TITLE	SERIAL No.

Special Publications. American Association of Economic Entomologists.
Melrose Highlands, Mass., &c.
See Index to Literature of American Economic Entomology.   E.S 2441

Special Publications of the American Committee on
International Wild Life Protection. Cambridge, Mass.
Spec.Publs Am.Comm.int.wild Life Prot. 1931-1940 (imp.)   Z.S 2590

Special Publications. American Fisheries Society. Ann Arbor.
Spec.Publs Am.Fish.Soc.   No.2 → 1960 →
                         No.2.   1960.   Z. Fish Section
                         No.3 → 1966 →   Z.S 2516 A

Special Publications. American Geographical Society. New York.
Spec.Publs Am.geogr.Soc. No.7, 9, 10, 18, 22, 23, 30,
1928 → (imp.)   S. 2358 B
No.21, 33, 36, 1938, 1956, 1964.   B. See Bot.Lbry Catalogue

Special Publications. American Society of Limnology and Oceanography.
Baltimore.
Spec.Publs Am.Soc.Limnol.& Oceanogr. No. 20 → 23 1949-1955.
Formerly Special Publications. Limnological Society of America.
Ann Arbor.   S. 2314 a

Special Publications. American Stomatopod Society. Eureka (Ca.)
Spec.Publs am.Stomatopod Soc. 1971 → (imp.)   S. 2548 A·

Special Publications. Australian Conservation Foundation. Eastwood.
Spec.Publs Aust.Conserv.Fdn No.1 → (1968 →)   S. 2122 a B

Special Publications. Bermuda Biological Station for Research.
St. George's West.
Spec.Publs Bermuda biol.Stn Res. 1969 →   S. 2297 a B

Special Publications. Bernice Pauahi Bishop Museum. Honolulu.
Spec.Publs Bernice Pauahi Bishop Mus.
No.1-5, 7, 11-12, 14-16, 19-22, 25-26, 30-31, 33-35, 37-39,
41-44, 48 → 1892 → (No.7, 1920-1922 Forms the Proceedings
of the first Pan-Pacific Scientific Conference at S. 2716.)   S. 2175 C
No.6, 1899-1913 See Fauna Hawaiiensis.   E. 77E q H & Z. 77E q S
No.13 & 40, 1929 & 1948   B. 581.9(969)NEA

Special Publications. Botanical Society of Bengal. Calcutta.
Spec.Publs bot.Soc.Beng. 1950 →   B.S 1600

Special Publications. Brigham Young University Geology
Studies. Provo.
Spec.Publs Brigham Young Univ.Geol.Stud. 1969 →   P.S 75 C o H

TITLE	SERIAL No.

Special Publications. British Columbia Provincial Museum. Victoria.
   Spec.Publs Br.Columb.prov.Mus.
      No.1 Supplement 1947.   B. 581.9(711)HEN
      No.2. 1947.   Z. 75B o V

Special Publications. Chekiang Provincial Fisheries
   Experiment Station. Tinghai, China.
   Spec.Publs Chekiang prov.Fish.Exp.Stn   1936-1937.   Z.S 1970 A

Special Publications of the Chicago Academy of Sciences.
   Spec.Publs Chicago Acad.Sci.   No.1-3 & 7 → 1902 →
   (Suspended 1912-1939.)   S. 2329 C
     No.4, 1940.   Z. Reptile Section
     No.5, 1944.   B. 582.28(73)GRA
     No.6, 1944.   R.R.Dic

Special Publications. Colorado Geological Survey. Denver.
   Spec.Publs Colo.geol.Surv.   1969 →   P.S 1879

Special Publications. Conchological Society of Southern Africa.
   (Kenilworth, Cape Province.).
   Spec.Publs conchol.Soc.sth.Afr.   No.1 → 1960 →
   No.1 was entitled   Memorandum.   Z. Mollusca Section

Special Publications of the Cushman Foundation for
   Foraminiferal Research. Ithaca.
   Spec.Publs Cushman Fdn   1952 →
   Formerly   Special Publications of the Cushman
   Laboratory of Foraminiferal Research.   Z. Protozoa Section

Special Publications of the Cushman Laboratory for
   Foraminiferal Research. Sharon, Mass.
   Later   Washington, D.C.
   Spec.Publs Cushman Lab.   1928-1949.
   Continued as   Special Publications of the Cushman
   Foundation for Foraminiferal Research. Ithaca.   Z. Protozoa Section

Special Publications. Department of Ichthyology,
   Rhodes University, Grahamstown.
   Spec.Publs Dep.Ichthyol.Rhodes Univ.   1967-1968.
   Continued as   Special Publications. J.L.B. Smith Institute
   of Ichthyology, Rhodes University. Grahamstown.   Z.S 2028 B

Special Publications. Fiji Museum. Suva.
   See   Fiji Museum Publication Series. Suva.   S. 2180 B

Special Publications. Florida Geological Survey. Tallahassee.
   Spec.Publs Fla geol.Surv.   No.1-4, 6 → 1956 →   P.S 1890 E

TITLE	SERIAL No.

Special Publications. Geological Society. London.
    Spec.Publs geol.Soc.Lond. 1964 →   P.S 102 B

Special Publications. Geological Society of Australia. Sydney.
    Spec.Publs geol.Surv.Aust. 1968 →   P.S 188 B

Special Publications. Geological Society of South Africa. Johannesburg.
    Spec.Publs geol.Soc.S.Afr. 1970 →   P.S 190 D

Special Publications. Geological Survey, Georgia. Atlanta.
    Spec.Publs geol.Surv.,Ga 1963 →   P.S 1891 A

Special Publications. Geological Survey of Iran. Tehran.
    Spec.Publs geol.Surv.Iran 1969 →   P.S 1745 B

Special Publications. Geological Survey of Kwangtung and Kwangsi. Canton.
    Spec.Publs geol.Surv.Kwantung 1929-1939
        (Wanting No.16)   P.S 1790 A

Special Publications. Indo-Pacific Fisheries Council.
    Spec.Publs Indo-Pacif.Fish.Coun. 1952 →   Z. 79 q I

Special Publications of the Institute of Jamaica. Kingston.
    Spec.Publs Inst.Jamaica 1891.   S. 2291 F

Special Publications. Institute of Oceanography, National Taiwan University. Taipei.
    Spec.Publs Inst.Oceanogr.Nat.Taiwan Univ. No.1 →   S. 2000 a G

Special Publications. International Commission for the Northwest Atlantic Fisheries. Halifax, N.S.
    Spec.Publs int.commn NW.Atlant.Fish. 1958 →   Z.S 2716

Special Publications. J.L.B. Smith Institute of Ichthyology, Rhodes University. Grahamstown.
    Spec.Publs J.L.B.Smith Inst.Ichthyol.Rhodes Univ. 1969 →
    Formerly Special Publications. Department of Ichthyology, Rhodes University. Grahamstown.   Z.S 2028 B

Special Publications. Kentucky Geological Survey. Lexington.
    Spec.Publs Ky geol.Surv. 1953 →   P.S 1925 D

Special Publications. Limnological Society of America. Ann Arbor.
    Spec.Publs limnol Soc.Am. 1939-1947.
    Continued as Special Publication. American Society of Limnology and Oceanography. Baltimore.   S. 2314 a

Special Publications. Lingnan Natural History Survey and Museum. Canton.
    Spec.Publs Lingnan nat.Hist.Surv.Mus. Nos.1-9, 12 & 13. 1942-1950.   S. 1963 B

TITLE	SERIAL No.

Special Publications. Manitoba Museum of Man and Nature. Winnipeg.
    Spec.Publs Manitoba Mus.Man Nat. 1970 →   S. 2683

Special Publications. Missouri Geological Survey and Water
    Resources. Rolla, Missouri.
    Spec.Publs Mo.geol.Surv.Wat.Resour. 1969 →   P.S 1952 F

Special Publications. Museum. Texas Tech University. Lubbock.
    Spec.Publs Mus.Texas Tech Univ. 1972 →   S. 2564 C

Special Publications. National Agricultural Research Bureau.
    Nanking.
    Spec.Publs natn.agric.Res.Bur.,Nanking 1934-1935.   E.S 1950

Special Publications. New England Museum of Natural History.
    Boston, Mass.
    Spec.Publs New Engl.Mus.nat.Hist. 1936-1937.   S. 2321 L

Special Publications of the New York Academy of Sciences.
    Spec.Publs N.Y.Acad.Sci. 1939 →   S. 2361 F

Special Publications. Ohio Herpetological Society.
    Columbus, Ohio.
    Spec.Publs Ohio herpet.Soc. 1958-1962.   Z. Reptile Section

Special Publications from the Osaka Museum of Natural History.
    Osaka.
    Spec.Publs Osaka Mus.nat.Hist. 1969 →   S. 1992 c C

Special Publications. Peak District Mines Historical Society.
    Spec.Publs Peak Distr.Mines hist.Soc. 1961 →   M.S 162 A

Special Publications. Public Library, Museum and Art Gallery
    of Western Australia. Perth.
    Spec.Publs publ.Libr.Mus.West.Aust. No.1, 1948.   Z. Bird Section

Special Publications. Pymatuning Laboratory of Ecology.
    University of Pittsburgh.
    Spec.Publ.Pymatuning Lab.Ecol. No.3 → 1965 →
    Formerly Special Publication. Pymatuning Laboratory
    of Field Biology. University of Pittsburgh.   S. 2327 a

Special Publications. Pymatuning Laboratory of Field Biology.
    Pittsburgh.
    Spec.Publs Pymatuning Lab.Fld Biol. 1956-1960.
    Continued as Special Publications. Pymatuning
    Laboratory of Ecology. University of Pittsburgh.   S. 2373 a

Special Publications. Ross Allen's Reptile Institute.
    Silver Springs, Florida.
    Spec.Publs Ross Allen's Rept.Inst. 1950 →   Z. Reptile Section

Special Publications. Royal Society of Canada. Toronto.
    Spec.Publs R.Soc.Canada No.3 → 1961 →   S. 2603 B

TITLE	SERIAL No.

Special Publications of the Royal Society of South Africa.
    Cape Town.
        Spec.Publs R.Soc.S.Afr. 1948.   S. 2002 B

Special Publications. S.W.A. (South West Africa) Scientific
    Society. Windhoek.
        Spec.Publs S.W.A.scient.Soc. No.6 → 1966 →   S. 2077 C

Special Publications. Saskatchewan Geological Society. Regina.
        Spec.Publs Sask.geol.Soc. 1973 →   P.S 1094 K

Special Publications from the Seto Marine Biological Laboratory.
    Sirahama.
        Spec.Publs Seto mar.biol.Lab. 1959 →
        Nos.1-17, 1959-1962 published as Biological Results
        of the Japanese Antarctic Research Expedition.   80.o.J
        Ser.II. Pt.1 → 1964 →   S. 1980 E

Special Publications. Society of Economic Paleontologists
    and Mineralogists. Tulsa.
        Spec.Publs Soc.econ.Paleont.Miner.Tulsa 5 → 1957 →   P.S 862 A

Special Publications. United States Coast and Geodetic Survey.
    Washington, D.C.
        Spec.Publs U.S.Cst geod.Surv. 1909-1938. (imp.)   S. 2428 B

Special Publications of the University of Nebraska State Museum.
    Lincoln.
        Spec.Publs Univ.Neb.St.Mus. 2 → 1961 →   S. 2474 B

Special Publications. Vermont Geological Survey. Montpelier.
        Spec.Publs Vt geol.Surv. 1962 →   P.S 1981 A

Special Publications. The Western Australian Museum. Perth.
        Spec.Publs W.Aust.Mus. 1948 →   S. 2156 C

Special Publications. Yokohama National University. Kamakura.
        Spec.Publs Yokohama natn.Univ. No.1 → 1959.   S. 1991 a B

Special Report. Alaska Cooperative Wildlife Research Unit.
    College, Alaska.
        Spec.Rep.Alaska Coop.Wildl.Res.Unit 1954 →   Z.S 2595

Special Report. Aquatic Research Institute. Stockton, California.
        Spec.Rep.aquat.Res.Inst. No.1 → 1963 →   Z.S 2474 A

Special Report. California State Mining Bureau, Division of Mines,
    San Francisco.
        Spec.Rep.Calif.St.Min.Bur. 1950 →   M.S 2635

Special Report. Department of Mines Victoria. Melbourne.
        Spec.Rep.Dep.Min.Vict. 1892-1902   M.S 2423

| TITLE | SERIAL No. |

Special Report. Division of Mines, California.
    See Special Report. California State Mining Bureau,
Division of Mines. San Francisco.   M.S 2635

Special Report. Food Investigation Board. Department of
    Scientific and Industrial Research. London.
    Spec.Rep.Fd Invest.Bd D.S.I.R. 1919 →   S. 205 Fa

Special Report. Formosa Agricultural Experiment Station.
    Spec.Rep.Formosa agric.Exp.Stn 1910-1917. (imp.).   E.S 1957 e

Special Report. Geological Society. London.
    Spec.Rep.geol.Soc.Lond. 1971 →   P.S 102 A

Special Report. Geological Survey of Alabama. Montgomery.
    Spec.Rep.geol.Surv.Ala. No.11 → 1920 →   P.S 1870 A

Special Report of the Geological Survey of China. Peking.
    Spec.Rep.geol.Surv.China 1921-1945
        (Wanting No.3)   P.S 1788

Special Report. Geological Survey Department, Swaziland.
    Johannesburg.
    Spec.Rep.geol.Surv.Dep.Swazild 1946-1949.   M.S 2106

Special Report. Geological Survey. Indiana Department of
    Conservation. Bloomington.
    Spec.Rep.geol.Surv.Indiana 1963 →   P.S 1906 A

Special Report. Geological Survey of Japan. Kawasaki.
    Spec.Rep.geol.Surv.Jap. No.3 → 1966 →   P.S 1766 A

Special Report of the Great Lakes Research Division, University
    of Michigan. Ann Arbor.
    Spec.Rep.Gt Lakes Res.Div. No.28 → 1967 → (imp.)   S. 2316 K/L

Special Report. Iron and Steel Institute. London.
    Spec.Rep.Iron St.Inst. No.26, 28, 32; 1939-1946 (imp.)   M.S 146 B

Special Report on the Mineral Resources of Great Britain. London.
    See Memoirs of the Geological Survey. Special Reports on the
    Mineral Resources of Great Britain. London.   P.S 1000 & M.S 154

Special Report. Misaki Marine Biological Institute, Kyoto
    University. Kyoto.
    Spec.Rep.Misaki mar.biol.Inst. No.2-4, 1965-1968.   S. 1980 E

Special Report Series. National Health Council. London.
    Spec.Rep.Ser.Natn Hlth Counc. No.59, 103, 107, 167, 1921 →   O.72Aa o G
                                             No.305, 1964.   P.A 7 o G
    Formerly National Health Insurance. Medical Research Committee.
    Special Report Series. London.

Special Report Series. Washington Department of Fisheries. Seattle.
    Spec.Rep.Ser.Wash.Dep.Fish. 1953.
    Continued as Fisheries Research Papers. Washington Department
    of Fisheries.   Z.S 2333

| TITLE | SERIAL No. |

Special Report. Wisconsin Geological and Natural History Survey.
   Madison.
      Spec.Rep.Wis.geol.nat.Hist.Surv. 1967 →   S. 2345 C
                Nos.1, 2, 3 & 5. 1906-1910.   T.R.S 4308

Special Report. Zoological Gardens, Gizeh.
      Spec.Rep.zool.Gdns Gizeh Nos.1-5, 1905-1910.   Z.O 74B o C

Special Scientific Report. Marine (Research) Laboratory.
   St.Petersburg, Florida.
      Spec.scient.Rep.mar.Lab.,St.Petersburg, Fla
      No. 5 → 1960 → (imp.)   Z.S 2316

Special Scientific Report. United States Department of the
   Interior, Fish & Wildlife Service, Washington.
   Spec.scient.Rep.U.S.Fish Wildl.Serv.
   Special Scientific Report, No.28, 58, 64, 1947-1948.
   Fisheries Series, No.6 → 1949 → (imp.)
   Wildlife Series, No.1 → 1949 → (imp.)   Z.S 2510 N-P

Special Soils Publications. National Geological Survey of China. Peiping.
      Spec.Soils Publs geol.Surv.China Nos.1-5 1934-1944:
      Ser.B. No.2-4. 1937-1938   P.S 1789 A and 1789 B

Special Studies. Utah Geological and Mineralogical Survey.
   Salt Lake City.
      Spec.Stud.Utah geol.min.Surv. 1962 →   P.S 1979 A

Special Zoology Leaflet. Field Museum of Natural History, Chicago.
      Spec.Zool.Leafl.Field Mus.nat.Hist. No.1, 1926.   Z.S 2330

Specialkartor med Beskrifningar. Sveriges Geologiska
   Undersökning. Stockholm.
      See Sveriges Geologisko Undersökning. Stockholm. Ser.Bb.   P.S 1400

Spectrochemical Abstracts. London.
      Spectrochem.Abstr. 1933 →   M.S 127 A

Spectrochimica Acta. Oxford, London, New York, Paris.
      Spectrochim.Acta Vol.22 → 1966; Part B,
      Vol.23 → 1967 →   M.S 3016 A LAB

Spectrum. London.
      Spectrum 1964 →   S. 173
               No.20 → 1966 → (imp.)   P.S 138
      Formerly British Science News. British Council. London.   S. 173

Speleological Abstracts. Settle.
      Speleol.Abstr. Vol.1 No.2 - No.6, 1965-1970.
      Continued as Current Titles in Speleology.   REF.ABS.43

| TITLE | SERIAL No. |

Speleon. Oviedo, etc.
    Speleon Tom.5 → 1954 → .... S. 1092

Spelunca (1961 →)
    See Bulletin. Spelunca & Mémoires. Spelunca. .... S. 944 C-D

Spelunca. Bulletin et Mémoires de la Spéléologie. Paris.
    Spelunca, Paris Tome iv, no.25. 1901-1913.
    Formerly Spelunca. Bulletin de la Société de Spéléologie
    and Mémoires de la Societe de Spéléologie. .... S. 944 B

Spelunca. Bulletin de la Société de Spéléologie. Paris.
    Spelunca, Paris 1895-1900.
    Continued as Spelunca. Bulletin et Mémoires de la
    Société de Spéléologie. .... S. 944 A

Spisanie na Bŭlgarskoto Geologichesko Druzhestvo. Sofia.
    Spis.bulg.geol.Druzh. 1927-1938, 1963 → .... P.S 490

Spisok Rastenii Gerbariya Flory SSSR. Moskva.
    Spisok Rast.Gerb.Flory SSSR Vyp.12 → 1953 →
    Formerly Spisok Rastenii Gerbariya Russkoi   B.Eur.(581.9(47):
    Flory. Petropoli.   579 AN

Spisok Rastenii Gerbariya Russkoi Flory. Petropoli.
    Spisok Rast.Gerb.russk.Flory 1898-1922.
    Continued as Spisok Rastenii Gerbariya   B.Eur.(581.9(47):
    Flory SSSR. Moskva.   579 AN

Spisok Roslin Gerbari Flori USRR. Kiev.
    Spisok Rosl.Gerb.Flori USRR 1934 →   B.S 581.9(477):579
                                                                              I.B

Spisuv Poctěných Jubilejní Cenou Král. C. Společnosti
    Náuk v Praze.
    Spisuv poct.jubil.Společn.Náuk Praze Nos.1, 3 & 4:
    1888-1890. .... S. 1703 G

Spisy Lékařské Fakulty Masarykovy University. Brno.
    Spisy lék.Fak.Masaryk.Univ. 1922-1948. .... S. 1709 A

Spisy Přírodovědecké Fakulty University J.E. Purkyně v Brne.
    Spisy přír.Fak.Univ.Brne 1959-1970.
    Formerly Spisy Vydávané Přírodovědeckou Fakultou Masarykovy
    University, Brno.
    Replaced by Scripta Facultatis Scientiarum Naturalium
    Universitatis Purkynianae Brunensis. Brno. .... S. 1709 B

Spisy Přírodovědeckého Klubu Severovýchodnich Cech v Hradci Králové.
    Spisy přír.Klubu Severovych.Cech Hradci Kralove 1948. ...... S. 1758 b

Spisy Vydávané Přírodovědeckou Fakultou Karlovy University. Praha.
    Spisy vydáv.přír.Fak.Karl.Univ. 1923-1949.
    Replaced by Universitas Carolina, Pragae. Biologica &
    Geologica. .... S. 1755 A

TITLE	SERIAL No.

Spisy Vydávané Přírodovědeckou Fakultou Masarykovy University, Brno.
    Spisy vydáv.přír.Fak.Masaryk.Univ. 1921-1958.
    (From 1948 in Rada A-M.)   S. 1709 B
    Nos.1-2 & 6. 1951-1957.   M.S 1732
    Continued as Spisy Přírodovědecké Fakulty University
    J.E. Purkyné v Brne.

Spisy Vysoké Skoly Veterinární. Brno.
    Spisy vys.Sk.vet., Brno 17-21, 1949-1952.
    Formerly Biologicke Spisy Vysoké Skoly Zvěrolekařske. Brno.
    Continued as Sbornik Vysoké Skoly Zemědelská a
    Lesnické Fakulty. Rada B.   S. 1710

Spolia Zeylanica, Colombo Museum. Colombo.
    Spolia zeylan. 1903 →   S. 1920 A & T.R.S 3012
                  1903-1915.   Z.S 1940
    (From 1924-1944 also styled Ceylon Journal of Science:
    Section B. Zoology & Geology.)

Spolia Zoologica Musei Hauniensis. Copenhagen.
    Spolia zool.Mus.haun. 1941 →   Z.S 575

Spomenik. Srpska Kral'evska Akademija. Beograd. Prvi Razred.
    Spomenik Prvi Razr. 1888-1936.   S. 1888 B

Sporovȳe Rasteniya.
    See Trudȳ Botanicheskogo Instituta Akademii Nauk SSSR.
    Leningrad, Moskva. Seriya 2.   B.C.S 29

Sport Fishery Abstracts. Fish & Wildlife Service, U.S.Department
    of the Interior. Washington.
    Sport Fishery Abstr. Vol.8 → 1963 →   Z.B 22 S

Sprawozdania Komisji Fizyograficznej oraz Materyaly do Fizyografii
    Kraju. Kraków.
    Spraw.Kom.fizyogr.Kraków 1866-1939.
    Continued as Materialy do Fizjografii Kraju. Polska
    Akademia Umiejetności. Kraków.   S. 1725 A

Sprawozdania Pánstwowego Instytutu Geologicznego. Warszawa.
    Spraw.pánst.Inst.geol. 1937-1938.
    Formerly Sprawozdania Polskiego Instytutu Geologicznego. Warszawa.
    Continued in Biuletyn. Panstwowego Instytutu Geologicznego.
    Warszawa.   P.S 1580

Sprawozdania Państwowego Muzeum Zoologicznego. Warszawa.
    Spraw.państ.Muz.zool.Warsz. 1929.   Z.S 1801 B

Sprawozdania Polskiego Instytutu Geologicznego, Warszawa.
    Spraw.pol.Inst.geol. 1920-1937
    Continued as Sprawozdania Pánstwowego Instytutu Geologicznego.
    Warszawa.   P.S 1580

Sprawozdania z Posiedzeń Towarzystwa Naukowego Warzawskiego.
    Warszawa.
    Spraw.Posied.Tow.Nauk.Warsz. 1908-1951. (imp.)   S. 1872 A

TITLE	SERIAL No.

Sprawozdania Poznańskiego Towarzystwa Przyjaciół Nauk. Poznań.
　　Spraw.poznań.Tow.przyj.Nauk.　No.13 → 1945 →　　　　S. 1729 C

Sprawozdania ze Stanu i Działalności Towarzystwa
　　Przyjaciól Nauk w Wilnie.
　　Spra.Stanu Dział.Tow.przyj.Nauk Wiln.　1936-1937.　　S. 1797 B

Sprawozdania Towarzystwa Naukowego w Toruniu.
　　Spraw.Tow.Nauk Torun　1947 (1949) →　　　　　　　　　S. 1874 a B

Sprawozdania Wrocławskiego Towarzystwa Naukowego. Wrocław.
　　Spraw.wrocł.Tow.nauk.　1946-1955. Ser A. 1956. Ser.B. 1957 →　S. 1798 A

Sprawozdanie.
　　See Sprawozdania.

Sredne-Volzhskiĭ Penzenskiĭ Oblastnoĭ Muzei. (Trudy.)
　　See Penzenskiĭ Gosudarstvennyĭ Oblastnoĭ Muzeĭ. (Trudy.)　　S. 1875

Sri Lanka Forester. Colombo.
　　Sri Lanka Forester　1972 →
　　Formerly Ceylon Forester. Colombo.　　　　　　　　　　B.S 1722

Stahlia. Miscellaneous Papers. Department of Biology.
　　University of Puerto Rico. Rio Piedras.
　　Stahlia　1961 →　　　　　　　　　　　　　　　　　　　S. 2282

Stain Technology. Baltimore, Md.
　　Stain Tech.　vol. 32 → 1957 →　　　　　　　　　　　　S. 2491

Stanford Ichthyological Bulletin.
　　Stanford ichthyol.Bull.　1938 →　　　　　　　　　　　Z.S 2408

Stanford Laboratory Guides, Biological Series. Stanford
　　University, Calif.
　　Stanford Lab.Guides　1926-1931.　　　　　　　　　　　Z.S 2407

Stanford University Publications. Palo Alto.
　　Stanf.Univ.Publs　Biological Sciences 1920-1953.　　S. 2407 B
　　　　　　　　　　　Geological Sciences 1924 →　　　　P.S 846
　　Formerly Leland Stanford Junior University Publications.
　　University Series. Palo Alto.

Starfish, The. London.
　　Starfish　No.6 → 1953 →
　　Formerly A.S.N.H.S.Journal. Carshalton.　　　　　　　S. 1 a

Staringia. Pinneberg.
　　Staringia　1971 →　　　　　　　　　　　　　　　　　　P.S 288

Starunia (Studia ad Poloniae Diluvium Cognoscendum Pertinentia.)
　　Polska Akademja Umiejetności. Kraków.
　　Starunia　1934-1953　　　　　　　　　　　　　　　　　P.S 586

TITLE	SERIAL No.

State Geological Survey of Kansas. Lawrence.
   See University Geological Survey of Kansas. Topeka.   P.S 1924

State Park Series Bulletin. West Virginia Geological and
   Economic Survey. Morgantown.
   St.Pk Ser.Bull.W.Va 1951 →   P.S 1986 B

State University of Montana Studies. Missoula.
   St.Univ.Mont.Stud. 1916-1926.   S. 2350 F

State University Studies in Natural History, Iowa City.
   St.Univ.Stud.nat.Hist.Iowa 1943 →
   Formerly Studies in Natural History. Iowa University.   S. 2341

State Wildlife Advisory News Service. Perth, W.A.
   See S.W.A.N.S. Department of Fisheries and Fauna.
   Perth, W.A.   S. 2157 a

Statement of the Permanent Wild Life Protection Fund. New York.
   Statem.perm.wild Life Prot.Fund 1914-1920.   Z. 10J o P

Statement Showing Quantities and Values of Minerals and Gems.
   Revenue and Agriculture Department, India. Calcutta.
   Statem.Quant.Values Miner.Gems India 1888-1890.
   Continued as Return of Mineral Production in India. Calcutta. M.S 1916

Statens Naturvetenskapliga Forskningsråds Arsbok. Stockholm.
   St.naturvet.ForskRåds Arsb. 1946-1954. (imp.)
   Continued as Svensk Naturvetenskap. Stockholm.   S. 572

Station Bulletin. Oregon Agricultural Experiment Station.
   Corvallis.
   Stn Bull.Ore.agric.Exp.Stn 1936-1938.   E.S 2488 a

Station Circular. Oregon Agricultural Experiment Station. Corvallis.
   Stn Circ.Ore.agric.Exp.Stn 1925-1927.   E.S 2488 a

Station Papers. Forest, Wildlife and Range Experiment Station.
   Moscow, Idaho.
   Stn Pap.Forest Wildl.Range Exp.Stn Moscow, Idaho 1966 →   S. 2350 d

Statistical Abstract of the Ministry of Agriculture and
   Forestry. Tokyo.
   Statist.Abstr.Minist.Agric.For.,Tokyo 1926-1937.   E.S 1927

Statistical Bulletin. International Commission for the Northwest
   Atlantic Fisheries. Halifax, Dartmouth, N.S.
   Statist.Bull.int.Commn NW.Atlant.Fish. Vol.2 → 1954 →
   (For Vol.1 See Report. International Commission for the
   Northwest Atlantic Fisheries. No.2.)   Z.S 2716 C

Statistical Bulletin. Minerals and Mineral Products. Canberra.
   Statist.Bull.Miner.miner.Prod. 1964-Sept.1971.
   Continued as Minerals and Mineral Products. Canberra.   M.S 2400 E

TITLE	SERIAL No.

Statistical Bulletin. U.S. Department of Agriculture. Washington.
    Statist.Bull.U.S.Dep.Agric. 1928-1960.   E.S 2452

Statistical Digest. Fish & Wildlife Service, U.S.
    Department of the Interior. Washington, D.C.
    Statist.Dig.Fish Wildl.Serv.U.S. 1942 →   Z.S 2510 B

Statistical Report. Ontario Department of Mines, Toronto.
    Statist.Rep.Ont.Dep.Mines 1968 →   M.S 2708 A

Statistical Summary, Mineral Industry of the British Empire
    and Foreign Countries. Imperial Institute. London.
    Statist.Summ.Miner.Ind.Br.Emp. 1913-1949.
    Continued as Statistical Summary of the Mineral Industry.
    Colonial (afterwards Overseas) Geological Surveys. London.   M.S 150 A

Statistical Summary of the Mineral Industry. Colonial (afterwards
    Overseas) Geological Surveys. London.
    Statist.Summ.Miner.Ind.colon.geol.Survs 1950 →
    Formerly Statistical Summary of the Mineral Industry of the
    British Empire and Foreign Countries Imperial
    Institute, London.   M.S 150 A

Statistical Summary on the Mineral Industry of the U.S.S.R. Leningrad.
    See Sbornik Statisticheskikh Svedenii po Gornoĭ i
    Gornozavodskoĭ Promyshlennosti SSSR. Leningrad.   M.S 1813

Statistics of the Mineral Production in India. Calcutta.
    Statist.Miner.Prod.India 1890-1903.   M.S 1916

Statistics of Science and Technology. London.
    Statist.Sci.Technol. 1967-1968.   S. 206 a

Status of Geological Research in the Caribbean. Mayagüez.
    Status geol.Res.Caribb. 1959 →   P.S 964 A

Status of Zoological Research in the Caribbean. Mayaguez.
    Status zool.Res.Caribb. 1961 →   Z.S 2285

Stavanger Museum Årbok.
    Stavanger Mus.Årb. 1946 →
    Formerly Stavanger Museums Årshefte.   S. 540

Stavanger Museums Aarsberetning.
    Stavanger Mus.Aarsb. 1890-1899.
    Continued as Stavanger Museums Aarshefte.   S. 540

Stavanger Museums Aarshefte.
    Stavanger Mus.Aarsh. 1900-1925.
    Formerly Stavanger Museum. Aarsberetning.
    Continued as Stavanger Museums Årshefte.   S. 540

Stavanger Museums Årshefte.
    Stavanger Mus.Årsh. 1925-1945.
    Formerly Stavanger Museums Aarshefte.
    Continued as Stavanger Museum Årbok.   S. 540

TITLE	SERIAL No.

Steenstrupia. Copenhagen.
    Steenstrupia 1970 →   E.S 510 & Z.S 575 A

Sterbeeckia. Orgaan van de Antwerpse Mycologische Kring. Antwerpen.
    Sterbeeckia 1961 →   B.S.M 105

Stereo-Atlas of Ostracod Shells. Leicester.
    Stereo-Atlas Ostracod Shells 1973 →   P.S 151

Sterkiana. Columbus, Ohio.
    Sterkiana 1959 →   Z. Mollusca Section

Sterna. Stavanger Museum.
    Sterna No.8 → 1953 →
    Formerly Småskrifter. Stavanger Museum.   T.B.S 622

Stettiner Entomologische Zeitung. Stettin.
    Stettin.ent.Ztg 1840-1944.   E.S 1317

Stockholm Contributions in Geology.
    Stockh.Contr.Geol. 1957 →   P.S 430

Stoke Park Series Bulletin. West Virginia Geological and Economic Survey. Morgantown.
    Stoke Pk Ser.Bull.W.Va geol.Surv. 1951 →   P.S 1986 B

Stomatopod. Eureka (Ca.)
    Stomatopod 1970 →   S. 2548

Straipsniu Rinkinys. Botanikos Institutas. Vilnius.
    Straips.Rink., bot.Inst.Vilnius 1961.
    Continued as Botanikos Klausimai. Botanikos Institutas.
    Vilnius.   B.S 1516

Strandloper. Pinelands, Cape.
    Strandloper 1972 →
    Formerly Circular. Conchological Society of Southern Africa.
    Cape Town.   Z. Mollusca Section

Stratigrafiya SSSR. Moskva.
    Stratigrafiya 1963 →   P.S 1500 B

Stratigraphic Sections. State of Israel Geological Survey. Jerusalem.
    Stratigr.Sect.Israel geol.Surv. No.4 → 1967 →   P.S 1740 B

Stray Feathers. Calcutta.
    Stray Feathers 1873-1888.   T.R.S 3001 & Z.S 1911

Stromatolite Newsletter. (Canberra City.)
    Stromatolite Newsl. 1972 →   P.S 989 L

Stromboli. Messina.
    Stromboli No.1, 3-5, 8 (1952-1958).
    Nuova Serie No.9 → (1964) →   M.S 1104

| TITLE | SERIAL No. |

Struktur und Mitgliederbestand. Deutsche Akademie der Naturforscher.
    Leopoldina. Halle Saale.
    Strukt.MitglBest.dt.Akad.Naturf.Leop. 1959 →   S. 1301 K

Student & Intellectual Observer of Science, Literature
    & Art. London.
    Student Intellect.Obs. 1868-1871.   S. 462 C
                   Vol.4, 5   1870-1871.   T.R.S 165
    Preceded by "The Intellectual Observer".

Student's Scientific Papers. Chernyshevsky Saratov
    State University.
    See Sbornik Nauchnykh Rabot Studentov Saratovskogo
    Gosudarstvennogo Universiteta. Saratov.   S. 1879 a B

Studi Entomologici. Trieste.
    Studi ent. 1925-1926.   E.S 1113

Studi Geologici Camerti. Camerino.
    Studi geol.camerti 1971 →   P.S 636

Studi Illustrativi della Carta Geologica d'Italia. Roma.
    Studi illust.Carta geol.Ital. 1968 →   P.S 1610 A

Studi Sassaresi. Annali della Facoltà di Agraria dell'Università
    di Sassari.
    Studi Sass. Sez.III. 1953 →   E.S 1131

Studi Speleologia e Faunistici sull'Italia Meridionale. Napoli.
    See Bolletino della Società dei Naturalisti in Napoli.   S. 1152 B

Studi Trentini. Rivista Trimestrale della "Società per
    gli Studi Trentini". Trento.
    Studi trentini Anno I, trim 2 - Ann XI. 1920-1930.
    (from 1928-1930, published in two series: Studi Trentini
    di Scienze Storiche & Studi Trentini di Scienze Naturali.)
    Continued as Studi Trentini di Scienze Naturali. Trento.   S. 1200

Studi Trentini di Scienze Naturali. Trento.
    Studi trent.Sci.nat. Ann XII-41. 1931-1964.
                    Sezione A-B & Supplements Ann 42 → 1965 →
    Formerly Studi Trentini. Rivista trimestrale della
    "Società per gli Studi Trentini". Trento.   S. 1200

Studia Biologica Hungarica. Academiae Scientiarum Hungaricae.
    Budapest.
    Studia biol.hung. 1964 →   S. 1721 J

Studia Botanica Cechica. Pragae.
    Studia bot.čech. 1938-1943.
    Continued as Studia Botanica Cechoslovaca. Pragae.   B.S 1230

Studia Botanica Cechoslovaca. Pragae.
    Studia bot.čsl. 1946-1952.
    Formerly Studia Botanica Cechica. Pragae.   B.S 1230

| TITLE | SERIAL No. |

Studia Botanica Hungarica. Budapest.
   Studia bot.hung. 1973 →
   Formerly Fragmenta Botanica Musei Historico-Naturalis
     Hungarici.   B.S 1272

Studia Entomologica. Petropolis.
   Studia ent. 1952 →   E.S 2357

Studia Entomologica Forestalia. Prague.
   Studia ent.for. 1970 →   E.S 1727

Studia Geologica. Salamanca.
   Studia geol. 1971 →   P.S 676

Studia Geologica Polonica. Warszawa.
   Studia geol.pol. 1958 →   P.S 573 A

Studia Helminthologica. Bratislava.
   Studia helminth. 1964 →   Z. Parasitic Worms Section

Studia Naturae. Zakład Ochrony Przyrody, Polska Akademii
   Nauk. Kraków.
   Studia Nat. Ser.A. 1967 →   S. 1726 G

Studia ad Poloniae Diluvium Cognoscendum Pertinentia.
   See Starunia.   P.S 586

Studia Societatis Scientiarum Torunensis. Torun-Polonia.
   Studia Soc.Sci.torun
     Sectio A (Matematica-Physica). Vol.3 → 1954 →
     Supplementum, 1949-1960. (imp.)   S. 1874 a A
     Sectio C (Geographia et Geologia) 1953 →   P.S 577
     Sectio D Botanica. Torun.   B.S 1295
     Sectio E (Zoologia) 1948 →   Z.S 1818

Studia Universitatis Babes-Bolyai. Cluj. Series II.
   Biologia. Bucuresti.
   Studia Univ.Babes-Bolyai Ser.II Biol. 1958 →
   Formerly Buletinul Universităților "V.Babeş" şi
   "Bolyai" Cluj.   S. 1886 B

Studia Zoologica. Budapest.
   Studia zool. 1929-1931.   Z.S 1741

Studie z Oboru Všeobecné Krasové Nauky, Védecké Spelaeologie
   a Sousedních Oboru. Brno.
   Studie Oboru všeob.kras.Nauky 1933-1941.   S. 1708 A

Studien aus dem Gebiet der Allgemeinen Karstforschung, der
   Wissenschaften Höhlenkunde und der Nachbargebieten. Brünn.
   See Studie z Oboru Všeobecné Krasové Nauky, Védecké
   Spelaeologie a Sousedních Oboru. Brno.   S. 1708 A

Studien des Goettingischen Vereins Bergmannischer Freunde. Göttingen.
   Stud.Goett.Vereins Vol.2, 1828-1858.   P.S 307

| TITLE | SERIAL No. |

Studiensammlung der Staats-Universität in Irkutsk.
   See Sbornik Trudov Professorov i Prepodavetelei
   Gosudarstvennogo Irkutskogo Universiteta. Irkutsk.   S. 1831 A

Studies in Bacteriology. Corvallis, Oregon.
   See Oregon State Monographs. Studies in Bacteriology.
   Corvallis.   B. 576.8 O.S.C

Studies from the Biological Laboratories of the Owens College,
   Manchester.
   Stud.biol.Labs.Owen Coll. 1886-1895.   Z.S 85 & S. 263 A

Studies from the Biological Laboratory, Johns Hopkins
   University. Baltimore.
   Stud.biol.Lab.Johns Hopkins Univ. 1883-1893.   Z.S 2315 B

Studies in the Biological Sciences. University of Minnesota.
   Minneapolis.
   Stud.biol.Sci.Univ.Minn.
      No.3, 1919.   E. 69 o M
      No.4-6, 1923-1927.   B.S 4261

Studies from the Biological Stations of the Biological
   Board of Canada. Toronto.
   Stud.biol.Stns Can. Nos. 15-43, 1922-1926. (imp.).
   Continued as Studies from the Stations of the
   Fisheries Research Board of Canada.   Z.S 2628 B

Studies in Biology. London.
   Stud.Biol. 1966 →   S. 178 C

Studies in Botany. Corvallis, Oregon.
   See Oregon State Monographs. Studies in Botany.   B.S 4360

Studies in Conservation. The Journal of the International
   Institute for the Conservation of Museum Objects. London, etc.
   Stud.Conserv. 1952 →   S. 2701 a

Studies of the East Siberian State University. Irkutsk.
   See Trudȳ Vostochno-Sibirskogo Gosudarstvennogo
   Universiteta. Moskva.   S. 1831 C

Studies in Entomology. Corvallis, Oregon.
   See Oregon State Monographs. Studies in Entomology.   E.S 2488 b

Studies on the Fauna of Curaçao, etc. The Hague.
   Stud.Fauna Curacao 1940 →
   From Vol.3 also published in Uitgaven van de Natuurweten-
   schappelijke Studiekring voor Suriname en de Nederlandse
   Antilles (formerly en Curaçao.)   S. 677 & Z. 75F o H

Studies on the Fauna of Suriname and other Guyanas. Utrecht.
   See Uitgaven van de Natuurwetenschappelijke Studiekring
   voor Suriname en de Nederlandse Antillen.   S. 677

TITLE	SERIAL No.

Studies. Fisheries Research of Canada. Ottawa.
    Stud.Fish.Res.Bd Can. 1963-1970.
    Formerly Studies from the Stations of the Fisheries Research
Board of Canada.   Z.S 2628 B

Studies on the Flora of Curaçao and other Caribbean Islands.
    See Uitgaven van de Natuurwetenschappelijke Studiekring voor
Suriname en de Nederlandse Antilles (formerly en Curaçao).
Utrecht.   S. 677

Studies in Genetics. University of Texas. Austin.
    See University of Texas Publications. Austin.   10 H.o.A

Studies from the Geological and Mineralogical Institute,
    Tokyo University of Education.
    Stud.geol.miner.Inst.Tokyo Univ.Educ. No.2-3, 1953-1954.
    Continued as Research Bulletin of the Geological and
Mineralogical Institute, Tokyo University of Education.   P.S 1773

Studies in Geology. Corvallis, Oregon.
    See Oregon State Monographs. Studies in Geology. Corvallis.   P.S 842

Studies in Geology. John Hopkins University. Baltimore.
    See John Hopkins University Studies in Geology. Baltimore.   P.S 845

Studies in History. A.H. Wright. Ithaca, N.Y.
    Stud.in Hist. 1937-1954.
        (Wanting No. 16.)   7.o.W

Studies in History and Philosophy of Science. London.
    Stud.Hist.Phil.Sci. 1970 →   S. 455 b

Studies from Institute for Medical Research. Federated
    Malay States. Singapore.
    Stud.Inst.med.Res.F.M.S. Nos.3-13. 1903-1916.   S. 1926 A

Studies of the Irkutsk State University.
    See Sbornik Trudov Professorov i Prepodavateleĭ
Gosudarstvennogo Irkutskogo Universiteta. Irkutsk.   S. 1831 A

Studies on Loch Lomond. H.D. Slack. Glasgow.
    Stud.Loch Lomond 1957 →   72 Ab.o.S

Studies in Micropaleontology. Moscow.
    See Etyudȳ po Mikropaleontologii.   P.S 518

Studies in Microscopical Science. London.
    Stud.micr.Sci. 1882-1887.   S. 418

Studies from the Morphological Laboratory, Cambridge
    University. London.
    Stud.morph.Lab.,Camb.Univ. 1880-1896.   Z.S 355

Studies from the Museum of Zoology in University College, Dundee.
    Stud.Mus.Zool.Univ.Coll.Dundee 1888-1890.   Z.S 16

| TITLE | SERIAL No. |

Studies in Natural History. Iowa University.
    <u>Stud.nat.Hist.Iowa Univ.</u> 1919-1940.
    <u>Formerly</u> Bulletin from the Laboratories of Natural History
    of the State University of Iowa.
    <u>Continued as</u> State University Studies in Natural History,
Iowa.   S. 2341

Studies in Natural Sciences. Eastern New Mexico University. Portales.
    <u>Stud.nat.Sci.east.New Mex.Univ.</u> 1971 →   S. 2569

Studies on the Neotropical Fauna. Amsterdam.
    <u>Stud.neotrop.Fauna</u> 1972 →
    <u>Formerly</u> Beiträge zur Neotropischen Fauna. Jena.   S. 70 o T

Studies. Palao Tropical Biological Station. Tokyo.
    <u>See</u> Palao Tropical Biological Station Studies.   S. 1991

Studies in Physical Anthropology. American Association of
    Physical Anthropology. Detroit.
    <u>Stud.phys.Anthrop.</u> No.1 1949.   P.A.S 772 B

Studies from the Physiological Laboratory of Cambridge
    University.
    <u>Stud.physiol.Lab.Camb.Univ.</u> 1873-1877.   Z.S 60

Studies from the Plant Physiological Laboratory of Charles
    University. Prague.
    <u>Stud.Pl.physiol.Lab.Charles Univ., Prague</u> 1923-1935.
        (Suspended 1939-1944)   B.S 1238

Studies and Reviews. General Fisheries Council for the
    Mediterranean, Food & Agriculture Organisation of
    the United Nations. Rome.
    <u>Stud.Rev.gen.Fish.Coun.Mediterr.</u> 1957 →   Z.S 2713 C

Studies. Southern Methodist University. Dallas, Texas.
    <u>See</u> Southern Methodist University Studies. Dallas, Texas.   S. 2462 C

Studies on Speciation. Institute for the Study of Natural
    Species. Washington.
    <u>Stud.Spec.</u> 1962.   E.COL.B.68a.

Studies in Speleology. Association of the Pengelly Cave
    Research Centre.
    <u>Stud.Speleol.</u> 1964 →   S. 21a B

Studies from the Stations of the Fisheries Research
    Board of Canada. Ottawa.
    <u>Stud.Stns Fish.Res.Bd Can.</u> 1951-1962.
    <u>Formerly</u> Studies from the Biological Stations of the
    Biological Board of Canada.
    <u>Continued as</u> Studies. Fisheries Research Board of Canada.   Z.S 2628 B

Studies in Systematic Botany. Provo, Utah.
    <u>Stud.Syst.Bot.</u> No.1, 1962.   B.S 4392

TITLE	SERIAL No.
Studies from the Tokugawa Institute. Tokyo. Stud.Tokugawa Inst. 1924 →	S. 1999
Studies in Tropical Oceanography. Miami, Florida. Stud.trop.Oceanogr. 1963 →	S. 2538 E
Studies in the Vegetation of the State. Nebraska University, Botanical Survey. Report on recent Collections. Lincoln. Stud.Veg.St.Neb.Univ.bot.Surv. 1901-1904.	B. 581.9 (782) U.N
Studies in Vermont Geology. Vermont Geological Survey. Montpelier. Stud.Vt Geol. 1970 →	P.S 1983
Studies from the Zoological Department, University of Birmingham. Stud.zool.Dep.Univ.Bgham Vol.2, 1910.	Z.S 59
Studies from the Zoological Laboratory, Mason College. See Studies from the Zoological Department, University of Birmingham.	Z.S 59
Studies from the Zoological Laboratory, University of Nebraska. Lincoln, Neb. Stud.zool.Lab.Univ.Neb. 1898-1909.	Z.S 2430
Studies in Zoology. Corvallis, Oregon. See Oregon State Monographs. Studies in Zoology.	Z.S 2400

Studii si Cercetări, Academia RPR. Bucuresti.
    Studii Cerc.Acad. RPR Nos.9, 12, 15, 24, 51-52.
    1925-1941.
	S. 1893 D

Studii si Cercetări de Agronomie. Filiala Cluj, Academia RPR.
    Studii Cerc.Agron.Cluj Anul 7-14, 1956-1963.
    Formerly Studii si Cercetări Stiintifice. Filiala Cluj, Academia RPR. Bucuresti. Seria II.
    Merged in Revue Roumaine de Biologie & Studii si Cercetări de Biologie.
	S. 1893 Td

Studii si Cercetări de Antropologie. Bucuresti.
    Studii Cerc.Antrop. Tom.9 → 1972 →
    Formerly Annuaire Roumain d'Anthropologie. Bucarest.
	P.A.S 361

Studii si Cercetări, Baza de Cercetări Stiintifice Timisoara, Academia RPR. Bucuresti.
    Stiinte Agricole then Biologie si Stiinte Agricole.
    Studii Cerc.Baza Cerc.Stiinte Timisoara 1958-1963.
    Formerly Studii si Cercetări Stiintifice. Baza de Cercetări Stiintifice. Timisoara, Academia RPR.
    Merged in Revue Roumaine de Biologie & Studii si Cercetări de Biologie.
	S. 1893 V

TITLE   SERIAL No.

Studii si Cercetări de Biologie. Bucuresti.
   Studii Cerc.Biol.
     Seria Biologie Animala. Tom.14-15, 1962-1963.
     Continued as Seria Zoologie. Tom.16-25, 1964-1973.
     Seria Biologie Vegetala. Tom.14-15, 1962-1963.
     Continued as Seria Botanică. Tom.16-25,1964-1973.
     Two series united Tom.26 → 1974 →
   Formerly Buletin Stiintific Academia Republicii Populare
Romîne. Bucuresti.   S. 1893 Fa-b

Studii Cercetari de Geologie. Bucuresti.
   Studii Cerc.Geol. 4-8, 1959-1963 (imp.)
   Formerly Buletin Stiintifice. Academia R.P.R. Bucuresti.
   Sect. Geol. Geogr.
   Continued as Studii si Cercetari de Geologie, Geofizica,
Geografie. Seria Geologie. Bucuresti.   P.S 468

Studii si Cercetari de Geologie Geofizica Geografie.
   Seria Geologie. Bucuresti.
   Studii Cerc.Geol.Geofiz.Geogr.Seria Geol. Tom.15 → 1970 →
   Formerly Studii si Cercetari de Geologie. Bucuresti.   P.S 468

Studii si Cercetări de Geologie-Geografie. Filiala Cluj,
   Academia RPR.
   Studii Cerc.Geol.-Geogr.Cluj 1956-1957.
   Formerly Studii si Cercetări Stiintifice. Filiala Cluj,
Academia RPR. Bucuresti.   S. 1893 Tb

Studii si Cercetări de Inframicrobiologie, (Microbiologie și
   Parazitologie). Bucuresti.
   Studii Cerc.Inframicrobiol. 1950-1971.
   Continued as Studii si Cercetări de Virusologie. Bucuresti. S. 1893 N

Studii si Cercetări. Institutului de Cercetări Forestiere. Bucuresti.
   Studii Cerc.Inst.Cerc.for. Vol.23A → 1963 →
   Formerly Analele. Institutului de Cercetări
Silvice. Bucuresti.   B.S 1521

Studii si Cercetări. Institutului de Cercetari Piscicole.
   Bucuresti.
   Studii Cerc.Inst.Cerc.pisc. Vol.II (V) 1960 →
   Formerly Analele Institutului de Cercetari Piscicole.   Z.S 1890

Studii si Cercetări. Institutului de Cercetări Silvice. Bucuresti.
   Studii Cerc.Inst.Cerc.Silv. Seria I. Vol.12-15. 1951-1954.
   Continued as Analele Institutului de Cercetări
Silvice. Bucuresti.   B.S 1521

Studii si Cercetări de Medicina. Filiala Cluj, Academia RPR.
   Studii Cerc.Med.Cluj 1956-1963.
   Formerly Studii si Cercetări Stiintifice. Filiala Cluj,
Academia RPR. Bucuresti.   S. 1893 Tc

TITLE   SERIAL No.

Studii si Cercetări Stiintifice. Baza de Cercetări Stiintifice,
    Timisoara, Academia RPR. Bucuresti.
    <u>Studii Cerc.stiint.Baza Timisoara</u> 1954.
    <u>Then continued in Series</u>: Seria II. Stiinte Biologice,
    Agricole si Medicale 1955.
    <u>Then continued in Sections</u>: Seria Stiinte Agricole 1956-1957.
    <u>Continued as</u> Studii si Cercetări. Baza de Cercetări
    Stiintifice Timisoara, Academia RPR. Bucuresti.   S. 1893 Dd

Studii si Cercetări Stiintifice. Filiala Cluj, Academia RPR.
    Bucuresti.
    <u>Studii Cerc.stiint.Cluj</u> 1950-1954.
        Seria II: Stiinte Biologice, Agricole si Medicale 1954-1955.
    <u>Continued as</u> Studii si Cercetări de Agronomie, Biologie,
    Geologie-Geografiche, Medicina, Filiala Cluj, Academia RPR. S. 1893 Ta

Studii si Cercetări Stiintifice. Filiala Iăsi, Academia RPR.
    Bucuresti.
    <u>Studii Cerc.stiint.Iăsi</u> Anul 2-5. 1951-1954.
        Seria II: Stiinte Biologice, Medicale si Agricole 1954-1955.
    <u>Then continued in Sections</u>: Biologie si Stiinte
    Agricole 1956-1963.
    <u>Merged in</u> Revue Roumaine de Biologie & Studii si
    Cercetări de Biologie.   S. 1893 Ua

Studii si Cercetări de Virusologie. Bucuresti.
    <u>Studii Cerc.Virusol.</u> 1972 →
    <u>Formerly</u> Studii si Cercetări de Inframicrobiologie,
    (Microbiologie si Parazitologie).   S. 1893 N

Studii si Comunicări. Muzeul Judetean Suceava.
    <u>Studii Comun.Muz.jud.Suceava</u> Stiinte Naturale. 1970 →   S. 1899 b

Studii si Comunicari. Muzeul de Stiintele Naturii. Bacau.
    <u>Studii Comun.Muz.Stiint.nat.Bacau</u> 1969-1970.   S. 1898 a
        Zoologie. 1971-1972.   S. 1898 a A
        Biologie Animale 1973 →   S. 1898 a A
        Botanica 1971-1972.   S. 1898 a b
        Biologie Vegetala 1973 →   S. 1898 a B
    <u>Formerly</u> Studii si Comunicari. Sectia Stiintele Naturii,
    Muzeul Judetean. Bacau.   S. 1898 a

Studii si Comunicări. Muzeul de Stiintele Naturii. Dorohoi.
    <u>See</u> Botosani. Studii si Comunicări. Muzeul de Stiintele
    Naturii. Dorohoi.   S. 1882 a

Studii si Comunicări. Sectia Stiintele Naturii, Muzeul Judetean. Bacau.
    <u>Studii Comun.Sect.Stiint.nat.Muz.Judetean Bacau</u> 1968.
    <u>Continued as</u> Studii si Comunicări. Muzeul de Stiintele
    Naturii. Bacau.   S. 1898 a

Studii si Monografii. Universitatea "V. Babes" din Cluj.
    <u>Studii Monogr.Univ.V.Babes</u> No.1. 1957.   S. 1886 C

TITLE	SERIAL No.

Studii Tehnice si Economice. Institutul Geologic. Bucuresti.
    Studii teh.econ.Inst.geol.Buc.
        Seria A. No. 7 → 1967 →
        Seria B. No.45 → 1970 →
        Seria C. No.16 → 1970 →
        Seria D. No. 7 → 1970 →
        Seria E. No. 1 → 1952 →
        Seria F. No. 1 → 1952 →
        Seria G. No. 1 → 1958 →
        Seria H. No. 3 → 1967 →
        Seria I. No. 3 → 1967 →
        Seria J. No. 1 → 1966 →   P.S 469

Study of Insects. Nagoya.
    Study Insects Vols.1-4, 1937-1940.   E.S 1902 a A

Stuttgarter Beiträge zur Naturkunde aus dem Staatlichen Museum
    für Naturkunde in Stuttgart. Stuttgart.
    Stuttg.Beitr.Naturk. 1957-1972.   S. 1614 A
    Continued in series:
        Ser.A, Biologie. 1973 →   S. 1614 A a
        Ser.B, Geologie und Paläontologie. 1972 →   S. 1614 A b

Stylops. Journal of Taxonomic Entomology. London.
    Stylops 1932-1935.
    Continued as Proceedings of the Royal Entomological Society
    of London. Series B.   E.S 17

Subalpino. Giornale di Scienze, Lettere ed Arti.
    (Rivista Italiana). Torino.
    Subalpino 1836-1839.   S. 1198

Subject Index of the Modern Works Added to the Library of the
    British Museum. London.   REF.
    Subj.Index Br.Mus.Lond. 1881 →   C.R.

Subject Index to Periodicals. Library Association. London.   REF.
    Subj.Index Periods, London 1917-1922, 1926-1948. (imp.) C.R.

Substitute. London.
    Substitute 1856-1857.   E.S 11

Succulenta. Amsterdam.
    Succulenta, Amst. 1947 →   B.S 300

Succulentarum Japonia. Tokyo.
    Succulent.jap. 1958 →   B.S 1994

Sudan Agricultural Journal. Khartoum.
    Sudan agric.J. Vol.1, No.2 → 1965 →   E.S 2155 a

Sudan Notes and Records. Khartoum.
    Sudan Notes Rec. Vol. 1, no.1 & vol. 3 → 1918 →   S. 2045

Sudan Research Information Bulletin. Khartoum.
    Sudan Res.Inf.Bull. 1965 →   S. 2047 D

Sudan Wild Life and Sport. Khartoum.
    Sudan Wild Life 1949-1950.   Z.S 2025

TITLE	SERIAL No.

Suffolk Natural History. Lowestoft.
    Suffolk nat.Hist. 1969 →
    Formerly Transactions of the Suffolk Naturalists' Society.   S. 299

Suid Afrikaanse Joernaal van Nywerheid & Arbeidskoerant.
    See South African Journal of Industries
    (& Labour Gazette). Pretoria.   S. 2004

Suid-Afrikaanse Joernaal vir Wetenskap.
    See South African Journal of Science.   S. 2001 A

Suid-Afrikaanse Wetenskap.
    See South African Science.   S. 2001 B

Suisan Gakkai Ho.
    See Proceedings of the Scientific Fisheries Association.
    Tokyo.   Z.S 1954

Suite des Mémoires (Rapports et Documents) de l'Académie
    Royale des Sciences. Paris.
    Suite des Mém.Acad.Sci.Paris 1718-1890.   S. 804 E

Summa Brasiliensis Biologiae. Rio de Janeiro.
    Summa bras.Biol. 1945-1948.   S. 2215 a

Summa Brasiliensis Geologiae. Fundacão Getúlio Vargas.
    Rio de Janeiro.
    Summa bras.Geol. Vol.1, Fasc.2 and 5. 1946-1947.   P.S 2011 A

Summaries of Theses for the Degree of Doctor of Philosophy.
    Harvard University. Cambridge, Mass.
    Summs Theses Ph.D.Harv. 1941.   S. 2482 F

Summaries of Theses approved for Higher Degrees in the Faculties
    of Science and Engineering. University of Glasgow.
    Summs Theses high.Degrees Fac.Sci.Engng Univ.Glasg. 1962/1963-1967/1968.
    Formerly Summaries of Theses approved for Higher Degrees
    in the Faculty of Science. University of Glasgow.   S. 98 a A

Summaries of Theses approved for Higher Degrees in the Faculty
    of Science. University of Glasgow.
    Summs Theses high.Degrees Fac.Sci.Univ.Glasg.
    1949/1950 - 1961/1962.
    Continued as Summaries of Theses approved for Higher Degrees
    in the Faculties of Science and Engineering. University
    of Glasgow.   S. 98 a A

Summarized Proceedings of the American Association for the
    Advancement of Science.
    Summd Proc.Am.Ass.Advmt Sci. 1911-1929.
    Formerly Proceedings of the American Association for
    the Advancement of Science.   S. 2301 A

TITLE	SERIAL No.

Summary of Activities. Geological Survey of Israel. Jerusalem.
    Summ.Act.geol.Surv.Israel 1949 →   P.S 1740 A

Summary of Annual Reports and Records of Boring Operations.
Geological Survey of Victoria, Melbourne.
    Summ.a.Rep.Rec.Boring Ops geol.Surv.Vict. 1919-1922.   P.S 1134 A

Summary of Current Literature. Water Pollution Research
Board. London.
    Summ.curr.Lit.Wat.Pollut.Res.Bd 1928-1948. (imp.)
    Continued as Water Pollution Abstracts.   S. 205 Wb

Summary of Florida Commercial Marine Landings. Miami.
    Summ.Fla comml mar.Landgs 1959 →   Z.S 2316 A

Summary, Notes and Papers of the Federal-Provincial Wildlife
Conference. Canadian Wildlife Service.
    Summ.Notes Pap.Fed.-Prov.Wildl.Conf. 29th. 1965.
    Continued as Transactions of the Federal-Provincial
Wildlife Conference.   Z.S 2635 E

Summary of Progress of the Geological Survey of Great Britain
and the Museum of Practical Geology. London.
    Summ.Progr.geol.Surv.Lond. 1923-1964.   P.S 1000
                             1923-1938.   M.S 155
Not issued for the years 1939-1944. For the years 1945-1950
See Report of the Geological Survey Board. London.
Formerly Memoirs of the Geological Survey. Summary of
Progress. London.
Continued in Report. Institute of Geological Sciences.

Summary of Progress of the Geological Survey of Uganda. Entebbe.
    Summ.Prog.geol.Surv.Uganda 1919-1929.   P.S 1179 c

Summary Report of the Geological Survey Branch, Department
of Mines, Canada. Ottawa.
    Summ.Rep.geol.Surv.Brch Can. 1900-1933.   P.S 1070
                             1912-1919. (imp.)   M.S 2706 A

Summary Report Mineral Resources Survey, Australia, Canberra.
    Summ.Rep.Miner.Resour.surv.Aust. 1945-1959.   M.S 2400

Summary Report. Mines Branch, Canada, Ottawa.
    Summ.Rep.Mines Brch Can. 1908-1920 (imp.)   M.S 2701

Summary Report. Minnesota Geological Survey. Minneapolis.
    Summ.Rep.Minn.geol.Surv. 1947 →   P.S 1943

Summary Report Series. Mineral Industry of New South Wales.
Department of Mines, Geological Survey of New South Wales, Sydney.
    See Mineral Industry of New South Wales. Department of Mines,
Geological Survey of New South Wales, Sydney.   M.S 2404 A

| TITLE | SERIAL No. |

Summary of Reports. American Association of Museums.
Washington, D.C.
Summ.Rep.Am.Ass.Mus. 1936-1948.
Formerly Report of the American Association of Museums.
Washington, D.C.   S. 2302 D

Sunbird. Brisbane.
Sunbird 1970 →   T.B.S 7202

Sunyatsenia. Journal of the Botanical Institute, Sun Yat Sen
University, Canton.
Sunyatsenia 1930-1943.   B.S 1870

Suomalaisen Eläin-ja Kasvitieteellisen Seuran Vanamon
Elaintieteellisiä Julkaisuja. Helsinki.
Suomal.eläin-ja kasvit.Seur.van.elain Julk 1932-1963.
Replaced by Annales Zoologici Fennici.
Formerly Suomalaisen Eläin-ja Kasvitieteellisen Seuran
Vanamon Julkaisuja. Helsinki.   Z.S 1861

Suomalaisen Eläin-ja Kasvitieteellisen Seuran Vanamon
Julkaisuja. Helsinki.
Suomal.eläin-ja kasvit.Seur.van.Julk. 1923-1931.
Continued as Suomalaisen Eläin-ja Kasvitieteellisen
Seuran Vanamon Elaintieteellisiä Julkaisuja &
Suomalaisen Eläin-ja Kasvitieteellisen Seuran Vanamon
Kasvitieteellisen Julkaisuja.   S. 1818 A

Suomalaisen Eläin-ja Kasvitieteellisen Seuran Vanamon
Kasvitieteellisiä Julkaisuja. Helsinki.
Suomal.eläin-ja kasvit.Seur.van.kasvit.Julk. 1931-1964.
Formerly Suomalaisen Eläin-ja Kasvitieteellisen Seuran
Vanamon Julkaisuja. Helsinki.
Replaced by Annales Botanici Fennici. Helsinki.   B.S 1487

Suomalaisen Eläin-ja Kasvitieteellisen Seuran Vanamon
Tiedonannot. Helsinki.
Suomal.eläin-ja kasvit.Seur.Van.Tiedon. 1951-1964.
Formerly Suomalaisen Eläin-ja Kasvitieteellisen Seuran
Vanamon Tiedonannot ja Pöytäkirjat.Helsinki.
Replaced by Annales Botanici Fennici & Annales
Zoologici Fennici.   S. 1818 C

Suomalaisen Eläin-ja Kasvitieteellisen Seuren Vanamon Tiedonannot
ja Pöytäkirjat. Helsinki.
Suomal.eläin-ja kasvit.Seur.Van.Tiedon.Pöytak. 1946-1950.
Continued as Suomalaisen Eläin-ja Kasvitieteellisen Seuran
Vanamon Tiedonannot. Helsinki.   S. 1818 C

Suomalaisen Tiedeakatemian Toimituksia. Helsinki.
Suomal.Tiedeakat.Toim. Ser.A. 1909-1946
   III Geol-Geogr. 1942 →
   IV Biologia 1945 →   S. 1827 A

TITLE	SERIAL No.

Suomen Eläimet. Helsinki.
    Suom.Eläim. 1931 →   Z. 72Q d H

Suomen Geologinen Kartta. Helsinki.
    Suom.geol.Kartta 1952 →   P.S 1472 A

Suomen Geologisen Seuren Julkaisuja.
    Suom.geol.Seur.Julk. 1929 →
    Contained in Bulletin de la Commission Géologique
    de Finlande.   P.S 1471

Suomen Hyönteistieteellinen Aikakauskirja. Helsinki.
    Suom.hyönt.Aikak. 1935 →   E.S 1821

Suomen Kalastuslehti. Helsinki.
    Suom.Kalastusl. Vuos 43 → 1936 →   Z.S 1866 A

Suomen Kalatalous. Helsinki.
    See Finlands Fiskerier. Helsinki.   Z.S 1866

Suomen Riista. Helsinki.
    Suom.Riista No.2-9, 1948-1954; Abstracts only,
    No.10 → 1956 →   Z.S 1864 A

Suplemento. Bibliografia Geologica y Cartografica de la
    Provincia de Buenos Aires.
    Supl.Biblfia geol.cartogr.,Prov.B.Aires 1964 →   P. 76 o B

Supplement to the International Hydrographic Review. Monaco.
    See International Hydrographic Review. Monte Carlo, etc.   M.S 3000 A

Supplementa Entomologica. Berlin.
    Supplta ent. 1912-1929.   E.S 1318

Supplementary Papers. Royal Geographical Society. London.
    Suppl.Pap.R.geogr.Soc. 1882-1893.   S. 211 C

Supplemento a "La Ricerca Scientifica". Consiglio Nazionale
    delle Ricerche. Rome.
    See Ricerca Scientifica. Consiglio Nazionale delle
    Ricerche. Roma. Supplemento.   P.S 1602 a

Supplemento das Memorias. Instituto Oswaldo Cruz. Rio de Janeiro.
    Supplto Mems Inst.Oswaldo Cruz 1928-1929.   S. 2211 A

Surrey Bird Report. Halsemere.
    Surrey Bird Rep. 1953 →   T.B.S 274 A

Surrey Garner. London.
    Surrey Garner 1886.   S. 480 a

Surrey Naturalist. Croydon.
    Surrey Nat. 1966 →   S. 174 a

TITLE	SERIAL No.

Surrey Naturalist Trust Newsletter. (Leatherhead.)
    See Newsletter. Surrey Naturalists' Trust. (Leatherhead.)   S. 174 a B

Surtsey Research Progress Report. Reykjavik.
    Surtsey Res.Prog.Rep. 1965 →   S. 509

Suruga no Konchū. Shizuoka.
    Suruga Konchū No.57 → 1967 →
    (Wanting No.61, 63 & 65.)   E.S 1908 a

Survey of Biological Progress. New York & London.
    Surv.biol.Prog. 1949-1962.   7.o.S

Survey Reports. New Hampshire Fish and Game Department. Concord.
    Sur.Rep.New Hamps.Fish Game Dep. 1936-1939.   S. 2332 b

Survey Reports of Volcanoes in Manchuria. Ryojun.
    Surv.Rep.Volc.Manchuria 1936-1938   M.S 1934

Surveyor, The. Columbus (Ohio).
    Surveyor 1972 →   S. 2332 a N

Surveys of Leicestershire Natural History. Loughborough
    Naturalists' Club.
    Surv.Leics.nat.Hist. 1962 →   S. 252

Sussex Mammal Report. Hove.
    Sussex Mamm.Rep. 1965 →   Z. Mammal Section

Süsswasserfauna Deutschlands, eine Exkursionsfauna.
    Herausgegeben von A. Brauer. Jena.
    Süsswasserfauna Dtl. 1909-1912.   Z. 72L o B

Süsswasser-Flora Deutschlands, Österreichs und der Schweiz. Jena.
    Süsswass.-Flora Dtl.Öst.Schweiz. 1913 →   B. See Bot.Lbry Catalogue

Svea. Folk-Kalender. Stockholm.
    Svea 1846-1858, 1878-1880 & 1891.   S. 590

Svensk Botanisk Tidskrift. Stockholm.
    Svensk bot.Tidskr. 1907 →   B.S 275

Svensk Faunistisk Revy. Stockholm.
    Svensk faun.Revy Argang 17, 1955.
    Continued as Zoologisk Revy. Stockholm.   Z.S 512

Svensk Fiskeri Tidskrift. Stockholm.
    Svensk Fisk.Tidskr. Arg.45 → 1936 →   Z.S 505

Svensk Insektfauna. Stockholm.
    Svensk Insektfauna 1902 →   E. 72 C.o.S

Svensk Naturvetenskap. Stockholm.
    Svensk.Naturv. Arg. 11 → 1957 →
    Formerly Statens Naturvetenskapliga Forskningsråds
    Arsbok. Stockholm.   S. 572

TITLE	SERIAL No.

Svenska Hydrografisk-Biologiska Kommissionens Skrifter. Göteborg.
    Svenska hydrogr.-biol.Kommn.Skr. 1925-1948.   S. 574
    Biologi Continued in Report. Institute of Marine
    Research. Lyseki.
    Hydrografi Continued in Report of the Fishery Board
    of Sweden. Series Hydrography.   S. 574

Svenska Linné-Sällskapets Arsskrift Uppsala.
    Svenska linnésällsk.Arsskr. 1918 →   S. 592

Svenska Trädgards-Föreningens Arsskrift. Stockholm.
    Svenska TrädgFören.Arsskr. 1842.   S. 573 A

Svenska Trädgardsföreningens Tidskrift. Stockholm.
    Svenska TrädgFör.Tidskr. 1880-1882, 1887-1899.   S. 573 B

Sveriges Geologiska Undersökning. Afhandlingar och Uppsatser.
    Stockholm. Ser.A-D.
    Sver.geol.Unders.Afh. 1862 →
    Ser.C. 1907 → Contains the Arsbok.   P.S 1400

Sveriges Natur, Stockholm.
    Sver.Nat. 1910-1912, 1927, 1952 →   S. 561

Swänska Mercurius, Stockholm.
    Swänsk.Merc. 1755-1761.   S. 589 a

Sweet Pea Annual. Burnley.
    Sweet Pea A. 1950 →   B.S 82

Swungs. Kensington.
    See Journal of the University of New South Wales
    Geological Society. Kensington.   P.S 1127

Syaboten. Tokyo Cactus Club. Tokyo.
    Syaboten No.1 → 1954 →   B.S 1993

Sydowia. Annales Mycologici. Horn. N.O.
    Sydowia 1947 →
    Formerly Annales Mycologici. Berlin.
    Beihefte 1957 →   B.M.S 20

Syesis. Victoria, British Columbia.
    Syesis 1968 →   S. 2658 D

Syllogeus. National Museum of Natural Sciences. Ottawa.
    Syllogeus 1972 →   S. 2632 F

Sylvia. Casopis Ornitologický. Praha.
    Sylvia, Praha 1936-1959.   Z.S 1755

Symbioses. Olivet.
    Symbioses 1969 →   S. 918

TITLE	SERIAL No.

Symbolae Asahikawenses. Journal of Asahikawa Women's
    Junior College. Asahikawa, Japan.
    Symb.Asahik. 1965 →   S. 1990 c

Symbolae Botanicae Upsalinses. Arbeten från Botaniska
    Institutionen i Uppsala.
    Symb.bot.upsal. 1932 →   B.S 283

Symposia in Biology. Brookhaven National Laboratory. Upton.
    See Brookhaven Symposia in Biology. Upton, N.Y.   S. 2413 a A

Symposia. British Ecological Society. Oxford.
    Symp.Br.ecol.Soc. 1960 →   See General Library Catalogue

Symposia of the British Society for Parasitology. Oxford.
    Symp.Br.Soc.Parasit. 1963 →   Z.S 329

Symposia on Development and Growth.
    Symp.Dev.Growth 1954 - 1957.   7.o.S

Symposia Genetica. Pavia.
    Symp.genet.Pavia 1951-1954.
    Continued as Symposia Genetica et Biologica Italica. Pavia.   S. 1153

Symposia Genetica et Biologica Italica. Pavia.
    Symp.genet.biol.Ital. Vol. 5 → 1957 →
    Formerly Symposia Genetica. Pavia.   S. 1153

Symposia & Guidebook. Field Trip. Permian Basin Section. Society of
    Economic Paleontologists & Mineralogists. Midland, Texas.
    See Guidebook. Field Trip. Permian Basin Section. Society of
    Economic Paleontologists & Mineralogists. Midland, Texas.   P.S 862 C

Symposia on Hydrobiology. Madison, Wisc.
    Symp.Hydrobiol. 1941.   10 I.o.M

Symposia of the Institute of Biology. London.
    Symp.Inst.Biol. Nos.1-11, 14 → 1952 →   7.o.I

Symposia. International Society for Cell Biology.
    Symp.int.Soc.Cell Biol. 1962 →   7.o.I

Symposia. Royal Entomological Society of London.
    Symp.R.ent.Soc.Lond. 1961 →   E. 72A o E

Symposia of the Society of Experimental Biology. Cambridge.
    Symp.Soc.exp.Biol. 1947 →   7.o.S

Symposia of the Society for General Microbiology. Cambridge.
    Symp.Soc.gen.Microbiol. No.2 → 1952 →   7.o.S

Symposia of the Society for the Study of Development and Growth.
    Princeton, N.J.
    See Symposia on Development and Growth.   7.o.S

| TITLE | SERIAL No. |

Symposia of the Underwater Association for Malta 1965.
    See Report. Underwater Association. Carshalton, Surrey. 1965.   S. 272

Symposia of the Zoological Society of London.
    Symp.zool.Soc.Lond. 1960 →   Z.S 1 N

Symposium.
    See Symposia.

Synopses of the British Fauna. Linnean Society of London.
    Synopses Br.Fauna 1944 →   Z. 72A o L

Syokubutu - Iho. Batavia.
    Syokubutu - Iho Vol.1, no.1 2603 (i.e. 1943).
    Also issued as Annals of the Botanic Gardens,
    Buitenzorg, Volume Hors Série, 1944.   B.S 1800 a

Syokubutu Oyobi Dôbutu. Tokyo.
    See Botany and Zoology. Tokyo.   S. 1999 a

Systema Helminthum. New York.
    Syst.helminthum 1958 →   Z. Parasitic Worm Section

Systematic Zoology. Washington, D.C.
    Syst.Zool. 1952 →   Z.S 2550
            1952 →   E.S 2412
            Vol.14 → 1965 →   P.S 863
            Vol.22 → 1973 →   T.R.S 9100

Systematische Beihefte. Internationale Revue der Gesamten
    Hydrobiologie und Hydrographie. Leipzig.
    See Internationale Revue der Gesamten Hydrobiologie
    und Hydrographie. Leipzig.   S. 1670 D

Systême Silurien du Centre de la Bohême. Prague & Paris.
    Syst.Silurien Centr.Bohêm. 1852-1911.   P.S 72 M.f.B

Szegedi Tudómanyegyetem Biológiai Intézetenek Evkönyve. Szeged.
    See Annales Biologicae Universitatis Szegediensis.   S. 1724 B

TITLE	SERIAL No.

TAAF. Paris.
    See Terres Australes et Antarctiques Françaises. Paris. S. 947 a

Tabellenserie van de Strandwerkgemeenschap. Hoogwoud.
    Tabellenser.Strandwerkgemeens. No.22 → 1968 →
    Formerly S.W.G. Tabellenserie. Hoogwoud. S. 700 H

Tableaux Analytiques de la Faune de l'URSS.
    See Opredeliteli po Faune SSSR. Z.S 1820 A

Tabulata. Santa Barbara.
    Tabulata 1967 → Z. Mollusca Section

Taetigkeit & Taetigkeitsbericht.
    See Tätigkeit & Tätigkeitsbericht.

Tageblatt der Versammlung Deutscher Naturforscher und Aerzte.
    Tagebl.dt.Naturf.u.Aerzte 1867-1900.
    Formerly Amtlicher Bericht über die Versammlung Deutscher
    Naturforscher und Aerzte.
    Continued as Verhandlungen der Gesellschaft Deutscher
    Naturforscher und Aerzte. S. 1302 A

Tagesberichte über die Fortschritte der Natur- und
    Heilkunde. Weimar.
    Tagesber.Fortschr.Nat.Heilk. 1850-1852.
    Formerly Notizen aus dem Gebiete der Natur- und Heilkunde.
    Continued as Froriep's Notizen aus dem Gebiete der
    Natur-und Heilkunde. Jena. Z. Tweeddale & S. 1638

Taiwan Tigaku Kizi. Taihoku.
    Taiwan Tig.Kizi 1930-1940. S. 2000 b

Taiwania. Laboratory of Systematic Botany, National Taiwan
    University. Taipei.
    Taiwania 1948 → B.S 1986 a

Talanta. London, etc.
    Talanta. 1958 → M.S 3015

Tallinna Botaanikaaia Uurimused. Tallinn.
    Tallinn Bot.Uurim. 1962 → B.S 1423

Tampereen Kaupungin Museolautakunnan Julkaisuja. Tampere.
    Tampereen Kaupungin Mus.Julk. 1972 → S. 1826 a A

Tampereen Kaupungin Museot Vuosikirja. Tampere.
    Tampereen Kaupungin Mus.Vuosikirja 1971 → S. 1826 a B

Tane. The Journal of the Auckland University College
    Field Club.
    Tane 1948 → S. 2170 B

TITLE	SERIAL No.

Tanganiyka Notes and Records. Dar-es-Salaam.
　　Tanganiyka Notes Rec. 1936-1965.
　　Continued as Tanzania Notes and Records. Journal of
　　the Tanzania Society. Dar-es-Salaam.　　　　　　　　　　S. 2027 a

Tanzania Notes and Records. Journal of the Tanzania Society.
　　Dar-es-Salaam.
　　Tanzania Notes Rec. No.65 → 1966 →
　　Formerly Tanganyika Notes and Records. Dar-es-Salaam.　S. 2027 a

Taprobanian, The. Bombay.
　　Taprobanian 1885-1888.　　　　　　　　　　　　　　　　S. 1931

Tarptautinis Metraštis Baltijos Jūros Kvartero Geologijos ir
　　Paleogeografijos, Krantu Morfologijos ir Dinamikos, Jūru
　　Geologijos ir Neotektonikos Klausimais.
　　See Baltica.Vilnius.　　　　　　　　　　　　　　　　　P.S 997 A

Tartu Riikliku Ulikooli Toimetised. Tartu.
　　See Uchenȳe Zapiski Tartuskogo Gosudarstvennogo Universiteta.
　　Tartu.　　　　　　　　　　　　　　　　　　　　　　　　S. 1857 A

Tartu Ulikooli Geoloogia - Instituudi Toimetused. Tartu.
　　Tartu Ulik.Geol.-Inst.Toim. 1924-1941 (imp.)
　　Formerly Geoloogia - Instituudi Toimetused.　　　　　P.S 590

Tartu Ulikooli Juures Oleva Loodusuurijate Seltsi Araunded.
　　Tartu.
　　See Protokolȳ Obshchestva Estestvȯispȳtateleĭ and
　　Annales Societatis Rebus Naturae.　　　　　　　　　　S. 1816 A

Taschenbuch für Freunde der Geologie in Allgemein Fasslicher
　　Weise Bearbeitet.
　　Taschenb.Fr.Geol. 1845-1847.　　　　　　　　　　　　　P.S 312

Taschenbuch für die Gesammte Mineralogie. Frankfurt am Main.
　　Taschenb.ges.Miner. 1807-1824.　　　　　　　　　　　　M.S 1301
　　　　　　　　Jahr.7 & 18, 1813-1824.　　　　　　　　　P.S 310
　　Continued as Zeitschrift für Mineralogie, Taschenbuch.
　　Frankfurt am Main.

Tasmanian Journal of Agriculture. Hobart.
　　Tasm.J.Agric. 1930 →　　　　　　　　　　　　　　　　E.S 2262 c

Tasmanian Journal of Natural Science, Agriculture, Statistics,
　　&c. Hobart.
　　Tasm.J.nat.Sci. 1841-1849.　　　　　　　　　　　　　　S. 2150

Tasmanian Naturalist. Hobart.
　　Tasm.Nat. 1907-1911, 1924-1928, 1946-1955, 1965 →
　　(From 1965 → Also Styled Supplement to the Bulletin
　　of the Tasmanian Field Naturalists' Club.)　　　　　S. 2153 A

| TITLE | SERIAL No. |

Tätigkeit der Geologischen Vereinigung Oberschlesiens. Glewitz.
  Tät.geol.Verein.Oberschles. 1924-1931.
  Continued as Jahresbericht der Geologischen Vereinigung
  Oberschlesiens. P.S 306

Tätigkeitsbericht der Forschungsgemeinschaft der
  Naturwissenschaftlichen, Technischen und Medizinischen
  Institute der Deutschen Akademie der Wissenschaften zu Berlin.
  TätBer.ForschGem.naturw.tech.med.Inst.dt.Akad.
  Wiss.Berlin 1959-1961. S. 1305 K

Tätigkeitsbericht der Naturforschenden Gesellschaft
  Baselland. Liestal.
  TätBer.naturf.Ges.Baselland 1900-1901; 1907-1911; 1917 → S. 1238

Tauschbörse und Mitteilungen der Arbeitsgemeinschaft
  Österreichischer Entomologen. Wien.
  Tauschbörse Mitt.ArbGem.öst.Ent. 1952-1954.
  Supplement to Entomologisches Nachrichtenblatt. Burgdorf. E.S 1711

Taxidermist News. Brooklyn, N.Y.
  Taxid.News Nos.1-14, 1938-1952. Z.S 2324

Taxometrics. A Newsletter dealing with Mathematical and
  Statistical Aspects of Classification. London.
  Taxometrics No.7- No.12, 1965-1969. S. 194 a

Taxon. Utrecht.
  Taxon 1951 → B.S 4600

Tay Estuary Research Report. Dundee.
  See Research Report. Tay Estuary Research Centre. S. 52 c

Tebiwa. Pocatello.
  Tebiwa Vol.12 → 1969 → S. 2457

Technical Bulletin. Agricultural Experiment Station. State
  College Mississippi.
  Tech.Bull.agric.Exp.Stn Miss. 1911-1918. E.S 2483 a

Technical Bulletin. Agricultural Experiment Station, Washington State
  Institute of Agricultural Sciences. Pullman.
  Tech.Bull.agric.Exp.Stn Wash.St. 1950 → E.S 2495

Technical Bulletin. Botanical Garden. University of
  British Columbia. Vancouver.
  Tech.Bull.bot.Gdn Univ.B.C. No.2 → 1973 → B.S 4501 b

Technical Bulletin. Bureau of Entomology of Chekiang
  Province. Hangchow.
  Tech.Bull.Bur.Ent.Chekiang Prov. 1933. E.S 1953

Technical Bulletin of California Agricultural Experiment
  Station. Berkeley.
  See University of California Publications in Entomology.
  Berkeley. E.S 2466

| TITLE | SERIAL No. |

Technical Bulletin. College of Agriculture, Bangalore.
    Tech.Bull.Coll.Agric.Bangalore No.1, 1952.   B. 581.9(548.21)GOV

Technical Bulletin. Commonwealth Institute of Biological Control,
    Ottawa. Farnham Royal.
    Tech.Bull.Commonw.Inst.biol.Control 1961 →   E.S 49

Technical Bulletin. Department of Agriculture and Fisheries,
    Siam. Bangkok.
    Tech.Bull.Dep.Agric.Fish.Siam 1936.   E.S 2026

Technical Bulletin. Directorate General of Agricultural Research
    and Projects, Ministry of Agriculture, Iraq. Baghdad.
    See Technical Bulletin. Ministry of Agriculture, Iraq.   E.S 2043

Technical Bulletin. Economic Commission for Asia and the Far East.
    Committee for Co-operation of Joint Prospecting for Mineral
    Resources in Asian Offshore Areas. Bangkok.
    Tech.Bull.econ.Commn Asia Far E.Comm.Co-op.jt Prosp.miner.
    Resour.asian off-shore Areas 1968 →   P.S 998 A & M.S 1943 A

Technical Bulletin. Hawaii Agricultural Experiment Station.
    Honolulu.
    Tech.Bull.Hawaii agric.Exp.Stn 1952-1956.   E.S 2276 b

Technical Bulletin. Illinois Department of Conservation.
    Springfield, Ill.
    See Illinois Department of Conservation Technical Bulletin.
    Springfield, Ill.   S. 2567

Technical Bulletin. Kansas Agricultural Experiment Station.
    Manhattan.
    Tech.Bull.Kans.agric.Exp.Stn 1918 → (imp.)   E.S 2477

Technical Bulletin. Lafayette Natural History Museum. Lafayette, La.
    Tech.Bull.Lafayette nat.Hist.Mus. 1969 →   S. 2514

Technical Bulletin. Maine Agricultural Experiment Station. Orono.
    Tech.Bull.Maine agric.Exp.Stn No.39 → 1969 →   E.S 2477 b

Technical Bulletin. Ministry of Agriculture, Iraq. Baghdad.
    Tech.Bull.Minist.Agric.Iraq No.1-3, 7, 13, 18, 20, 1960-1965.
    Continued as Bulletin. Ministry of Agriculture, Iraq.   E.S 2043

Technical Bulletin. Ministry of Agriculture. United Arab
    Republic. Cairo.
    Tech.Bull.Minist.Agric.U.A.R. 1967 →   E.S 2154 a

Technical Bulletin. Ministry of Agriculture, Fisheries and Food.
    London.
    Tech.Bull.Minist.Agric.Fish.Fd 1954 →   E.S 58 a

Technical Bulletin. Minnesota Agricultural Experiment Station. St. Paul.
    Tech.Bull.Minn.Agric.Exp.Stn 1932 →   E.S 2482

| TITLE | SERIAL No. |

Technical Bulletin. Minnesota Department of Conservation,
    Division of Fish and Game. St.Paul.
    Tech.Bull.Minn.Dep.Conserv.Div.Fish Game 1944-1945.   S. 2469

Technical Bulletin. Mississippi Agricultural Experiment Station.
    See Technical Bulletin. Agricultural Experiment Station.
    State College of Mississippi.   E.S 2483 a

Technical Bulletin. New Hampshire Agricultural Experiment Station.
    Durham.
    Tech.Bull.New Hamps.agric.Exp.Stn 1935-1937.   E.S 2486 a

Technical Bulletin. New York State Agricultural Experiment
    Station. Geneva.
    Tech.Bull.N.Y.St.agric.Exp.Stn 1908-1947. (imp.)   E.S 2487 h

Technical Bulletin. Royal Botanical Gardens. Hamilton.
    Tech.Bull.R.bot.Gdns Hamilton 1963 →   B.S 4515

Technical Bulletin of Stephen F. Austin State Teachers College.
    Nacogdoches. Texas.
    Tech.Bull.S.F.Austin St.Coll. 1939-1941.
        (Negative Photostat.)   S. 2315 a

Technical Bulletin. Taiwan Agricultural Research Institute. Taipeh.
    Tech.Bull.Taiwan agric.Res.Inst. 1946.   E.S 1957 C

Technical Bulletin. U.S. Department of Agriculture. Washington.
    Tech.Bull.U.S.Dep.Agric. 1927 →
    Replacing Department Bulletin. U.S. Department
    of Agriculture. Washington.   E.S 2455

Technical Bulletin. Virginia Agricultural Experiment Station.
    Blacksburg.
    Tech.Bull.Va agric.Exp.Stn 1921-1937. (imp.)   E.S 2491 f

Technical Bulletin. Washington State Agricultural Experiment
    Station. Pullman.
    See Technical Bulletin. Agricultural Experiment Station,
    Washington State Institute of Agricultural Sciences. Pullman. E.S 2495

Technical Bulletin. West African Cocoa (formerly Cacao) Research
    Institute. Tafo.
    Tech.Bull.W.Afr.Cocoa Res.Inst. 1954 →   E.S 2174 c

Technical Bulletin. West African Timber Borer Research Unit. London.
    Tech.Bull.W.Afr.Timb.Borer Res.Unit 1959 →   E.S 2173 b

Technical Communications. Commonwealth Institute of
    Biological Control. Farnham Royal.
    Tech.Commun.Commonw.Inst.biol.Control 1960 →   E.S 49 a

TITLE	SERIAL No.

Technical Communications. Department of Agricultural Technical
    Services, South Africa. Pretoria.
    Tech.Commun.Dep.agric.tech.Serv.S.Afr.
    No.1-84, 103 → 1960 → (imp.)   E.S 2167 a

Technical Communications. National Botanic Gardens. Lucknow.
    Tech.Commun.natn.bot.Gdns Lucknow 1968.   B.S 1624 b

Technical Education Series. Technological Museum. Sydney.
    Tech.Educ.Ser.Tech.Mus.Sydney Nos. 7-28 (imp.) 1891-1931.   S. 2135 A

Technical Memorandum. CSIRO Division of Wildlife Research. Canberra.
    Tech.Memo.CSIRO Div.Wildl.Res. 1969 →   Z.S 2141

Technical Monographs. Texas Agricultural Experiment Station.
    College Station.
    Tech.Monogr.Texas agric.Exp.Stn No.2 → 1965 →   S. 2476

Technical Newsletter. Forest Products Research Institute, Ghana.
    Kumasi.
    Tech.Newsl.For.Prod.Res.Inst.Ghana Vol.4 → 1970 →   B.S 2317 b

Technical Notes. Federal Department of Forest Research,
    Nigeria. Ibadan.
    Tech.Notes Fed.Dep.For.Res.Nigeria No.39 → 1969 →   B.S 2294 b

Technical Notes. Forest Products Research Institute, Ghana.
    Kumasi.
    Tech.Notes For.Prod.Res.Inst.Ghana No.11 → 1969 →   B.S 2317 c

Technical Papers. Agricultural Experiment Station, Puerto Rico.
    Rio Piedras.
    Tech.Pap.agric.Exp.Stn P.Rico 1950 →   E.S 2393 b

Technical Papers. Arctic Institute of North America. Montreal.
    Tech.Pap.Arct.Inst.N.Am. 1956 →   S. 2640 D

Technical Papers. Bureau of Mines, United States. Washington.
    Tech.Pap.Bur.Mines, Wash. No.5-381; 1912-1927 (imp.)   M.S 2621 A

Technical Papers of the Bureau of Sport Fisheries and Wildlife.
    Washington, D.C.
    Tech.Pap.Bur.Sport Fish.Wildl. 1966 →   Z.S 2510 I

Technical Papers. California Agricultural Experiment Station. Berkeley.
    Tech.Pap.Calif.agric.Exp.Stn 1923-1925.
    Continued as Hilgardia. California Agricultural Experiment
    Station. Berkeley.   E.S 2467

Technical Papers. Canadian Museum Association. Ottawa.
    Tech.Pap.Can.Mus.Ass. 1966.   S. 2604

Technical Papers. Division of Biology and Horticulture.
    New Zealand Department of Agriculture. Wellington.
    Tech.Pap.Div.Biol.Hort. No.1. 1905.   E.S 2268

TITLE	SERIAL No.

Technical Papers. Division of Fisheries C.S.I.R.O. Australia, Melbourne.
Tech.Pap.Div.Fish.C.S.I.R.O.Aust. No.1-3, 1951-1955.
Continued as Technical Papers. Division of Fisheries & Oceanography. C.S.I.R.O. Australia.   Z.S 2112 A

Technical Papers. Division of Fisheries & Oceanography. C.S.I.R.O. Australia, Melbourne.
Tech.Pap.Div.Fish.Oceanogr.C.S.I.R.O.Aust. No.4 → 1958 →
Formerly Technical Papers. Division of Fisheries. C.S.I.R.O. Australia.   Z.S 2112 A

Technical Papers. Divison of Wildlife Research. C.S.I.R.O. Australia, Melbourne.
Tech.Pap.Div.Wildl.Res.C.S.I.R.O.Aust. No.2 → 1962 →
Formerly Technical Papers. Wildlife Survey Section. C.S.I.R.O. Australia.   Z.S 2112 B

Technical Papers. Forest Research Institute, New Zealand. Wellington.
Tech.Pap.Forest Res.Inst.N.Z. 1954 →   B.S 2482

Technical Papers. Geological Survey of Korea. Seoul.
Tech.Pap.geol.Surv.Korea 1958 →   P.S 1794 B

Technical Papers. Great Lakes Research Institute, University of Michigan. Ann Arbor.
See Publications. Great Lakes Research Institute, University of Michigan.   S. 2316 J

Technical Papers. Iowa Geological Survey. Des Moines.
Tech.Pap.Iowa geol.Surv. No.3, 1936.   P.S 1915

Technical Papers. London Shellac Research Bureau. London.
Tech.Pap.Lond.Schellac Res.Bur. 1934-1946.   E.S 75

Technical Papers. Mineragraphic Investigations. Commonwealth Scientific and Industrial Research Organization. Melbourne.
See Mineragraphic Investigations.Technical Paper. C.S.I.R.O.   M.S 2426

Technical Papers. United States Bureau of Mines. Washington.
Tech.Pap.Bur.Min.,Wash. No.5-381; 1912-1927 (imp.)   M.S 2621 A

Technical Papers. Water Pollution Research D.S.I.R. London.
Tech.Pap.Wat.Pollut.Res.D.S.I.R. 1929 →   S. 205 Wa

Technical Papers. Wildlife Survey Section, Commonwealth Scientific & Industrial Research Organisation, Australia, Melbourne.
Tech.Pap.Wildl.Surv.Sect.C.S.I.R.O.Aust. No.1, 1958.
Continued as Technical Papers. Division of Wildlife Research C.S.I.R.O. Australia, Melbourne.   Z.S 2112 B

TITLE	SERIAL No.

Technical Publications. Moss Landing Marine Laboratories.
    Moss Landing.
        Tech.Publs Moss Landing mar.Lab. 71-1 → 1971 →   S. 2390 a D

Technical Report. Bell Museum of Natural History. Minneapolis.
    Tech.Rep.Bell Mus.nat.Hist. 1963-1968.   S. 2467 C

Technical Report. Department of Biological Sciences, University of
    Southern California. Los Angeles.
        Tech.Rep.Dep.biol.Sci.Univ.sth.Calif. 1970 →   S. 2404 b G

Technical Report. Department of Mines, New South Wales. Sydney.
    Tech.Rep.Dep.Mines N.S.W. 1953-1960.   M.S 2403 A

Technical Report. International Pacific Halibut Commission. Seattle.
    Tech.Rep.int.Pacif.Halibut Commn 1969 →   Z.S 2712 C

Technical Report. Museum of Anthropology, University of Michigan.
    Ann Arbor.
        Tech.Rep.Mus.Anthrop.Univ.Mich. 1971 →   P.A.S 792 E

Technical Report. Ohio Department of Natural Resources. Columbus.
    Tech.Rep.Ohio Dep.nat.Res. No.4-9, 1960-1961.   S. 2332 c B

Technical Report and Scientific Papers. Imperial Institute.
    London.
        Tech.Rep.imp.Inst.,Lond. 1903.   S. 187 D

Technical Report Series. Resource Management Branch, Fisheries
    and Marine Service, Canada. Winnipeg.
        Tech.Rep.Ser.Resour.Mgmt Brch Fish.mar.Serv.Can. 1973 →   Z.S 2636

Technical Reports. Fisheries Research Board of Canada. Winnipeg.
    Tech.Rep.Fish.Res.Bd Can. No.33 → 1967 →   Z.S 2628 I

Technical Reports. Marine Sciences Research Laboratory,
    Memorial University of Newfoundland. St.John's.
    See MSRL Technical Report.   S. 2645

Technical Series Bulletin. University of Missouri School of
    Mines and Metallurgy. Rolla.
    See Bulletin of the School of Mines and Metallurgy;
    University of Missouri.   P.S 858

Technical Series. Bureau of Entomology, United States Department
    of Agriculture. Washington.
        Tech.Ser.Bur.Ent.U.S. 1895-1914   E.S 2447 b

Technical Series. Bureau of Plant Industry Pennsylvania
    Department of Agriculture.
    See Bulletin of the Pennsylvania Department of Agriculture.   B.S 4421

TITLE	SERIAL No.

Technical Series. Florida State Board of Conservation. Tallahassee.
  Tech.Ser.Fla St.Bd Conserv.
  No.5-57, 1952-1968.
  Continued as Technical Series. Florida Department of
  Natural Resources.   Z.S 2316

Technical Series. Florida State Department of Natural Resources.
  St.Petersburg, Fla.
  Tech.Ser.Fla St.Dep.nat.Resour. No.58-69, 1969-1972.
  Formerly Technical Series. Florida State Board of
  Conservation.   Z.S 2316
  Superseded by Florida Marine Research Publications.
  St. Petersburg, Florida.   S. 2480

Technical Series. Intergovernmental Oceanographic Commission.
  UNESCO.
  Tech.Ser.Intergovtl oceanogr.Comm. 1965 →   S. 2714 D

Technical Series. Royal Geographical Society.
  See R.G.S. Technical Series.   S. 211 I

Technical Series. Scripps Institution of Oceanography. Berkeley.
  See Bulletin.Scripps Institution of Oceanography.
  Technical Series.   S. 2319 M

Technical Translations. Clearinghouse for Federal, Scientific
  & Technical Information. Washington, D.C.
  Tech.Transl.Wash.,D.C. 1967.   S. 2431 - REF

Technology and the Environment. London.
  Technol.Envir. 1971 →   S. 171 a

Tectonophysics. International Journal of Geotechtonics and
  the Geology and Physics of the Interior of the Earth.
  Amsterdam, etc.
  Tectonophysics 1964 →   M.S 3020

Tektonika SSSR. Geologicheskiĭ Institut. Akademiya Nauk SSSR. Moskva.
  Tektonika 1948 → Imp.   P. 72 Q.q.L

Tematicheskiĭ Sbornik. Otdel Fiziologii i Biofiziki Rasteniĭ,
  Akademiya Nauk Tadzhikskoĭ SSR. Dushanbe.
  Tematich.Sb.Otd.Fiziol.Biofiz.Rast.Akad.Nauk
  tadzhik SSR 1962 →   B.S 1353 a

Tematicheskiĭ Sbornik Rabot po Gel'mintologii Sel'skokhozyaĭstvennȳkh
  Zhivotnȳkh. Moskva.
  Tematich.Sb.Rab.gel'mint.sel'khoz.Zhivot. Vyp.12 → 1966 → Z.S 1847 B

Temminckia. Leiden.
  Temminckia 1936 →   Z.S 625
  1936-1960.   T.R.S 761

| TITLE | SERIAL No. |

Tenthredo. Acta Entomologica. Takeuchi Entomological Laboratory.
Kyoto.
    Tenthredo   1936-42.   E.S 1913

Természetrajzi Füzetek. Budapest.
    Természetr.Füz. 1877-1902.
    Continued as Annales Historico-Naturales Musei
Nationalis Hungarici. Budapest.   S. 1716 & T.R.S 1832

Természettudományi Közlöny. Budapest.
    Természettud.Közl. 1869-1907.
    Formerly Közlönye. Magyar Természettudományi
Társulat. Pesten.   S. 1720 A

Természettudományi Pályamunkák. Budapest.
    Természettud.Pályamunk. 1837-1858.   S. 1721 A

Terra. Bucuresti.
    Terra, Bucuresti 1969 →
    Formerly Natura. Seria Geografie-Geologie. Bucuresti.   S. 1893 a B

Terra. Los Angeles.
    Terra Los Angeles Vol.9, No.2 → 1970 →
    Formerly Quarterly. Los Angeles County Museum of Natural
History. Los Angeles.   S. 2309 A

Terra Australis. Canberra.
    Terra Aust. 1971 →   P.A.S 870 a

Terrae Incognitae. Amsterdam.
    Terrae Incognit. 1969 →   S. 644

Terre Malgache. Tananarive.
    Terre malg. 1966 →   S. 2084 C

Terre (La) et la Vie. Paris.
    Terre Vie 1931-1940: 1947 →   S. 940
                    1931-1937: 1947 →   T.R.S 827

Terres Australes et Antarctiques Françaises. Paris.
    Terres aust.antarct.fr. 1957 →   S. 947 a

Tertiary Times. Bromley.
    Tertiary Times 1970 →   P.S 143

Tesis del Museo de La Plata.
    Tesis Mus.La Plata 1940-1945.   S. 2231 J

Téthys. Marseille.
    Téthys 1969 →   S. 899 E

Texas Journal of Science. San Marcos.
    Tex.J.Sci. 1949 →
    Formerly Proceedings and Transactions of the Texas
Academy of Sciences. Austin.   S. 2315 B

TITLE	SERIAL No.

Teysmannia. Batavia.
   Teysmannia 1890-1922 (wanting Vol.7, 28, 32)   B.S 1821

Tezisy Dokladov. Pribaltiĭskaya Ornitologicheskaya Konferentsiya.
   See Trudy Pribaltiĭskoĭ Ornitologicheskoĭ Konferentsii.   Z. 72Q q C

Thai Forest Bulletin. Bangkok. Botany.
   Thai Forest Bull. 1954 →   B.S 1761

Thai National Scientific Papers. Applied Scientific Research
   Corporation of Thailand. Bangkok.
   Thai natn.scient.Pap. Fauna Series 1968 →   Z.S 1988

Thai National Scientific Papers. Bangkok.
   Thai natn.scient.Pap. Biol. Series 1971.   S. 1914 c D

Thai Science Bulletin. Bangkok.
   Thai Sci.Bull. 1939-1941; 1949-1959.
   Formerly Siam Science Bulletin. Bangkok,
   which title it had from 1947-1948.   S. 1913 a

Thalassia. Venezia.
   Thalassia 1932-1944.
   Continued as Nova Thalassia.   S. 1161 A

Thalassia Jonica. Taranto.
   Thalassia jonica 1958 →   S. 1171

Thalassia Jugoslavica. Zagreb.
   Thalassia jugosl. 1956 →   S. 1705 a

Thanet Panorama. Ramsgate.
   See Panorama. Ramsgate.   S. 336 a

Theoretical Population Biology. New York.
   Theoret.Pop.Biol. 1970 →   S. 2554

Thermal Analysis Review. London.
   Thermal Analysis Rev. 1962-1966.   M.S 163

Theses Abstracts. Graduate School of Arts and Sciences.
   University of the Philippines.
   Thes.Abstr.Grad.Sch.Arts Sci.Univ.Philipp. 1947-1965.   S. 1976 a D

Theses and Dissertations accepted for Higher Degrees.
   University of London.
   Thes.Diss.higher Degrees Univ.Lond.
   1962/1963 → (Wanting 1963/1964)   S. 215 a

Thiergarten. Stuttgart.
   Thiergarten Jahrg.1, 1864.   Z.S 1302

| TITLE | SERIAL No. |

Thomas Cawthron Memorial Lecture. Cawthron Institute. Nelson, N.Z.
    Thomas Cawthron Meml Lect. No.33 → 1958 → (imp.)
    Formerly Cawthron Lecture Series. Cawthron Institute.
Nelson, N.Z.   S. 2182 D

Thompson Yates (and Johnston) Laboratories Report. Liverpool.
    Thompson Yates Johnston Labs Rep. 1898-1906 (imp.)
    Continued in Annals of Tropical Medicine and Parasitology.
Liverpool.   Z.S 322 A

Tidsskrift for Historisk Botanik. København.
    Tidsskr.hist.Bot. 1918-1919.   B.S 208

Tidsskrift for Naturvidenskaberne. Kjøbenhavn.
    Tidsskr.Naturv. 1822-1828.   S. 531

Tidsskrift for Physik og Chemi Samt disse Videnskabers
    Anvendelse. Kjøbenhavn.
    Tidsskr.Phys.Chem. 1862-1894.   M.S 602

Tidsskrift for Populaere Fremstillinger af Naturvidenskaben.
    Kjøbenhavn.
    Tidsskr.pop.Fremst.Naturw. 1855-1883.   S. 533

Tiedonantoja. Riista-ja Kalatalouden Tutkimuslaitos
    Kalantutkimusosasto. Helsinki.
    Tiedon.Riista Kalatal.Tutkim.Kalantut. 1972 →   Z.S 1866 c

Tier- und Naturphotographie. Baden-Baden.
    Tier-u.NatPhotogr. Bd.1-3 (2), 1958-1960.   Z.S 1342

Tier und Umwelt. Hamburg.
    Tier Umwelt Neue Folge 1964 →   Z.S 1307

Tierärztliches Archiv für die Sudetenländer. Troppau.
    A.- Wissenschaftlicher Teil.
    Tierärztl.Arch.Sudetenl.(A) Bd. 3 1923.   Z. 17D o D

Tierreich. Berlin.
    Tierreich 1896 →   T.R.S 1359 & Z. 1 q L

Tierwelt Deutschlands und der Angrenzenden Meeresteile. Jena.
    Tierwelt Dtl. 1925 → (Excluding entomology.)   Z. 72L o D
                1927 → (Entomology only.)   E. 72L o D

Tierwelt Mitteleuropas. Ein Handbuch zu ihrer Bestimmung
    als Grundlage für Faunistisch-Zoogeographische
    Arbeiten. Herausgegeben von P. Brohmer &c. Leipzig.
    Tierwelt Mitteleur. 1927 → (Excluding entomology.) Z. 72L o B
                    1927-1936 (Entomology only.)   E. 72L o B

| TITLE | SERIAL No. |

Tierwelt der Nord- und Ostsee. Leipzig.
    Tierwelt N.-u.Ostsee 1925 →   Z. 72L o G

Tijdschrift voor Entomologie. 's Gravenhage, Amsterdam.
    Tijdschr.Ent. 1857 →   E.S 603

Tijdschrift voor Indische Taal-Land-en Volkenkunde. Batavia.
    Tijdschr.indische Taal-Land-en Volkenk. 1853-1955.   S. 1951 A

Tijdschrift voor Natuurlijke Geschiedenis en Physiologie.
    Amsterdam.
    Tijdschr.Natuurl.Gesch.Physiol. 1834-1845.   S. 613 & T.R.S 759
                                                             Z.S 660 & Tweeddale

Tijdschrift der Nederlandsche Dierkundige Vereeniging. Leiden.
    Tijdschr.ned.dierk.Vereen. 1874-1933.   Z.S 615
                                  1874-1927.   T.R.S 752
    Continued as Archives Néerlandaises de Zoologie. Leiden.

Tijdschrift voor de Wisen Natuurkundige Wetenschappen,
    uigegeven door de Eerste Klasse van het Koninklijk
    Nederlandsch Instituut van Wetenschappen, Letterkunde en
    Schoone Kunsten. Amsterdam.
    Tijdschr.Wisen Natuurk.Wet.Amst. 1848-1852.   S. 602 D

Timehri. Demerara and Georgetown.
    Timehri 1822 →   S. 2205 A

Times Literary Supplement. London.
    Times lit.Suppl. No. 3,540 → 1970 →   Librarian's Room

Time's Telescope. London.
    Time's Telescope 1814-1834.   S. 421

Tinea, Tokyo.
    Tinea, Tokyo 1953 →   E.S 1915 a

Tiscia. Dissertationes Biologicae a Collegio Exploratorum
    Fluminis Tisciae Editae. Szeged.
    Tiscia 1965 →   S. 1724 a

Tissue and Cell. Edinburgh.
    Tissue Cell 1968 →   S. 420 a

Titles of Dissertations approved for the Ph.D., M.Sc., and M. Litt.
    Degrees in the University of Cambridge.
    Titles Diss.Univ.Camb. 1957 →
    Formerly Abstracts of Dissertations approved for the Ph.D.,
    M.Sc., and M. Litt. Degrees in the University of Cambridge.   S. 35 B

| TITLE | SERIAL No. |

Toelichting. Geologische Kaart van Sumatra.
    Toelich.geol.Kaart Sumatra Blad 2, 4, 5 and 10. 1931-1932. P.S 1299

Tohoku Journal of Agricultural Research. Sendai.
    Tohoku J.agric.Res. 1950 → S. 1988 C

Tohoku Konchu Kenkyu. Morioka City.
    Tohoku Konchu Kenkyu 1964 → E.S 1905 a

Toimetised. Eesti NSV Teaduste Akadeemia.
    See Eesti NSV Teaduste Akadeemia Toimetised. Tallinn. S. 1817 A - B

Tokyo Journal of Climatology. Tokyo.
    Tokyo J.Clim. 1965 → S. 1988 d B

Tombo. Tokyo.
    Tombo 1959 → E.S 1916b

Tomurcuk. Istanbul.
    Tomurcuk 1952-1958.
    Continued as Koruma. Istanbul. E.S 1888

Tori. Bulletin of the Ornithological Society of Japan. Tokyo.
    Tori 1915 → T.R.S 3801 A & Z.S 1952 a A

Torreia. Direccion Nacional de Zoologicos y Acuarios. Habana.
    Torreia Nueva Serie 1967 →
    (Previous series issued by the Museo Poey.) S. 2287 B

Torreia. Museo Poey. Habana.
    Torreia Nos.1, 12-14, 16-19 & 22. 1939-1954.
    (Nueva Serie started in 1967 by the Direccion Nacional
      de Zoologicos y Aquarios.) S. 2287 B

Torreya. Torrey Botanical Club. New York.
    Torreya 1901-1945.
    Continued in Bulletin of the Torrey Botanical Club
    (and Torreya). New York. B.S 4317 b

Torry Research on the Handling and Preservation of Fish and
    Fish Products. Edinburgh.
    Torry Res.Handl.Preserv.Fish 1959 → Z.S 482

Tortuga Gazette. California Turtle and Tortoise Club. Los Angeles.
    Tortuga Gaz. Vol.4 → 1968 → Z. Reptile Section

Torvmarkskartor med Beskrivningar. Sveriges Geologiska
    Undersökning. Stockholm.
    See Sveriges Geologiska Undersökning. Stockholm. Ser.D. P.S 1400

Townsville Naturalist. Townsville, N.Qd.
    Townsville Nat. 1953 → (imp.) S. 2144

| TITLE | SERIAL No. |

Toxicology. Amsterdam.
   Toxicology 1973 →   S. 2759

Toxicon. Oxford, London, &c.
   Toxicon 1962 →   S. 2719 a

Trabajos. Catedra de Artrópodos, Departamento de Zoologia. Madrid.
   Trab.Cated.Artrop.Dep.Zool.Madrid 1973 →   E.S 1009

Trabajos del Cuarto Congreso Cientifico (1º Pan-Americano). Santiago de Chile.
   Trab.Cuarto Congr.cient.Santiago 1909-1910.   S. 2208

Trabajos del Departamento de Botanica y Fisiologia Vegetal, Universidad de Madrid. Madrid.
   Trab.Dep.Bot.Fisiol.veg.Univ.Madrid 1968 →   B.S 583

Trabajos de Divulgacion. Direccion General de Pesca e Industrias Conexas. Mexico.
   Trab.Divulg.Dir.gen.Pesca, Mex. Vol.13, num.128 → 1967 →   Z.S 2261 A

Trabajos de Divulgacion. Museo "Felipe Poey" de la Academia de Ciencias de Cuba. La Habana.
   Trab.Divulg.Mus."Felipe Poey" Acad.Cienc.Cuba No.41 → 1967 → S. 2289 B

Trabajos de la Estación Agrícola Experimental de León. León.
   Trab.Estac.agric.exp.León 1964 →   E.S 1007

Trabajos. Estación Limnológica de Pátzcuaro.
   Trab.Estac.limnol.Patzcuaro No.1 & 4. 1940-1941.   S. 2265 A

Trabajos de Geologia, Facultad de Ciencias, Universidad de Oviedo.
   Trab.Geol.Fac.Cienc.Univ.Oviedo 1967 →   P.S 658 a

Trabajos del Instituto de Biologia Animal. Madrid.
   Trab.Inst.Biol.anim.,Madr. 1933-1936.   Z.S 1015

Trabajos del Instituto de Botánico y Farmacología (Julio A. Roca). Buenos Aires.
   Trab.Inst.Bot.Farmac., B.Aires 1914-1940. (imp.)   B.S 3003

Trabajos del Instituto de Ciencias Naturales José de Acosta. Madrid.
   Trab.Inst.cienc.nat.,Madr. Ser.Biol. 1942-1955.   S. 1031 B
                                   Ser.Geol. 1943-1945.   P.S 665
   Continued in Boletin de la R. Sociedad Española de la Historia Natural. Madrid.

Trabajos del Instituto Español de Oceanografia. Madrid.
   Trab.Inst.esp.Oceanogr. 1929 →   Z.S 1001 B

TITLE	SERIAL No.

Trabajos e Investigaciones. Estación de Biología Marina.
   Instituto Tecnológico de Veracruz. Mexico.
   <u>Trab.Invest.Estac.Biol.mar</u>. 1961-1966.
   (Wanting No.4.)   S. 2259 B

Trabajos sobre Islas de Lobos y Lobos Marinos. Montevideo.
   <u>Trab.Isl.Lobos</u> 1956 →   Z.S 2220 A

Trabajos del Jardin Botánico. Universidad de Santiago.
   Santiago de Compostela.
   <u>Trab.Jard.bot.Univ.Santiago</u> 1950 →   B.S 590

Trabajos del Laboratorio de Historia Natural. Valencia.
   <u>See</u> Anales del Instituto General y Técnico de Valencia.   S. 1026 B

Trabajos del Museo Botánico, Universidad Nacional de Córdoba.
   <u>Trab.Mus.bot.Univ.nac.Córdoba</u> 1947-1964.   B.S 3011

Trabajos del Museo de Ciencias Naturales de Barcelona.
   <u>Trab.Mus.Cienc.nat.Barcelona</u> Nueva Ser.
     Serie Geologica. Vol.1. No.1. 1947.   S. 1008 A
     Serie Zoologica. 1947-1957.   "
     "   " 1947-1954.   T.R.S 1108
   <u>Formerly</u> Musei Barcinonensis Scientiarum Naturalium
   Opera. Barcelona.

Trabajos Museo Nacional de Ciencias Naturales. Madrid.
   <u>Trab.Mus.nac.Cienc.nat.Madr</u>.
     Serie Botanica, 1912-1934.   B.S 582
     Serie Geologica, 1912-1935.   P.S 660
     Serie Zoologica, 1912-1932.   Z.S 1010

Trabajos del Museo de Zoologica. Barcelona.
   <u>See</u> Trabajos del Museo de Ciencias Naturales de Barcelona.   S. 1008 A

Trabalhos do Centro de Biologia Piscatoria. Lisboa.
   <u>See</u> Memorias da Junta de Investigações do Ultramar. Lisboa.   S. 1053 F

Trabalhos do Centro de Botânico da Junta de Investigacões do
   Ultramar. Lisboa.
   <u>Trabhs Cent.Bot.Jta Invest.Ultramar</u> 1961-1970.
   (Nos.1-3 are reprints from Memorias. Junta Investigacões
     do Ultramar.)   B.S 609

Trabalhos do Centro de Investigação Cientifica Algodoeira.
   Lourenço Marques.
   <u>Trabhs Cent.Invest.cient.algod.Lourenco Marq</u>. 1949-1951.
   <u>Continued as</u> Memórias y Trabalhos. Centro de Investigação
   Cientifica Algodoeira. Lourenço Marques.   S. 2088

Trabalhos. Conferência Internacional dos Africanistas
   Ocidentais. Bissau, 1947. Lisboa.
   <u>Trabhs Conf.Int.Afr.Ocid</u>. 2nd Conf. vols. 2-3. 1950-1951.   74.o.P

| TITLE | SERIAL No. |

Trabalhos do Instituto de Biologia Marítima e Oceanografia,
 Universidade do Recife.
  Trabhs Inst.Biol.marít.Oceanogr.Univ.Recife Vol.2, No.1, 1960.
  Continued as Trabalhos do Instituto Oceanográfico da
  Universidade do Recife.   S. 2214 b

Trabalhos Instituto de Botânico Dr. Goncalo Sampaio. Porto.
  Trabhs Inst.Bot.Dr.Goncalo Sampaio 1943 →   B.S 611 b

Trabalhos do Instituto Botânico da Faculdade de Ciências de Lisboa.
  Trabhs Inst.bot.Fac.Ciênc.Lisb. 1925-1942.   B.S 613

Trabalhos do Instituto de Investigacão Cientifica de Mocambique.
 Lourenco Marques.
  Trabhs Inst.Invest.cient.Mocamb. 1961 →   S. 2089 D

Trabalhos do Instituto Oceanografico da Universidade do Recife.
  Trabhs Inst.Oceanogr.Univ.Recife Vol.3-8, 1963-1967.
  Formerly Trabalhos do Instituto de Biologia Marítima e
  Oceanografia, Universidade do Recife.
  Continued as Trabalhos Oceanográficos da Universidade
  Federal de Pernambuco.   S. 2214 b

Trabalhos Oceanográficos da Universidade Federal de Pernambuco. Recife.
  Trabhs oceanogr.Univ.fed.Pernambuco No.9 → 1970 →
  Formerly Trabalhos do Instituto Oceanográfico da
  Universodade do Recife.   S. 2214 b

Tracts. Bibliothèque des Jeunes Naturaliste. Montréal.
  See Bibliothèque des Jeunes Naturalistes. Société
  Canadienne d'Histoire Naturelle. Montréal.   S. 2666

Trade Circular. Division of Forest Products. Council
 (Commonwealth) for Scientific and Industrial Research
 (Organization). Commonwealth of Australia. Melbourne.
  Trade Circ.Div.Forest Prod.C.S.I.R.O. Nos.6, 11, 14, 18,
  25, 28, 38-46, 48 & 50. 1935-1954.   S. 2113 D

Traduction. Bureau de Recherches Géologiques et Minières.
 Service d'Information Géologique. Paris.
  Traduct.Bur.Rech.géol.min.Paris No.2685 → 1960 →   P.S 514

Transactions of the Aberdeen Working Men's Natural History
 and Scientific Society.
  Trans.Aberd.wkg Men's nat.Hist.scient.Soc. 1901-1916.   S. 6 c

Transactions of the Academy of Science of St.Louis. Mo.
  Trans.Acad.Sci.St.Louis 1856-1958.   S. 2381 A
                           1856-1900.   T.R.S 5144

Transactions of the Albany Institute.
  Trans.Albany Inst. 1828-1893.   S. 2312 A

| TITLE | SERIAL No. |

Transactions of the Allelodidactic Society. London.
    Trans.Allelodidactic Soc. 1847-1848.   S. 163

Transactions of the All-Union Scientific Research
    Institute of Economic Mineralogy. Moscow.
    See Trudy Vsesoyuznogo Nauchno-Issledovatelskogo Instituta
    Mineral'nogo Sy'rya. Moskva.   M.S 1809

Transactions of the All-Union Scientific Research Institute
    of Marine Fisheries and Oceanography. Moscow.
    See Trudy Vsesoyuznogo-Nauchno Issledovatel'skogo
    Instituta Morskogo Rybnogo Khozyaĭstva i Okeanografii.
    Moskva.   S. 1946 b

Transactions of the American Entomological Society. Philadelphia.
    Trans.Am.ent.Soc. 1867 →   E.S 2422

Transactions of the American Fisheries Society. St.Paul,
    Minnesota.
    Trans.Am.Fish.Soc. Vol. 89 → 1960 →   Z.S 2516

Transactions of the American Game Conference. Washington, D.C.
    Trans.Am.Game Conf. 16th-21st. 1929-1935.
    Replaced by Proceedings of the North American Wildlife
    Conference.   Z.S 2709

Transactions of the American Geophysical Union. Washington.
    Trans.Am.geophys.Un. 1920-1958. (No.1-2, 4, 6-9
    published in Reprint & Circular Series and Bulletin of the
    National Research Council, q.v.; No.12 wanting.)   S. 2419 B
                      1967-1968.   M.S 2657
    Continued as Eos. Transactions of the American
    Geophysical Union.

Transactions of the American Institute of Mining Engineers. New York.
    Trans.Am.Inst.Min.Engrs 1871-1918.
    Continued as Transactions of the American Institute of
    Mining and Metallurgical Engineers. New York.   M.S 2692

Transactions of the American Institute of Mining and
    Metallurgical Engineers. New York.
    Trans.Am.Inst.Min.metall.Engrs 1919-1921.
    Formerly Transactions of the American Institute of Mining
    Engineers. New York.   M.S 2692

Transactions of the American Microscopical Society.
    Lancaster, Pa.
    Trans.Am.microsc.Soc. 1895 →
    Formerly Proceedings of the American Microscopical
    Society. Menasha, Wis.   S. 2500

Transactions of the American Philosophical Society.
    Philadelphia, Pa.
    Trans.Am.phil.Soc. 1769 →   S. 2304 B

TITLE	SERIAL No.

Transactions and Annual Report. Liverpool Geographical Society.
    Trans.a.Rep.Lpool geogr.Soc. 1896-1914.
    Formerly Report (of the Council) of the Liverpool
Geographical Society.   S. 162 A

Transactions and Annual Report of the Manchester Microscopical
Society.
    Trans.a.Rep.Manchr micr.Soc. 1884-1900.
    Formerly Report. Manchester Microscopical Society.
    Continued as Report and Transactions of the Manchester
Microscopical Society.   S. 262

Transactions and Annual Report. North Staffordshire Field Club.
    Trans.a.Rep.N.Staffs.Fld Club 1915-1960.
    Formerly Report and Transactions. North Staffordshire
(Naturalists') Field Club.   S. 371 B

Transactions of the Anthropological Society of London.
    See Journal of the Anthropological Society of London.   P.A.S 1 C

Transactions of the Arctic Institute. Leningrad.
    See Trudy Arkticheskogo Nauchno-Issledovatel'skogo
Instituta. Leningrad.   S. 1842 A

Transactions of the Armenian Anti-Pest Station. Erevan.
    See Trudy Armyanskoĭ Protivochumnoĭ Stantsii. Erevan.   S. 1858 d

Transactions of the Armenian Branch of the Academy of Sciences
of the USSR. Biol.Ser. Erevan.
    See Trudy Armyanskogo Filiala Akademii Nauk SSSR.   S. 1802 Ca A

Transactions of the Asiatic Society of Japan. Yokohama.
    Trans.Asiat.Soc.Japan 1872-1907.   S. 1996

Transactions. Atlantic Scientific Research Institute of Fisheries
and Oceanography. Kaliningrad.
    See Trudy. Atlanticheskiĭ Nauchno-Issledovatel'skiĭ
Institut Rybnogo Khozyaĭstva i Okeanografii. Kaliningrad.   S. 1847 a

Transactions of the Barnsley Naturalists' Society.
    Trans.Barnsley Nat.Soc. vol. 5. 1885-1886.
    Formerly Quarterly Transactions of the Barnsley
Naturalists' Society.   S. 8

Transactions of the Barrow Naturalists' Field Club.
(1876-1879 cover only).
    See Report (and Proceedings) of the Barrow Naturalists'
Field Club.   S. 9

Transactions of the Biological Club of Nippon. Tokyo.
    See Biologica. Tokyo.   S. 1964 a

| TITLE | SERIAL No. |

Transactions of the Biological Institute. The Armenian Branch
of the Academy of Sciences of the USSR. Erevan.
See Trudy Biologicheskogo Instituta. Armyanskii Filial.   S. 1802 Ca A

Transactions &c. Birmingham & Midland Institute. Walsall.
Archaeological Section. 1870-1894.
Trans.Bgham Midl.Inst.Arch.Sect. 1871-1895.   S. 17 b A

Transactions of the Bishop's Stortford and District
Natural History Society.
Trans.Bishop's Stortford Distr.nat.Hist.Soc. 1950-1953.   S. 17 d

Transactions of the Bose Research Institute. Calcutta.
Trans.Bose Res.Inst. 1918 →   B.S 1607

Transactions of the Botanical Society of Edinburgh.
Trans.bot.Soc.Edinb. 1839-1890; 1970 →
From 1890-1970 See Transactions and Proceedings of the
Botanical Society of Edinburgh.   B.S 6 b

Transactions of the Botanical Society of Pennsylvania.
See Transactions & Proceedings of the Botanical Society
of Pennsylvania. Philadelphia.   B.S 4374

Transactions. British Bryological Society. London.
Trans.Br.bryol.Soc. 1946-1971.
Formerly Report British Bryological Society.
Continued as Journal of Bryology. Oxford.   B.B.S 4

Transactions. British Cave Research Association. Bridgwater.
Trans.Br.Cave Res.Ass. 1974 →   S. 156 B

Transactions of the British Ceramic Society, Stoke-on-Trent.
Trans.Br.Ceram.Soc. 1938 →
Formerly Transactions of the Ceramic Society, Stoke-on-Trent.   M.S 124

Transactions of the British Institute of Preventive Medicine.
London.
Trans.Br.Inst.prev.Med. 1897.   B.S.S 13

Transactions of the British Mycological Society. London.
Trans.Br.mycol.Soc. 1896 →   B.M.S 7

Transactions of the Buchan Club. Peterhead.
Trans.Buchan Club 1926-1940. imp.   S. 323 B

Transactions of the Burnley Literary & Scientific Club.
Trans.Burnley lit.scient.Club Vols.1-10, 20-27, 33, 35-36.
1874-1919.   S. 33

Transactions of the Burton-on-Trent Natural History and
Archaeological Society. London.
Trans.Burton-on-Trent nat.Hist.archaeol.Soc. 1889-1933.   S. 23 B

Transactions of the Buteshire Natural History Society. Rothesay.
Trans.Butesh.nat.Hist.Soc. 1907-1935, 1938 → (imp.)   S. 344

| TITLE | SERIAL No. |

Transactions of the Cambridge Philosophical Society.
    Trans.Camb.phil.Soc. 1820-1928. — S. 6 B

Transactions of the Canadian Institute. Toronto.
    See Transactions of the Royal Canadian Institute. Toronto. — S. 2605 A

Transactions. Caradoc & Severn Valley Field Club. Shrewsbury.
    Trans.Caradoc Severn Vall.Fld Club 1893-1945. — S. 360 B

Transactions of the Cardiff Naturalists' Society.
    Trans.Cardiff Nat.Soc. Vol. 35 → 1903 → — S. 25
        Vol. 35-36, 1903. — T.R.S 463
    Formerly Report and Transactions. Cardiff
    Naturalists' Society. — S. 25

Transactions of the Carlisle Natural History Society.
    Trans.Carlisle nat.Hist.Soc. 1909 → — S. 37 A

Transactions of the Cave Research Group of Great Britain.
    Leamington Spa.
    Trans.Cave Res.Grp Gt Br. 1948 → — S. 15 a B

Transactions of the Central Geological and Prospecting Institute.
    Leningrad.
    See Trudȳ Tsentral'nogo Nauchno-Issledovatel'skogo Geologo-
    Razvedochnogo Instituta (TSNIGRI). Leningrad. — P.S 1528

Transactions of the Ceramic Society. Stoke on Trent.
    Trans.Ceram.Soc. 1917-38
    Formerly Transactions of the English Ceramic Society. Tunstall.
    Continued as Transactions of the British Ceramic Society.
    Stoke on Trent. — M.S 124

Transactions of the Chartered Surveyors' Institution. London.
    Trans.chart.Surv.Instn 1930-1943.
    Formerly Transactions of the Surveyors' Institution. London.
    Continued as Transactions. Royal Institution of
    Chartered Surveyors. London. — S. 219 A

Transactions. Chesterfield and Derbyshire Institute of Mining,
    Civil and Mechanical Engineers. London
    Trans.Chester.Derby.Inst.Min.Engrs Vol. 13-14; 1884-86
    Continued as Transactions. Chesterfield and Midland.
    Counties Institution of Engineers Chesterfield. London. — M.S 309

Transactions. Chesterfield and Midland Counties Institution
    of Engineers, Chesterfield. London.
    Trans.Chester.Midland Count.Inst.Engrs Vol. 15-17; 1886-89
    Formerly Transactions. Chesterfield and Derbyshire
    Institute of Mining, Civil and Mechanical Engineers. London — M.S 309

Transactions of the Chicago Academy of Sciences.
    Trans.Chicago Acad.Sci. 1867-1870. — S. 2329 K

| TITLE | SERIAL No. |

Transactions of the Chichester and West Sussex Natural History
    and Microscopical Society.
        Trans.Chichester nat.Hist.micr.Soc. 1882-1889.
        Formerly Report. Chichester and West Sussex Natural
    History and Microscopical Society.   S. 27 a A

Transactions of the City of London Entomological & Natural
    History Society.
        Trans.Cy Lond.ent.nat.Hist.Soc. 1890-1913.   E.S 29 & S. 170
        Continued as Transactions of the London Natural History
    Society.

Transactions of the Clifton College Scientific Society.
        Trans.Clifton Coll.scient.Soc. 1871-1875.   S. 27 b A

Transactions of the Committee of the Turkmen Government for
    Protection of Nature and the Development of the Riches
    of Nature. Ashkhabad.
        See Izvestiya Turkmenskogo Mezhduvedomstvennogo
    Komiteta po Okhrane Prirodȳ i Razvitiyu Prirodnȳkh
    Bogatstv. Ashkhabad.   S. 1859 c

Transactions of the Conference on Petroleum. Academy of
    Sciences of the Ukrainian SSR.
        See Trudy Neftyanoĭ Konferentsii. Akademiya Nauk
    USSR. Kiev.   P. 72 Q.o.K

Transactions of Conferences. Josiah Macy Junior Foundation, New York.
        Trans.Confs Josiah Macy jr Fdn 1954 → (imp.)
            Polysaccharides in Biology. 1955 →
            Cold Injury.   1957.
            Group Processes.   1955.
            Nerve Impulse.   1954.
            Neuropharmacology.   1955.
            Shock & Circulatory
            Homeostasis.   1955.   7.o.N
            Gestation.   1957-1958.   Z. 12C o V

Transactions of the Connecticut Academy of Arts and Sciences.
    New Haven.
        Trans.Conn.Acad.Arts Sci. 1866 →   S. 2351 B
                                1866-1903.   T.R.S 5124

Transactions of the County of Middlesex Natural History
    & Science Society.
        Trans.Middlx nat.Hist.sci.Soc. 1886-1891.   T.R.S 143 & S. 269

Transactions of the Croydon Natural History and Scientific Society.
        See Proceedings and Transactions of the Croydon Natural
    History and Scientific Society. Croydon.   S. 28

Transactions of the Cumberland Association for the Advancement
    of Literature and Science. Keswick.
        Trans.Cumberland Ass.Advanc.Sci. 1875-1893.   S. 29

TITLE	SERIAL No.

Transactions of the Derby Natural History Society.
    Trans.Derby nat.Hist.Soc. 1960 →   S. 90

Transactions of the Dublin Society.
    Trans.Dublin Soc. 1799-1810.   S. 46 D

Transactions (Proceedings) of the Dudley and Midland Geological
    and Scientific Society and Field Club. Dudley.
    Trans.Dudley Mid.geol.sci.Soc. 1862-1867.
    Continued as Proceedings of the Dudley and Midland
    Geological and Scientific Society and Field Club.   P.S 135

Transactions of the Dumfriesshire and Galloway Natural History
    and Antiquarian Society.
    See Transactions and Journal of the Proceedings of the
    Dumfriesshire and Galloway Natural History Society.
    Edinburgh.   S. 51

Transactions on the Dynamics of Development. Moscow.
    See Trudy po Dinamike Razvitiya. Moskva.   S. 1846

Transactions of the East Kent Natural History Society. Canterbury.
    Trans.E.Kent nat.Hist.Soc. 1885-1889.   S. 24 B

Transactions of the East Lothian Antiquarian & Field
    Naturalists' Society. Edinburgh.
    Trans.E.Loth.Antiq.Fld Nat.Soc. 1924 →   S. 105

Transactions of the Eastbourne Natural History
    (Scientific & Literary) Society.
    Trans.Eastbourne nat.Hist.Soc. N.S. 1-4. 1881-1912.
    Continued as Transactions & Journal of the Eastbourne
    Natural History, Photographic, Literary &
    (Archaeological) Society.   S. 54 A

Transactions of the East-Siberian Geological and Prospecting
    Trust. Irkutsk.
    See Trudy Vostochno-Sibirskogo Geologo-Razvedochnogo Tresta.
    Irkutsk.   P.S 1531

Transactions of the Edinburgh Field Naturalists &
    Microscopical Society.
    Trans.Edinb.Fld Nat.microsc.Soc. 2-7, 1886-1915.
    Formerly Transactions of the Edinburgh Naturalists'
    Field Club.   S. 61

Transactions of the Edinburgh Geological Society.
    Trans.Edinb.geol.Soc. 1866-1963.   P.S 163
                           Vol.15, 1952.   M.S 103 A
    Amalgamated with the Transactions of the Geological Society
    of Glasgow to form the Scottish Journal of Geology.

| TITLE | SERIAL No. |

Transactions of the Edinburgh Naturalists' Field Club.
    Trans.Edinb.nat.Fld Club   1881-1886.
    Continued as Transactions of the Edinburgh Field
    Naturalists' and Microscopical Society.   S. 61

Transactions of the English Ceramic Society. Tunstall.
    Trans.Engl.Ceram.Soc.   1905-1917
    Formerly Transactions of the North Staffordshire
    Ceramic Society. Hanley.
    Continued as Transactions of the Ceramic Society.
    Stoke on Trent.   M.S 124

Transactions of the Entomological Society of Japan. Kyoto.
    Trans.ent.Soc.Japan   1907-1909.   E.S 1909

Transactions of the Entomological Society of London.
    Trans.ent.Soc.London   1807-1812: 1834-1933.
    (The first Society ceased to exist sometime after 1812,
    the second Society was founded in 1833.)
    Continued as Transactions of the Royal Entomological
    Society of London.   E.S 17

Transactions of the Entomological Society of New South Wales. Sydney.
    Trans.ent.Soc.N.S.W.   1862-1873.   E.S 2251

Transactions of the Entomological Society of the South of
    England. Southampton.
    Trans.ent.Soc.S.Engl.   1929-1932.
    Formerly Transactions of the Hampshire Entomological
    Society. Southampton.
    Continued as Transactions of the Society for British
    Entomology. Southampton.   E.S 30

Transactions of the Epping Forest and County of Essex
    Naturalists' Field Club. Buckhurst Hill.
    Trans.Epping Forest Essex Nat.Fld Cl.   1880-1882.
    Continued as Transactions of the Essex Field Club.   S. 22 A,
    Buckhurst Hill.   E.S 26 & T.R.S 167

Transactions of the Essex Field Club. Buckhurst Hill.
    Trans.Essex Fld Cl.   Vol.3-4.   1882-1887.
    Formerly Transactions of the Epping Forest and County of
    Essex Naturalists' Field Club. Buckhurst Hill.   S. 22 A,
    Continued as Essex Naturalist.   E.S 26 & T.R.S 167

Transactions of the Ethnological Society of London. New Series.
    London.
    Trans.ethnol.Soc.Lond.   1861-1869.
    Formerly and Continued as Journal of the Ethnological
    Society of London.   P.A.S 4 B & Z.S 180 B

Transactions of the Far-East Geological and Prospecting Trust
    of the USSR.
    See Materialy̅ po Geologii i Polezny̅m Iskopaemy̅m Dal'nyago
    Vostoka. Vladivostok.   P.S 1540

TITLE	SERIAL No.

Transactions. Far-Eastern Association of Tropical Medecine.
   Trans.Far-East.Ass.trop.Med. 6th. Tokyo 1925, Vol.1
   7th. British India 1927, Vol.1-3.
   9th. Nanking 1934, Vol.1-2.   Z. 69 o F

Transactions of the Federal-Provincial Wildlife Conference.
   Canadian Wildlife Service.
   Trans.Fed.-Prov.Wildl.Conf. 31st. → 1967 →
   Formerly Summary, Notes and Papers of the Federal-
   Provincial Wildlife Conference.   Z.S 2635 E

Transactions of the Federated Institution of Mining Engineers.
   Newcastle/Tyne.
   Trans.Fed.Instn.Min.Engrs 1889-98
   Continued as Transactions of the Institution of Mining
   Engineers. London.   M.S 310

Transactions of the Folkestone Natural History Society.
   Trans.Folkestone nat.Hist.Soc. 1949-1950.   S. 86 E

Transactions of the Geological, Hydrological and Geodetical
   Trust of Ukraine.
   See Trudȳ Ukrainskogo Geologo-Gidro-Geodezicheskogo
   Tresta i Ukrnigri. Moskva.   P.S 1538

Transactions of the Geological and Mining Faculties. Beograd.
   See Zbornik Geoloshkog i Rudarskog Fakulteta,
   Tehnička Velika Skola. Beograd.   P.S 457

Transactions of the Geological Oil Institute, Moscow, Leningrad, &c.
   See Trudȳ Neftyanogo Geologo-Razvedochnogo Instituta. Moskva,
   Leningrad, &c.   P.S 525 & P.S 525 A

Transactions of the Geological and Prospecting Service of the
   U.S.S.R. Moscow.
   See Trudȳ Glavnogo-Geologo Razvedochnogo Upravleniya V.S.N.
   Kh.SSSR, and Trudȳ Vsesoyuznogo Geologo-Razvedochnogo
   Ob"edineniya NKTP.SSSR. Leningrad.   P.S 1525

Transactions of the Geological Society. London.
   Trans.geol.Soc.Lond. 1811-1856.   P.S 110

Transactions of the Geological Society of Australia. Melbourne
   Trans.geol.Soc.Austral. 1886-1892.   P.S 188

Transactions of the Geological Society of Glasgow.
   Trans.geol.Soc.Glasg. 1860-1963.
   Amalgamated with the Transactions of the Edinburgh
   Geological Society to form the Scottish Journal of Geology.   P.S 165

| TITLE | SERIAL No. |

Transactions of the Geological Society of Glasgow.
　　Palaeontological Series.
　　　Trans.geol.Soc.Glasg.Pal.Ser. (1868).　　　　　　P.S 166

Transactions of the Geological Society of Pennsylvania.
　　Philadelphia.
　　　Trans.geol.Soc.Pa. 1834-1835.　　　　　　　　　　P.S 868

Transactions of the Geological Society of South Africa. Johannesburg.
　　　Trans.geol.Soc.S.Afr. 1896-1945.
　　　Continued in Transactions and Proceedings of the Geological
　　Society of South Africa.　　　　　　　　　　　　　　P.S 190

Transactions of the Glasgow Archaeological Society New Series.
　　Glasgow.
　　　Trans.Glasg.archaeol.Soc. 1885-1899.　　　　　P.A.S 15 A

Transactions of the Guernsey Society of Natural Science and
　　Local Research.
　　　See Report and Transactions of the Guernsey Society of
　　Natural Science(and Local Research).Guernsey.　　S. 113 & T.R.S 582

Transactions. Gulf Coast Association of Geological Societies.
　　　Trans.Gulf Cst Ass.geol.Socs 1953 →
　　　Formerly Annual Meeting of Gulf Coast Association of
　　Geological Societies.　　　　　　　　　　　　　　　P.S 937

Transactions of the Hampshire Entomological Society. Southampton.
　　　Trans.Hamps.ent.Soc. 1924-1928.
　　　Continued as Transactions of the Entomological Society of
　　the South of England.　　　　　　　　　　　　　　　E.S 30

Transactions of the Hawick Archaeological Society.
　　　Trans.Hawick archaeol.Soc. 1906-1908.　　　　　S. 119 A

Transactions of the Hertfordshire Natural History Society &
　　Field Club.
　　　Trans.Herts.nat.Hist.Soc.Fld Club 1879 →
　　　Formerly Transactions of the Watford Natural History
　　Society & Hertfordshire Field Club.　　　　　　　S. 391 A & T.R.S 137

Transactions of the Highland and Agricultural Society of
　　Scotland. Edinburgh.
　　　Trans.Highl.agric.Soc.Scotl. 1944-1948.　　　　E.S 57 a
　　　　　　　　　　　　Vol.25-26, 1913-1914.　　　　T.R.S 224
　　　Continued as Transactions of the Royal Highland and
　　Agricultural Society of Scotland. Edinburgh.　　E.S 57 a

Transactions of the Historic Society of Lancashire & Cheshire.
　　　Trans.Hist.Soc.Lancs.& Chesh. 1854-1883.
　　　Formerly Proceedings and Papers. Historic Society
　　of Lancashire & Cheshire.　　　　　　　　　　　　　S. 155

TITLE	SERIAL No.

Transactions of the Hull Geological Society.
   Trans.Hull geol.Soc. 1893-1937.   P.S 150

Transactions of the Hull Scientific & Field Naturalists' Club.
   Trans.Hull scient.Fld Nat.Club 1898-1919.   S. 126 A

Transactions of the Illinois Natural History Society. Springfield.
   Trans.Ill.nat.Hist.Soc. Vol.1, 2nd.edn. 1861.   S. 2408 a

Transactions of the Illinois State Academy of Science. Springfield.
   Trans.Ill.St.Acad.Sci. 1908 →   S. 2408

Transactions. Institute of Biology, Latvian Academy of Sciences.
   See Trudy Instituta Biologii. Akademiya Nauk Latviĭskoĭ SSR. Riga.   S. 1852 a C

Transactions of the Institute of Economic Mineralogy. Moscow.
   See Trudy Instituta Prikladnoi Mineralogii, Moskva.   M.S 1809

Transactions of the Institute of Fisheries and Scientific Explorations.
   See Trudy Instituta Rybnogo Khozyaĭstva i Promyslovykh Issledovaniĭ. Leningradskoe Otdelenie.   Z.S 1821

Transactions of the Institute of Geography. Moscow, Leningrad.
   See Trudy Instituta Geografii. Akademiya Nauk SSSR. Moskva, Leningrad.   P.S 512

Transactions of the Institute of Geology of Ore Deposits, Petrography, Mineralogy and Geochemistry. Moscow.
   See Trudy Instituta Geologii Rudnyk Mestorozhdenii, Petrografii, Mineralogii i Geokhimii. Moskva.   M.S 1816

Transactions of the Institute of Marine Fisheries and Oceanography. Moscow.
   See Trudy Vsesoyuznogo Nauchno-Issledovatel'skogo Instituta Morskogo Rybnogo Khozyaĭstva i Okeanografii. Moskva.   S. 1846 b

Transactions of the Institute of Oceanography. Moscow.
   See Trudy Instituta Okeanologii. Akademiya Nauk SSSR. Moskva.   S. 1802 d

Transactions of the Institute for Scientific Exploration of the North. Moskow.
   See Trudy Nauchno-Issledovatel'skogo Instituta po Izucheniyu Severa. Moskva.   S. 1842 A

Transactions. Institution of Applied Mineralogy, Moscow.
   See Trudy Instituta Prikladnoi Mineralogii, Moskva.   M.S 1809

| TITLE | SERIAL No. |

Transactions of the Institution of Mining Engineers.
  Newcastle/Tyne.
  Trans.Instn Min.Engrs 1897-1921 (imp.)
  Formerly Transactions of the Federated Institution
  of Mining Engineers. Newcastle/Tyne.   M.S 310

Transactions of the Institution of Mining and Metallurgy. London.
  Trans.Instn Min.Metall. 1892-1961; 1966 →
  (Section B only)
  Issued as part of Bulletin of the Institution of Mining
  and Metallurgy 1950 →   M.S 118

Transactions of the International Conference of the Association
  for the Study of the Quaternary Period in Europe.
  See Conference of the International Association on
  Quaternary Research.   P.S 997

Transactions. International Congress of Prehistoric Archaeology.
  Trans.Int.Congr.prehist.Archaeol. Sess.3 1868 (1869.)   P.A.S 901

Transactions of the International Symposia of the International
  Society for Plant Geography and Ecology. The Hague.
  See Bericht über die Internationalen Symposia der
  Internationalen Vereinigung für Vegetationskunde.
                                         See Gen.Lbry Catalogue

Transactions of the International Union of Game Biologists.
  See International Union of Game Biologists.   Z.S 2711

Transactions of the Inverness Scientific Society and Field Club.
  Trans.Inverness scient.Soc.Fld Club 1875-1925.   S. 130

Transactions of the Isle of Man Natural History & Antiquarian
  Society. Douglas.
  Trans.Isle Man nat.Hist.Soc. 1879-1884.   S. 335 C

Transactions of the Jamaica Society of Arts. Kingston, etc.
  Trans.Jamaica Soc.Arts 1854-1855.
  Continued as Transactions of the Royal Society of Arts
  (& Agriculture) of Jamaica. Kingston.   S. 2293

Transactions & Journal of the Eastbourne Natural History &
  Archaeological Society.
  Trans.J.Eastbourne nat.Hist.Soc. Vol.12, No.2-4. 1939-1946.
  Formerly Transactions & Journal of the Natural History
  Photographic, Literary & Archaeological Society.
  Continued as Journal & Transactions of Eastbourne Natural
  History & Archaeological Society.   S. 54 A

Transactions & Journal of the Eastbourne Natural History,
  Photographic, Literary & (Archaeological) Society.
  Trans.J.Eastbourne nat.Hist.Soc. Vol.5-12, No.1. 1927-1938.
  Formerly Transactions of the Eastbourne Natural History,
  Scientific & Literary Society.
  Continued as Transactions of the Eastbourne Natural History
  & Archaeological Society.   S. 54 A

TITLE	SERIAL No.

Transactions and Journal of the Proceedings of the
  Dumfriesshire and Galloway Natural History and Antiquarian
  Society. Edinburgh.
  Trans.J.Proc.Dumfries.Galloway nat.Hist.Antiq.Soc. 1862 →   S. 51

Transactions of the Kansai Entomological Society. Osaka.
  Trans.Kansai ent.Soc. 1930-1950 (wanting Vol.8, Vol.10,
  Vol.11, Pt.2, Vol.12, Pt.1.)   E.S 1918

Transactions of the Kansas Academy of Science. Topeka.
  Trans.Kans.Acad.Sci. 1868 →   S. 2410

Transactions of the Karelian Fisheries Station.
  See Trudȳ Karel'skoĭ Nauchno-Issledovatel'skoĭ
  Rȳbokhozyaĭstvennoĭ Stantsii.   Z.S 1857 A

Transactions. Kaspian Scientific Research Institute of Fisheries.
  Moskva.
  See Trudȳ. Kaspiĭskii Nauchno-Issledovatel'skiĭ Institut
  Rybnogo Khozyaĭstva. Moskva.   Z.S 1839

Transactions of the Kent Field Club. Maidstone.
  Trans.Kent Fld Club 1957 →   S. 256

Transactions of the Kentucky Academy of Science. Lexington.
  Trans.Ky Acad.Sci. 1914 →   S. 2527

Transactions of the Kinki Coleopterological Society. Japan.
  Trans.Kinki coleopt.Soc. 1946-1949.
  Merged with Entomological Review, Japan, 1949.   E.S 1911

Transactions of the Laboratory of Experimental Biology
  of the Zoopark of Moscow.
  See Trudy Laboratorii Eksperimental'noĭ Biologii
  Moskovskogo Zooparka. Moskva.   S. 1846

Transactions of the Lancaster Philosophical Society.
  Trans.Lancaster Phil.Soc. 1886-1887.   S. 142

Transactions of the Leeds Geological Association. Leeds, etc.
  Trans.Leeds geol.Ass. 1883-1967.
  Continued as Journal of Earth Sciences. Leeds.   P.S 140

Transactions of the Leeds Naturalists' Club & Scientific
  Association.
  Trans.Leeds Nat.Cl.scient.Ass. 1886-1890.
  Continued as List of Papers and Meetings. Leeds Naturalists'
  Club and Scientific Association.   S. 145 B & E.S 93

TITLE	SERIAL No.

Transactions of the Leicester Literary and Philosophical
    Society. Leicester.
        Trans.Leicester lit.phil.Soc. 1835-1879, 1889 →   S. 151 B

Transactions of the Leningrad Geological Hydrogeological and
    Geodetic Trust. Leningrad.
        See Trudȳ Leningradskogo Geologo-Gidrogeodezicheskogo
    Tresta. Leningrad.   P.S 1522

Transactions of the Lepidopterological Society of Japan.
        See Butterflies and Moths. Kyoto.   E.S 1909 a

Transactions. Lincolnshire Naturalists' Union.
        Trans.Lincs.Nat.Un. 1893-1894, 1905 →   S. 153

Transactions of the Linnaean Society of New York.
        Trans.Linn.Soc.N.Y. 1882 →   Z.S 2483
                1882-1884.   T.R.S 5130

Transactions of the Linnean Society of London.
        Trans.Linn.Soc.Lond. 1791 →   Z.S 20 A & T.R.S 132 C
                1791-1875. 2nd.Ser.Botany 1875-1922.   B.S 58 b
                2nd. Ser.Zoology. 1875-1884.   E.S 14

Transactions of the Literary & Historical Society of Quebec.
        Trans.lit.hist.Soc.Quebec 1824-1905.
            (Wanting Vols. 3 & 4, pt. 4 & NS pt 5.)   S. 2635

Transactions of the Literary & Philosophical Society
    of New York.
        Trans.lit.phil.Soc.N.Y. 1815-1825.   S. 2521

Transactions of the Liverpool Biological Society.
        See Proceedings & Transactions of the Liverpool
    Biological Society.   S. 157 A

Transactions Liverpool Botanical Society. Liverpool.
        Trans Lpool bot.Soc. 1909.   B.S 21 a

Transactions. Liverpool Geographical Society.
        See Transactions and Annual Report. Liverpool Geographical
    Society.   S. 162 A

Transactions. Liverpool Geological Association. Liverpool.
        Trans.Lpool geol.Ass. 1880-1890.
        Continued as Journal. Liverpool Geological Association.   P.S 132

Transactions of the Lomonossov Institute of Geochemistry,
    Crystallography and Mineralogy. Moscow and Leningrad.
        See Trudy Lomonosovskogo Instituta Geokhimii, Kristallografii
    i Mineralogii. Moskva.   M.S 1806

TITLE	SERIAL No.

Transactions of the London Natural History Society.
    Trans.Lond.nat.Hist.Soc. 1914-1920.
    Formerly Transactions of the City of London Entomological
    & Natural History Society.
    Continued as London Naturalist.   S. 174 B

Transactions of the Maidstone & Mid-Kent Natural History
    & Philosophical Society.
    Trans.Maidstone nat.Hist.& phil.Soc. 1870.   S. 255

Transactions of the Malvern Naturalists' Field Club. Worcester.
    Trans.Malvern Nat.Fld Cl. 1855-1879.   S. 399

Transactions. Manchester Entomological Society.
    See Report and Transactions of the Manchester
    Entomological Society. Manchester.   E.S 38

Transactions of the Manchester Geological and Mining Society.
    See Manchester Geological and Mining Society Transactions.   P.S 153

Transactions of the Manchester Geological Society.
    See Manchester Geological(and Mining Society)Transactions.   P.S 153

Transactions of the Manchester Microscopical Society.
    See Transactions and Annual Report of the Manchester
    Microscopical Society.   S. 262

Transactions. Manitoba Historical & Scientific Society. Winnipeg.
    Trans.Manitoba hist.scient.Soc. 1883-1885.
    Formerly Publication. Manitoba Historical & Scientific
    Society. Winnipeg.
    Continued as Report. Manitoba Historical & Scientific
    Society. Winnipeg.   S. 2681

Transactions of the Maryland Academy of Science and Literature.
    Baltimore.
    Trans.Md Acad.Sci.Lit. 1837.
    Continued as Transactions of the Maryland Academy of
    Sciences. Baltimore.   S. 2317 A

Transactions of the Maryland Academy of Sciences. Baltimore.
    Trans.Md Acad.Sci. 1888-1908.
    Formerly Transactions of the Maryland Academy of Science
    & Literature.
    Continued as Bulletin of the Maryland Academy of
    Sciences. Baltimore.   S. 2317 B

Transactions of the Medico-Botanical Society of London.
    Trans.med.-bot.Soc.Lond. Vol.1 pt.1 & 3. 1827-1834.   B.S 108

Transactions of the Microscopical Society of London.
    Trans.microsc.Soc.Lond. 1844-1852.   S. 415 A

TITLE	SERIAL No.

Transactions of the Midland Scientific Association.
    Burton-on-Trent.
    Trans.Midl.scient.Ass. 1864-1870.   S. 22 b

Transactions of the Mineralogical Museum. Leningrad.
    See Trudȳ Mineralogicheskogo Muzeya (im.A.E. Fersmana)
    Akademiya Nauk SSSR. Leningrad.   M.S 1806

Transactions of the Mining Association and Institute of
    Cornwall. Cambourne.
    Trans.Min.Ass.Inst.Cornwall 1885-95   M.S 1590

Transactions of the Mining and Geological and Metallurgical
    Institute of India. Calcutta.
    Trans.Min.geol.metall.Inst.India 1906 → (imp.)   M.S 1901

Transactions. Mining Institution of Scotland. London. etc.
    Trans.Min.Instn Scotl. Vol.15-22, 23 (i.-iii); 1894-1902   M.S 306

Transactions of the Missouri Academy of Science. Columbia.
    Trans.Mo.Acad.Sci. Vol.6 → 1972 →   S. 2484 a

Transactions of the Moscow Geological-Prospecting Institute of
    Orjonikhidze.
    See Trudȳ Moskovskogo Geologo-Razvedochnogo Instituta imeni
    S. Ordzhonikidze.   P.S 522

Transactions of the Moscow Society of Naturalists. Biological
    Series: (Histology & Embryology).
    See Trudȳ Moskovskogo Obshchestva Ispȳtateleĭ Prirodȳ.
    Otdel Biologicheskiĭ. Moskva.   S. 1838 C

Transactions of the National Institute of Sciences of India. Calcutta.
    Trans.natn.Inst.Sci.India 1935-1958.   S. 1918 B

Transactions of the Natural History and Antiquarian Society
    of Penzance.
    Trans.nat.Hist.antiq.Soc.Penzance
    1845-1865.   T.R.S 145
    1845-1846, 1848-1865.   S. 321 A
    (Styled "Report" on the wrappers; none published 1856-1861.)
    Continued as Transactions of the Penzance Natural History
    and Antiquarian Society.   S. 321 A & T.R.S 145

Transactions of the Natural History Society of Aberdeen.
    Trans.nat.Hist.Soc.Aberdeen 1878 & 1885.   S. 6 a

Transactions of the Natural History Society of Formosa. Taihoku.
    Trans.nat.Hist.Soc.Formosa 1925-1942.   S. 2000
    Vol.23-33, 1933-1943.   T.R.S 3606
    Formerly Journal of the Natural History Society
    of Taiwan (Formosa). Taihoku.
    Continued as Transactions of the Natural History Society
    of Taiwan. Taipei.   S. 2000

| TITLE | SERIAL No. |

Transactions of the Natural History Society of Glasgow.
    Trans.nat.Hist.Soc.Glasg. 1892-1911.   S. 101 A
                          1892-1899.   T.R.S 217
    Formerly Proceedings (& Transactions) of the Natural
    History Society of Glasgow.
    Continued in Glasgow Naturalist.

Transactions of the Natural History Society of Hartford.
    Trans.nat.Hist.Soc.Hartford 1836.   S. 2339 & T.R.S 5221

Transactions of the Natural History Society of Northumberland,
    Durham & Newcastle-upon-Tyne. Newcastle, &c.
    Trans.nat.Hist.Soc.Northumb. 1831-1838; 1904 →   S. 282 A - B
                             1831-1838; 1904-1932.   T.R.S 151
    (From 1865 to 1913 see Natural History Transactions of
    Northumberland, Durham and Newcastle-upon-Tyne. Newcastle.

Transactions of the Natural History Society of Queensland.
    Brisbane.
    Trans.nat.Hist.Soc.Qd. 1895.   S. 2138

Transactions of the Natural History Society of Taiwan. Taipei.
    Trans.nat.Hist.Soc.Taiwan 1943-1944.
    Formerly Transactions of the Natural History Society of
    Formosa. Taihoku.   S. 2000

Transactions of the New York Academy of Sciences. New York.
    Trans.N.Y.Acad.Sci. 1881-1897; 1938 →
    (1898-1937 published in Annals of the New York Academy
       of Sciences S. 2361 A.)   S. 2361 C

Transactions of the New York Microscopical Society.
    See American Quarterly Microscopical Journal. New York.   S. 2507

Transactions New York State Agricultural Society. Albany.
    Trans.N.Y.St.agric.Soc. 1843-1871 (imp.).   E.S 2487 m

Transactions of the New Zealand Institute
    Trans.N.Z.Inst. 1908-1911.
    Formerly & Continued as Transactions and Proceedings
    of the New Zealand Institute.   S. 2161 A & T.R.S 7302

Transactions of the Newbury District Field Club.
    Trans.Newbury Distr.Fld Club 1870-1911; 1930 →   S. 278 A

Transactions of the Nippon Lepidopterological Society. Kyoto.
    Trans.Nippon lepid.Soc. Vol.1, Pt.1. 1945.
    Continued as Butterflies and Moths. Kyoto.   E.S 1909 a

Transactions of the Norfolk & Norwich Naturalists' Society.
    Norwich.
    Trans.Norfolk Norwich Nat.Soc. 1869 →   S. 296 & T.R.S 138
                                1869-1914.   E.S 23

| TITLE | SERIAL No. |

Transactions of the North American Wildlife Conference.
    Washington, D.C.
    Trans.N.Am.Wildl.Conf. 1937-1940; 1944 →
    Formerly Proceedings of the North American Wildlife
    Conference.   Z.S 2709

Transactions of the North Staffordshire Ceramic Society. Hanley.
    Trans.N.Staffs.Ceram.Soc. 1901-05
    Continued as Transactions of the English Ceramic Society.
    Tunstall.   M.S 124

Transactions of the North Staffordshire Field Club.
    See Transactions and Annual Report. North Staffordshire
    Field Club.   S. 371 B

Transactions of the Northern Association of Literary &
    Scientific Societies.
    Trans.nth.Ass.lit.scient.Socs 1887-1899.   S. 131

Transactions. Northern Cavern and Mine Research Society. Keighley.
    Trans.nth.Cavern Mine Res.Soc. 1961-1963.   M.S 167 A

Transactions. Northern Naturalists Union. Newcastle-upon-Tyne.
    Trans.nth.Nat.Un. 1931-1953.   S. 286

Transactions of the Northern Scientific and Economic
    Expedition. Moscow.
    See Trudȳ Severnoĭ Nauchno-Promyslovoĭ Ekspeditsii VSNKh.
    Petrograd.

Transactions of the Northumberland, Durham and Newcastle-upon-Tyne
    Natural History Society.
    See Transactions of the Natural History Society of Northumberland,
    Durham and Newcastle-upon-Tyne.   S. 282 A-B

Transactions of the Nottingham Naturalists' Society.
    See Report and Transactions of the Nottingham
    Naturalists' Society.   S. 301

Transactions of the Nova Scotian Institute of Natural Science.
    Halifax.
    Trans.N.S.Inst.nat.Sci. 1863-1864.
    Continued as Proceedings and Transactions of the Nova
    Scotian Institute of Science. Halifax.   S. 2621

Transactions of the Oceanographic Commission. Moscow.
    See Trudȳ Okeanograficheskoi Komissii. Akademiya Nauk
    SSSR. Moskva.   S. 1802 i B

Transactions of the Oceanographical Institute. Moscow.
    See Trudȳ Gosudarstvennogo Okeanograficheskogo
    Instituta. Moskva; Trudȳ Morskogo Nauchnogo Instituta,
    Moskva & Trudȳ Plovuchego Morskogo Nauchnogo Instituta,
    Moskva.   S. 1845 A

TITLE	SERIAL No.

Transactions of the Odontological Society of Great Britain. London.
    Trans.odont.Soc.Gt Br. 1856-1907. S. 201

Transactions of the Oil-Geological Institute, Moscow, Leningrad, &c.
    See Trudy Neftyanogo Geologo-Razvedochnogo Instituta. Moskva, Leningrad, &c. P.S 525 & P.S 525 A

Transactions of the Optical Society. London.
    Trans.opt.Soc. 1907-1908. S. 202

Transactions. Oswestry Offa Field Club.
    Trans.Oswestry Offa Fld Club 1923-1930. S. 308

Transactions. Ottawa Field-Naturalists' Club.
    Trans.Ottawa Fld Natsts Club 1879-1887.
    Continued as Ottawa Naturalist. S. 2630

Transactions of the Oxford University Junior Scientific Club.
    Trans.Oxf.Univ.jr scient.Club N.S.1 → Ser.5 No.9. 1897-1938.
    Formerly Journal of the Oxford University Junior Scientific Club. S. 314

Transactions of the Pacific Committee of the Academy of Sciences of the USSR.
    See Trudy Tikhookeanskogo Komiteta. Akademiya Nauk SSSR. Leningrad. S. 1802 Z

Transactions of the Paisley Naturalists' Society.
    Trans.Paisley Nat.Soc. 1912-1942. S. 317

Transactions of the Palaeontological Society of Japan.
    See Transactions and Proceedings of the Palaeontological Society of Japan. Tokyo. P.S 752

Transactions. Papua and New Guinea Scientific Society. Port Moresby.
    Trans.Papua New Guinea scient.Soc. 1960-1969.
    Replaced by Proceedings. Papua and New Guinea Scientific Society. Port Moresby. S. 2143 b

Transactions of the Penzance Natural History and Antiquarian Society.
    Trans.Penzance nat.Hist.antiq.Soc. 1880-1898. S. 321 A
                                          1880-1888. T.R.S 145
    (Styled "Report & Transactions" on the wrappers.)
    Formerly Transactions of the Natural History and Antiquarian Society of Penzance.

Transactions of the Perthshire Society of Natural History.
    See Transactions and Proceedings of the Perthshire Society of Natural History, Perth. S. 324

Transactions of the Philosophical Institute of Victoria. Melbourne.
    Trans.phil.Inst.Vict. 1857-1858.
    Formerly Transactions of the Philosophical Society of Victoria. Melbourne.
    Continued as Transactions & Proceedings of the Royal Society of Victoria. Melbourne. S. 2105 A

TITLE	SERIAL No.

Transactions of the Philosophical & Literary Society of Leeds.
    Trans.phil.lit.Soc.Leeds 1837.   S. 146 B

Transactions of the Philosophical Society of New South Wales. Sydney.
    Trans.phil.Soc.N.S.W. 1862-1865.
    Continued as Transactions (and Proceedings) of the Royal Society of New South Wales. Sydney.   S. 2107 A

Transactions of the Philosophical Society of Victoria. Melbourne.
    Trans.phil.Soc.Vict. 1854-1855.
    Continued as Transactions of the Philosophical Institute of Victoria. Melbourne.   S. 2105 A

Transactions of the Plinian Society. Edinburgh.
    Trans.Plinian Soc. 1828-1829.
    Formerly Abstracts of the Proceedings of the Plinian Society of Edinburgh.   S. 59

Transactions of the Plumstead & District Natural History Society.
    Trans.Plumstead Distr.nat.Hist.Soc. 1927-1932.   S. 330

Transactions of the Plymouth & District Field Club.
    Trans.Plymouth Distr.Fld Club 1912-1917.   S. 327

Transactions of the Plymouth Institution.
    Trans.Plymouth Instn 1830.   S. 326 A

Transactions of the Plymouth Institution and Devon and Cornwall Natural History Society.
    See Report and Transactions of the Plymouth Institution and Devon and Cornwall Natural History Society, Plymouth.   S. 326 B

Transactions and Proceedings of the Botanical Society of Edinburgh.
    Trans.Proc.bot.Soc.Edinb. Vol.19-40, 1890-1970.
    Formerly & Continued as Transactions of the Botanical Society of Edinburgh.   B.S 6 b

Transactions and Proceedings of the Botanical Society of Pennsylvania. Philadelphia.
    Trans.Proc.bot.Soc.Pa 1897-1911.   B.S 4374

Transactions and Proceedings of the Geological Society of South Africa. Johannesburg.
    Trans.Proc.geol.Soc.S.Afr. 1946 →
    Formerly Proceedings of the Geological Society of South Africa and Transactions of the Geological of South Africa.   P.S 190

| TITLE | SERIAL No. |

Transactions and Proceedings of the New Zealand Institute.
    Wellington.
        Trans.Proc.N.Z.Inst.   1868-1934.   S. 2161 A & T.R.S 7302
                                    1868-1883.   Z.S 2160
        From 1908-1911 See Proceedings of the New Zealand
        Institute and Transactions of the New Zealand Institute.
        Continued as Transactions and Proceedings of the Royal
        Society of New Zealand. Dunedin.

Transactions and Proceedings of the Palaeontological Society
    of Japan. Tokyo.
        Trans.Proc.palaeont.Soc.Japan N.S. 1 → 1951 →
        Formerly Published in the Journal of the Geological
        Society of Japan.   P.S 752

Transactions & Proceedings of the Perthshire Society of
    Natural Science. Perth.
        Trans.Proc.Perthsh.Soc.nat.Sci. 1886 →
        Formerly Proceedings of the Perthshire Society of
        Natural Science. Perth.   S. 324

Transactions and Proceedings of the Royal Microscopical Society.
    See Journal of the Royal Microscopical Society. London.

Transactions and Proceedings of the Royal Society of
    New Zealand. Dunedin.
        Trans.Proc.R.Soc.N.Z. 1934-1952.
        Formerly Transactions and Proceedings of the New Zealand
        Institute. Wellington.
        Continued as Transactions of the Royal Society of
        New Zealand. Dunedin.   S. 2161 A & T.R.S 7302

Transactions and Proceedings of the Royal Society of Victoria.
    Melbourne.
        Trans.Proc.R.Soc.Vict. Vol.5-24, 1860-1888.
        (Vol.5 styled Transactions.)
        Formerly Transactions of the Philosophical Institute
        of Victoria. Melbourne.
        Continued as Proceedings of the Royal Society of Victoria.
        Melbourne.   S. 2105 A

Transactions and Proceedings of the South London Entomological
    and Natural History Society. London.
        Trans.Proc.S.Lond.ent.nat.Hist.Soc. 1932-1933.
        Formerly Proceedings of the South London Entomological
        and Natural History Society.
        Continued as Proceedings and Transactions of the
        South London Entomological and Natural History Society. S. 216 & E.S 28

Transactions & Proceedings. Torquay Natural History Society. Torquay.
        Trans.Torquay nat.Hist.Soc. 1922 →   S. 379 A
                                           1922-1938.   T.R.S 166
        Formerly Journal of the Torquay Natural History Society.

TITLE	SERIAL No.

Transactions & Proceedings of the Victorian Institute for
the Advancement of Science. Melbourne.
Trans.Vict.Inst.Adv.Sci. 1854-1855.   S. 2105 C

Transactions and Report of the Manchester Microscopical Society.
See Transactions and Annual Report of the Manchester
Microscopical Society.   S. 262

Transactions and Reports. Geological Congress of FNR Yugoslavia.
See Kongres Geologa Jugoslavije.   P.S 459

Transactions. Research Institute for Experimental Morphogenesis.
See Trudȳ Nauchno-Issledovatel'skogo Instituta
Eksperimental'nogo Morfogeneza.   Z.S 1928 C

Transactions of the Rochdale Literary & Scientific Society.
Trans.Rochdale lit.scient.Soc. 1878-1949.   S. 342 A

Transactions of the Royal Asiatic Society of Great Britain
and Ireland. London.
Trans.R.Asiat.Soc. 1827-1835.   S. 206 B

Transactions of the Royal Canadian Institute. Toronto.
Trans.R.Can.Inst. 1889 →
Formerly Proceedings of the Canadian Institute.   S. 2605 A

Transactions of the Royal Dublin Society.
See Scientific Transactions of the Royal Dublin Society.   S. 46 E

Transactions of the Royal Entomological Society of London.
Trans.R.ent.Soc.Lond. 1933 →
Formerly Transactions of the Entomological Society
of London.   E.S 17

Transactions of the Royal Geographical Society of Australasia,
Queensland Branch. Brisbane.
Trans.R.geogr.Soc.Australasia Qd Brch 1924.
(Forms Vol.1 of the Reports of the Great Barrier
Reef Committee.)   S. 2137

Transactions of the Royal Geological Society of Cornwall. Penzance.
Trans.R.geol.Soc.Corn. 1818 →   P.S 170

Transactions of the Royal Highland and Agricultural Society of
Scotland. Edinburgh.
Trans.R.Highld agric.Soc.Scotl. 1949 →
Formerly Transactions of the Highland and Agricultural
Society of Scotland. Edinburgh.   E.S 57 a

Transactions Royal Horticultural Society London.
Trans.R.Hort.Soc. 1808-1848.
1st. series. Vols. 1-7. 1808-1830.
Third Edition. Vol. 1 & 2. 1820-1822.
2nd. series. Vols. 1-3. 1835-1848.   B.S 50 c

| TITLE | SERIAL No. |

Transactions. Royal Institution of Chartered Surveyors. London.
   Trans.R.Instn chart.Surv. 1943-1948.
   Formerly Transactions of the Chartered Surveyors'
   Institution. London.   S. 219 A

Transactions of the Royal Irish Academy. Dublin.
   Trans.R.Ir.Acad. 1787-1906   S. 5 B

Transactions of the Royal Society of Arts (& Agriculture)
of Jamaica. Kingston.
   Trans.R.Soc.Arts Jamaica 1856-1868,(imp.)
   Formerly Transactions of the Jamaica Society of Arts.
Kingston, etc.   S. 2293

Transactions of the Royal Society of Arts and Sciences of Mauritius.
Port Louis.
   Trans.R.Soc.Arts Sci.Maurit.
   1846-1850: 1860-1865: 1872-1875: 1878-1887:
   Sec.C. Nos.1-15. 1932-1949. (Suspended 1853-1859 & 1890-1932.)
                                                                                    S. 2091 C
   1847.   T.R.S 7701
   Continued as Proceedings of the Royal Society of Arts
and Science of Mauritius.   S. 2091 C

Transactions of the Royal Society of Canada.
   Trans.R.Soc.Can. 1882 - All sections: Forms part
of Proceedings and Transactions of the Royal Society of Canada. S. 2603

Transactions of the Royal Society of Edinburgh.
   Trans.R.Soc.Edinb. 1788 →   S. 4 C
                  1788-1897; 1902-1914 (imp.)   T.R.S 175 A

Transactions (and Proceedings) of the Royal Society of
New South Wales. Sydney.
   Trans.R.Soc.N.S.W. 1867-1875.
   Formerly Transactions of the Philosophical Society
of New South Wales. Sydney.
   Continued as Journal of the Proceedings of the Royal
Society of New South Wales. Sydney.   S. 2107 A

Transactions of the Royal Society of New Zealand. Dunedin.
   Trans.R.Soc.N.Z.
      Vol.80-88, 1952-1961.   S. 2161 A
      General   1962-1970.   S. 2161 A a
      Botany   1961-1968.   S. 2161 A b
      Zoology   1961-1968.   T.R.S 7302 & S. 2161 A c
      Biological Sciences 1968-1970.   T.R.S 7302 & S. 2161 A d
      Geology (Earth Sciences) 1961-1970.   P.S 1156 a
   Formerly Transactions and Proceedings of the Royal
Society of New Zealand. Dunedin.

| TITLE | SERIAL No. |

Transactions of the Royal Society of South Africa. Cape Town.
  Trans.R.Soc.S.Afr. 1909 →
  Formerly Transactions of the South African Philosophical
  Society. Cape Town.   S. 2002

Transactions of the Royal Society of South Australia. Adelaide.
  Trans.R.Soc.S.Aust. 1877 →   S. 2102 A & T.R.S 7203

Transactions of the Royal Society of Tropical Medicine
  and Hygiene. London.
  Trans.R.Soc.trop.Med.Hyg. 1920 →
  Formerly Transactions of the Society of Tropical Medicine
  & Hygiene.   S. 222

Transactions of the Royal Society of Victoria. Melbourne.
  Trans.R.Soc.Vict. 1888-1914.   S. 2105 B

Transactions of the San Diego Society for Natural History.
  Trans.S.Diego Soc.nat.Hist. 1905 →   S. 2399 B
               Vol.3 → 1917 →   T.R.S 5141 C

Transactions of the San Francisco Microscopical Society.
  Trans.S.Francisco microsc.Soc. 1893.   S. 2403

Transactions of the Sapporo Natural History Society.
  Trans.Sapporo nat.Hist.Soc. 1905-1949.   S. 1987 a A

Transactions of the Science Society of China. Shanghai.
  Trans.Sci.Soc.China Vol.1 & 4-8, 1922-1934.   S. 1981 D

Transactions of the Scientific Society of Turkestan. Tashkent.
  See Trudȳ Turkestanskogo Nauchnogo Obshchestva. Tashkent.   S. 1859 b

Transactions of the Scottish Natural History Society. Edinburgh.
  Trans.Scott.nat.Hist.Soc. 1898-1902.   S. 62

Transactions of the Second Moscow University.
  See Trudȳ Vtorogo Moskovskogo Universiteta. Moskva.   S. 1840

Transactions of the Seismological Society of Japan. Yokohama.
  Trans.seismol.Soc.Japan 1880-1890.   P.S 760

Transactions of the Severn Valley Naturalists' Field Club.
  Wellington.
  Trans.Severn Valley Nat.Fld Cl. 1865-1870.   S. 360 B

Transactions of the Shikoku Entomological Society. Matsuyama.
  Trans.Shikoku ent.Soc. 1950 →   E.S 1910

Transactions of the Shropshire Archaeological & Natural History
  Society. Shrewsbury.
  Trans.Shrops.archaeol.nat.Hist.Soc. 1877-1935.   S. 361
                                      1877-1892.   T.R.S 139
  Continued as Transactions. Shropshire Archaeological Society.   S. 361

TITLE	SERIAL No.

Transactions of the Shropshire Archaeological Society. Shrewsbury.
    Trans.Shrops.archaeol.Soc. 1936 →
    Formerly Transactions of the Shropshire Archaeological
    & Natural History Society.   S. 361

Transactions of the Sigenkagaku Kenkyusyo. Tokyo.
    Trans.Sigenkag.Kenk. 1943.   S. 1984 a A

Transactions of Sikhote-Alin State Reserve. Moscow.
    See Trudȳ Sikhote-Alinskogo Gosudarstvennogo Zapovednika.
    Moskva.   S. 1805 a C

Transactions. Société Guernesiaise.
    See Report and Transactions.Société Guernesiaise. Guernsey.   S. 113

Transactions of the Society for British Entomology. Southampton.
    Trans.Soc.Br.Ent. 1934 →
    Formerly Transactions of the Entomological Society of
    the South of England. Southampton.   E.S 30

Transactions of the Society Instituted at London for the
    Encouragement of Arts, Manufactures, & Commerce. London.
    Trans.Soc.Arts, Lond. 46. 1827-1828.   S. 215 A

Transactions of the Society of Tropical Medicine and Hygiene.
    London.
    Trans.Soc.trop.Med.Hyg. 1907-1920.
    Continued as Transactions of the Royal Society of Tropical
    Medicine and Hygiene.   S. 222

Transactions of the Soil and Geobotanical Institution of Middle
    Asiatic State University. Taschkent. Turkmenstanian Series.
    See Acta Universitatis Asiae Mediae. Tashkent. Ser.Vlld.
    Pedologia. Fasc.1-2. 1930.   S. 1859 B

Transactions of the South African Philosophical Society. Cape Town.
    Trans.S.Afr.phil.Soc. 1877-1909.
    Continued as Transactions of the Royal Society of
    South Africa. Cape Town.   S. 2002

Transactions of the South-Eastern Union of Scientific
    Societies. London.
    Trans.S.-E. Un.scient.Socs 1897.
    Continued as Report & Transactions of the South-Eastern
    Union of Scientific Societies.   S. 218 A

Transactions. Southwestern Federation of Geological Societies.
    Abilene.
    Trans.SWest.Fed.geol.Socs. 1960 →   P.S 841 A

Transactions of the Soviet Section of the International
    Association for the Study of the Quaternary. Leningrad, Moscow.
    See Trudȳ Sovetskoĭ Sektsii Mezhdunarodnoĭ Assotsiatsii
    po Izucheniyu Chetvertichnogo Perioda.   P.S 997

| TITLE | SERIAL No. |

Transactions of the Soviet Union Geological Institute (VSEGEI) Leningrad.
   See Trudȳ Vsesoyusnogo Nauchno Issledovatel'skogo Geologicheskogo Instituta (VSEGEI). Leningrad.   P.S 1528

Transactions of the State Oceanographical Institute. Leningrad.
   See Trudȳ Gosudarstvennogo Okeanograficheskogo Instituta. Moskva.   S. 1845 A

Transactions of the Stirling Field Club.
   Trans.Stirling Fld Cl. 1878-1882.
   Continued as Transactions of the Stirling Natural History & Archaeological Society.   S. 373

Transactions of the Stirling Natural History & Archaeological Society.
   Trans.Stirling nat.Hist.archaeol.Soc. 1882-1932.
   Formerly Transactions of the Stirling Field Club.   S. 373

Transactions of the Suffolk Naturalists' Society.
   Trans.Suffolk Nat.Soc. 1929-1969.
   Continued as Suffolk Natural History.   S. 299

Transactions of the Sukhumi Botanical Garden. Sukhumi.
   See Trudȳ Sukhumshogo Botanicheskogo Sada. Sukhumi.   B.S 1430 e

Transactions of the Surveyors' Institution. London.
   Trans.Surv.Instn 1893-1930.
   Continued as Transactions of the Chartered Surveyors' Institution. London.   S. 219 A

Transactions of the Tennessee Academy of Science. Nashville.
   Trans.Tenn.Acad.Sci. 1912-1917.
   Continued as Journal of the Tennessee Academy of Science. Nashville.   S. 2535

Transactions of the Texas Academy of Sciences. Austin.
   Trans.Tex.Acad.Sci. 1892-1940.
   Continued as Proceedings and Transactions of the Texas Academy of Sciences. Austin.   S. 2315 A

Transactions of Tomsk State University.
   See Izvestiya Tomskogo (Gosudarstvennogo) Universiteta.   S. 1867 A

Transactions. Torquay Natural History Society.
   See Transactions and Proceedings. Torquay Natural History Society.   S. 379 A

Transactions of the Tyneside Naturalists' Field Club. Newcastle-upon-Tyne.
   Trans.Tyneside Nat.Fld Cl. 1846-1864.   S. 281 & T.R.S 150

| TITLE | SERIAL No. |

Transactions of the United Geological and Prospecting Service
of the U.S.S.R. Moscow.
See Trudy̆ Glavnogo Geologo-Razvedochnogo Upṛavleniya
V.S.N.Kh. SSSR. Moskva.   P.S 1525

Transactions of the Utah Academy of Sciences. Provo.
Trans.Utah Acad.Sci. 1908-1921.
Continued as Abstracts of Papers. Utah Academy of Sciences.
Provo.   S. 2396

Transactions of the Uzbekistan Institute of Tropical Medicine.
See Trudy̆ Uzbekistanskogo Instituta Tropicheskoĭ
Meditsiny im. Faĭzully Khodzhaeva.   73 A.o.K

Transactions of the Vale of Derwent Naturalists' Field Club.
Rowlands Gill.
Trans.Vale Derwent Nat.Fld Club 1908-1913.
Formerly Notes on the History, Geology and Entomology
of the Vale of Derwent being papers read before the
Burnopfield Vale of Derwent Naturalists' Field Club.   S. 30

Transactions of the Wagner Free Institute of Science of
Philadelphia.
Trans.Wagner free Inst.Sci.Philad. 1887-1927.   S. 2371 A
                                    1887, 1890-1899.   T.R.S 5149

Transactions of the Watford Natural History Society &
Hertfordshire Field Club.
Trans.Watford nat.Hist.Soc. 1875-1880.
Continued as Transactions of the Hertfordshire Natural
History Society & Field Club.   S. 391 & T.R.S 137

Transactions. Weardale Naturalists' Field Club.
Trans.Weardale Nat.Fld Club 1900-1904.   S. 392

Transactions of the West Kent Natural History, Microscopical
& Photographic Society.
Trans.W.Kent nat.Hist.microsc.photogr.Soc. 1900-1904.
Formerly President's Address, Papers & Reports.
West Kent Natural History, Microscopical & Photographic
Society.   S. 112

Transactions of the Wisconsin Academy of Sciences, Arts and
Letters. Madison.
Trans.Wis.Acad.Sci.Arts Lett. 1870 →   S. 2346 B
                               1870-1899.   T.R.S 5150

Transactions Wisconsin State Agricultural Society. Madison.
Trans.Wis.St.agric.Soc. 1852-1853.   E.S 2496 b

Transactions of the Woolhope Naturalists' Field Club. Hereford.
Trans.Woolhope Nat.Fld Club 1852 →   S. 120
                             1852-1904.   T.R.S 257

TITLE	SERIAL No.

Transactions of the Worcestershire Naturalists' Club. Worcester.
    <u>Trans.Worcs.Nat.Club</u> 1847-1966.   S. 401 A
                          1847-1899.   T.R.S 159

Transactions of the Yorkshire Naturalists' Union. Leeds.
    <u>Trans.Yorks.Nat.Un.</u> 1877-1946.   S. 474 B
                          1877-1907.   E.S 21

Transactions of the Zoological Institute. Kiev.
    <u>See</u> Trudȳ Instȳtutu Zoolohiyi. Kȳyiv.   Z.S 1827 C

Transactions of the Zoological Society of London.
    <u>Trans.zool.Soc.Lond.</u> 1833 →   T.R.S 130 B & Z. 1 B
                            1833 → (imp.)   E.S 18

Transfusion. Philadelphia, Pa.
    <u>Transfusion</u> Vol.10, No.4 → 1970 →
    (Wanting Vol.10, No.5; Vol.11, No.1 & 2; Vol.12, No.1)   P. SBG Unit

Translation News. Delft.
    <u>Transl.News</u> 1971 →   S. 633 B

Translation of the Russian Game Reports. Canadian Wildlife
    Service, Ottawa.
    <u>Transl.russ.Game Rep.</u> Vols.1-6, 1957-1959.   Z. 67 o C

Translation Series. Fisheries Research Board of Canada.
    Nanaimo, B.C.
    <u>Transl.Ser.Fish.Res.Bd Canada</u> No.194 → 1960 → (<u>imp.</u>).   Z.S 2628 E

Transunti. Reale Accademia dei Lincei. Roma.
    <u>See</u> Atti della Reale Accademia dei Lincei. Roma. Transunti. S. 1107 B

Transvaal Agricultural Journal. Pretoria.
    <u>Transv.agric.J.</u> 3-8, 1905-1910.
    <u>Continued as</u> Agricultural Journal of Union of South Africa.
    Pretoria.   E.S 2164

Transvaal Museum Memoirs. Pretoria.
    <u>Transv.Mus.Mem.</u> No.1, 1943.   Z. Reptile Section
      Nos. 2, 4, 6, 9 & 10. 1946 →   P.A. <u>See</u> Pal.Lbry Catalogue
      Nos. 3, 5, 7 & 8. 1949 →   E. <u>See</u> Ent.Lbry Catalogue

Travail de l'Institut de Botanique de l'Université de Montpellier
    et de la Station Zoologique de Cette. Cette.
    <u>Travail Inst.Bot.Univ.Montpellier</u>
    Série Mixte, No.2, 4 & 5, 1905 & 1916.
    (<u>See also</u> Travaux de l'Institut de Zoologie de l'Université
    de Montpellier et de la Station Maritime (Zoologique)
    de Cette. Série Mixte.   B. <u>See</u> Bot.Lbry Catalogue

| TITLE | SERIAL No. |

Travail du Laboratoire d'Histologie de la Faculté de Médecine
de Montpellier et de la Station Zoologique de Cette.
  Travail Lab.Histol.Univ.Montpellier No.1, 1903.
  Forms part of Travaux de l'Institut de Zoologie de
  l'Université de Montpellier et de la Station
  Zoologique de Cette. Série Mixte.   Z.S 975 A

Travailleur et Talisman.
  See Expéditions Scientifiques du Travailleur et du Talisman. Z. 78 q F

Travaux de l'Académie (Impériale, Nationale) de Reims.
  Trav.Acad.natn.Reims   1851-1884.
  Formerly Séances et Travaux de l'Académie de Reims.   S. 956

Travaux Algologiques. Paris.
  Trav.algol. 1942.
  Formerly and Continued as Revue Algologique. Paris.   B.A.S 14

Travaux de l'Association de l'Institut Scientifique du Caucase
  du Nord. Rostoff-sur Don.
  See Trudy Severo-Kavkazskoĭ Assotsiatsii Nauchno-Issledovatel'
  skikh Institutov. Rostov-na-Donu.   S. 1877

Travaux de l'Association International pour l'Etude du Quaternaire.
  See Conference of the International Association on
  Quaternary Research.   P.S 997 A

Travaux. Association Internationale de Limnologie Théoretique
  et Appliquée.
  See Verhandlungen der Internationalen Vereinigung für
  Theoretische und Angewandte Limnologie.   S. 1573 D

Travaux de Biologie. Université de Montréal.
  Trav.Biol.Univ.Montréal 1966-1970.
  Replaces Contributions de l'Institut de Biologie de
  l'Université de Montréal & Travaux de l'Institut de Biologie
  Génerale et de Zoologie de l'Université de Montréal.   Z.S 2650 A

Travaux Biologiques del'Institut J.B. Carnoy. Louvain.
  Trav.biol.Inst.J.B. Carnoy 1929 →   B.S 380

Travaux Bryologiques. Paris.
  See Revue Bryologique et Lichenologique. Paris.   B.B.S 12

Travaux du Bureau pour l'Assèchement des Marais de Polésie
  Polonaise. Brzésc nad Bugiem.
  See Prace Biura Meljoracji Polesia.   S. 1796

TITLE	SERIAL No.

Travaux de Bureau Géologique, Service Géologique de Madagascar
et Dépendances. Tananarive.
 Trav.Bur.geol.Madagascar   No.38 → 1952 → (imp.)   P.S 1240 A

Travaux du Centre d'Océanographie et des Pêches de Nosy-Bé.Madagascar.
 See Cahiers de l'Office de la Recherche Scientifique et
 Technique Outre-Mer.   S. 953

Travaux du Centre Océanographique de Pointe-Noire.
 See Cahiers de l'Office de la Recherche Scientifique et
 Technique Outre-Mer. Océanographie.   S. 953

Travaux du Centre de Recherches Anthropologiques Préhistoriques
et Ethnographiques. Alger.
 Trav.Cent.Rech.Anthrop.Préhist.Ethnogr.Alger   1964 →   P.A.S 484 A

Travaux du Centre de Recherches et d'Etudes Océanographiques.
 Paris and Boulogne.
 Trav.Cent.Rech.Etud.océanogr.   Vol.8 → 1968 →   S. 835

Travaux des Collaborateurs. Service de la Carte Géologique
de l'Algérie.
 See Publications du Service de la Carte Géologique de
 l'Algérie (Nouvelle Série).Bulletin.   P.S 1216

Travaux de la Commission pour l'Etude du Lac Bajkal. Petrograd.
 See Trudȳ Komissii po Izucheniyu Ozera Baĭkala. Petrograd.   S. 1802 N

Travaux de la Commission pour l'Etude du Quaternaire de l'Académie
des Sciences de l'U.R.S.S.
 See Trudy Komissii po Izucheniyu Chetvertichnogo Perioda.   P.S 502

Travaux de la Commission pour l'Etude de la République
Autonome Soviétique Socialiste Iakoute. Leningrad.
 See Trudȳ Komissii po Izucheniyu Yakutskoĭ ASSR.
 Leningrad.   S. 1802 Pb

Travaux et Comptes Rendus. Institut Polonais des Recherches
Forestières. Kraków.
 See Rozprawy i Sprawozdania. Instytut Badawczy Lasòw
 Państwowych w Warszawie.   S. 1874 A

Travaux et Comptes Rendus. Institut de Recherches de Forêts
d'Etat à Varsovie.
 See Rozprawy i Sprawozdania. Zakład Doświadczalny Lasów
 Państwowych w Warszawie.   S. 1874 A

Travaux et Documents de l'O.R.S.T.O.M. Paris.
 Trav.Docum.O.R.S.T.O.M. 1969 →   S. 953 I

| TITLE | SERIAL No. |

Travaux de l'Expedition Aralo-Caspienne.
   See Trudy Aralo-Kaspiĭskoĭ Ekspeditsii. S.-Peterburg.   S. 1855 B

Travaux de l'Expedition-Complexe pour l'Etude de la
République des Kirhgiz.
   See Trudȳ Kirgizskoĭ Kompleksnoi Expeditsii.   S. 73 A.o.L

Travaux de l'Expédition Scientifique d'Olonetz. Leningrad.
   See Trudy Olonetskoĭ Nauchnoĭ Ekspeditsii. Leningrad.   S. 1804 A

Travaux de la Faculté des Sciences. Université de Dakar.
   Trav.Fac.Sci.Univ.Dakar 1963 →   S. 2038 a C

Travaux de la Faculté des Sciences, Université de Rennes.
   Trav.Fac.Sci.Univ.Rennes Série Oceanogr.biol. 1968-1970.
   Continued as Travaux du Laboratoire de Biologie Halieutique. S. 958 a B

Travaux de la Filiale de Tadjikistan de l'Académie des
Sciences de l' URSS.
   See Trudȳ Tadzhikskoĭ Bazȳ, Akademiya Nauk SSSR.
Moskva, Leningrad.   Z. Fish Section

Travaux sur la Géologie de Bulgarie. Sofia.
   See Trudove Vărkhu Geologiyata na Bulgariya. Sofiya.   P.S 488 A-B

Travaux de Géologie. Laboratoire de Géologie. Université de Huî.
   Trav.Géol.Huî 1962 →   P.S 1781

Travaux Géologiques. Comité des Publications Silésiennes.
Académie Polonaise des Sciences et des Lettres. Cracovie.
   Trav.géol.Com.Publs silés.Acad.pol.Sci.Lett.
No.1 & 3, 1934 & 1937. For Nos.2, 4 & 5
   See also Prace Geologiczne, Wydawnictwa Slaskie.
Polska Akademiya Umiejetności. Kraków.   P.S 587

Travaux de l'Institut de Biologie Générale de l'Université de Vilno.
   See Prace Zakładu Biologji Ogólnej, Uniwersytetu St. Batorego
w Wilnie. Wilno.   Z.S 1305

Travaux de l'Institut de Biologie Génerale et de Zoologie
de l'Université de Montréal.
   Trav.Inst.Biol.gen.Zool.Univ.Montréal No.2-85, 1939-1962 (imp.)
   Replaced by Travaux de Biologie. Université de Montréal.   Z.S 2650 A

Travaux de l'Institut de Biologie de Peterhof. Leningrad.
   See Trudy Petergofskogo Biologicheskogo Instituta. Leningrad. S. 1865

Travaux de l'Institut Botanique. Kharkov.
   See Trudȳ Instytutu Botaniky. Khar'kiv.   B.S 1370

| TITLE | SERIAL No. |

Travaux de l'Institut Botanique de Tbilisi.
   See Trudȳ Tbilisskogo Botanicheskogo Instituta. Tbilisi.   B.S 1430

Travaux de l'Institut Botanique de l'Université de Neuchâtel.
   Trav.Inst.bot.Univ.Neuchâtel 1950 →   B.S 830

Travaux. Institut Botanique, Université de Stockholm.
   See Meddelanden från Stockholms Högskolas Botaniska Institut.   B.S 277

Travaux de l'Institut Chérifien. Tanger.
   See Travaux de l'Institut Scientifique Chérifien. Tanger.

Travaux de l'Institut Français d'Etudes Andines. Paris-Lima.
   Trav.Inst.fr.Etud.andines 1949 → (Wanting No.6.)   S. 2207 b

Travaux de l'Institut de Géographie de l'Université de Cluj.
   See Lucrările Institutului de Geografie al Universității
din Cluj.   S. 1886 A

Travaux de l'Institut de Géologie et d'Anthropologie
Préhistorique de la Faculté des Sciences de Poitiers.
   Trav.Inst.Géol.Anthrop.préh.Fac.Sci.Poitiers 1959-1968.
   Continued as Bulletin des Sciences de la Terre de l'Université
de Poitiers.   P.S 238

Travaux de l'Institut de Géologie de l'Université de Vilno.
   See Prace Zakładu Geologicznego (i Geograficznego)
Uniwersytetu St. Batorego w Wilnie.   P.S 575

Travaux de l'Institut Géologique. Académie des Sciences de la
RSS Géorgienne. Tiflis.
   See Trudȳ Geologicheskogo Instituta. Akademiya Nauk
Gruzinskoĭ SSR.

Travaux de l'Institut Géologique de l'Académie des Sciences
de l'U.R.S.S. Léningrad.
   See Trudȳ Geologicheskogo Instituta. Akademiya Nauk S.S.S.R.   P.S 508

Travaux. Institut Geologique de Pologne.
   See Prace Panstwowy Instytut Geologiczny. Warszawa.   P.S 1585

Travaux de l'Institut Hongrois de Recherches Biologiques. Tihany.
   See Archiva Biologica Hungarica. Tihany.   S. 1716 C

Travaux de l'Institut Lomonossoff de Géochemie Cristallographie
et Minéralogie. Moscow and Leningrad.
   See Trudȳ Lomonosovskogo Instituta Geokhimii,
Kristallografii i Mineralogii. Moskva.   M.S 1806

TITLE   SERIAL No.

Travaux de l'Institut Minéralogique de l'Académie des Sciences
  de l'U.R.S.S. Leningrad.
    See Trudȳ Mineralogicheskogo Instituta.   M.S 1806

Travaux de l'Institut Océanologique de l'URSS. Moscou.
    See Trudȳ Instituta Okeanologii. Akademiya Nauk. Moskva.   S. 1802 d

Travaux de l'Institut Paléontologique de l'Académie des Sciences
  de l'U.R.S.S.
    See Trudȳ Paleontologicheskogo Instituta. Akademiya
    Nauk SSSR.   P.S 509

Travaux de l'Institut Paléozoologique de l'Académie des Sciences
  de l'U.R.S.S.
    See Trudȳ Paleozoologicheskogo Instituta. Akademiya
    Nauk SSSR.   P.S 509

Travaux de l'Institut Pétrographique (Loewinson-Lessing)
  près l'Académie des Sciences de l'U.R.S.S. Leningrad.
    See Trudȳ Petrograficheskogo Instituta. Akademiya Nauk
    SSSR. Leningrad.   M.S 1804

Travaux de l'Institut des Recherches Biologiques et de la
  Station Biologique à l'Université de Perm (Molotov).
    See Trudȳ Biologicheskogo Nauchno-Issledovatel'skogo
    Instituta i Biologicheskoĭ Stantsii pri Permskom
    Gosudarstvennom Universitete. Perm (Molotov).   S. 1849 B & T.R.S 2008 B

Travaux de l'Institut de Recherches Scientifiques sur la
  Pêche et les Industries s'y Rattachant - Varna.
    See Trudove na Nauchnoizsledovatelskiya Institut po
    Ribarstvo i Ribna Promishlenost - Varna. Sofiya.   Z.S 1884

Travaux de l'Institut des Recherches Scientifiques à
  l'Université d'Etat. Voronèje.
    See Trudȳ Nauchno-Issledovatel'skogo Instituta pri
    Voroneshskom Gosudarstvennom Universitete. Voronezh.   S. 1868 A

Travaux de l'Institut des Sciences Géologiques de l'Académie des
  Sciences de l'U.R.S.S.
    See Trudȳ Instituta Geologicheskikh Nauk. Akademiya
    Nauk SSSR.   P.S 508

Travaux l'Institut des Sciences Naturelles de Peterhof.
    See Trudȳ Petergofskogo Estestvenno-Nauchnogo Instituta.
    Petergof.   S. 1865

Travaux de l'Institut Scientifique de Biologie. Tomsk.
    See Trudȳ Biologicheskogo Nauchno-Issledovatel'skogo
    Instituta Tomskogo Gosudarstvennogo Universiteta.   S. 1867 B

TITLE	SERIAL No.

Travaux de l'Institut Scientifique Chérifien. Tanger.
   Trav.Inst.scient.chérif.
      Série Générale 1953-1954.   S. 2037
      Série Botanique 1952 →   B.S 2310
      Série Géologie et Géographie Physique 1951 →   P.S 1226
      Série Zoologique 1951 →   T.R.S 4308 & Z.S 2010

Travaux de l'Institut de Spéologie "Emile Racovitza". Bucharest.
   Trav.Inst.Spéol. Emile Racovitza   Vol.9 → 1970 →
   Formerly Lucrările Institutului de Spéologie "Emil Racovita".
   Bucharest.   S. 1894 c

Travaux de l'Institut de Zoologie. Académie des Sciences
   de las RSS Georgienne.
   See Trudȳ Zoologicheskogo Instituta. Akademiya Nauk
   Gruzinskoi SSR.   Z.S 1837

Travaux de l'Institut de Zoologie et Biologie. Kieff.
   See Trudȳ Instytutu Zoolohiyi ta Biolohiyi. Kȳyiv.   S. 1834 a F

Travaux de l'Institut de Zoologie de l'Université de Montpellier
   et de la Station Maritime (Zoologique) de Cette.
   Trav.Inst.Zool.Univ.Montpellier 1891-1918   Z.S 975
     Série Mixte. No.1. 1903.   Z.S 975 A
                No.2, 4 & 5, 1905-1916.   B. See Bot.Lbry Catalogue
                No.3. See Annales.Institut
     Océanographique. Paris. Tom.1, fasc.10, 1911.   S. 934
   Formerly Travaux Originaux de Laboratoire Zoologique de la
   Faculté des Sciences de Montpellier et de la Station
   Maritime de Cette.

Travaux de l'Institut de Zoologie de l'Université de Wilno.
   See Prace Zakładu Zoologicznego, Uniwersytetu Stefana
   Batorego w Wilnie. Wilno.   Z.S 1805 A

Travaux de l'Institut Zoologique de l'Académie des Sciences
   de l'URSS.
   See Trudȳ Zoologicheskogo Instituta. Akademiya Nauk
   SSSR. Leningrad.   T.R.S 2004 & Z.S 1820

Travaux de l'Institut Zoologique de Lille et de la Station
   Maritime de Wimereux. Paris.
   Trav.Inst.zool.Lille   1879-1925 (imp.).   Z.S 935 &
   Continued as Travaux de la Station Zoologique de Wimereux.   Sections

Travaux des Instituts de Géologie et de Géographie de
   l'Université de Wilno.
   See Prace Zakładu Geologicznego (i Geograficznigo)
   Uniwersytetu St. Batorego w Wilnie & Prace Towarzystwa
   Przyjaciól Nauk w Wilnie. Wydział Nauk Matematyznych
   i Przyrodniczych.   P.S 575

| TITLE | SERIAL No. |

Travaux du Jardin Botanique de Tiflis.
    See Trudy Tiflisskago Botanicheskago Sada. Tiflis.   B.S 1430

Travaux des Jeunes Scientifiques. Association des Jeunes
    Scientifiques. Montréal.
    Trav.Jeunes Scients   1963 →   S. 2629

Travaux du Laboratoire Arago. Banyuls-sur-Mer.
    Trav.Lab.Arago   N.S. 1 → 1946 →   S. 928 a B

Travaux du Laboratoire de Biologie Halieutique. Université
    de Rennes.
    Trav.Lab.Biol.halieut.   1971-1972.
    Formerly Travaux de la Faculté des Sciences, Université
    de Rennes. Série Oceanogr.biol.   S. 959 a B

Travaux du Laboratoire de Botanique Systematique et de
    Phytogéographie de Université Libre de Bruxelles.
    Trav.Lab.Bot.syst.Phytogéogr.Univ.Brux.   1953 → (imp.)   B.S 358

Travaux du Laboratoire d'Entomologie. Muséum National d'Histoire
    Naturelle. Paris.
    Trav.Lab.Ent.Mus.natn.Hist.nat.   1932-1935.   E.S 814

Travaux du Laboratoire Forestier de Toulouse.
    Trav.Lab.for.Toulouse   1928 →   B.S 467

Travaux. Laboratoire de Géologie, Ecole Normale Supérieure. Paris.
    Trav.Lab.Géol.Ecole norm.Sup.   1967 →   P.S 234

Travaux du Laboratoire de Géologie de la Faculté des Sciences
    de Grenoble. Mémoires.
    Trav.Lab.Géol.Grenoble.Mém.   1 → 1960 →   P.S 222 a

Travaux du Laboratoire de Géologie de la Faculté des Sciences
    de l'Université D'Aix-Marseille. Marseille.
    Trav.Lab.Géol.Univ.Aix-Marseille   Tome 3-8, 1940-1965.   P.S 223
    Replaced by Travaux du Laboratoire de Géologie Historique
    et de Paléontologie. Université de Provence. Marseille.   P.S 219 a

Travaux du Laboratoire de Géologie de la Faculté des Sciences
    de L'Université de Bordeaux.
    Trav.Lab.Geol.Univ.Bordeaux   1949 →   P.S 215

Travaux du Laboratoire de Géologie de la Faculté des Sciences
    de L'Université de Grenoble.
    Trav.Lab.Géol.Univ.Grenoble   Tome 5-41, 1899-1965.
    Continued as Géologie Alpine. Travaux du Laboratoire
    de Géologie de la Faculté des Sciences de Grenoble.   P.S 222

TITLE   SERIAL No.

Travaux du Laboratoire de Géologie de la Faculté des Sciences
    de l'Université de Lyon.
    Trav.Lab.Géol.Univ.Lyon 1921-1943: N.S. No.2-14, 1955-1967.   P.S 224

Travaux du Laboratoire de Géologie Historique et de Paléontologie.
    Université de Provence. Marseille.
    Trav.Lab.Géol.hist.Paléont.Univ.Provence 1971 →   P.S 219 a
    Replaces Travaux du Laboratoire de Géologie de la Faculté
    des Sciences de l'Université d'Aix-Marseille. Marseille.   P.S 223

Travaux du Laboratoire d'Hydrobiologie et Pisciculture
    de l'Université de Grenoble.
    Trav.Lab.Hydrobiol.Piscic.Univ.Grenoble 1926 →
    Formerly Travaux du Laboratoire de Pisciculture
    de l'Université de Grenoble.   S. 872

Travaux du Laboratoire Ichtyologique d'Astrakhan de L'Administration
    des Pêcheries du Volga et de la Mer Caspienne.
    See Trudy Astrakhanskoĭ Ikhtīologicheskoĭ Laboratorīi pri
    Upravlenii Kaspiĭsko-Volzhskikh Rybnykh i Tyulen'ikh
    Promyslov.   Z.S 1330

Travaux du Laboratoire de La Jaysinia à Samoëns (Haute-Savoie).
    Trav.Lab. La Jaysinia 1957 →
    (No.1 and 2 published as Publications du Muséum National
    d'Histoire Naturelle, Paris, No.17 & 18.)   S. 931 K

Travaux du Laboratoire de Morphologie Evolutive.
    See Trudy Laboratorii Evolyutsionnoĭ Morfologii.   Z.S 1822

Travaux du Laboratoire du Muséum d'Histoire Naturelle de
    Saint-Servan.
    Trav.Lab.Mus.Hist.nat.St.Servan 1928.
    Continued as Bulletin du Laboratoire Maritime du Muséum
    d'Histoire Naturelle de Saint-Servan.   S. 968

Travaux du Laboratoire de Paléontologie. Faculté des Sciences,
    Université de Paris. Orsay.
    Trav.Lab.Paléont.Univ.Paris 1971 →   P.S 232

Travaux de Laboratoire Parasitologique de l'Université d'Etat
    de Moscou sous la Redaction de Professeur K.I. Skrjabin.
    See Raboty Parazitologicheskoĭ Laboratorii.
    Moskovskii Gosudarstvennyi Universitet.   S. 1844 D

Travaux du Laboratoire de Pisciculture de l'Université
    de Grenoble.
    Trav.Lab.Piscic.Univ.Grenoble 1909-1914 (imp.)
        1921-1926.
    Continued as Travaux du Laboratoire d'Hydrobiologie et de
    Pisciculture de l'Université de Grenoble.   S. 872

TITLE   SERIAL No.

Travaux du Laboratoire de Zoologie Expérimentale et de
    Morphologie des Animaux.
    See Trudȳ Laboratorii Eksperimental'noĭ Zoologii
    i Morfologii Zhivotnȳk.   Z.S 1820 B

Travaux du Laboratoire de Zoologie de la Faculté des Sciences
    de Caen et du Laboratoire Maritime de Luc-sur-Mer. Caen.
    Trav.Lab.Zool.Fac.Sci.Caen Lab.marit.Luc-sur-Mer 1967-1969 →
    Formerly Bulletin du Laboratoire Maritime de Luc-sur-Mer.
    Travaux du Laboratoire Maritime et du Laboratoire de Zoologie
    de la Faculté des Sciences de Caen.   Z.S 826

Travaux du Laboratoire de Zoologie et de la Station Aquicole
    Grimaldi de la Faculté des Sciences de Dijon.
    Trav.Lab.Zool.Stn aquic.Grimaldi Dijon 1952 →   Z.S 885

Travaux du Laboratoire Zoologique et de la Station
    Biologique de Sébastopol.
    See Trudȳ Osoboĭ Zoologicheskoĭ Laboratorĭi i
    Sevastopol'skoi Biologicheskoĭ Stantsĭi.   Z.S 1815 A

Travaux des Laboratoires. Institut de Biologie Marine de
    l'Université de Bordeaux.
    See Bulletin de la Station Biologique d'Arcachon.   S. 817

Travaux des Laboratoires. Société Scientifique et Station
    Zoologique (Biologique) d'Arcachon. Paris.
    Trav.Labs Soc.scient.Stn zool.Arcachon 1896-1908.
    Continued as Bulletin.Station Biologique d'Arcachon. Bordeaux. S. 817

Travaux et Mémoires publiés par la Société Ramond.
    Bagnères-de-Bigorre.
    Trav.Mém.Soc.Ramond 1919 - (1930).   S. 826 B

Travaux sur la Morphologie des Animaux. Kieff.
    See Zbirnȳk Prats z Morfolohiyi Tvarȳn. Kyiv.   Z.S 1827 B

Travaux du Musée Botanique de l'Académie Impériale des
    Sciences de St.-Pétersbourg (de l'Académie des Sciences
    de Russie).
    Trav.Mus.bot.Acad.Sci.Russ. 1902-1932.
    Continued in Trudȳ Botanicheskogo Instituta. Akademiya
    Nauk SSSR. Leningrad, Moskva.   B.S 1401

Travaux du Musée Géologique Académie des Sciences de l'U.R.S.S.
    See Trudȳ Geologicheskogo Muzeya. Akademiya Nauk S.S.S.R.   P.S 506

TITLE								SERIAL No.

Travaux du Musée Géologique (et Mineralogique) Pierre le Grand
    près l'Académie Impériale des Sciences de St-Petersbourg.
    <u>See</u> Trudy̆ Geologicheskago (i Mineralogicheskago) Muzeya
    imeni Petra Velikago Imperatorskoĭ Akademii Nauk'
    S.-Peterburg'.   P.S 505

Travaux du Musée Minéralogique de l'Académie des Sciences
    de l'U.R.S.S. Leningrad.
    <u>See</u> Trudy̆ Mineralogicheskogo Muzeya. Leningrad.   M.S 1806

Travaux du Musée de la Terre.
    <u>See</u> Prace Muzeum Ziemi.   P.S 571 B

Travaux du Musée Zoologique. Kiev.
    <u>See</u> Zbirny̆k Prats Zoolohichnoho Muzeyu.   T.R.S 2007 & Z.S 1827

Travaux du Muséum de Géorgie. Tiflis.
    <u>Trav.Mus.Géorgie</u> 1920-1933.
    <u>Formerly</u> Izvestiya Kavkazskago Muzeya. Tiflis.   S. 1858 G

Travaux du Muséum d'Histoire Naturelle "Gr.Antipa". Bucuresti.
    <u>Trav.Mus.Hist.nat."Gr.Antipa"</u> 1957 →   S. 1894 b
                                1957-1962.   T.R.S 1860

Travaux des Naturalistes de la Vallée du Loing. Moret-sur-Loing.
    <u>Trav.Nat.Vall.Loing</u> 1927 →   S. 915 C

Travaux Originaux du Laboratoire Zoologique de la Faculté
    des Sciences de Montpellier et de la Station Maritime
    de Cette. Nouvelle (deuxième) série.
    <u>Trav.Lab.zool.Montpellier</u> 1885.
    <u>Continued as</u> Travaux de l'Institut de Zoologie de
    l'Université de Montpellier et de la Station Maritime
    (Zoologique) de Cette.   Z.S 975

Travaux sur les Pêcheries du Québec.
    <u>Trav.Pêch.Québ.</u> 1964 →
    <u>Formerly</u> Contributions du Ministère de la Chasse et des
    Pêcheries, Québec.   Z.S 2656

Travaux sur le Radium et les Minerais Radioactifs. Léningrad.
    <u>See</u> Trudy̆ po Izucheniyu Radiya i Radioaktivny̆kh
    Rud. Leningrad.   M.S 1808

Travaux Récents des Collaborateurs. Service de la Carte Géologique
    de l'Algérie.
    <u>See</u> Bulletin du Service de la Carte Géologique de l'Algérie.
    Alger. Travaux Récents des Collaborateurs.   P.S 1216 A

Travaux et Recherches. Fédération Tarnaise de Spéléo-Archeologie. Tarn.
    <u>Trav.Rech.Fed.tarn.Spel.-Archeol.</u> 1962 →   S. 808

| TITLE | SERIAL No. |

Travaux de la Réserve d'Etat d'Altaï. Moscow.
   See Trudȳ Altaĭskogo Gosudarstvennogo Zapovednika. Moskva. S. 1805 a B

Travaux de la Réserve d'Etat d'Astrakhan.
   See Nauchnye Trudȳ Goszapovednikov. Moskva. S. 1805 a A

Travaux de la Réserve d'Etat de Noursome. Moscow.
   See Trudȳ Naurzumskogo Gosudarstvennogo Zapovednika.
Moskva. S. 1805 a D

Travaux Scientifiques. Institut de Recherche Scientifique de la
Pêche. Varna.
   See Trudove. Nauchnoizsledovatelskiya Institut po
Ribarstvo i Ribna Promishlenost. Varna. Z.S 1884

Travaux Scientifiques du Parc National de la Vanoise. Chambéry.
   Trav.scient.Parc natn.Vanoise 1970 → S. 914

Travaux Scientifiques de l'Université Populaire Russe de Prague.
   See Nauchnye Trudȳ. Russkiĭ Narodnȳĭ Universitet v Prague. S. 1762 a

Travaux Scientifiques de l'Université de Rennes.
   Trav.scient.Univ.Rennes 1902-1924. S. 958 a

Travaux de la Section Géologique du Cabinet de Sa Majésté.
St.-Petersbourg.
   Trav.Sect.géol.Cab.St.-Pétersb. 1895-1915. P.S 1520

Travaux de la Section de Pédologie de la Société des Sciences
Naturelles (et Physiques) du Maroc. Rabat.
   Trav.Sect.Pédol.Soc.Sci.nat.Maroc Tom.14, 1959. S. 2036 C
                                                            1950-1954; 1958-1959. T.R.S 4305 E

Travaux de la Section Scientifique et Technique.
Institut Français de Pondichéry.
   Trav.Sect.scient.tech.Inst.fr.Pondichéry 1957 → S. 1912 c

Travaux de la Section Soviétique de l'Association Internationale
pour l'Etude du Quaternaire. Leningrad, Moscou.
   See Trudȳ Sovetskoĭ Sektsii Mezhdunarodnoĭ Assotsiatsii
po Izucheniyu Chetvertichnogo Perioda (INQUA).
Leningrad, Moskva. P.S 997

Travaux de la Section de Zoologie. Académie des Sciences
de l'URSS - Filiale Georgienne.
   See Trudȳ Zoologicheskogo Sektora. Akademiya Nauk
SSSR, Gruzinskoe Otdelenie, Zakavkazskiĭ Filial. Tiflis. Z.S 1837

Travaux du Service Geologique de Pologne. Varsovie.
   Trav.Serv.géol.Pol. 1921-1950.
   Continued as Prace. Panstwowy Instytut Geologiczny. Warszawa. P.S 1585

| TITLE | SERIAL No. |

Travaux. Service des Réserves Naturelles Domaniales et de la
   Conservation de la Nature. Brussels.
   Trav.Serv.Rés.nat.dom. No.2 → 1966 →   S. 718

Travaux de la Société Botanique de Genève.
   Trav.Soc.bot.Genève 1952-1966/1967.
   Formerly Bulletin de la Société Botanique de Genève.
   Continued as Saussurea. Genève.   B.S 822

Travaux de la Société Bulgare des Sciences Naturelles. Sofia.
   See Trudove na Bŭlgarskoto Prirodoispitatelno
   Druzhestvo. Sofiya.   S. 1898 A

Travaux de la Société d'Histoire Naturelle de l'Ile Maurice.
   Port Louis.
   See Procès-Verbaux de la Société d'Histoire Naturelle
   de l'Ile Maurice. Port Louis.   S. 2091 B

Travaux de la Société Impériale des Naturalistes de St.-Pétersbourg.
   See Trudȳ Imperatorskago Sankt-Peterburgskago Obshchestva
   Estestvoispȳtateleĭ.   S. 1855 A

Travaux de la Société des Naturalistes de Charkow.
   See Trudȳ Khar'kovskogo Obshchestva Ispytateleĭ Prirodȳ
   pri Ukrglavnauke. Khar'kov.   S. 1833 A

Travaux de la Société des Naturalistes de l'Université
   Impériale de Kasan.
   See Trudȳ Obshchestva Estestvoispȳtateleĭ pri Imperatorskom
   Kazanskom Universitete. Kazan.   S. 1832 A

Travaux de la Société des Naturalistes de Varsovie.
   See Trudȳ Varshavskago Obshchestva Estestvoispȳtateleĭ.
   Varshava.   S. 1870 A - C

Travaux de la Société des Sciences et des Lettres de Wilno.
   See Prace Towarzystwa Przyjaciol Nauk w Wilnie.
   Wydział Nauk Matematyeznych i Przyrodniczych.   S. 1797 A

Travaux de la Station Biologique du Caucase du Nord.
   See Rabotȳ Severo-Kavkazskoĭ Gidrobiologicheskoĭ Stantsīi
   pri Gorskom Sel'sko-Khozyaĭstvennom Institute. Vladikavkaz.   Z.S 1835 A

Travaux de la Station Biologique du Dniepre. Kiev.
   See Zbirnȳk Prats' Dniprovs'koyï Biolohichnoyï
   Stantsiyï. Kȳyiv.   S. 1834 a M

Travaux de la Station Biologique à Karadagh.
   See Trudȳ Karadahs'koyi Nauchnoyi Stantsiyi imeni T.I.
   Vyazems'koho. Simferopol.   S. 1838 I

| TITLE | SERIAL No. |

Travaux de la Station Biologique du Lac d'Oredon. Université
de Toulouse.
   See Recueil. Travaux de la Station Biologique du Lac
d'Oredon. Toulouse.   S. 970 A

Travaux de la Station Biologique Maritime de Stalin.
   See Trudove na Morskata Biologichna Stantsiya v gr. Stalin.   S. 1899 B

Travaux de la Station Biologique de Murman. Leningrad.
   See Raboty Murmanskoĭ Biologicheskoĭ Stantsii. Murmansk.   S. 1855 D

Travaux de la Station Biologique de Roscoff. Paris.
   Trav.Stn biol.Roscoff 1923 →   S. 947

Travaux de la Station Biologique de Sébastopol.
   See Trudy Sevastopol'skoĭ Biologicheskoĭ Stantsii.   Z.S 1815 A

Travaux de la Station Hydrobiologique. Kiev.
   See Trudy Hidrobiolohichnoyi Stantsiyi. Akademiya
Nauk URSR. Kyyiv.   S. 1834 a G

Travaux de la Station Limnologique de Lac Bajcal. Leningrad.
   See Trudy Baĭkal'skoĭ Limnologicheskoĭ Stantsii. Leningrad.   S. 1802 N

Travaux de la Station des Sciences Naturelles à Karadagh
(Crimée). Moscow.
   See Trudy Karadahs'koyi Nauchnoyi Stantsiyi imeni
T.I. Vyazems'koho. Simferopol.   S. 1838 I

Travaux de la Station Zoologique d'Arcachon.
   See Travaux des Laboratoires. Société Scientifique et Station
Zoologique (Biologique) d'Arcachon.   S. 817

Travaux de la Station Zoologique Maritime d'Agigea.
   See Lucrările ale Statiei Zoologice Maritime
"Regele Ferdinand I" dela Agigea.   Z.S 1889

Travaux de la Station Zoologique de Wimereux. Paris.
   Trav.Stn zool.Wimereux 1925-1938.
   Formerly Travaux de l'Institut Zoologique de Lille   Z.S 935 &
et de la Station Maritime de Wimereux.   Sections

Travaux de l'Université de Poznan.
   See Prace Naukowe Uniwersytetu Poznańskiego. Poznań.   S. 1728

Travaux Zoologiques. Académie des Sciences de l'URSS,
Filiale Transcaucasienne. Section Géorgienne.
   See Trudy Zoologicheskogo Sektora. Akademiya Nauk
SSSR, Gruzinskoe Otdelenie.   Z.S 1837

TITLE	SERIAL No.

Travel. London.
    Travel, Lond. 1896-1902.   S. 497

Travels of Airborne Pollen. Albany (N.Y.)
    See Progress Report. New York State Museum and Science Service. Albany (N.Y.)   B.S 4303

Treballs de la Institució Catalana d'Historia Natural. Barcelona.
    Treb.Inst.catal.Hist.nat. 1915-1923.   S. 1004 B

Treballs del Museu de Ciències Naturals de Barcelona.
    See Musei Barcinonensis Scientiarum Naturalium Opera. Barcelona.   S. 1008 A & T.R.S 1108
    For Série Botanica See Publicacions de l'Institut Botánic. Barcelona.   B.S 570

Treballs de la Societat de Biologia de Barcelona.
    Treb.Soc.Biol.Barcelona 1913-1934.   S. 1004 D

Trees. Journal of the Men of the Trees. London.
    Trees, Lond. 1936-1956. (imp.)   B.S 76

Trees and Life. Southampton.
    Trees Life 1956 →   B.S 76

Trees in South Africa. Johannesburg.
    Trees S.Afr. 1949 →   B.S 2276

Treganza Anthropology Museum Papers. San Francisco State College.
    Treganza Anthrop.Mus.Pap. 1971 →
    Formerly Occasional Papers. Anthropology Museum, San Francisco State College.   P.A.S 814

Trematodȳ Zhivotnȳkh i Cheloveka. Osnovȳ Trematodologii. Moskva.
    Tremat.Zhivot.Cheloveka 1947 →   Z. Platyhelminth Section

Trencsén Megyei Természettudomanyi Tarsulat Evokönyvei.
    See Jahresheft des Naturwissenschaftlichen Vereins des Trencsiner Komitäts. Trencsin.   S. 1765

Treubia. Buitenzorg.
    Treubia 1919 →   T.R.S 7602 & Z.S 2195

Trias. Utrecht.
    Trias 1973 →   S. 699 c

Tribolium Information Bulletin. Chazy, New York.
    Tribolium Inf.Bull. 1958 →   E.S 2439 A

Tribuna Farmacêutica. Orgao Oficial da Faculdade de Farmácia da Universidade do Paraná. Curitiba.
    Tribuna Farm. Vol.33 → 1965 →   B.S 3052 b

TITLE	SERIAL No.

Triennial Report. Botany Division, D.S.I.R., New Zealand. Wellington.
    Trienn.Rep.Bot.Div.D.S.I.R., N.Z. 1957 →   B.S 2483

Triennial Report. Cawthron Institute. Nelson, N.Z.
    Trienn.Rep.Cawthron Inst. 1963 →
    Formerly Biennial Report. Cawthron Institute.   S. 2182 A

Trilobite, The. Milwaukee. Wis.
    Trilobite 1944-1957.   P.S 835

Trilobite News. Oslo.
    Trilobite News 1971 →   P.S 412

Trimonthly Report. Ohio Herpetological Society. Columbus.
    Trimon.Rep.Ohio herpet.Soc. 1958-1959.
    Continued as Journal. Ohio Herpetological Society.
    Columbus.   Z. Reptile Section

Trinidad Official and Commercial Register and Almanack. Port of Spain.
    Trinidad Almanack 1879-80, 1888-89.   S. 75 F.o.G

Tri-ology Technical Report, Division of Plant Industry. Gainesville, Florida.
    Tri-ology tech.Rep.Div.Pl.Ind.Fla Vol.9 → 1970 →   S. 2436 c D

Tromsø Museums Aarsberetning.
    Tromsø Mus.Aarsberetn. 1873-1935; 1952 →   S. 541 B

Tromsø Museums Aarshefter.
    Tromsø Mus.Aarsh. 1878-1951.   S. 541 A

Tromsø Museums Skrifter.
    Tromsø Mus.Skr. 1925 → (except Vol.6.)   S. 541 D
    Vol.6,1958.   B. 581.9(481)BEN

Trondhiemske Selskabs Skrifter.
    See Skrifter. Trondhiemske Selskabs. Kiöbenhavn.   S. 554 A

Tropenpflanzer. Berlin.
    Tropenpflanzer 1897-1912.
    For Beihefte See Beihefte zum Tropenpflanzer. Berlin.   B.S 914

Tropical Agriculture. The Journal of the Imperial College of Agriculture. Trinidad.
    Trop.Agric.,Trin. 1924 →   E.S 2380

Tropical Agriculturist and Magazine of the Ceylon Agricultural Society. Peradeniya.
    Trop.Agric.Mag.Ceylon agric.Soc. 1881 → (imp.)   E.S 2016

TITLE	SERIAL No.

Tropical Diseases Bulletin. London.
    Trop.Dis.Bull. 1912 →
    Formerly Bulletin. Sleeping Sickness Bureau. London.
    Supplement: See Sanitation Supplements.   Z.B 69 T

Tropical Ecology.
    Trop.Ecol. 1961 →
    Formerly Bulletin of the International Society for
    Tropical Ecology.   B.S 4604

Tropical Fish Hobbyist. New York.
    Trop.Fish Hobby. 1952 →   Z.S 2473

Tropical Life. London.
    Trop.Life 1905-1908.   S. 496

Tropical Pest Bulletin. London.
    Trop.Pest Bull. 1972 →   E.S 41 a

Tropical Science. Tropical Products Institute. London.
    Trop.Sci. 1959 →
    Formerly Colonial Plant & Animal Products. London.   B.S 91

Tropical Veterinary Bulletin. London.
    Trop.vet.Bull. 1912-1930.   Z.B 69 V

Tropical Woods. Yale University School of Forests. New Haven, Conn.
    Trop.Woods 1925 →   B.S 4190 a

Tropische Natuur. Nederlandsch-Indische Natuurhistorische
    Vereeniging. Weltevreden.
    Trop.Natuur. 1912-1941, 1952-1953.
        (Wanting Jaarg. 31.)
    Continued as Penggemar Alam. Bogor.   S. 1957

Trout and Salmon. Peterborough, London.
    Trout Salm.
    Vol.2, No.14 - Vol.7, No.75, 1956-1961 (imp.)   Z.S 319

Trudi.
    See Trudȳ

Trudove na Bŭlgarskoto Prirodoispitatelno Druzhestvo. Sofiya.
    Trud.bulg.prir.Druzh. 1900-1936.   S. 1898 A

Trudove ot Chernomorskata Biologichna Stantsiya v Gr.Varna.
    Trud.chernomorsk.biol.Sta.Varna 1933-1948. 1955.
    From 1949-1954 See Trudove na Morskata Biologichna Stantsiya
    v Stalin.   S. 1899 B

TITLE	SERIAL No.

Trudove na Instituta po Zoologiya. Bŭlgarska Akademiya na
    Naukite. Sofiya.
    Trud.Inst.Zool.Sof. 1950 →   Z.S 1885 A

Trudove na Morskata Biologichna Stantsiya v gr. Stalin.
    Trud.morsk.biol.Sta.Stalin 1949-1954. (imp.)
    Formerly and continued as Trudove ot Chernomorskata
    Biologichna Stantsiya v Gr.Varna. Sofia.   S. 1899 B

Trudove na Nauchnoizledovatelskiya Institut po Ribarstvo
    i Ribna Promishlenost - Varna. Sofiya.
    Trud.nauchnoizsled.Inst.Ribarst.ribna Prom. Tom 2, 1960.
    Continued as Trudove na Tsentralniya Nauchnoizsledovatelski
    Institut po Ribovŭdstvo i Ribolov. Varna.   Z.S 1884

Trudove na Tsentralniya Nauchnoizsledovatelski Institut
    po Ribovŭdstvo i Ribolov - Varna.
    Trud.tsent.nauchno-izsled.Inst.Ribov.Ribol.Varna Tom.3, 1960.
    Formerly Trudove na Nauchnoizsledovatelskiya Institut
    po Ribarstvo i Ribna Promishlenost - Varna.
    Continued as Izvestiya na Tsentralniya Nauchno-
    izsledovatelski Institut po Ribovŭdstvo i Ribolov - Varna.   Z.S 1884

Trudove Vărkhu Geologiyata na Bŭlgariya. Sofiya.
    Trudove Vărkhu geol.Bŭlg.
    Seriya Stratigrafiya i Tektonika. 1960-1965.
    Seriya Paleontologiya. 1959-1966.   P.S 488 A-B

Trudove. Vissh Pedagogickeski Institut. Plovdiv. Matematika
    Fizika, Chimiya, Biologiya.
    Trudove Vissh pedag.Inst. 1963.
    Continued as Nauchni Trudove. Vissh Pedagogicheskii
    Institut. Plovdid.   S. 1897 a

Trudovi na Geološki Zavod na Narodna Republika Makedonija. Skopje.
    Trudovi geol.Zav.Skopje 1947 →   P.S 453

Trudȳ Akademii Nauk Litovskoĭ SSR.
    See Lietuvos TSR Mokslu Akademijos Biologijos
    Instituto Darbai.   S. 1799 A

Trudȳ Alma-Atinskogo Botanicheskogo Sada.
    Trudȳ alma-atin.bot.Sada Vol.3-7. 1956-1963.
    Continued as Trudȳ Botanicheskikh Sadov Akademii
    Nauk Kazakskoi SSR.   B.S 1346

Trudȳ Altaĭskogo Gosudarstvennogo Zapovednika. Moskva.
    Trudȳ altaĭsk.gos.Zapov. Fasc. 1-2. 1938.   S. 1805 a B

| TITLE | SERIAL No. |

Trudy Amurskoĭ Ikhtiologicheskoĭ Ekspeditsii 1945-1949.
  Tom. I-IV, 1950-1958.
  See Materialy k Poznaniyu Fauny i Flory SSSR,
  Novaya Seriya. Otdel Zoologicheskiĭ. Vp.16, 24, 32, 26.   Z.S 1854

Trudy Antropologicheskago Otdela Imperatorskago Obshchestva
  Lyubitelei Estestvoznaniya, Antropologii i Etnografii, Moskva.
  See Izvestiya Imperatorskago Obshchestva Lyubiteleĭ
  Estestvoznaniya, Antropologii i Etnografii pri Imperatorskom
  Moskovskom Universitete. Moskva.   S. 1841 A

Trudy Aralo-Kaspiĭskoĭ Ekspeditsii. S.-Peterburg.
  Trudy aralo-kasp.Eksped. 1875-1905.   S. 1855 B

Trudy Arkticheskogo i Antarkticheskogo Nauchno-Issledovatel'skogo
  Instituta. Moskva.
  Trudy arkt.antarkt.nauchno-issled.Inst. Vol.259 → 1964 →   S. 1842 A

Trudy Arkticheskogo Nauchno-Issledovatel'skogo Instituta. Leningrad.
  Trudy arkt.nauchno-issled.Inst. 1931-1937.
    (Wanting nos. 20, 48, 57, 62, 70, 72 & 79.)
  Formerly Trudy Instituta po Izucheniyu Severa. Moskva.
  Continued as Trudy Vsesoyuznogo Arkticheskogo
  Instituta. Leningrad.   S. 1842 A

Trudy Armyanskogo Filiala Akademii Nauk SSSR. Erevan.
  Seriya Biologicheskaya.
  Trudy armyansk.Fil.Akad.Naùk SSSR vol. 2, 1937.   S. 1802 Ca A

Trudy Armyanskoĭ Protivochumnoĭ Stantsii. Erevan.
  Trudy armyansk.Protivoch.Stn Vyp.2-3. 1963-1964.   S. 1858 d

Trudy Astrakhanskogo Gosudarstvennogo Zapovednika. Moskva.
  See Nauchnye Trudy Goszapovednikov. Moskva.   S. 1805 a A

Trudy Astrakhanskoĭ Ikhtiologicheskoĭ Laboratorii pri
  Upravlenii Kaspiĭsko-Volzhskikh Rybnykh i Tyulen'ikh Promyslov.
  Trudy astrakh.ikhtiol.Lab. Vol.3 No.6 - Vol.7 No.1,
    1918-1926.
  Formerly Trudy Ikhtiologicheskoĭ Laboratorii. Astrakhan.
  Continued as Trudy Astrakhanskoĭ Nauchnoĭ
  Rybokhozyaĭstvennoĭ Stantsii.   Z.S 1830

Trudy Astrakhanskoĭ Nauchnoĭ Rybokhozyaĭstvennoĭ Stantsii.
  Trudy astrakh.nauch.rybokhoz.Sta. Vol.7 Nos. 2-4, 1931.
  Formerly Trudy Astrakhanskoĭ Ikhtiologicheskoĭ
  Laboratorii.   Z.S 1830

TITLE	SERIAL No.

Trudȳ. Atlanticheskiĭ Nauchno-Issledovatel'skiĭ Institut
　　Rȳbnogo Khozyaĭstva i Okeanografii. Kaliningrad.
　　Trudȳ atlant.nauchno-issled.Inst.Rȳb.Khoz.Okeanogr.
　　Vȳp.11, 1964; 16 → 1965 →　　　　　　　　　　　　　　　S. 1847 a

Trudȳ Azerbaĭdzhanskogo Filiala. Akademiya Nauk SSSR. Baku.
　　Trudȳ azerb.Fil.Akad.Nauk SSSR　nos. 7, 20, 26-27,
　　29, 38-41, 47, 53-55 & 58.　1934-1939.　　　　　　　　　S. 1810 a C
　　No.39, (Geol.Ser.) 1938.　　　　　　　　　　　　　　　　P. 16 o B

Trudȳ Azovsko-Chernomorskogo Nauchno-Issledovatel'skogo
　　Instituta Rȳbnogo Khozyaistva i Okeanografii. Simferopol.
　　Trudȳ azov.-chernomorsk.nauch.-issl.Inst.Rȳb.Khozyaist.
　　Okeanogr.　Vp.11, 1938; Vp.22-24, 27 → 1964 →
　　Formerly　Trudy Azovsko-Chernomorskoĭ Nauchnoĭ
　　Rȳbokhozyaĭstvennoĭ Stantsĭi.　　　　　　　　　　　　　Z.S 1838

Trudȳ Azovsko-Chernomorskoĭ Nauchnoĭ Rybokhozyaĭstvennoĭ Stantsii.
　　Trudȳ azov.-chernomorsk.nauch.rybokhoz.Sta.　Vȳp.4-8. 1930.
　　Formerly　Trudy Kerchenskoĭ Nauchnoĭ Rybokhozyaĭstvennoĭ
　　Stantsii.
　　Continued as　Trudȳ Azovsko-Chernomorskogo Nauchno-
　　Issledovatel'skogo Instituta Rȳbnogo Khozyaistva i
　　Okeanografii. Simferopol.　　　　　　　　　　　　　　　Z.S 1838

Trudȳ Baĭkal'skoĭ Limnologicheskoĭ Stantsii. Leningrad.
　　Trudȳ baĭkal.limnol.Sta　1931-1961.　(imp.)
　　Formerly　Trudȳ Komissii po Izucheniyu Ozera Baĭkala. Petrograd.
　　Continued as　Trudȳ Limnologicheskogo Instituta. Moskva
　　& Leningrad.　　　　　　　　　　　　　　　　　　　　　S. 1802 N

Trudȳ Belomorskoĭ Biologicheskoĭ Stantsii. M.G.U.
　　Trudȳ belom.biol.Sta M.G.U.　1962 →　　　　　　　　　　S. 1844 E

Trudȳ Belorusskogo Gosudarstvennogo Universiteta v g. Minske.
　　Trudȳ beloruss.gos.Univ.　1922-1930.　　　　　　　　　　S. 1814

Trudȳ Biologicheskogo Fakul'teta Tomskogo Gosudarstvennogo
　　Universita.
　　Trudȳ biol.Fak.tomsk.gos.Univ.　(83.)　(1930.)
　　Formerly & Continued as　Izvestiya Tomskogo (Gosudarstvennogo)
　　Universiteta.　　　　　　　　　　　　　　　　　　　　　S. 1867 A

Trudȳ Biologicheskogo Instituta　Akademii Nauk Kirgizskoĭ
　　S.S.R. Frunze.
　　Trudȳ biol.Inst.Frunze　vol. 2-4.　1947-1951.　　　　　S. 1820 a A

Trudȳ Biologicheskogo Instituta, Armyanskiĭ Filial.
　　Akademiya Nauk SSSR. Erevan.
　　Trudȳ biol.Inst.,Erevan　1939.　　　　　　　　　　　　　S. 1802 Ca B

| TITLE | SERIAL No. |

Trudy Biologicheskogo Nauchno-Issledovatel'skogo Instituta
i Biologicheskoĭ Stantsiĭ pri Permskom Gosudarstvennom
Universitete. Perm (Molotov).
 Trudy biol.nauchno-issled.Inst.perm.gos.Univ. 1927-1940. S. 1849 B
                   1927-1930. T.R.S 2008 B

Trudy Biologicheskogo Nauchno-Issledovatel'skogo Instituta Tomskogo
Gosudarstvennogo Universiteta.
 Trudy biol.nauchno-issled.Inst.tomsk.gos.Univ. 1935-1939.
 Formerly Trudy Tomskogo Gosudarstvennogo Universiteta. S. 1867 B

Trudy Biologicheskoĭ Stantsii "Borok". Moskva.
 Trudy biol.Sta."Borok" Nos.2-3. 1955-1958.
 Continued as Trudy Instituta Biologii Vodokhranilishch.
Moskva. S. 1802 l. B

Trudy Biologo-Pochvennogo Instituta, ANSSSR. Vladivostok.
 Trudy biol.-pochvenn.Inst., Vladivostok 1969 → S. 1802 m C

Trudy Borodinskoĭ Biologicheskoĭ Stantsii v Karelii. Leningrad.
 See Trudy Prěsnovodnoĭ Biologicheskoĭ Stantsiĭ Imp.
S.-Peterburgskago Obshchestva Estestvoispȳtateleĭ.
S.-Peterburg. S. 1855 E

Trudy Botanicheskikh Sadov Akademii Nauk Kazakskoĭ SSR.
 Trudy bot.Sadov Akad.Kazaks. Vol.8 → 1965 →
 Formerly Trudy Alma-Atinskogo Botanicheskogo Sada. B.S 1346

Trudy Botanicheskogo Instituta Akademii Nauk SSSR. Leningrad, Moskva.
 Trudy bot.Inst.Akad.Nauk SSSR
  Seriya 1 1933-1964. Seriya 3-7 1933 → B.S 1377 a-f
  Seriya 2 1933-1959. B.C.S 29
  Seriya 8 Paleobotanika 1956 → P.S 509 A

Trudy Botanicheskogo Instituta. Akademiya Nauk Armyanskoĭ SSR. Erevan.
 Trudy bot.Inst.Erevan No.4 & 14 → 1946 & 1964 →
 Formerly Trudy Botanicheskogo Instituta, Armyanskii Filial.
Akademiya Nauk SSSR. Erevan. B.S 1355 a

Trudy Botanicheskogo Instituta. Akademiya Nauk Tadzhikskoi SSR.
Dushanbe.
 Trudy bot.Inst.Dushanbe Tom 18, 1962.
 Continued as Trudy Pamirskoĭ Biologicheskoĭ Stantsii
Botanicheskogo Instituta. Akademiya Nauk Tadzhikskoi SSR. B.S 1353

Trudy Botanicheskogo Instituta, Armyanskiĭ Filial. Akademiya
Nauk SSSR. Erevan.
 Trudy bot.Inst., Erevan Nos.1-3, 1941.
 Continued as Trudy Botanicheskogo Instituta. Akademiya
Nauk Armyanskoĭ SSR. Erevan. B.S 1355 a

TITLE	SERIAL No.

Trudȳ Botanicheskogo Instituta. Azerbaĭdzhanskiĭ Filial, Akademiya
  Nauk SSSR. Baku.
  <u>See</u> Trudȳ Instituta Botaniki im. V.L. Komarova. Akademiya
  Nauk Azerbaĭdzhanskoĭ SSR. Baku.   B.S 1351

Trudȳ Botanicheskogo Sada. Akademiya Nauk Ukrainskoĭ SSR. Kiev.
  <u>Trudȳ bot.Sada Kiev</u> 1949-1958. (imp.)
  <u>Continued as</u> Pratsi Botanichnoho Sadu.   B.S 1372

Trudȳ Botanicheskogo Sada Latviĭskogo Gosudarstvennogo
  Universiteta im. Petra Stuchki. Riga.
  <u>See</u> Pētera Stučkas Latvijas Valsts Universitātes
  Botaniskā Dārza Raksti. Rigā.   B.S 1500

Trudȳ Botanicheskogo Sada Moskovskogo Gosudarstvennogo
  Universiteta. Moskva.
  <u>Trudȳ bot.Sada mosk.gos.Univ.</u> 1937 →   B.S 1380

Trudȳ Botanicheskogo Sada im. Prof. B.M. Kozo - Polyanskogo.
  Voronezh.
  <u>Trudȳ bot.Sada Voronezh</u> Tom.2 → 1963 →   B.S 1405

Trudȳ Byuro po Entomologii. S.-Peterburg.
  <u>Trudȳ Byuro Ent.</u> 1894-1923
  <u>Continued as</u> Trudȳ Otdela Prikladnoĭ Entomologii.   E.S 1809

Trudȳ Byuro Kol'tsevaniya. Moskva.
  <u>Trudȳ Byuro Kol'tsev.</u> Vȳp. 8 → 1955 →
  (Transl. of 'selected articles' filed with original.)   Z. 18 q R

Trudȳ Byuro po Prikladnoĭ Botanike. S.-Peterburg.
  <u>Trudȳ Byuro prikl.Bot.</u> 1908-1917. (imp.)
  <u>Continued as</u> Trudȳ po Prikladnoĭ Botanike, Genetike i
  Selektsii. Leningrad.   B.S 1400

Trudȳ Dal'nevostochnogo Filiala imeni V.L. Komarova, Akademiya
  Nauk SSSR. Vladivostok.
  <u>Trudȳ dal'nevost.Fil.Akad.Nauk SSSR</u>
    Seriya Zoologicheskaya, Tom 1(5) → 1951 →   Z.S 1824
    Seriya Botanicheskaya, Vol.2 → 1956 →   B.S 1379

Trudȳ po Dinamike Razvitiya. Moskva.
  <u>Trudȳ Din.Razv.</u> 1931-1935.
  <u>Formerly</u> Trudȳ Laboratorii Eksperimental'noĭ Biologii
  Moskovskogo Zooparka. Moskva.   S. 1846

Trudy Ekspeditsii Pamirskaya Ekspeditsiya 1928g. Leningrad.
  <u>Trudy Eksped.pamir.Eksped.1928</u> Vyp.1-8, 1929-1931.   73 A.o.S

TITLE　　　　　　　　　　　　　　　　　　　　　　　SERIAL No.

Trudy̆ Estestvenno-Istoricheskogo Fakul'teta Universiteta
　　Im. J.E. Purkyné. Brno.
　　See Spisy Přírodovědecké Fakulty University
　　J.E. Purkyné v Brne.　　　　　　　　　　　　　　　　　　S. 1709 B

Trudy̆ Estestvenno-Istoricheskogo Muzeya im. G.Zardabi. Akademiya
　　Nauk Azerbaidzhanskoĭ SSR. Baku.
　　Trudy̆ estest.-istor.Muz.G.Zardabi
　　Vy̆p.4, 1951; 6-8, 1953-1954; 10-12, 1955-1960.　　　　　S. 1810 a D

Trudy̆ Etnograficheskago Otdela Imperatorskago Obshchestvo Lyubitelei
　　Estestvoznaniya, Antropologii i Etnografii, Moskva.
　　See Izvestiya Imperatorskago Obshchestva Lyubiteleĭ
　　Estestvoznaniya, Antropologii i Etnografii pri Imperatorskom
　　Moskovskom Universitete. Moskva.　　　　　　　　　　　　S. 1841 A

Trudy̆ Gel'mintologicheskoĭ Laboratorii. Akademiya Nauk
　　SSSR. Moskva and Leningrad.
　　Trudy̆ gel'mint.Lab. 1948 →　　　　　　　　　　　　　　Z.S 1847

Trudy̆ Geograficheskago Otdeleniya Imperatorskago Obshchestva
　　Lyubitelei Estestvoznaniya, Antropologii i Etnografii, Moskva.
　　See Izvestiya Imperatorskago Obshchestva Lyubiteleĭ
　　Estestvoznaniya, Antropologii i Etnografii pri Imperatorskom
　　Moskovskom Universitete. Moskva.　　　　　　　　　　　　S. 1841 A

Trudy̆ Geologicheskago Komiteta. S.-Peterburg.
　　Trudy̆ geol.Kom. 1883-1930.
　　Continued as Trudy̆ Glavnogo Geologo-Razvedochnogo
　　Upravleniya V.S.N.Kh. SSSR. Leningrad.　　　　　　　　　P.S 1505

Trudy̆ Geologicheskago (i Mineralogicheskago) Muzeya imeni
　　Petra Velikago Imperatorskoĭ Akademii Nauk'. S.-Peterburg'.
　　Trudy̆ geol.miner.Muz. 1907-1926.
　　Continued as Trudy̆ Geologicheskogo Muzeya Akademii
　　Nauk SSSR and Trudy̆ Mineralogicheskogo Muzeya Akademii
　　Nauk SSSR.　　　　　　　　　　　　　　　　　　　　　　　　P.S 505

Trudy̆ Geologicheskogo Instituta. Akademiya Nauk Gruzinskoĭ SSR.
　　Tbilisi.
　　Trudy̆ geol.Inst.Tbilisi
　　Geologicheskaya Seriya 1 (VI) - 14 (XIX), 1942-1965 (imp.)
　　Nov.Ser. 1965 →　　　　　　　　　　　　　　　　　　　　　P.S 556
　　Min.-Petrogr.Ser. No.1-2, 4 & 6, 1943-1962.　　　　　　M.S 1827
　　Formerly Bulletin de l'Institut Géologique de Géorgie.

Trudy̆ Geologicheskogo Instituta. Akademiya Nauk Kirgizskoĭ SSR. Frunze.
　　Trudy̆ geol.Inst., Frunze Vyp. 2. 1951.
　　Continued as Trudy̆ Instituta Geologii. Akademiya Nauk
　　Kirgizskoĭ SSR. Frunze.　　　　　　　　　　　　　　　　　P.S 527

TITLE	SERIAL No.

Trudy Geologicheskogo Instituta. Akademiya Nauk SSSR. Leningrad.
    <u>Trudy geol.Inst.Leningr.</u> 1932-1939: 1956 →
    <u>Formerly</u> Trudy Geologicheskogo Muzeya. Akademiya Nauk SSSR.
    <u>From 1939-1956 Contained in</u> Trudy Instituta Geologicheskikh
    Nauk. Akademiya Nauk SSSR.   P.S 508

Trudy Geologicheskogo Instituta. Akademiya Nauk SSSR.
    Kazanskiĭ Filial. Moskva.
    <u>Trudy geol.Inst.kazan.Fil.</u> Vol.3. 1956.   P.S 1546

Trudy Geologicheskogo Instituta. Kirgizskiĭ Filial Akademii
    Nauk SSSR. Frunze.
    <u>See</u> Trudy Instituta Geologii. Akademiya Nauk Kirgizskoĭ
    SSR. Frunze.   P.S 527

Trudy Geologicheskogo Instituta Narodnoĭ Respubliki Makedonii.
    <u>See</u> Trudovi na Geološki Zavod na Narodna Republika
    Makedonija. Skopje.   P.S 453

Trudy Geologicheskogo Muzeya. Akademiya Nauk SSSR. Leningrad.
    <u>Trudy geol.Muz.</u> 1926-1931.
    <u>Formerly</u> Trudy Geologicheskago (i Mineralogicheskago)
    Muzeya imeni Imperatora Petra Velikago.
    <u>Continued as</u> Trudy Geologicheskogo Instituta. Akademiya
    Nauk SSSR.   P.S 506

Trudy Geologicheskogo Muzeya imeni A.P. Karpinskogo.
    (Akad.Nauk SSSR)
    <u>Trudy geol.Muz.Karpinskogo</u> 1957 →   P.S 516

Trudy Geologicheskoĭ Chasti Kabineta ego Imperatorskago Velichestva.
    S.-Petersburg.
    <u>See</u> Travaux de la Section Géologique du Cabinet de Sa Majésté.
    St.-Petersbourg.   P.S 1520

Trudy Geologichesko-Razvedochnogo Instituta Narodnoĭ
    Respubliki Makedonii.
    <u>See</u> Trudovi Geološki Zavod na Narodna Republika
    Makedonija. Skopje.   P.S 453

Trudy Glavnogo Botanicheskogo Sada. Petrograd, Moskva.
    <u>Trudy glav.bot.Sada Petrogr.</u> 1918-1930, Vuip 3-9. 1953-1963.
    <u>Formerly</u> Trudy Imperatorskago S.-Peterburgskago
    Botanicheskago Sada. S.-Peterburg.   B.S 1403

Trudy Glavnogo Geologo-Razvedochnogo Upravleniya V.S.N.Kh. SSSR. Moskva.
    <u>Trudy glav.geol.-razv.Uprav.V.S.N.Kh.</u>
    1930-1931. (Wanting Nos.2 and 8.)
    <u>Formerly</u> Trudy Geologicheskago Komiteta. St.-Peterburg.
    <u>Continued as</u> Trudy Vsesoyuznogo Geologo-Razvedochnogo
    Ob"edineniya NKTP SSSR. Leningrad.   P.S 1525

TITLE	SERIAL No.

Trudȳ Gorno-Geologicheskogo Instituta. Akademiya Nauk SSSR.
  Ural'skiĭ Filial. Sverdlovsk.
  <u>Trudȳ gorno-geol.Inst.ural'.Fil</u>. No.22-56, (imp.) 1953-1961.
  <u>Continued as</u> Trudȳ Instituta Geologii. Akademiya
  Nauk SSSR. Ural'skiĭ Filial. Sverdlovsk.   P.S 1537

Trudȳ Gornotaezhnoĭ Stantsii. Dal'nevostochnȳĭ Filial
  Akademii Nauk SSSR. Vladivostok.
  <u>Trudȳ gornotaezh.Sta.Vladiv</u>. vols.2-3. 1938-1939.   S. 1802 Cb

Trudȳ Gosudarstvennogo Muzeya Tsentral'no-Promyshlennoĭ
  Oblasti. Moskva.
  <u>Trudȳ gos.Muz.tsent.prom.Obl</u>. Nos.1-2. 1925.   S. 1839

Trudȳ Gosudarstvennogo Nauchno-Issledovatel'skogo Instituta
  Eksperimental'nogo Morfogeneza. Moskva.
  <u>Trudȳ gos.nauchno-issled.eksp.Morfogen</u>. Tom.2, 1934.
  <u>Continued as</u> Trudȳ Nauchno-Issledovatel'skogo Instituta
  Eksperimental'nogo Morfogeneza. Moskva.   Z.S 1928 C

Trudȳ Gosudarstvennogo Nauchno-Issledovatel'skogo Neftyanogo
  Instituta. Moskva.
  <u>Trudȳ gos.nauchno-issled.neft.Inst</u>. 4. 1929.   P. 72 Q.q.S

Trudȳ Gosudarstvennogo Nikitskogo Botanicheskogo Sada. Yalta.
  <u>Trudȳ gos.nikit.bot.Sada</u> Tom.19-36, 1935-1962: Tom.39, 1967 →
  For Tom.37-38, 1964-1967 <u>See</u> Sbornik Nauchnȳkh Trudov.
  Gosudarstvennȳĭ Ordena Trudovogo Krasnogo Znameni Nikitskiĭ
  Botanicheskiĭ Sad & Nauchnȳe Trudȳ. Gosudarstvennȳĭ
  Ordena Trudovogo Krasnogo Znameni Nikitskiĭ Botanicheskiĭ
  Sad. Moskva.
  <u>Formerly</u> Zapiski Gosudarstvennogo Nikitskogo Opȳtnogo
  Botanicheskogo Sada. Leningrad.   B.S 1445

Trudȳ Gosudarstvennogo Okeanograficheskogo Instituta. Moskva.
  <u>Trudȳ gos.okeanogr.Inst</u>. 1931-1934; 1955 →
  <u>Formerly</u> Trudȳ Morskogo Nauchnogo Instituta. Moskva.   S. 1845 A

Trudȳ. Groznenskiĭ Neftyanoĭ Nauchno-Issledovatel'skiĭ
  Institut. Moskva.
  <u>Trudȳ grozn.neft.nauchno-issled.Inst</u>. Vyp.8, 1960.   P.S 545

Trudȳ Hidrobiolohichnoyi Stantsiyi. Akademiya Nauk URSR. Kȳyiv.
  <u>Trudȳ hidrobiol.Sta.Kȳyiv</u> 1934-1939.   S. 1834 a G

Trudȳ Ikhtīologicheskoĭ Laboratorīi Upravleniya Kaspīĭsko-
  Volzhskikh Rȳbnykh i Tyulen'ikh Promȳslov. Astrakhan.
  <u>Trudȳ ikhtīol.Lab.Uprav.kasp.volzh.rȳb.tyul.Promȳs</u>.
  Vol.2 No.2 - Vol.3 No.5, 1912-1914
  <u>Continued as</u> Trudȳ Astrakhanskoĭ Ikhtīologicheskoĭ
  Laboratorīi.   Z.S 1830

TITLE   SERIAL No.

Trudȳ Il'menskogo Gosudarstvennogo Zapovednika (im V.I. Lenina).
   Moskva.
   Trudȳ Il'mensk.gos.Zapov. No.1, 4-5, 7, 1938-1959.
   Continued as Trudȳ Zapovednika. Ural'skiĭ Filial.
   Akademiya Nauk SSSR. Sverdlovsk.   S. 1805 a F

Trudȳ Imperatorskago S.-Peterburgskago Botanicheskago Sada.
   S.-Peterburg.
   Trudȳ imp.S.-Peterb.bot.Sada 1871-1915.
   Continued as Trudȳ Glavnogo Botanicheskogo Sada. Petrograd.   B.S 1403

Trudȳ Imperatorskago Sankt-Peterburgskago Obshchestva
   Estestvoispȳtateleĭ. S.-Peterburg. (Leningrad.)
   Trudȳ imp.S.-Peterb.Obshch.Estest. 1870-1914.
   Continued as Trudȳ Leningradskogo Obshchestva
   Estestvoispȳtateleĭ. Leningrad.   S. 1855 A

Trudȳ Instituta Biologicheskoĭ Fiziki. Moskva.
   Trudȳ Inst.biol.Fiz. 1955.   S. 1802 f

Trudȳ Instituta Biologii. Akademiya Nauk Latviĭskoĭ SSR. Riga.
   Trudȳ Inst.Biol.Riga 1953-1963.
   (Numbers 8 and 13 wanting.)   S. 1852 e C

Trudȳ Instituta Biologii.Ural'skiĭ Filial. Akademiya Nauk SSR.
   Sverdlovsk.
   Trudȳ Inst.Biol.Sverdlovsk No.5-56, 1954-1966 (imp.)
   Continued as Trudȳ Instituta Ekologii Rasteniĭ i Zhivotnykh.
   Ural'skiĭ Filial. Akademiya Nauk SSR.   S. 1802 j

Trudȳ Instituta Biologii. Vilnius.
   See Lietuvos TSR Mokslu Akademijos Biologijos
   Instituto Darbai. Vilnius.   S. 1799 A

Trudȳ Instituta Biologii Vnutrennikh Vod. Moskva.
   Trudȳ Inst.Biol.Vnutrenn.Vod 6 (9) → 1963 →
   Formerly Trudȳ Instituta Biologii Vodokhranilishch.
   Moskva.   S. 1802 l. B

Trudȳ Instituta Biologii Vodokhranilishch. Moskva.
   Trudȳ Inst.Biol.Vodokhran. Vyp.1-5 (4-8), 1959-1963.
   Formerly Trudȳ Biologicheskoĭ Stantsii "Borok". Moskva.
   Continued as Trudȳ. Institut Biologii Vnutrennikh
   Vod. Moskva.   S. 1802 l. B

Trudȳ Instituta Biologii. Yakutskiĭ Filial Akademiya Nauk. SSSR.
   Trudȳ Inst.Biol.Yakutsk. 1955-1962.
   (Vuip 1-2 & 5 Microfilm)   S. 1802 Pd

Trudȳ Instituta Botaniki. Akademiya Nauk Gruzinskoĭ SSR. Tbilisi.
   See Trudȳ Tbilisskogo Botanicheskogo Instituta.   B.S 1430

| TITLE | SERIAL No. |

Trudy Instituta Botaniki. Akademiya Nauk Kazakhskoĭ SSR. Alma-Ata.
  Trudy Inst.Bot.Alma-Ata Vol.2 → 1955 →   B.S 1346 e

Trudy Instituta Botaniki. Akademiya Nauk Kirgizskoĭ SSR. Frunze.
  Trudy Inst.Bot.Frunze   Nos.1-3, 1955-(1958)
  Formerly Trudy Instituta Botaniki i Rastenievodstva.
  Akademiya Nauk Kirgizskoĭ SSR. Frunze.   B.S 1359 a

Trudy Instituta Botaniki. Akademiya Nauk Tadzhikskoĭ SSR. Stalinabad.
  Trudy Inst.Bot.,Stalinabad Tom.47 → 1956 →   B.S 1414

Trudy Instituta Botaniki. Khar'kov.
  See Trudy Instytutu Botaniky. Khar'kiv.   B.S 1370

Trudy Instituta Botaniki im. V.L. Komarova. Akademiya Nauk
    Azerbaidzhanskoĭ SSR. Baku.
  Trudy Inst.Bot.Baku 1936-1939, 1949-1964. (imp.) B.S 1351

Trudy Instituta Botaniki i Rastenievodstva. Akademiya Nauk
    Kirgizskoĭ SSR. Frunze.
  Trudy Inst.Bot.Rasteniev, Frunze 1954.
  Continued as Trudy Instituta Botaniki. Akademiya Nauk
  Kirgizskoĭ SSR. Frunze.   B.S 1359 a

Trudy Instituta Ekologii Rastenii i Zhivotnykh. Ural'skiĭ Filial,
    Akademiya Nauk SSR. Sverdlovsk.
  Trudy Inst.Ekol.Rast.Zhivot.:Sverdlovsk 1967 →
  Formerly Trudy Instituta Biologii. Ural'skiĭ Filial.
  Akademiya Nauk SSR. Sverdlovsk.   S. 1802 j

Trudy Instituta Eksperimental'noĭ Biologii. Akademiya
    Nauk Estonskoĭ SSR. Tallinn.
  See Eksperimentaalbioloogia Instituudi Uurimused. Tallinn.   S. 1817 C

Trudy Instituta Eksperimental'noĭ Biologii. Akademiya Nauk
    Kazakhskoi SSR. Alma-Ata.
  Trudy Inst.eksp.Biol.Alma-Ata 1964 →   S. 1832 b B

Trudy Instituta Evolyutsionnoĭ Fiziologii i Patologii Vysshe
    Nervnoĭ Deyatel'nosti. im. I.P.Pavlova.
  Trudy Inst.evol.Fiziol.Patol.vyssh.nerv.Deyat. 1947.   13.o.S

Trudy Instituta Fiziologii Rastenii imeni K.A. Timiryazeva. Moskva.
  Trudy Inst.Fiziol.Rast. 1937-1955. (imp.)
  Continued as Trudy Laboratorii Evolyutsionnoĭ i
  Ekologicheskoĭ Fiziologii im. B.A. Kellera. Moskva.   B.S 1381

Trudy Instituta Genetiki. Akademiya Nauk SSSR. Leningrad.
  Trudy Inst.Genet. No.21-33 & 35. 1954-1965.   S. 1802 e

| TITLE | SERIAL No. |

Trudȳ Instituta Genetiki i Selektsii Azerbaĭdzhanskoĭ SSR. Baku.
　　Trudȳ Inst.Genet.Selek., Baku　Tom 4 → 1966 →　　　　B.S 1352

Trudȳ Instituta Geografii. Akademiya Nauk SSSR.Moskva,Leningrad.
　　Trudȳ Inst.Geogr.Leningr.　53 → 1953 → (imp.)
　　(Only issues containing the Series Materialȳ po Geomorfologii
　　i Paleogeografii SSSR retained)　　　　　　　　　　　　P.S 512

Trudȳ Instituta Geologicheskikh Nauk. Akademiya Nauk
　　Belorusskoĭ SSR. Minsk.
　　Trudȳ Inst.geol.Nauk Minsk　1958.　　　　　　　　　　P.S 1551 A

Trudȳ Instituta Geologicheskikh Nauk. Akademiya Nauk
　　Kazakhskoi SSR. Alma-Ata.
　　Trudȳ Inst.geol.Nauk Alma-Ata
　　Vol.1. Geol.Ser.No.1. 1956　　　　　　　　　　　　　　P.S 1550

Trudȳ Instituta Geologicheskikh Nauk. Akademiya Nauk SSSR. Moskva.
　　Trudȳ Inst.geol.Nauk Mosk.　1939-1956 (Imp.) (Published in Series.)
　　Formerly & Continued as　Trudȳ Geologicheskogo Instituta.
　　Akademiya Nauk SSSR.
　　Also Continued as　Trudȳ Instituta Geologii Rudnȳkh
　　Mestorozhdeniĭ, Petrografii, Mineralogii i Geokhimii.　　P.S 508

Trudȳ Instituta Geologicheskikh Nauk. Akademiya Nauk Ukrainskoĭ
　　SSR. Kiev.
　　Trudȳ Inst.geol.Nauk Kiev
　　　Seriya Petrografii, Mineralogii i Geokhimii.
　　　Vȳp.14-21, 1962-1964. (imp.)　　　　　　　　　　　　P.S 531 H
　　　Seriya Geomorfologii i Chetvertichnoĭ
　　　Geologii. Vȳp.1, 1957.　　　　　　　　　　　　　　　P.S 531 F
　　　For Seriya Stratigrafiĭ i Paleontologiĭ
　　　See　Trudȳ Instytutu Heolohichnȳkh Nauk. Kȳyiv.　　　P.S 531 C

Trudȳ Instituta Geologii. Kiev.
　　See　Trudȳ Instȳtutu Heolohiyi. Ukrayins'ka Akademia
　　Nauk. Kȳyiv.　　　　　　　　　　　　　　　　　　　　P.S 531 A

Trudȳ Instituta Geologii. Akademiya Nauk Azerbaĭdzhanskoĭ SSR. Baku.
　　Trudȳ Inst.Geol., Baku　Tom 21-23, 1961-1964.　　　　P.S 1556 A

Trudȳ Instituta Geologii Akademii Nauk Estonskoĭ SSR. Tallinn.
　　See　Geoloogia Instituudi Uurimused. Eesti NSV Teaduste
　　Akadeemia. Tallinn.　　　　　　　　　　　　　　　　　P.S 1549

Trudȳ Instituta Geologii. Akademiya Nauk Kirgizskoĭ SSR. Frunze.
　　Trudȳ Inst.Geol.Akad.Nauk kirgiz.　Vyp.4, 5 & 10.　1953-1958.
　　Formerly　Trudȳ Geologicheskogo Instituta. Akademiya
　　Nauk Kirgizskoĭ SSR. Frunze.　　　　　　　　　　　　　P.S 527

TITLE	SERIAL No.

Trudy Instituta Geologii. Akademiya Nauk Latviĭskoĭ SSR. Riga.
    Trudy Inst.Geol.Akad.Nauk Riga 1947-1963.
    (Wanting Tom 3)   P.S 559

Trudy Instituta Geologii. Akademiya Nauk SSSR. Ural'skiĭ
    Filial. Sverdlovsk.
    Trudy Inst.Geol.ural'.Fil. Vyp.62 → 1962 →
    Formerly Trudy Gorno-Geologicheskogo Instituta. Akademiya
    Nauk SSSR. Ural'skiĭ Filial. Sverdlovsk.   P.S 1537

Trudy Instituta Geologii. Akademiya Nauk Tadzhikskoĭ SSR.
    Stalinabad. Dunshanbe.
    Trudy Inst.Geol.Dushanbe Tom.4 and 6 → 1961 →   P.S 1567

Trudy Instituta Geologii i Geofiziki. Sibirskoe Otdelenie.
    Novosibirsk.
    Trudy Inst.Geol.Geofiz.sib.Otd. 1960 →   P.S 1516

Trudy Instituta Geologii i Geografii Akademii Nauk Latviĭskoĭ
    SSR. Riga.
    See Trudy Instituta Geologii. Akademiya Nauk Latviĭskoĭ SSR.
    Riga.   P.S 559

Trudy Instituta Geologii. Gosudarstvennyĭ Geologicheskii
    Komitet SSSR. Vilnius.
    Trudy Inst.Geol., Vilnius 1965-1969.
    Continued as Trudy Litovskogo Nauchno-Issledovatel'skogo
    Geologorazvedochnogo Instituta. Vilnius.   P.S 575

Trudy Instituta Geologii i Poleznykh Iskopaemykh. Akademiya
    Nauk Latviĭskoĭ SSR. Riga.
    See Trudy Instituta Geologii. Akademiya Nauk Latviĭskoĭ SSR.
    Riga.   P.S 559

Trudy Instituta Geologii Rudnykh Mestorozhdeniĭ Petrografii,
    Mineralogii i Geokhimii. Moskva.
    Trudy Inst.Geol.rudn.Mestorozh. 1956-1963 (imp.)   M.S 1816

Trudy Instituta Ikhtiologii i Rybnogo Khozyaĭstva.
    Akademiya Nauk Kazakhskoĭ SSR. Alma-Ata.
    Trudy Inst.Ikhtiol.ryb.Khoz.Akad.Nauk Kazakhskoĭ SSR.
    Tom.4 → 1963 →   Z.S 1814

Trudy Instituta Istorii Estestvoznaniya. Akademiya Nauk SSSR.
    Moskva, Leningrad.
    Trudy Inst.Istor.Estest. 1947-1953.
    Continued as Trudy Instituta Istorii Estestvoznaniya i
    Tekhniki. Akademiya Nauk SSSR. Moskva.   S. 1804 b A

TITLE   SERIAL No.

Trudy Instituta Istorii Estestvoznaniya i Tekhniki. Akademiya
   Nauk SSSR. Moskva.
     Trudy Inst.Istor.Estest.Tekh. Tom.2-42, 1954-1962.
     Formerly Trudy Instituta Istorii Estestvoznaniya. Akademiya
     Nauk SSSR. Moskva, Leningrad.   S. 1804 b A

Trudy Instituta po Izucheniyu Severa. Moskva.
     Trudy Inst.Izuch.Sev. No.31-49. 1926-1931.   S. 1842 A
                        1926.   M.S 1817 B
     Formerly Trudy Nauchno-Issledovatel'skogo Instituta
     po Izucheniyu Severa. Moskva.
     Continued as Trudy Arkticheskogo Nauchno-Issledovatel'skogo
     Instituta. Leningrad.

Trudy Instituta Kristallografii. Moskva.
     Trudy Inst.Kristall. Vyp.8-12; 1953-1956.   M.S 1812

Trudy Instituta Lesa. Akademiya Nauk SSSR. Moskva, Leningrad.
     Trudy Inst.Lesa, Mosk. Tom.7-49, 1951-1959. (imp.)
     Continued as Trudy Instituta Lesa i Drevesiny.
     Akademiya Nauk SSSR. Moskva, Leningrad.   B.S 1383 a

Trudy Instituta Lesa i Drevesiny. Akademiya Nauk SSSR.
   Moskva, Leningrad.
     Trudy Inst.Lesa Drev. Tom.51-66, 1962-1963.
     Formerly Trudy Instituta Lesa. Akademiya Nauk SSSR.
     Moskva, Leningrad.   B.S 1383 a

Trudy Instituta Mikrobiologii. Akademiya Nauk Latviĭskoĭ
   SSR. Riga.
     Trudy Inst.Mikrobiol.Akad.Nauk Latv.SSR 1952-1963.   S. 1852 e D

Trudy Instituta Mikrobiologii. Akademiya Nauk SSSR. Moskva.
     Trudy Inst.Mikrobiol.Mosk. Vol.3-11, 1954-1961.   B.M.S 58 b

Trudy Instituta Mikrobiologii i Virusologii. Akademiya Nauk
   Kazakhskoi SSR. Alma-Ata.
     Trudy Inst.Mikrobiol.Virus., Alma-Ata 1956 →   B.M.S 59

Trudy Instituta Mineralogii, Geokhimii i Kristallokhimii
   Redkikh Elementov. Moskva.
     Trudy Inst.Miner.Geokhim.Kristallokhim.redk.Elem. 1957 →   M.S 1806 C

Trudy Instituta Morfologii Zhivotnykh. Akademiya Nauk SSSR.
   Moskva & Leningrad.
     Trudy Inst.Morf.Zhivot. Tom.1-2, 8-9, 11-42. 1949-1962.   Z.S 1851

Trudy Instituta Okeanologii. Akademiya Nauk SSSR. Moskva.
     Trudy Inst.Okeanol. 1946 →   (imp.)   S. 1802 d

TITLE	SERIAL No.

Trudȳ Instituta Paleobiologii. Akademiya Nauk Gruzinskoĭ SSR. Tbilisi.
    <u>Trudȳ Inst.Paleobiol.Tbilisi</u>   1954 →   P.S 558

Trudȳ Instituta Pochvovedeniya. Akademiya Nauk Kazakhskoĭ SSR.
    Alma-Ata.
    <u>Trudȳ Inst.Pochv., Alma-Ata</u>   Vol.10 → 1960 →   B.S 1346 f

Trudȳ Instituta Pochvovedeniya i Geobotaniki Sredne-Aziatskogo
    Gosudarstvennogo Universiteta. Tashkent. Turkmenistanskaya
    Seriya.
    <u>See</u> Acta Universitates Asiae Mediae. Tashkent. Ser.Vlld.
    Pedologia, Fasc.1-2, 1930.   S. 1959 B

Trudȳ Instituta Prikladnoĭ Mineralogii. Moskva.
    <u>Trudȳ Inst.prikl.Miner.</u> Vyp.59-67, 69, 70: 1933-1935.
    <u>Formerly</u> Trudȳ Nauchno-Issledovatel'skikh Institutov
    Promȳshlennosti. Moskva.
    <u>Continued as</u> Trudȳ Vsesoyuznogo Nauchno-Issledovatel'skogo
    Instituta Mineral'nogo Syr'ya. Moskva.   M.S 1809

Trudȳ Instituta Rȳbnogo Khozyaĭstva i Promȳslovykh
    Issledovaniĭ. Leningradskoe Otdelenie.
    <u>Trudȳ Inst.rȳb.Khoz.promȳsl.Issled.leningr.</u> 1929.   Z.S 1821

Trudȳ Instituta Zashchitȳ Rasteniĭ. Akademiya Nauk Gruzinskoĭ
    SSR. Tbilisi.
    <u>Trudȳ Inst.Zashch.Rast.,Tbilisi</u> 1948, 1957.   E.S 1785

Trudȳ Instituta Zoologii. Akademiya Nauk Azerbaĭdzhanskoĭ
    SSR. Baku.
    <u>Trudȳ Inst.Zool., Baku</u> Tom.11-27, 1946-1968.
    (Wanting Tom.19.)
    <u>Formerly</u> Trudy Zoologicheskogo Instituta. Azerbaĭdzhanskiĭ
    Filial, Akademiya Nauk SSSR. Baku.   Z.S 1853

Trudȳ Instituta Zoologii. Akademiya Nauk Gruzinskoĭ SSR. Tbilisi.
    <u>Trudȳ Inst.Zool.Tbilisi</u> Tom.9 →   1950 →
    <u>Formerly</u> Trudȳ Zoologicheskogo Instituta.
    Akademiya Nauk Gruzinskoĭ SSR.   Z.S 1837

Trudȳ Instituta Zoologii. Akademiya Nauk Kazakhskoĭ SSR.
    Alma Ata.
    <u>Trudȳ Inst.Zool., Alma Ata</u> Tom 4, 6 → 1955 →   Z.S 1813 A

Trudȳ Instituta Zoologii i Parazitologii im. Akademika
    E.N. Pavlovskogo. Akademiya Nauk Tadzhikskoĭ SSR.
    Dushanbe (Stalinabad).
    <u>Trudȳ Inst.Zool.Parazit.Akad.Nauk tadzhik.SSR</u>
    Tom 21-25, 1954-1962.   Z.S 1856

Trudȳ Instituta Zoologii i Parazitologii. Akademiya Nauk
    Kirgizakoĭ SSR. Frunze.
    <u>Trudȳ Inst.Zool.Parazit.Akad.Nauk Kirgiz.SSR.</u> 1954-1959.   Z.S 1844

TITLE	SERIAL No.

Trudȳ Instȳtutu Botanikȳ. Khar'kiv.
    Trudȳ Inst.Bot.Khar'kiv. 1936-1938 (imp.)   B.S 1370

Trudȳ Instȳtutu Heolohichnȳkh Nauk. Kȳyiv.
    Trudȳ Inst.heol.Nauk, Kȳyiv Seriya Stratȳhrafiyi
    i Paleontolohiyi. No.8-49, 1956-1964. (imp.)   P.S 531 C

Trudȳ Instȳtutu Heolohiyi po Donbasu. Akademiya Nauk URSR. Kiev.
    Trudȳ Inst.Heol.Donbasu 1939.   P.S 531 B

Trudȳ Instȳtutu Heolohiyi. Ukrayins'ka Akademia Nauk. Kȳyiv.
    Trudȳ Inst.Heol.Kȳyiv 1935-1939. (imp.)   P.S 531 A

Trudȳ Instȳtutu Zoolohiyi. Kȳyiv.
    Trudȳ Inst.Zool.Kȳyiv 1948-1957.
    Formerly Trudȳ Instȳtutu Zoolohiyi ta Biolohiyi.
    Continued as Pratsi Instȳtutu Zoolohiyi. Kȳyiv.   Z.S 1827 C

Trudȳ Instytutu Zoolohiyi ta Biolohiyi. Kyyiv.
    Trudȳ Inst.zool.Biol.Kyyiv Vols.2, 8, 14, &
    Nos.5 & 12.1934-1939. For Vol.3, 1934, Vol.10, 1936.
    Vol.16, 1937, See Zbirnyk Prats' z Morfolohiyi Tvarȳn. Kȳyiv.
    Continued as Trudȳ Instȳtutu Zoolohiyi. Kȳyiv.   S. 1834 a F

Trudȳ po Izucheniyu Radiya i Radioktivnȳkh Rud. Leningrad.
    Trudȳ Izuch.Radiya radioakt.Rud. 1924-1928.   M.S 1808

Trudȳ Karadahs'koyi Biolohichnoyi Stantsiyi. Kȳyiv.
    See Trudȳ Karadahs'koyi Nauchnoyi Stantsiyi imeni
    T.I. Vyazems'koho. Simferopol.   S. 1838 I

Trudȳ Karadahs'koyi Nauchnoyi Stantsiyi imeni T.I. Vyazems'koho.
    Simferopol.
    Trudȳ karadah.nauch.Sta.T.I.Vyazems'koho
    1917-1940; 1952; 1957-1962.   S. 1838 I

Trudȳ Karagandinskogo Botanicheskogo Sada.
    Trudȳ karagand.bot.Sada 1960 →   B.S 1346 d

Trudȳ Karel'skoi Nauchno-Issledovatel'skoĭ Rȳbokhozyaĭstvennoĭ
    Stantsii. Leningrad.
    Trudȳ karel.nauchno-issled.rȳbokhoz.Sta. Tom.1 1935.   Z.S 1857 A

Trudȳ. Kaspiĭskii Nauchno-Issledovatel'skiĭ Institut Rybnogo
    Khozyaĭstva. Moskva.
    Trudȳ kasp.nauchno-issled.Inst.ryb.Khoz. Tom 23-26, 1967-1971.Z.S 1839
    Continued in Trudȳ Vsesoyoznogo-Nauchno-Issledovatel'skogo
    Instituta Morskogo Rȳbnogo Khozyaistva i Okeanografii.   S. 1846 b

TITLE	SERIAL No.

Trudy̅ Kazakhstanskogo Filiala. Akademiya Nauk SSSR. Moskva.
  Trudy̅ kazakhstan.Fil.Akad.Nauk SSSR. Vol.11. 1938. 73 A.o.S

Trudy̅ Kerchenskoĭ Ikhtiologicheskoĭ Laboratorii. Kerch.
  Trudy̅ Kerch.ikhtiol.Lab. Tom 1 Vp.1, 1926.
  Continued as Trudy̅ Kerchenskoĭ Nauchnoi Rybo-
  khozyaĭstvennoĭ Stantsii. Z.S 1838

Trudy̅ Kerchenskoĭ Nauchnoĭ Rybokhozyaĭstvennoĭ Stantsii.
  Trudy̅ kerch.nauch.rybokhoz.Sta. Tom 1,Vp.2-5, 1927-1930.
  Formerly Trudy̅ Kerchenskoĭ Ikhtiologicheskoĭ Laboratorii.
  Continued as Trudy Azovsko-Chernomorskoĭ Nauchnoĭ
  Rybokhozyaĭstvennoĭ Stantsii. Z.S 1838

Trudy̅ Kharkivs'kogo Tovaristva Doslidnikiv Prirody. Kharkiv.
  Trudy̅ kharkiv.Tov.Dosl.Prir. 1927-1930.
    (Wanting Tom.54)
  Formerly Trudy̅ Khar'kovskogo Obshchestva Ispy̅tateleĭ
  Prirody pri Ukrhlavnauke. Khar'kov. S. 1833 A

Trudy̅ Khar'kovskogo Obshchestva Ispy̅tateleĭ Prirody̅ pri
  Ukrhlavnauke. Khar'kov.
  Trudy̅ khar'kov.Obshch.Ispyt.Prir. 1925.
  Formerly Trudy Obshchestva Ispy̅tateleĭ Prirody̅ pri
  Imperatorskom Khar'kovskom Universitetê. Khar'kov.
  Continued as Trudy̅ Kharkivs'kogo Tovaristva Doslidnikiv
  Prirody̅. Kharkiv. S. 1833 A

Trudy̅ Kirgizskoĭ Kompleksnoĭ Ekspeditsii. Moskva, etc.
  Trudy̅ kirgiz.kompleks.Eksped. vol.1-3, pt.1. 1935-1936. 73 A.o.L

Trudy̅ Komissii po Izucheniyu Chetvertichnogo Perioda.
  Trudy̅ Kom.Izuch.chetv.Perioda 1932 → (imp.) P.S 502

Trudy̅ Komissii po Izucheniyu Ozera Baĭkala. Petrograd.
  Trudy̅ Kom.Izuch.Ozera Baĭkala 1918-1930.
  Continued as Trudy Baĭkal'skoĭ Limnologicheskoĭ
  Stantsii. Leningrad. S. 1802 N

Trudy̅ Komissii po Izucheniyu Yakutskoĭ ASSR. Leningrad.
  Trudy Kom.Izuch.yakut.ASSR 1926-1930.
    (Wanting Tom.13.) S. 1802 Ph

Trudy̅ Komissii po Opredeleniyu Absolyutnogo Vozrasta
  Geologicheskikh Formatsiĭ. Leningrad.
  Trudy̅ Kom.Opred.absol.Vozr.geol.Form. No.4 → 1955 → P.S 1502

Trudy̅ Kompleksnoĭ Yuzhnoĭ Geologicheskoĭ Ekspeditsii.
  Akademiya Nauk SSSR. Leningrad.
  Trudy̅ kompleks.yuzh.geol.Eksped. 1958 → P.S 1563

TITLE	SERIAL No.

Trudȳ Kosinskoĭ Biologicheskoĭ Stantsii. Moskva.
    Trudȳ kosin.biol.Sta. 1924-1930.
    Continued in Trudȳ Limnologicheskoĭ Stantsii v Kosine.
Moskva.   S. 1838 D

Trudȳ Kostromskogo Nauchnogo Obshchestva po Izucheniyu
    Mestnogo Kraya. Kostroma.
    Trudȳ kostrom.nauch.Obshch.Izuch.mest.Kr. No.37. 1926.   S. 1853

Trudȳ Laboratorii Eksperimental'noĭ Biologii Moskovskogo
    Zooparka. Moskva.
    Trudȳ lab.éksp.Biol.mosk.Zoopk. 1926-1929.
    Continued as Trudȳ po Dinamike Razvitiya. Moskva.   S. 1846

Trudȳ Laboratorii Eksperimental'noĭ Zoologii i Morfologii
    Zhivotnȳkh. Leningrad.
    Trudȳ Lab.eksp.Zool.Morf.Zhivot. 1930-1935.   Z.S 1820 B

Trudȳ Laboratorii Evolyutsionnoĭ i Ekologicheskoĭ Fiziologii
    im. B.A. Kellera. Moskva.
    Trudȳ Lab.Evol.ekol.fiz.Kellera Tom 4 → 1962 →
    Formerly Trudȳ Instituta Fiziologii Rasteniĭ im.
K.A. Timiryazeva. Moskva.   B.S 1381

Trudȳ Laboratorii Evolyutsionnoĭ Morfologii. Leningrad.
    Trudȳ Lab.evol.Morf. Tom.1-2, 1933-1935.   Z.S 1822

Trudȳ Laboratorii Geologii Dokembriya. Moskva.
    Trudȳ Lab.Geol.Dokembr. Vol.5-7, 9, 11 → 1955 →   P.S 1547

Trudȳ Laboratorii Geologii Uglya. Moskva.
    Trudȳ Lab.Geol.Uglya Vol.4, 6-7, 9 → 1956 →   P.S 1545

Trudȳ Laboratorii Gidrogeologicheskikh Problem imeni Akademika
    F.P. Savarenskogo.
    Trudȳ Lab.gidrogeol.Probl. Tom.14, 19 → 1957 →   P.S 1548

Trudȳ Laboratorii Lesovedeniya. Akademiya Nauk SSSR. Moskva.
    Trudȳ Lab.Lesov. 1960 →   B.S 1389

Trudȳ Laboratorii Osnov Rȳbovodstva. Leningrad.
    Trudȳ Lab.Osnov.Rȳbov. 1947-1949.   Z.S 1832

Trudȳ Laboratorii Ozerovedeniya. Moskva.
    Trudȳ Lab.Ozerov. Tom.1-15, 20 → 1950 →   S. 1802 h

Trudȳ Laboratorii Sapropelevȳkh Otlozheniĭ. Akademiya
    Nauk SSSR. Moskva.
    Trudȳ Lab.sapropel.Otlozh. 7 → 1959 →   P.S 1500 A

TITLE	SERIAL No.

Trudȳ Laboratorii Vulkanologii. Akademiya Nauk. SSSR. Moskva.
    Trudȳ Lab.Vulk. Vyp.12-15; 1956-1958 (imp.)   M.S 1825

Trudy Latviĭskogo Nauchno-Issledovatel'skogo Instituta
    Zhivotnovodstva i Veterinarii. Riga.
    See Nauka-Zhivotnovodstvu. Riga.   Z.S 1826

Trudȳ Leningradskogo Geologo-Gidrogeodezicheskogo Tresta.
    Leningrad.
    Trudȳ leningr.geol.-gidrogeod.Tresta No.9. 1935.   P.S 1522

Trudȳ Leningradskogo Obshchestva Estestvoispȳtateleĭ. Leningrad.
    Trudȳ leningr.Obshch.Estest. 1914 →
    Formerly Trudȳ Imperatorskago Sankt-Peterburgskago
    Obshchestva Estestvoispȳtateleĭ. S.-Peterburg. (Leningrad.)   S. 1855 A

Trudȳ po Lesnomu Opȳtnomu Delu Zasek. Moskva.
    Trudȳ les.opȳt.Delu Zasek 1939 →   B.S 1382

Trudy Limnologicheskogo Instituta. Moskva & Leningrad.
    Trudȳ limnol.Inst. 1962 →
    Formerly Trudȳ Baĭkal'skoĭ Limnologicheskoĭ Stantsii.
    Leningrad.   S. 1802 N

Trudȳ Limnologicheskoĭ Stantsii v Kosine. Moskva.
    Trudȳ limnol.Sta.Kosine 1931-1939.
    Formerly Trudȳ Kosinskoĭ Biologicheskoĭ Stantsii.
    Moskva.   S. 1838 D

Trudȳ Litovskogo Nauchno-Issledovatel'skogo Geologorazvedochnogo
    Instituta. Vilnius.
    Trudȳ litov.nauchno-issled.geologorazved.Inst. 1971 →
    Formerly Trudȳ Instituta Geologii. Gosudarstvennyĭ
    Geologicheskiĭ Komitet SSSR. Vilnius.   P.S 575

Trudȳ Lomonosovskogo Instituta, Geokhimii, Kristallografii i
    Mineralogii. Moskva and Leningrad.
    Trudȳ lomonosov.Inst.Geokhim.Kristallogr.Miner. 1932-1938.
    Formerly Trudȳ Mineralogicheskogo Instituta.
    Akademiya Nauk S.S.S.R. Leningrad.
    Continued as Trudȳ Mineralogicheskogo Muzeya.
    Akademiya Nauk S.S.S.R.   M.S 1806

Trudȳ Mezhdunarodnoĭ Konferentsii Assotsiatsii po Izucheniyu
    Chetvertichnogo Perioda Evropȳ. Leningrad, Moskva.
    See Conference of the International Association on
    Quaternary Research.   P.S 997

TITLE	SERIAL No.

Trudȳ Mineralogicheskogo Instituta. Akademiya Nauk SSSR. Leningrad.
   Trudȳ miner.Inst. 1931.
   Formerly Trudȳ Mineralogicheskogo Muzeya. Akademiya
   Nauk SSSR. Leningrad.
   Continued as Trudȳ Lomonosovskogo Instituta, Geokhimii,
   Kristallografii i Mineralogii. Moskva.   M.S 1806 A

Trudȳ Mineralogicheskogo Muzeya (im.A.E. Fersmana). Akademiya
   Nauk SSSR. Leningrad.
   Trudȳ miner.Muz. 1926-1930, 1949 →
   (1931 contained in Trudȳ Mineralogicheskogo Instituta.
   1932-1938 contained in Trudȳ Lomonosovskogo Instituta
   Geokhimii, Kristallografii i Mineralogii.   M.S 1806

Trudȳ Mineralogicheskago Obshchestva. Sanktpeterburg.
   Trudȳ miner.Obshch. 1830-1831.   M.S 1818

Trudȳ Ministerstva Geologii i Okhranȳ Nedr SSSR. Glavgeologiya
   USSR. Moskva.
   Trudȳ Minist.geol.Okhr.Nedr SSSR. Vol.1-2. 1959   P.S 1564

Trudȳ Mleevskoĭ Sadovo-Ogorodnoĭ Opȳtnoi Stantsii. Mleev.
   Trudȳ mleev.sad.-ogorod.opȳt.Sta. 1928-1931 (imp.)   S. 1836 a

Trudȳ Molodykh Uchenȳkh. Vsesoyuznyi Nauchno-Issledovatel'skiĭ
   Institut Morskogo Rybnogo Khozyaĭstva i Okeanografii
   (VNIRO). Moskva.
   Trudȳ molod.Uchen. 1969-1970.   S. 1846 b B
   Continued in Trudȳ Vsesoyuzhogo-Nauchnogo-Issledovatel'skogo
   Instituta Morskogo Rybnogo Khozyaistva i Okeanografii.
   Moskva.   S. 1846 b

Trudȳ Mongol'skoĭ Komissii. Komitet Nauk Mongol'skoĭ
   Narodnoĭ Respubliki. Akademiya Nauk SSSR. Moskva.
   Trudȳ mongol'.Kom. Vuip 51, 53, 1953-1954.   Z.S Mammal Section
                        Vuip 59, 1954.   P. 73 H.q.S

Trudȳ Morskogo Nauchnogo Instituta. Moskva.
   Trudȳ morsk.nauch.Inst. 1928-1930.
   Formerly Trudȳ Plovuchego Morskogo Nauchnogo Instituta. Moskva.
   Continued as Trudȳ Gosudarstvennogo Okeanograficheskogo
   Instituta. Moskva.   S. 1845 A

Trudȳ. Moskovskiĭ (Ordena Trudovogo Krasnogo Znameni)
   Institut Neftekhimicheskoĭ i Gazovoĭ Promȳshlennosti
   imeni I.M. Gubkina.
   Trudȳ mosk.Inst.neftekhim.gaz.Prom. 25. 1959.
   Formerly Trudȳ. Moskovskiĭ (Ordena Trudovogo Krasnogo Znameni)
   Neftyanoĭ Institut imeni Akad. I.M. Gubkina. Leningrad.   P. 72 Q.o.M

| TITLE | SERIAL No. |

Trudȳ. Moskovskiĭ (Ordena Trudovogo Krasnogo Znameni) Neftyanoĭ
  Institut imeni Akad. I.M. Gubkina. Leningrad.
  <u>Trudȳ mosk.neft.Inst.</u> 19. 1957.
  <u>Continued as</u> Trudȳ. Moskovskiĭ (Ordena Trudovogo Krasnogo
  Znameni) Institut Neftekhimicheskoĭ i Gazovoĭ
  Promȳshlennosti imeni I.M. Gubkina.   P. 72 Q.o.M

Trudȳ Moskovskogo Geologo-Razvedochnogo Instituta imeni S.
  Ordzhonikidze. Moskva, Leningrad.
  <u>Trudȳ mosk.geol-razv.Inst.</u> Tom 3, 1936 → (imp.)   P.S 522

Trudȳ Moskovskogo Obshchestva Ispȳtateleĭ Prirodȳ. Moskva.
  <u>Trudȳ mosk.Obshch.Ispȳt.Prir.</u> 1961 →   S. 1838 C

Trudȳ Moskovskogo Zooparka.
  <u>Trudȳ mosk.Zoopk.</u> 1940-1946.   Z.S 1812

Trudȳ Murmanskogo Biologicheskogo Instituta. Murmanskiĭ Morskiĭ
  Biologicheskiĭ Institut. Akademiya Nauk SSSR.
  <u>Trudȳ murmansk.biol.Inst.</u> 1960 →
  <u>Formerly</u> Trudȳ Murmanskoĭ Biologicheskoĭ Stantsii.
  Kol'skiĭ Filial im. S.M. Kirova. Akademiya Nauk SSSR.   S. 1855 D

Trudȳ Murmanskoĭ Biologicheskoĭ Stantsii. Kol'skiĭ Filial im.
  S.M. Kirova. Akademiya Nauk SSSR.
  <u>Trudȳ murmansk.biol.Stn.</u> 1948-1958.
  <u>Formerly</u> Rabotȳ Murmanskoĭ Biologicheskoĭ Stantsii. Leningrad.
  <u>Continued as</u> Trudȳ Murmanskogo Biologicheskogo Instituta.
  Murmanskiĭ Morskoĭ Biologicheskiĭ Institut.
  Akademiya Nauk SSSR.   S. 1855 D

Trudȳ Nauchno-Issledovatel'skikh Institutov Promȳshlennosti. Moskva.
  <u>Trudȳ nauchno-issled.Inst.Prom.</u> Vyp.52-55, 1932.
  <u>Continued as</u> Trudȳ Instituta Prikladnoĭ Mineralogii. Moskva. M.S 1809

Trudȳ Nauchno-Issledovatel'skogo Instituta Biologii i Biologicheskogo
  Fakul'teta. Kharkovskiĭ Gosudarstvennȳĭ Universitet. Khar'kov.
  <u>Trudȳ nauchno-issled.Inst.Biol.khar'kov.gos.Univ.</u>
  Tom.22, 25, 33-34, 37. 1955-1956, 1962-1963.
  <u>Formerly</u> Pratsi Naukovo-Doslidnoho Zooloho-
  Biolohichnoho Instȳtutu. Kȳȳiv. Karkiv.   S. 1833 B

Trudȳ Nauchno-Issledovatel'skogo Instituta Eksperimental'nogo
  Morfogeneza. Moskva.
  <u>Trudȳ nauchno-issled.Inst.eksp.Morfogen.</u> Tom.6, 1938.
  <u>Formerly</u> Trudȳ Gosudarstvennogo Nauchno-Issledovatel'skogo
  Instituta Eksperimental'nogo Morfogeneza. Moskva.   Z.S 1828 C

Trudȳ Nauchno-Issledovatel'skogo Instituta Geografii. Moskva.
  <u>Trudȳ Nauchno-Issled.Inst.geogr.</u> No.2 1926.   73 C.o.S

TITLE   SERIAL No.

Trudȳ Nauchno-Issledovatel'skogo Instituta Geologii Arktiki. Leningrad.
    Trudȳ nauchno-issled.Inst.Geol.Arkt. Tom.67 → imp. 1958 →   P.S 1553

Trudȳ Nauchno-Issledovatel'skogo Instituta po Izucheniyu
    Severa. Moskva.
    Trudȳ nauchno-issled.Inst.Izuch.Sev. Nos. 25-30, 1925-1926.
    Formerly Trudȳ Severnoĭ Nauchno-Promyslovoĭ Ekspeditsii.
    VSNKh. Petrograd.
    Continued as Trudȳ Instituta po Izucheniyu Severa. Moskva.   S. 1842 A

Trudȳ Nauchno-Issledovatelskogo Instituta Rȳbnogo
    Khozyaĭstva y Promȳshlennosti-Varna.
    See Trudove na Nauchnoizsledovatelskiya Institut
    po Ribarstvo i Ribna Promishlenost.   Z.S 1884

Trudȳ Nauchno-Issledovatel'skogo Instituta pri Voronezhskom
    Gosudarstvennom Universitete. Voronezh.
    Trudȳ nauchno-issled.Inst.voronezh.gos.Univ. 1927-1930.   S. 1868 A

Trudȳ Nauchno-Issledovatel'skogo Instituta Zoologii. Moskva.
    Trudȳ nauchno-issled.Inst.Zool. 1925-1931.   Z.S 1828

Trudȳ Nauchno-Issledovatel'skogo Protivochumnogo Instituta
    Kavkaza i Zakavkaz'ya. Stavropol.
    Trudȳ nauchno-issled.protiv.Inst.Kavk.Zakavk. 1956-1961.   S. 1860 a

Trudȳ Naurzumskogo Gosudarstvennogo Zapovednika. Moskva.
    Trudȳ naurzum.gos.Zapov. No.1 1937.   S. 1805 a D

Trudȳ Neftyanogo Geologo-Razvedochnogo Instituta. Moskva.
    Leningrad, &c.
    Trudȳ neft.geol.razv.Inst.
    Serija A, Vyp 7, 1932-1939 (imp.)
    Serija B, Vyp 18, 1932-1936 (imp.)
    Continued as Trudȳ Vsesoyuznogo Neftyanogo Nauchno-Issledovatel'skogo
    Geologo-Razvedochnogo Instituta (VNIGRI)   P.S 525 & P.S 525 A

Trudȳ Neftyanoĭ Konferentsii. Akademiya Nauk USSR. Kiev.
    Trudȳ neft.Konf.Kiev 1938.   P. 72 Q.o.K

Trudȳ Nizhne-Volzhskogo Kraevogo Muzeya. Saratov.
    Trudȳ nizhne-volzh.kraev.Muz. 1929.   S. 1879

Trudȳ Obshchestva Estestvoispȳtateleĭ pri Imperatorskom
    Kazanskom Universitete. Kazan.
    Trudȳ Obshch.Estest.imp.kazan.Univ. 1871-1922.   S. 1832 A

Trudȳ Obshchestva Estestvoispȳtateleĭ pri Imperatorskom
    Varshavskom Universitete. Varshava.
    Trudȳ Obshch.Estest.imp.Varshav.Univ. 1891-1896.   S. 1870 D

TITLE	SERIAL No.

Trudȳ Obshchestva Estestvoispȳtateleĭ pri Imperatorskom
    Yur'evskom Universitete.
    <u>See</u> Schriften der Naturforscher-Gesellschaft bei der
    Universität Jurjeff (Dorpat, Tartu).   S. 1816 D

Trudȳ Obshchestva Ispȳtateleĭ Prirodȳ pri Imperatorskom
    Khar'kovskom Universitete. Kharkov.
    <u>Trudȳ Obshch.Ispyt.Prir.imp.Khar'kov</u>. 1869-1918.
    <u>Continued as</u> Trudȳ Khar'kovskogo Obshchestva Ispȳtateleĭ
    Prirodȳ pri Ukrhlavnauke. Khar'kov.   S. 1833 A

Trudȳ Obshchestva Izucheniya Man'chzhurskogo Kraya.
    <u>Trudȳ Obshch.Izuch.man'chzhur.Kraya</u> 1927.   S. 1986 A

Trudȳ Okeanograficheskoĭ Komissii. Akademiya Nauk SSSR. Moskva.
    <u>Trudȳ okeanogr.Kom</u>. 1957-1961.   S. 1802 i B

Trudȳ Olonetskoĭ Nauchnoĭ Ekspeditsii. Leningrad.
    <u>Trudȳ Olonets.nauch.Eksped</u>. Parts 1,3,5-6 & 8, 1921-1928.   S. 1804 A

Trudȳ Osoboĭ Zoologicheskoĭ Laboratorĭi i Sevastopol'skoĭ
    Biologicheskoĭ Stantsii. Petrograd.
    <u>Trudȳ osob.zool.Lab.Sevastop.biol.Sta</u>. 1915-1929.
    <u>Continued as</u> Trudȳ Sevastopol'skoĭ Biologicheskoĭ
    Stantsii.   Z.S 1315 A

Trudȳ. Otdel Fiziologii i Biofiziki Rasteniĭ, Akademiya Nauk
    Tadzhikskoĭ SSR. Dushanbe.
    <u>Trudȳ Otd.Fiziol.Biofiz.Rast.Akad.Nauk tadzhik.SSR</u> 1962 →   B.S 1353 b

Trudȳ Otdela Prikladnoĭ Entomologii. Petrograd.
    <u>Trudȳ Otd.prikl.Ent</u>. 1923-1925.
    <u>Formerly</u> Trudȳ Byuro po Entomologii.
    <u>Continued as</u> Trudȳ po Prikladnoĭ Entomologii.   E.S 1809

Trudȳ Otdeleniya Fizicheskikh Nauk Imperatorskago Obshchestva
    Lyubiteleĭ Estestvoznaniya, Antropologii i Etnografii, Moskva.
    <u>See</u> Izvestiya Imperatorskago Obshchestva Lyubiteleĭ
    Estestvoznaniya, Antropologii i Etnografii pri Imperatorskom
    Moskovskom Universitete. Moskva.   S. 1841 A

Trudȳ Paleontologicheskogo Instituta. Akademiya Nauk SSSR.
    Moskva, Leningrad.
    <u>Trudȳ paleont.Inst.</u> Tom 7 → 1937 →
    <u>Formerly</u> Trudȳ Paleozoologicheskogo Instituta. Akademiya
    Nauk SSSR.   P.S 509

Trudȳ Paleozoologicheskogo Instituta. Akademiya Nauk SSSR.
    Moskva, Leningrad.
    <u>Trudȳ paleozool.Inst</u>. 1932-1937.
    <u>Continued as</u> Trudȳ Paleontologicheskogo Instituta.
    Akademiya Nauk SSSR.   P.S 509

| TITLE | SERIAL No. |

Trudȳ Pamirskoĭ Biologicheskoĭ Stantsii Botanicheskogo Instituta.
  Akademiya Nauk Tadzhikskoi SSR. Dushanbe.
  Trudȳ pamir.biol.Sta. Vol.19 → 1962 →
  Formerly Trudȳ Botanicheskogo Instituta. Akademiya
  Nauk Tadzhikskoi SSR.   B.S 1353

Trudȳ Penzenskago Obshchestva Lyubiteleĭ Estestvoznaniya
  (i Kraevedeniya). Penza.
  Trudȳ penz.Obshch.Lyub.Estest.   1913-1928.   S. 1848 A

Trudȳ Pervogo Vsesoyuznogo S'ezda po Okhrane Prirodȳ v SSSR.
  Moskva.
  Trudȳ Perv.Vses.Okhr.prir.SSSR   1935.   72 Q.o.R

Trudȳ Petergofskogo Biologicheskogo Instituta. Petergof.
  Trudȳ petergof.biol.Inst.   1932-1939.
  (No.17, 1939 in Uchenȳe Zapiski Leningradskogo Gosudarstvennogo
  Universiteta. No.35, Ser.Biol. No.9 S. 1803 A.)
  Formerly Trudȳ Petergofskogo Estestvenno-Nauchnogo
  Instituta. Petergof.   S. 1865

Trudȳ Petergofskogo Estestvenno-Nauchnogo Instituta. Petergof.
  Trudȳ petergof.estest.-nauch.Inst.   1925-1932.
  Continued as Trudȳ Petergofskogo Biologicheskogo
  Instituta. Petergof.   S. 1865

Trudȳ Petrogradskogo Obshchestva Estestvoispȳtateleĭ. Petrograd.
  See Trudȳ Leningradskogo Obshchestva Estestvoispȳtateleĭ.
  Leningrad.   S. 1855 A

Trudȳ Petrograficheskogo Instituta. Akademiya Nauk S.S.S.R.
  Leningrad.
  Trudȳ petrogr.Inst.   1931-1939.   M.S 1804

Trudȳ Plovuchego Morskogo Nauchnogo Instituta. Moskva.
  Trudȳ plov.morsk.nauch.Inst.   1923-1927.
  Continued as Trudȳ Morskogo Nauchnogo Instituta. Moskva.   S. 1845 A

Trudȳ Poltavskoĭ Sel'khozyaĭstvennoĭ Opȳtnoĭ Stantsii. Poltava.
  Trudȳ poltav.sel'.-khoz.opȳt.Sta.   1911-1930.   E.S 1808

Trudȳ. Polyarnyĭ Nauchno-Issledovatel'skiĭ i Proektnyĭ Institut
  Morskogo Rybnogo Khozyaĭstva i Okeanografii im N.M.
  Knipovicha (PINRO). Murmansk.
  Trudȳ polyar.nauchno-issled.proekt.Inst.morsk.rȳb.Khoz.Okeanogr.
  Vyp.21 → 1967 →   Z.S 1816

Trudȳ Prĕsnovodnoĭ Biologicheskoĭ Stantsii Imp. S.-Peterburgskago
  Obshchestva Estestvoispȳtateleĭ. S.-Peterburg.
  Trudȳ presnov.biol.Sta.S-peterb.Obshch.Estest.
  Tom 8, no.2, & Tom 9, no. 1. 1936.   S. 1855 E

TITLE   SERIAL No.

Trudȳ Pribaltiĭskoĭ Ornitologicheskoĭ Konferentsii.
    Trudȳ Pribalt.Orn.Konf. 3rd. - 5th.
    1954-1963 (1957-1967).
    Formerly Sbornik Dokladov Ornitologicheskoĭ Konferentsii.
    Continued as Materialȳ Pribaltiĭskoĭ Ornitologicheskoĭ
    Konferentsii.   Z. 72Q q C

Trudȳ po Prikladnoĭ Botanike, Genetike i Selektsii (Prikladnoi
    Botanike i Selektsii) Leningrad.
    Trudȳ prikl.Bot.Genet.Selek. 1918-1937: 1957 → (imp.)
    Formerly Trudȳ Byuro po Prikladnoĭ Botanike. S.-Peterburg.   B.S 1400

Trudȳ po Prikladnoĭ Entomologii. Leningrad.
    Trudȳ prikl.Ent. 1926-1930.
    Formerly Trudȳ Otdela Prikladnoĭ Entomologii.
    Continued as Trudȳ po Zashchite Rasteniĭ. Leningrad. Ser.Ent. E.S 1809

Trudȳ Problemnykh i Tematicheskikh Soveshchaniĭ. Zoologicheskiĭ
    Institut, Akademiya Nauk SSSR.
    Trudȳ probl.temat.Soveshch.zool.Inst. Vyp.4-7. 1954-1957.   Z. 72Q q S

Trudȳ Prȳrodnȳcho-Tekhnichnoho Viddȳlu. Vseukrayins'ka
    Akademiya Nauk. Kȳyiv.
    Trudȳ prȳr.-tekh.Vidd.Kȳyiv 1930-1932.   S. 1834 a C

Trudȳ Respublikanskoĭ Stantsii Zashchitȳ Rasteniĭ. Vsesoyuznaya
    Akademiya Sel'skokhozyaĭstvennȳkh nauk im. V.I. Lenina,
    Kazakhskiĭ Filial. Alma-Ata.
    Trudȳ respub.Sta.Zashch.Rast.Alma-Ata Tom.3. 1956.   B.M.S 53

Trudȳ Russkago Entomologicheskago Obshchestva. S. Peterburg (Leningrad).
    Trudȳ russk.ent.Obshch. 1861-1932.
    (Suspended 1917-1923).
    Continued as Trudȳ Vsesoyuznogo Entomologicheskogo
    Obshchestva.   E.S 1802

Trudȳ Russkikh Estestvoispȳtateleĭ v S.-Peterburge.
    Trudȳ Russk.Estestv. 1867-73, 1883-90.
      (Wanting Sess.5-6, 1876 & 1879.)   S. 1801

Trudȳ Saratovskago Obshchestva Estestvoispȳtateleĭ
    i Lyubiteleĭ Estestvoznaniya.
    1902-1906 - See Otchet Volzhskoĭ Biologicheskoĭ
                   Stantsii
    1906-1926 - See Rabotȳ Volzhskoĭ Bīologicheskoĭ
                   Stantsii   Z.S 1840

Trudȳ Saratovskogo Otdeleniya.
    (1951-54)   Kaspiĭskogo Filiala VNIRO.
    (1956-58)   VNIORKh.
    (1960 →)   GosNIORKh.
    Trudy saratov.Otdel.kasp.Fil.VNIRO Tom 1-3, 1951-54
                    VNIORKh.   Tom 4-5, 1956-58
                    GosNIORKh.   Tom 6 →   1960 →   Z.S 1840

| TITLE | SERIAL No. |

Trudȳ Sed'moĭ Sessii Komissii po Opredeleniyu Absolyutnogo Vozrasta
  Geologicheskikh Formatsiĭ. Moskva.
  Trudȳ sed'moĭ Sessii Kom.Opred.absol.Vozr.geol.Form.
    4, 5 & 7. 1957-1960.   P. 72 Q.o.L

Trudȳ Sektora Agrobotaniki. Akademiya Nauk Kazakhskoĭ SSR.
  Alma-Ata.
  Trudȳ Sekt.Agrobot.Alma-Ata Vol.5-8, 1957-1960.   B.S 1346 b

Trudȳ Sektora Paleobiologii. Akademiya Nauk Gruzinskoĭ SSR.
  See Trudȳ Instituta Paleobiologii. Akademiya Nauk
  Gruzinskoĭ SSR.   P.S 558

Trudȳ Sektsii po Mikologii i Fitopatologii Russkogo Botanicheskogo
  Obshchestva. Petrograd.
  Trudȳ Sekts.Mikol.Fitopat.russk.bot.Obshch. 1923.   B.M.S 63

Trudȳ Sessii Vsesoyuznogo Paleontologicheskogo Obshchestva. Moskva.
  Trudȳ Sess.vses.paleont.Obshch. 1957 →   P.S 541

Trudȳ Sevanskoĭ Gidrobiologicheskoĭ Stantsii. Erevan.
  Trudȳ sevan.gidrobiol.Sta. 1938 →
  Formerly Trudȳ Sevanskoĭ Ozernoĭ Stantsii.   Z.S 1905

Trudȳ Sevanskoĭ Ozernoĭ Stantsii. Erevan.
  Trudȳ sevan.ozer.Sta. 1927-1932.
  Continued as Trudȳ Sevanskoĭ Gidrobiologicheskoĭ Stantsii.   Z.S 1905

Trudȳ Sevastopol'skoĭ Biologicheskoĭ Stantsii. Leningrad.
  Trudȳ Sevastopol.biol.Sta. Tom.1-5, 8-17, 1929-1964.
  Formerly Trudȳ Osoboĭ Zoologicheskoĭ Laboratorii
  i Sevastopol'skoĭ Biologicheskoĭ Stantsii.   Z.S 1915 A

Trudy Severnoĭ Nauchno-Promyslovoĭ Ekspeditsii. VSNKh. Petrograd.
  Trudy sev.nauchno-prom.Eksped. Nos.3-5, 9-13, 16-24.
    1920-1924.   S. 1842 A
    1923-1925 (imp.)   M.S 1817 B
  Continued as Trudȳ Nauchno-Issledovatel'skogo Instituta
  po Izucheniyu Severa. Moskva.

Trudȳ Severo-Kavkazskogo Instituta Zashchitȳ Rasteniĭ. Rostov.
  Trudȳ sev.-kavk.Inst.Zashch.Rast. 1932.   E.S 1806 a

Trudȳ Severo-Kavkazskoĭ Assotsiatsii Nauchno-Issledovatel'skikh
  Institutov. Rostov-na-Donu.
  Trudȳ sev.-kavk.Ass.nauchno-issled.Inst. 1926-1930. (imp.)   S. 1877

Trudȳ Sibirskogo Nauchno-Issledovatel'skogo Instituta Geologii,
  Geofiziki i Mineral'nogo Sȳr'ya (SNIIGGIMS). Leningrad,
  Novosibirsk.
  Trudȳ sib.nauchno-issled.Inst.Geol.Geofiz.miner.Syr'ya
    2 → 1959 → (imp.)   P.S 1516 a

| TITLE | SERIAL No. |

Trudȳ Sikhote-Alinskogo Gosudarstvennogo Zapovednika. Moskva.
    Trudȳ sikhote-alin.gos.Zapov. Vols.1-2. 1938.   S. 1805 a C

Trudy SNIIGGIMS.
    See Trudy Sibirskogo Nauchno-Issledovatel'skogo Instituta
    Geologii, Geofiziki i Mineral'nogo Sȳr'ya (SNIIGGIMS).
    Leningrad.   P.S 1516 a

Trudȳ Soveshchaniĭ. Ikhtiologicheskaya Komissiya, Akademiya
    Nauk SSSR. Moskva.
    Trudȳ Soveshch.ikhtiol.Kom. Vȳp.5-13,1955-1961 (imp.)   Z. 22 q S

Trudȳ Soveta po Izucheniyu Prirodnȳkh Resursov. Seriya
    Yakutskaya. Leningrad.
    Trudȳ Sov.Izuch.Prir.Res.Ser.Yakutsk. 1935.
    (Wanting Nos. 20-22.)
    Formerly Trudȳ Soveta po Izucheniyu Proizvoditel'nȳkh
    Sil. Seriya Yakutskaya. Leningrad.   S. 1802 Pc

Trudȳ Soveta po Izucheniyu Proizvoditel'nȳkh Sil. Seriya
    Yakutskaya. Leningrad.
    Trudȳ Sov.Izuch.proizv.Sil.Ser.yakutsk. 1931-1934.
    Continued as Trudȳ Soveta po Izucheniyu Prirodnȳkh
    Resursov. Seriya Yakutskaya. Leningrad.   S. 1802 Pc

Trudȳ Sovetskoĭ Sektsii Mezhdunarodnoĭ Assotsiatsii po Izucheniyu
    Chetvertichnogo Perioda (INQUA). Leningrad, Moskva.
    Trudȳ sov.Sekts.mezhd.Ass.Izuch.chetv.Perioda 1937-1939.   P.S 997

Trudȳ. Sovmestnaya Sovetsko-Mongol'skaya Nauchno-Issledovatel'skaya
    Geologicheskaya Ekspeditsiya. Moskva.
    Trudȳ sovmest.sov.-mongol'.nauchno-issled.geol.Eksped.
    Vȳp.2 → 1970 →   P.S 1565

Trudȳ Sredne-Aziatskogo Gosudarstvennogo Universiteta. Tashkent.
    See Acta Universitatis Asiae Mediae. Tashkent.   S. 1859 B

Trudy Sredne-Aziatskogo Industrial'nogo Instituta, Gornȳĭ
    Fakul'tet. Tashkent.
    Trudȳ sred.-aziat.ind.Inst.gorn.Fak. Vyp.2. (10) 1938.
    Formerly Materialȳ po Geologii Sredneĭ Asiĭ.   P.S 552

Trudȳ Sredne-Aziatskogo Issledovatel'skogo Instituta
    Zashchitȳ Rasteniĭ. Tashkent.
    Trudȳ sred.-aziat.Inst.Zashch.Rast. 1931-1937 (imp.).
    Formerly Trudȳ Uzbekistanskoĭ Opȳtnoĭ Stantsii Zashchitȳ
    Rasteniĭ. Tashkent.   E.S 1811 a

Trudȳ Sredne-Aziatskogo Nauchno-Issledovatel'skogo
    Protivochumnogo Instituta. Alma-Ata.
    Trudȳ sred.-aziat.nauchno-issled.protiv.Inst. 1951-1959.   E.S 1811

Trudȳ Stavropol'skago Sel'skokhozyaĭstvennago Instituta. Stavropol.
    Trudȳ stavropol.sel'.-khoz.Inst. 1921-1922.   Z.S 1825

TITLE	SERIAL No.

Trudȳ Sukhumskogo Botanicheskogo Sada. Sukhumi.
    Trudȳ sukhum.bot.Sada Bd 8 → 1955 →   B.S 1430 e

Trudȳ Sungariiskoĭ Rechnoĭ Biologicheskoĭ Stantsiĭ. Kharbin.
    See Otdel'noe Izdanie. Obshchestvo Izucheniya
    Man'chzhurskogo Kraya. Ser.B.   S. 1986 B

Trudȳ Sverdlovskogo Gornogo Instituta imeni V.V.
    Vakhrusheva. Moskva.
    Trudȳ sverdlovsk.gorn.Inst. 26. 1956.   P. 72 Q.q.S

Trudȳ Tadzhikskoĭ Bazȳ, Akademiya Nauk SSSR. Moskva, Leningrad.
    Trudȳ tadzhik.Bazȳ Akad.Nauk SSSR Tom.7, 1938.   Z. Fish Section

Trudȳ Tallinskogo Botanicheskogo Sada. Tallin.
    See Tallinn Botaanikaaia Uurimused.   B.S 1423

Trudȳ Tashkentskogo Gosudarstvennogo Universiteta im. V.I. Lenina.
    See Nauchnye Trudȳ. Tashkentskii Gosudarstvennȳi Universitet
    im. V.I. Lenina. Tashkent.   S. 1859 B

Trudȳ Tbilisskogo Botanicheskogo Instituta. Tbilisi.
    Trudȳ tbiliss.bot.Inst. 1934 → (Wanting Vol.9 & 14.)
    Formerly Trudȳ Tiflisskago Botanicheskago Sada. Tiflis.   B.S 1430

Trudȳ Tiflisskago Botanicheskago Sada. Tiflis.
    Trudȳ tiflis.bot.Sada 1895-1899, 1901, 1914-1917, 1920-1930.
    Continued as Trudȳ Tbilisskogo Botanicheskogo Instituta.
    Tbilisi.   B.S 1430

Trudȳ Tikhookeanskogo Komiteta. Akademiya Nauk SSSR. Leningrad.
    Trudȳ tikhookèan.Kom. Nos.1,2, & 4. 1930-1937.   S. 1802 Z

Trudȳ Tomskogo Gosudarstvennogo Universiteta, Tomsk.
    Trudȳ tomsk.gos.Univ. Tom.85-87, 1932-1935.
    Formerly Izvestiya Tomskogo(Gosudarstvennogo) Universiteta.
    Continued as Trudȳ Biologicheskogo Nauchno-Issledovatel'skogo
    Instituta, Tomskogo Gosudarstvennogo Universitata.   S. 1967 A

Trudȳ Tret'eĭ Zakavkazskoĭ Konferentsii Molodȳkh Nauchnȳkh
    Rabotnikov Geologicheskikh Institutov. Akademii Nauk
    Azerbaĭdzhanskoi, Armyanskoĭ i Gruzinskoĭ SSR.
    Trudȳ tret'ei zakavkaz.Konf.molod.nauch.Rabot.geol.Inst.
        3. 1960. (1962).   P. 72 Q.o.T

Trudȳ Tsentral'nogo Lesnogo Gosudarstvennogo Zapovednika.
    Smolensk.
    Trudȳ tsent.les.gos.Zapov. No.2. 1937.   S. 1805 a G

| TITLE | SERIAL No. |

Trudȳ Tsentral'nogo Nauchno-Issledovatel'skogo Geologo-Razvedochnogo
    Instituta (TSNIGRI). Leningrad.
    <u>Trudȳ tsent.nauchno-issled.geologo-razv.Inst.</u>
    No.2 - 130 (imp.) 1934 - 1939
    <u>Continued as</u> Trudȳ Vsesoyuznogo Nauchno-Issledovatel'skogo
    Geologicheskogo Instituta (VSEGEI). Leningrad.   P.S 1528

Trudȳ Tsentral'nogo Nauchno-Issledovatel'skogo Instituta
    Rȳbovodstva i Rȳbolovstva. Varna.
    <u>See</u> Trudove na Tsentralniya Nauchnoizsledovatelski
    Institut po Ribovŭdstvo i Ribolov.   Z.S 1884

Trudȳ Tsentral'nogo Sibirskogo Botanicheskogo Sada. Akademiya
    Nauk SSSR. Sibirskoe Otdelenie. Novosibirsk.
    <u>Trudȳ tsent.sib.bot.Sada</u> No.2-10, 1964-1965.   B.S 1415

Trudȳ TSNIGRI.
    <u>See</u> Trudȳ Tsentral'nogo Nauchno-Issledovatel'skogo
    Geologo-Razvedochnogo Instituta (TSNIGRI). Leningrad.   P.S 1528

Trudȳ Turkestanskogo Nauchnogo Obshchestva. Tashkent.
    <u>Trudȳ turkest.nauch.Obshch.</u> 1923-1925.   S. 1859 b

Trudȳ Turkmenskogo Botanicheskogo Sada. Ashkhabad.
    <u>Trudȳ turkmen.bot.Sada</u> 2 → 1956 →   B.S 1348

Trudȳ Turkmenskogo Sel'skokhozyaĭstvennogo Instituta. Ashkhabad.
    <u>Trudȳ turkmen.sel'khoz.Inst.</u> 1935.   S. 1859 a

Trudȳ. Ukraĭns'ka Akademiya Nauk, Fizicho-Matematichnoho Viddilu.
    <u>See</u> Zbirnȳk Prats Zoolohichnoho Museyu. Kȳyiv.   Z.S 1827

Trudȳ Ukrainskogo Geologo-Gidro-Geodezicheskogo Tresta i
    Ukrnigri. Moskva.
    <u>Trudȳ Ukrain.geol.-gidro-geodez.Tresta</u> 1934.   P.S 1538

Trudȳ Ukrainskogo Nauchno-Issledovatel'skogo Geologorazvedochnogo
    Institute. (UkrNIGRI.) Moskva.
    <u>Trudȳ ukr.nauchno-issled.geologorazv.Inst.</u> Vyp.1-2, 1959.   P.S 1564

Trudȳ Ukrayins'koho Instȳtutu Prȳkladnoyi Botanikȳ. Khar'kiv.
    <u>Trudȳ ukr.Inst.prȳkl.Bot.</u> 1930.   B.S 1360

Trudȳ Upravleniya Geologii Soveta Ministrov Tadzhikskoĭ SSR.
    Moskva.
    <u>Trudȳ Upravl.Geol.Sov.Minist.tadzhik.SSR</u>
    Paleontologiya i Stratigrafiya. vȳp.2 → 1966 →   P.S 1568

Trudȳ Uzbekistanskogo Instituta Tropicheskoĭ Meditsinȳ
    im.Faĭzullȳ Khodzhaeva.
    <u>Trudȳ uzbekist.Inst.Trop.Med.</u> Tom 1. No.1. 1930.   73 A.o.K

TITLE	SERIAL No.

Trudȳ Uzbekistanskoĭ Opȳtnoĭ Stantsii Zashchitȳ Rasteniĭ. Tashkent.
    Trudȳ uzbekist.opȳt.Sta.Zashch.Rast. 1930.
    Continued as Trudȳ Sredne-Aziatskogo Issledovatel'skogo
    Instituta Zashchitȳ Rasteniĭ.   E.S 1811 a

Trudȳ Varshavskago Obshchestva Estestvoispȳtateleĭ.Varshava.
    Trudȳ varsh.Obshch.Estest. Protokol.Obshch.Sobr.
    Nos. 3-12, 15, 21-25, 1891-1914:
    Otdel Biol. 1889-1908 (1910);
    Otdel Fiz.i.Khim. 1889-1904 (1906).   S. 1870 A - C

Trudȳ VIGIS
    See Trudȳ Vsesoyuznogo Instituta Gel'mintologii imeni
    Akademika K.I. Skryabina. Moskva.   Z.S 1847 A

Trudȳ VNIGNI.
    See Trudȳ. Vsesoyuznȳi Nauchno-Issledovatel'skiĭ
    Geologorazvedochnȳĭ Neftyanoĭ Institut (VNIGNI). Leningrad.   P.S 1562

Trudȳ VNIGRI
    See Trudȳ Vsesoyuznogo Neftyanogo Nauchno-Issledovatel'skogo
    Geologo-Razvedochnogo Instituta (VNIGRI)   P.S 525

Trudȳ VNII.
    See Trudȳ.Vsesoyuznȳĭ Neftegazovȳĭ Nauchno-Issledovatel'skiĭ
    Institut (VNII). Leningrad.   P. 72Q o G

Trudȳ VNIIGAZ.
    See Trudȳ. Vsesoyuznȳi Nauchno-Issledovatel'skiĭ Institut
    Prirodnȳkh Gazov (VNIIGAZ.) Moskva.   P. 72Q q M

Trudȳ VNIRO.
    See Trudȳ Vsesoyuznogo-Nauchnogo-Issledovatel'skogo
    Instituta Morskogo Rȳbnogo Khozyaĭstva i Okeanografii.
    Moskva.   S. 1846 b

Trudȳ Voronezhskogo Gosudarstvennogo Universiteta. Voronezh.
    See Acta Universitatis Voronegiensis. Voronezh.   S. 1868 B

Trudȳ Voronezhskogo Gosudarstvennogo Zapovednika. Moskva.
    Trudȳ voronezh.gos.Zapov. No.1. 1938.   S. 1805 a E

Trudȳ Voronezhskogo Otdeleniya Vsesoyuznogo Nauchno-
    Issledovatel'skogo Instituta Prudovogo i Rȳbnogo
    Khozyaĭstva.
    Trudȳ voronezh.Otd.Vses.nauchno-issled.
    Inst.prud.rȳb.Khoz. 1935-1936.   Z.S 1842

| TITLE | SERIAL No. |

Trudy Vostochno-Sibirskogo Filiala. Akademiya Nauk SSSR. Moskva.
    Seriya Geologicheskaya.
    <u>Trudy vost.-sib.Fil.Akad.Nauk SSSR</u> No.2. 1955   P.S 1531 B

Trudy Vostochno-Sibirskogo Geologo-Razvedochnogo Tresta. Irkutsk.
    <u>Trudy vost.-sib.geol.razv.Tresta</u> Nos.1-5, 8-9. 1932-1934   P.S 1531

Trudy Vostochno-Sibirskogo Gosudarstvennogo Universiteta. Moskva.
    <u>Trudy vost.-sib.gos.Univ.</u> 1932-1942 (imp.)   S. 1831 C

Trudy VSEGEI.
    <u>See</u> Trudy Vsesoyuznogo Nauchno-Issledovatel'skogo
    Geologicheskogo Instituta (VSEGEI). Leningrad.   P.S 1528

Trudy. Vserossiiskii Nauchno-Issledovatel'skii Institut
    Prudovogo Rybnogo Khozyaistva. Moskva.
    <u>Trudy vseross.nauchno-issled.Inst.prud.ryb.Khoz.</u>
    Tom.10-13, 1961-1965.
    <u>Continued as</u> Trudy.Vsesoyuznyi Nauchno-Issledovatel'skii
    Institut Prudovogo Rybnogo Khozyaistva. Moskva.   Z.S 1834

Trudy Vserossiiskogo Entomo-Fitopatologicheskogo S'ezda. Leningrad.
    <u>Trudy vseross.ent.-fitopatol.S'ezda</u> (No.) 4, 1922.   E.S 1819 a

Trudy Vserossiiskogo S'ezda Zoologov, Anatomov i Gistologov.
    Leningrad.
    <u>Trudy vseross.S'ezda Zool.Anat.Gistol.</u> 1923-1928.   Z.S 1845

Trudy. Vsesoyuznoe Paleontologicheskoe Obshchestvo. Moskva.
    <u>Trudy vses.paleont.Obshch.</u> 1957 →   P.S 541

Trudy Vsesoyuznogo Aerogeologicheskogo Tresta. Moskva.
    <u>Trudy vses.aerogeol.Tresta</u> Vol.3. 1957   P.S 1552

Trudy Vsesoyuznogo Arkticheskogo Instituta. Leningrad.
    <u>Trudy vses.arkt.Inst.</u> 1937 (imp.)
    <u>Formerly</u> Trudy Arkticheskogo Nauchno-Issledovatel'skogo
    Instituta. Leningrad.   S. 1842 A

Trudy Vsesoyuznogo Entomologicheskogo Obshchestva. Akademiya
    Nauk SSSR. Moskva.
    <u>Trudy vses.ent.Obshch.</u> 1951 →
    <u>Formerly</u> Trudy Russkago Entomologicheskago Obshchestva.   E.S 1802

Trudy Vsesoyuznogo Geologo-Razvedochnogo Ob'edineniya
    NKTP SSSR. Leningrad.
    <u>Trudy vses.geol.-razv.Ob"ed.NKTP.</u> 1932-1934. (imp.)
    <u>Formerly</u> Trudy Glavnogo Geologo-Razvedochnogo Upravleniya
    V.S.N.Kh. SSSR. Moskva.   P.S 1525

TITLE	SERIAL No.

Trudȳ Vsesoyuznogo Gidrobiologicheskogo Obshchestva. Moskva.
    Trudȳ vses.gidrobiol.Obshch. Tom 5 → 1953 →   S. 1802 c

Trudȳ Vsesoyuznogo Instituta Gel'mintologii imeni Akademika
    K.I. Skryabina. Moskva.
    Trudȳ vses.Inst.Gel'mint. Tom 6 → 1959 →   Z.S 1847 A

Trudȳ Vsesoyuznogo Instituta Zashchitȳ Rasteniĭ. Stavropol
    and Leningrad.
    Trudȳ vses.Inst.Zashch.Rast. Vȳp.8 - 16. 1957-1961.
    Continued as Trudȳ Vsesoyuznogo Nauchno-Issledovatel'skogo
    Instituta Zashchitȳ Rasteniĭ. Leningrad.   E.S 1800

Trudȳ Vsesoyuznogo-Nauchnogo-Issledovatel'skogo Instituta Morskogo
    Rȳbnogo Khozyaĭstva i Okeanografii. Moskva.
    Trudȳ vses.nauchno.-issled.Inst.morsk.rȳb.Khoz.Okeanogr.
    1935 → (imp.)   S. 1846 b
    Incorporates Trudȳ Molodykh Uchenykh. Moskva.
        Tom 4 → 1971 →
    and Trudȳ Kaspiiskii Nauchno-Issledovatel'skii Institut
        Rybnogo Khozyaistva. Moskva.
        Tom 27 → 1972 →   S. 1846 b
    Tom 35, 1958 - selected articles (transl.) only   Z. 66 o M

Trudȳ Vsesoyuznogo Nauchno-Issledovatel'skogo Geologicheskogo
    Instituta (VSEGEI).Leningrad.
    Trudȳ vses.nauchno-issled.geol.Inst.
    No.132, 134-135. 1941. N.S. 1954 →
    (From 1955-56 styled Materialȳ VSEGEI.)
    Formerly Trudȳ Tsentral'nogo Nauchno-Issledovatel'skogo
    Geologo-Razvedochnogo Instituta (Ts NIGRI). Leningrad.   P.S 1528

Trudȳ Vsesoyuznogo Nauchno-Issledovatel'skogo Instituta
    Mineral'nogo Sȳr'ya. Moskva.
    Trudȳ vses.nauchno-issled.Inst.miner.Sȳr'ya
    Vyp.68, 70-111. 1935-1936 (imp.)
    Formerly Trudȳ Instituta Prikladnoĭ Mineralogii. Moskva.   M.S 1809

Trudȳ Vsesoyuznogo Nauchno-Issledovatel'skogo Instituta Zashchitȳ
    Rasteniĭ. Leningrad.
    Trudȳ vses.nauchno-issled.Inst.Zashch.Rast. Vȳp.17 → 1963 →
    Formerly Trudȳ Vsesoyuznogo Instituta Zashchitȳ Rasteniĭ.   E.S 1800

Trudȳ Vsesoyuznogo Neftyanogo Nauchno-Issledovatel'skogo
    Geologo-Razvedochnogo Instituta (VNIGRI). Leningrad, Moskva.
    Trudȳ vses.neft.nauchno-issled.geol.-razv.Inst.
    Novaya Seriya 1950 → (imp.)
    -------------- Mikrofauna SSSR. Sbornik 4 → 1950 →
    -------------- Geologicheskiĭ Sbornik 2 → 1956 →
    -------------- Ocherki po Geologii 1956 →
    -------------- Paleontologicheskiĭ Sbornik 2 → 1960 →
    (The designation "Novaya Seriya" is discontinued from Vyp.101)
    Formerly Trudȳ Neftyanogo Geologo-Razvedochnogo Instituta.
    Moskva, Leningrad, &c.   P.S 525

| TITLE | SERIAL No. |

Trudy̆ Vsesoyuznogo Ornitologicheskogo Zapovednika Gassan-Kuli. Moskva.
    Trudy̆ vses.orn.Zap.Gassan-Kuli No.1, 1940.   Z.S 1841

Trudy̆ III Vsesoyuznogo S'ezda Geologov. Tashkent.
    Trudy̆ III Vses.S'ezda Geol. 1928 (1929)   P.S 1535

Trudy̆ Vsesoyuznogo Soveshchaniya po Razrabotke Unifitsirovannoĭ
    Skhemy̆ Stratigrafii Mezozoĭskikh Otlozheniĭ Russkoĭ Platformy̆.
    Trudy̆ vses.Sovesh.Razr.Unifit.Skhemy Stratigr.mezoz.
    Otlozh.Russk.Platf. 1954. (1956).   P. 72 Q.q.C

Trudy̆. Vsesoyuzny̆i Arkticheskiĭ Institut. Leningrad.
    See Trudy̆ Arkticheskogo Instituta. Leningrad.   S. 1842 A

Trudy̆. Vsesoyuzny̆i Nauchno-Issledovatel'skii Geologorazvedochny̆ĭ
    Neftyanoĭ Institut (VNIGNI). Leningrad.
    Trudy̆ vses.nauchno-issled.geol.Neft.Inst.
    No.8 → 1957 → (imp.)   P.S 1562

Trudy̆. Vsesoyuzny̆ĭ Nauchno-Issledovatel'skiĭ Institut Prirodny̆kh
    Gazov (VNIIGAZ).Moskva.
    Trudy̆ vses.nauchno-issled.Inst.prir.Gaz. 1959.   P. 72 Q.q.M

Trudy̆. Vsesoyuznyĭ Nauchno-Issledovatel'skiĭ Institut
    Prudovogo Ry̆bnogo Khozyaĭstva. Moskva.
    Trudy̆ vses.nauchno-issled.Inst.prud.ry̆b.Khoz. Tom.14 → 1966 →
    Formerly Trudy̆. Vserossiiskii Nauchno-Issledoval'skii
    Institut Prudovogo Ry̆bnogo Khozyaĭstva. Moskva.   Z.S 1834

Trudy̆.Vsesoyuzny̆ĭ Neftegazovy̆ĭ Nauchno-Issledovatel'skiĭ
    Institut (VNII). Leningrad.
    Trudy̆ vses.neftegaz.nauchno-issled.Inst. 9. 1956.   P. 72 Q.o.G

Trudy̆ Vtorogo Moskovskogo Universiteta. Moskva.
    Trudy̆ vtor.mosk.Univ. 1927-1928.   S. 1840

Trudy̆ Yakutskogo Filiala. Akademiya Nauk SSSR. Seriya
    Geologicheckaya.Sbornik. Moskva.
    Trudy̆ yakutsk.Fil.Akad.Nauk SSSR 6. 1961.   P. 73 A.o.L

Trudy̆ Zapovednika. Ural'skiĭ Filial. Akademiya Nauk SSSR. Sverdlovsk.
    Trudy̆ Zapov. No.8, 1961.
    Formerly Trudy̆ Il'menskogo Gosudarstvennogo Zapovednika
    (im V.I. Lenina). Moskva.   S. 1805 a F

Trudy̆ po Zashchite Rasteniĭ. Leningrad.
    Trudy̆ Zashch.Rast. 1930-1936. (4 Series.)
    (Ser.Ent. Formerly Trudy̆ Prikladnoĭ Entomologii.)   E.S 1816

Trudy̆ po Zashchite Rasteniĭ Sibiri. Novosibirsk.
    Trudy̆ Zashch.Rast.Sib. 1931.   E.S 1810

Trudy̆ po Zashchite Rasteniĭ Vostochnoĭ Sibiri. Irkutsk.
    Trudy̆ Zashch.Rast.vost.Sib. 1933-1935   E.S 1810

| TITLE | SERIAL No. |

Trudȳ Zoologicheskogo Instituta. Akademiya Nauk
   Gruzinskoĭ SSR. Tbilisi.
   <u>Trudȳ zool.Inst.Tbilisi</u> Tom 4-5, 1941-1943.
   <u>Formerly</u> Trudȳ Zoologicheskogo Sektora. Akademiya
   NaukSSSR, Gruzinskoe Otdelenie.
   <u>Continued as</u> Trudȳ Instituta Zoologii. Akademiya Nauk
   Gruzinskoĭ SSR.   Z.S 1837

Trudȳ Zoologicheskogo Instituta. Akademiya Nauk SSSR. Leningrad.
   <u>Trudȳ zool.Inst.Leningr.</u> 1932 →   Z.S 1820
                              1932-1940.   T.R.S 2004
   <u>Formerly</u> Ezhegodnik Zoologicheskago Muzeya
   Imperatorskoĭ Akademii Nauk. S.-Peterburg.

Trudȳ Zoologicheskogo Instituta. Azerbaĭdzhanskiĭ Filial,
   Akademiya Nauk SSSR. Baku.
   <u>Trudȳ zool.Inst.Baku</u> Tom 8-9, 1938.
   <u>Continued as</u> Trudȳ Instituta Zoologii. Akademiya Nauk
   Azerbaĭdzhanskoĭ SSR. Baku.   Z.S 1853

Trudȳ Zoologicheskogo Sektora. Akademiya Nauk SSSR. Gruzinskoe
   Otdelenie, Zakavkazskiĭ Filial. Tiflis.
   <u>Trudȳ zool.Sekt.Tbilisi</u> 1934-1941.
   <u>Continued as</u> Trudȳ Zoologicheskogo Instituta. Akademiya
   Nauk Gruzinskoĭ SSR.   Z.S 1837

Trudȳ Zoolohichnoho Muzeyu. Kȳyiv.
   <u>Trudȳ zool.Muz.Kȳyiv</u> Vol.1 (wanting pp.150-168),
     1939 (1941.)   E.S 1787

Trudȳ Zoopsikhologicheskoĭ Laboratorii. Gosudarstvennyĭ
   Darvinovskiĭ Muzeĭ. Moskva.
   <u>Trudȳ zoopsikhol.Lab.gos.darvin.Muz.</u> 1928, 1935.   S. 1805

Tschermaks Mineralogische und Petrographische Mitteilungen. Wien.
   <u>Tschermaks miner.petrogr.Mitt.</u> 1889-1927, 1948 →
   <u>Formerly</u> Mineralogische und Petrographische
   Mitteilungen. Wien. (In 1928 the title reverted to
   Mineralogische und Petrographische Mitteilungen, and
   from 1929-1943 styled Zeitschrift für Kristallographie,
   Mineralogie und Petrographie, Abt.B.)   M.S 1310

Tsitologiya. Leningrad.
   <u>Tsitologiya</u> Vol.13, No.6 → 1971 →   S. 1803 a

Tsitologiya i Genetika. Akademiya Nauk Ukrainskoĭ SSR. Kiev.
   <u>Tsitol.Genet.</u> 1967 →   S. 1834 a B

Tuatara. Wellington, N.Z.
   <u>Tuatara.</u> 1947 →   S. 2163

Tübinger Naturwissenschaftliche Abhandlungen. Stuttgart.
   <u>Tübinger naturw.Abh.</u> 1922-1935.   S. 1596

| TITLE | SERIAL No. |

Tudómányos Közleményei. Szeged.
   See Acta Litterarum ac Scientiarum regiae Universitatis Hungaricae Francisco-Josephinae. Szeged.   S. 1724 A

Tufts College Studies. Scientific Series. Medford, Mass.
   Tufts Coll.Stud. 1894-1937.   S. 2435

Tulane Studies in Geology. New Orleans.
   Tulane Stud.Geol. 1962 →   P.S 881

Tulane Studies in Zoology. New Orleans.
   Tulane Stud.Zool. 1953-1968.
   Continued as Tulane Studies in Zoology and Botany.   Z.S 2415

Tulane Studies in Zoology and Botany. New Orleans.
   Tulane Stud.Zool.Bot. 1968 →
   Formerly Tulane Studies in Zoology.   Z.S 2415

Tulsa Geological Society Digest.
   Tulsa geol.Soc.Dig. 1934 →   P.S 877

Türk Biologi Dergisi. Türk Biologi Dernegi'nin Yayin Organi. Istanbul.
   Türk Biol.Derg. 1958 →
   Formerly Biologi. Türk Biologi Dernegi'nin Yayin Organi. Istanbul.   S. 1887 b

Türk Fizikî ve Tabiî Ilimler Sosyetesi Yillik Bildiriğleri ve Arsivi. Istanbul.
   Türk.fiz.tabiî Ilim.Sosyet.yill.Bild.Ars. 1934-1935.   S. 1887 a

Turkish Journal of Biology. Istanbul.
   See Türk Biologi Dergisi. Türk Biologi Dernegi'nin Yayin Organi. Istanbul.   S. 1887 b

Türkiye Cümhuriyetinde Jeolojik Görümler. Ankara.
   Türk.Cumh.jeol.Gor. 1936 →
   (Internal Series of Yüksek Ziraat Enstitüsü Çalismalari.)   P.S 497

Türkiye Jeoloji Kurumu Bülteni. Istanbul.
   Türk.Jeol.Kur.Bült. 1947 →   P.S 498

Turrialba. Revista Interamericana de Ciencies Agricolas. Turrialba, Costa Rica.
   Turrialba Vol.15 → 1965 →   E.S 2394 a

Turtox News. Chicago.
   Turtox News Vols.9-27; 35 → 1931 → (imp.)   S. 2501

| TITLE | SERIAL No. |

Turun Suomalaisen Yliopiston. Turku.
    See Annales Universitatis Fennicae Aboensis.   S. 1823 b

Turun Yliopiston Julkaisuja. Turku.
    See Annales Universitatis Turkuensis.   S. 1823 b

Tydskrif vir Natuurwetenskappe. Suid Afrikaanse Akademie
    vir Wetenskap en Kuns. Pretoria.
    Tydskr.Natuurwet. 1961 →
    Formerly Tydskrif vir Wetenskap en Kuns. Bleomfontein.   S. 2009 B

Tydskrif van die Suid-Afrikaanse Biologiese Vereniging. Pretoria.
    See Journal of the South African Biological Society.
    Pretoria.   S. 2066 B

Tydskrif van die Suid-Afrikaanse Bosbouvereniging
    See Journal of the South African Forestry Association.   B.S 2307

Tydskrif vir Wetenskap en Kuns. Bleomfontein.
    Tydskr.Wet.Kuns N.S. Deel 1-20. 1940-1960.
    Continued as Tydskrif vir Natuurwetenskappe. Pretoria.   S. 2009 A

Tyo To Ga (Transactions. Lepidopterological Society of Japan)
    See Butterflies and Moths. Kyoto.   E.S 1909 a

U.B.I. Wendingen. Utrecht.
    U.B.I. Wendingen 9-10, 1972.   S. 678
                     New Series. 1974 →   S. 678 A

U.L. Science Magazine. Monrovia.
    U.L.Sci.Mag. 1972 →   S. 2098

U.M.R. Journal. University of Missouri at Rolla.
    U.M.R.Jl 1968 →   S. 2392

UNISIST Newsletter. Paris.
    UNISIST Newsl. 1973 →   S. 2714 K

U.P. Research Digest. University of the Philippines. Quezon City.
    U.P.Res.Dig.Univ.Philipp. Vol.5-6, 1965-1967.   S. 1976 a C

U.S.D.A. Forest Service Research Papers PNW. Portland, Oregon.
    See Research Papers. Pacific Northwest Forest and Range
    Experiment Station.   B.S 4363

| TITLE | SERIAL No. |

Uchenȳe Zapiski Imperatorskago Moskovskago Universiteta.
   Moskva. Otdel Estestvennoistoricheskiĭ.
   Uchen.Zap.imp.mosk.Univ. 1880-1916.
   <u>Continued as</u> Uchenȳe Zapiski Moskovskogo
   Gosudartesvennogo Universiteta.   S. 1844 C

Uchenȳe Zapiski Kazakhskogo Gosudarstvennogo Universiteta im
   S.M. Kirova. Alma-Ata.
   Uchen.Zap.kazakh.gos.Univ. 1938-1961 (imp.)
   (Published in series.)   S. 1832 a

Uchenȳe Zapiski Kazanskogo Gosudarstvennogo Universiteta imeni
   V.I. Ul'yanova-Lenina. Kazan'.
   Uchen.Zap.kazan.gos.Univ. Tom 95, Kniga 3-4 & Tom 96, Kniga 3.
      1935-1936. Geologiya 5-7.   P.S 529
      Tom.92, Kniga 5-6, Tom.94, Kniga 4, Tom.95, Kniga 8,
      Tom.96, Kniga 7, Tom.98, Kniga 8, Tom.99, Kniga 4-5,
      Zoologiya 1-4, 6-8, 1932-1939.   Z.S 1823

Uchenȳe Zapiski. Khar'kovskiĭ Gosudarstvennyĭ Universitet
   imeni A.M. Gor'kogo.
   <u>See</u> Trudȳ Nauchno-Issledovatel'skogo Instituta Biologii
   i Biologicheskogo Fakul'teta. Khar'kovskiĭ Gosudarstvennyĭ
   Universitet. Khar'kov.   S. 1833 B

Uchenȳe Zapiski. Latviĭskiĭ Gosudarstvennyĭ Universitet imeni
   Petra Stuchki. Riga.
   Uchen.Zap.latv.gos.Univ. Tom.67 → 1965 → (imp.)   S. 1851 a G

Uchenȳe Zapiski Leningradskogo Ordena Lenina Gosudarstvennogo
   Universiteta. Leningrad.
   Uchen.Zap.leningr.gos.Univ.
   Ser.Biol.Nauk, No.1-4, 6, 9, 40-46, 48-49. 1935-1962.
   Ser.Fiz. No.1-3, 1935-1937:
   Ser.Fiz.i Geol.Nauk, No.14 → 1963 →
   Ser.Geogr.Nauk, No.11, 1956:
   Ser.Geol.Pochv.(Geogr.) No.1-3 & 11. 1935-1944:
   Ser.Geol.Nauk, No.7 & 9, 1956-1957:
   Ser.Khim. No.1-2, 1936-1937:
   Ser.Mat.(Astron.) No.1-2, 1936-1937.   S. 1803 A

Uchenȳe Zapiski Moskovskogo Gosudarstevennogo Universiteta.
   Uchen.Zap.mosk.gos.Univ. 1933-1961, (imp.)
   <u>Formerly</u> Uchenȳe Zapiski Imperatorskago Moskovskago
   Universiteta. Moskva. Otdel Estestvennoistoricheskiĭ.   S. 1844 C

Uchenȳe Zapiski. Nauchno-Issledovatel'skiĭ Institut Geologii
   Arktiki. Leningrad.
   Uchen.Zap.nauchno-issled.Inst.Geol.Arkt.
   Regional'naya Geologiya, Vyp.7 → 1965 →   P.S 1553 C
   Paleontologiya i Biostratigrafiya, 1960 → (imp.)   P.S 1553 D

| TITLE | SERIAL No. |

Uchenye Zapiski. Permskiĭ Gosudarstvennyĭ Universitet
  im M. Gor'kogo. Perm.
  Uchen.Zap.perm.gos.Univ. 1935-1936.   S. 1849 a

Uchenye Zapiski Saratovskogo Gosudarstvennogo Universiteta
  imeni N.G. Chernyshevskogo. Saratov.
  Uchen.Zap.saratov.gos.Univ. Tom. 7-8, 10-14. 1929-1939.   S. 1879 a A

Uchenye Zapiski Severo-Kavkazskogo Instituta Kraevedeniya.
  Vladikavkaz.
  Uchen.Zap.sev.-kavk.Inst.Kraev. 1926.   Z.S 1835

Uchenye Zapiski Tartuskogo Gosudarstvennogo Universiteta. Tartu.
  Uchen.Zap.tartu.gos.Univ. No.75 → 1959 →   S. 1857 A

Uebersicht der Arbeiten und Veränderungen der Schlesischen
  Gesellschaft für Vaterländische Kultur. Breslau.
  Uebers.Schles.Ges.Vaterl.Kult. 1824-1849.
  Continued as Jahresbericht der Schlesischen Gesellschaft
  für Vaterländische Kultur. Breslau.   S. 1376 A

Uganda Journal. Kampala.
  Uganda J. 1934-1942, 1946 →
  For 1943-1945 Styled Bulletin of the Uganda Society.   S. 2025 a A

Uganda Wild Life and Sport. Entebbe.
  Uganda Wildl.Sport 1956-1960.
  Continued as Wild Life and Sport.   Z.S 2060

Uitgaven van de Natuurwetenschappelijke Studiekring voor Suriname
  en de Nederlandse Antillen (formerly en Curacao). Utrecht, etc.
  Uitg.natuurw.StudKring Suriname 1945 →   S. 677

Uitgaven van de Natuurwetenschappelijke Werkgroep Nederlandse
  Antillen. Curaçao.
  Uitg.natuurw.Werkgrp ned.Antillen 1951 →   S. 2280
  Nos.5 & 12, 1955 & 1960 (Also styled Fauna Nederlandse   Z. Bird Section
  Antillen.)   & Z. 75F o H

Uitgezogte Verhandelingen uit de Nieuwste Werken von de
  Societeiten der Wetenschappen in Europa, Amsterdam.
  Uitgez.Verh.Amst. 1757-1765.   S. 681

Ukrainian Botanical Journal.
  See Ukrayins'kyĭ Botanichnyĭ Zhurnal. Kyyiv.   B.S 1371

Ukrayins'kyĭ Botanichnyĭ Zhurnal. Kyyiv.
  Ukr.bot.Zh. Tom.13 → 1956 →
  Formerly Botanichnyi Zhurnal. Kyyiv.   B.S 1371

Umi. La Mer. Tokyo.
  See Mer. Tokyo.   S. 1991 d

Umbelliferae Newsletter. Kew.
  Umbellif.Newsl. 1971 →   B.S 64 c

| TITLE | SERIAL No. |

Umwelt. Zeitschrift der Biologischen Station. Wilhelminenberg. Vienna.
    Umwelt. 1946-1947.   S. 1683

Unasylva. Washington.
    Unasylva 1947-1971.   B.S 4590

Underground Water Supply Paper. Department of Mines, Tasmania. Hobart.
    Undergr.Wat.Supply Pap.Tasm. 1921 →   M.S 2448

Underwater Information Bulletin. Guildford.
    Underwat.Inf.Bull. Vol.6 → 1974 →
    Formerly Underwater Journal and Information Bulletin.   S. 412 a C

Underwater Journal and Information Bulletin. Guildford.
    Underwat.J.Inf.Bull. Vol.3-5, 1971-1973.
    Formerly Underwater Science and Technology Information Bulletin, and Underwater Science and Technology Journal. Guildford.
    Continued as Underwater Information Bulletin.   S. 412 a C

Underwater Naturalist. Bulletin of the American Littoral Society. Highlands, N.J.
    Underwat.Nat. 1962 →   S. 2472

Underwater Science and Technology Information Bulletin. Guildford.
    Underwat.Sci.Technol.Inf.Bull. 1970.
    Continued as Underwater Journal and Information Bulletin. Guildford.   S. 412 a C

Underwater Science and Technology Journal. Guildford.
    Underwat.Sci.Technol.J. 1970.
    Continued as Underwater Journal and Information Bulletin. Guildford.   S. 412 a C

Underwater Technology (& Inner Space News). Poole, Dorset.
    Underwat.Technol. No.1-3, 1967.   S. 410 a B

Underwater World. New Malden.
    Underwat.Wld 1966-1967.   S. 410 a A

Unesco Bulletin for Libraries. Paris.
    Unesco Bull.Libr. 1947 → (imp.)   S. 2714

Unesco Technical Papers in Marine Science.
    Unesco tech.Pap.mar.Sci. No.5 → 1966 →   S. 2714 C

Ungarische Botanische Blätter.
    See Magyar Botanikai Lapok. Budapest.   B.S 1262

Ungarische Entomologische Zeitschrift.
    See Rovartani Lapok. Budapest.   E.S 1730

Ungarische Rundschau für Geologie und Paläontologie.
    See Földtani Szemle.   P.S 367

TITLE	SERIAL No.

Ungava Bay Papers. Scientific Results of the Oxford University
Hudson Strait Expedition. 1931. London.
　Ungawa Bay Papers 1932-1938.　　　　　　　　　　　　　　75 B.q.O

Union of Burma Journal of Life Sciences. Rangoon.
　Un.Burma J.Life Sci. 1968 →　　　　　　　　　　　　　　S. 1938 c

Union of Burma Journal of Science and Technology. Rangoon.
　Un.Burma J.Sci.Tech. 1968 →　　　　　　　　　　　　　　S. 1938 b

United Arab Republic Journal of Pharmaceutical Sciences. Cairo.
　U.A.R.J.pharm.Sci. Vol.11-12, 1970-1971.
　Formerly Journal of Pharmaceutical Sciences of the
　United Arab Republic.
　Continued as Egyptian Journal of Pharmaceutical Sciences.　B.S 2263

United Empire. Royal Colonial Institute Journal. London.
　Unit.Emp. 1910-1945.
　Formerly Journal of the Royal Colonial Institute. London.　S. 209 A

United Kingdom Mineral Statistics. London.
　U.K.miner.Statist. 1973 →　　　　　　　　　　　　　　　M.S 132

United Kingdom Research Vessels Cruise Programmes. London.
　See Cruise Programmes. United Kingdom Research Vessels.
　London.　　　　　　　　　　　　　　　　　　　　　　　　S. 224 a F

United States Catalog. Minneapolis & New York.
　U.S.Cat. 1912. Supplement (- Cumulative Book Index). 1912-1927.
　Continued as Cumulative Book Index. New York.　　　　　REF. CR.

United States Exploring Expedition 1838-1842. Philadelphia.
　U.S.Explor.Exped. Vols. 1-5, 9 & 15. 1845-1863.　　　　70.f.U

Universitas Carolina. Universita Karlova v Praze. Praha.
　Univ.carol.　Biologica 1955-1957.　　　　　　　　　　　　S. 1755 B
　　　　　　　Geologica 1955-1957.　　　　　　　　　　　　M.S 1715
　Replaces Spisy Vydávané Přírodovědeckou Fakultou
　Karlovy University. Praha.
　Continued in Acta Universitatis Carolinae. Biologica.

Universitetet i Bergen Arbok. Avhandlingar og Arsberetning.
　Univ.Bergen Arb. 1948-1960.
　　(From 1898-1948 the Aarsberetning was issued separately
　　See S.552.B.)
　Formerly Bergens Museums Aarbog (Aarbok) Afhandlingar og
　Aarsberetning.
　Continued as Arbok for Universitet i Bergen.
　Mat.-Naturv.Ser.　　　　　　　　　　　　　　　　　　　　S. 552 A

Universitetet i Bergen Arsberetning.
　Univ.Bergen Arsberetn. 1948-1949.
　Formerly Bergens Museums Aarsberetning.
　Continued as Universitetet i Bergen Arsmelding.　　　　　S. 552 B

Universitetet i Bergen Arsmelding.
　Univ.Bergen Arsmeld. 1949 →
　Formerly Universitetet i Bergen Arsberetning.　　　　　　S. 552 B

| TITLE | SERIAL No. |

Universitetet i Bergen Skrifter.
    Univ.Bergen Skr. 1948 →
    Formerly Bergens Museums Skrifter.   S. 552 G

University of British Columbia Publications.
    Biological Sciences. Vancouver.
    Univ.Br.Columb.Publs biol.Sci. 1945-1946.   S. 2660

University Bulletin. University of Michigan. Ann Arbor.
    Univ.Bull.Mich. N.S. Vol.7, No.5, 9 and 14
    (Bot. Ser. No.1-2 and 4), 1906.   B.S 4250(B)

University of California Publications in Agricultural Sciences.
    Berkeley.
    Univ.Calif.Publs agric.Sci. 1912-1939.   E.S 2466 a

University of California Publications. American Archaeology
    and Ethnology.
    Univ.Calif.Publs Am.Archaeol.Ethnol. 1903 →   P.A.S 776 A

University of California Publications in Anthropology. Berkeley
    and Los Angeles.
    Univ.Calif.Publs Anthrop. 1964 →   P.A.S 776 D

University of California. Publications in Biological Sciences.
    Los Angeles.
    See Publications of the University of California at
    Los Angeles in Biological Sciences.   S. 2319 a A

University of California Publications in Botany. Berkeley.
    Univ.Calif.Publs Bot. 1902 →   B.S 4166

University of California Publications. Bulletin of the Department
    of Geology. Berkeley.
    Univ.Calif.Publs Bull.Dep.Geol. 1893-1921.
    Continued as University of California Publications in
    Geological Sciences. Berkeley.   P.S 1884

University of California Publications in Entomology. Berkeley.
    Univ.Calif.Publs Ent. 1906 →   E.S 2466

University of California Publications. Geography. Berkeley.
    Univ.Calif.Publs Geogr. 1913 →   S. 2319 F

University of California Publications in Geological Sciences. Berkeley.
    Univ.Calif.Publs geol.Sci. 1922 →
    Formerly University of California Publications. Bulletin
    of the Department of Geology. Berkeley.   P.S 1884

University of California Publications in Oceanography.
    See Bulletin. Scripps Institution of Oceanography.
    Technical and Non-Technical Series.   S. 2319 M-N

| TITLE | SERIAL No. |

University of California Publications. Pathology. Berkeley.
    Univ.Calif.Publs Path. 1903-1919.   S. 2319 D

University of California Publications. Physiology. Berkeley.
    Univ.Calif.Publs Physiol. 1902-1956.   S. 2319 B

University of California Publications in Zoology. Berkeley.
    Univ.Calif.Publs Zool. 1902 →   T.R.S 5121 & Z.S 2319 C

University of Chicago Science Series.
    Univ.Chicago Sci.Ser. 1914-1940.   S. 2330 a F

University of Colorado Bulletin. Boulder.
    Univ.Colo.Bull. Vol.13, Nos, 1, 4, 1913.   S. 2323 a C

University of Colorado Studies. General Series. Boulder.
    Univ.Colo.Stud.gen.Ser. Vol.1, Nos.3 - Vol.29, No.4
    1903-1957.   S. 2323 a A

University of Colorado Studies. Physical and Biological
    Sciences. Boulder.
    Univ.Colo.Stud.phys.biol.Sci. 1940-1947.
    Continued as University of Colorado Studies.
    Series in Biology. Boulder.   S. 2323 a A

University of Colorado Studies. Series in Anthropology. Boulder.
    Univ.Colo.Stud.Ser.Anthrop. 1948-1971.   P.A.S 784

University of Colorado Studies. Series in Biology. Boulder.
    Univ.Colo.Stud.Ser.Biol. 1950-1970.
    Formerly University of Colorado Studies. Physical and
    Biological Sciences. Boulder.   S. 2323 a B

University of Colorado Studies. Series in Earth Sciences. Boulder.
    Univ.Colo.Stud.Ser.Earth Sci. 1965-1968.
    Formerly University of Colorado Studies. Series
    in Geology. Boulder.   P.S 805

University of Colorado Studies. Series in Geology. Boulder.
    Univ.Colo.Stud.Ser.Geol. 1-3 1963-1964.
    Continued as University of Colorado Studies. Series in
    Earth Sciences.   P.S 805

University of Florida Publication. Biological Science Series.
    Gainesville.
    Univ.Fla Publ.biol.Sci.Ser. 1930-1950. (imp.)   S. 2436 A

University of Florida Publications and Theses.
    Gainesville, Fla.
    Univ.Fla Publs & Theses 1955-1965.   S. 2436 B

| TITLE | SERIAL No. |

University Geological Survey of Kansas Topeka.
    Univ.geol.Surv.Kans. 1896-1908, 1937 →   P.S 1924

University of Illinois Bulletin. Urbana.
    Univ.Ill.Bull.
        Vol.13, No.45. 1916   B. 582.4pl48 TRE
        Vol.17, No.41. 1920   P. 75 C o B
        Vol.20, No.50. 1923   E. Hymenoptera Room
    From 1914-1916 & 1934-1940 numbers of the Bulletin were
    also styled Illinois Biological Monographs.

University of Iowa Studies in Natural History.
    See State University Studies in Natural History.
    Iowa City.   S. 2341

University of Kansas Geological Survey Bulletin.
    See Bulletin of the Kansas University Geological
    Survey. Topeka.   P.S 1923

University of Kansas Paleontological Contributions.
    See Paleontological Contributions. University of
    Kansas. Topeka.   P.S 1920

University of Kansas Publications. Museum of Natural History.
    Lawrence.
    Univ.Kans.Publs Mus.nat.Hist. 1946-1971.   S. 2344 D & T.R.S 5129 B

University of Kansas. Science Bulletin.
    See Kansas University Science Bulletin.   S. 2344 B

University of Michigan Studies. Scientific Series. Ann Arbor.
    Univ.Mich.Stud.scient.Ser. Vols.5, 6 & 12. 1931-1941.   S. 2316 G
        Vol.9 & 13, 1937-1957.   B. See Bot.Lbry Catalogue

University of Minnesota Studies. Biological Sciences.
    See Studies in the Biological Sciences. University
    of Minnesota. Minneapolis.

University of Missouri Studies. A Quarterly of Research. Columbia.
    Univ.Mo.Stud. 1902, 1926 → (nat.hist.only.)   S. 2384

University of Montana Studies. Missoula.
    See State University of Montana Studies.   S. 2350 F

University of Nebraska Studies. Lincoln.
    Univ.Neb.Stud. N.S. No.1 → 1946 → (imp.)
    Formerly University of Nebraska Studies. Studies in Science
    and Technology. Lincoln.   S. 2473 A

University of Nebraska Studies. Studies in Science & Technology.
    Lincoln.
    Univ.Neb.Stud.Sci.Technol. 1941-1942.
    Formerly University Studies of the University of Nebraska. Lincoln.
    Continued as University of Nebraska Studies. Lincoln.   S. 2473 A

| TITLE | SERIAL No. |

University of New Mexico Publications in Biology. Albuquerque.
    <u>Univ.New Mex.Publs Biol</u>. 1946 →
    <u>Formerly</u> Bulletin of the University of New Mexico.
    Albuquerque. Biological Series.   S. 2313 a B

University of New Mexico Publications in Geology. Albuquerque.
    <u>Univ.New Mex.Publs Geol</u>. 1945-1966.   P.S 843 A

University of New Mexico Publications in Meteoritics. Albuquerque.
    <u>Univ.New Mex.Publs Meteorit</u>. 1946-1969.   M.S 2602 B

University of Nottingham. Abbott Memorial Lecture.
    <u>See</u> Abbott Memorial Lecture. University of Nottingham. P.S 141

University of Oklahoma Biological Survey Publications.
    <u>See</u> Publications of the University of Oklahoma Biological
    Survey. Norman.   S. 2386 B

University of Oklahoma Bulletin. Norman.
    <u>Univ.Okla.Bull</u>. N.S. No.20, 268, 299 & 314. 1923-1925. S. 2386 B

University of Oregon Monographs. Studies in Botany. Eugene.
    <u>Univ.Ore.Monogr.Stud.Bot</u>. No.1. 1936.
    <u>Continued from</u> University of Oregon Publications. Eugene. B.S 4361

University of Oregon Monographs. Studies in Entomology. Eugene.
    <u>Univ.Ore.Monogr.Stud.Ent</u>. 1939 →   E.S 2488 d

University of Oregon Publications. Eugene.
    <u>Univ.Ore.Publs</u> 1920-1925 (nat.hist.only.)   S. 2335
      Geology Series 1926-1932.   P.S 867
      Plant Biology Series Vol.1, No.1, 1929.   B.S 4361 a
    <u>Replaced by</u> University of Oregon Monographs.

University Publications in Botany. Seattle.
    <u>Univ.Wash.Publs Bot</u>. Vol.1, No.1, 1915.   B.S 4413

University of Queensland Papers. Department of Biology.
    <u>See</u> Papers from the Department of Biology. University
    of Queensland. Brisbane.   S. 2139 A

University of Queensland Papers. Department of Botany.
    <u>See</u> Papers from the Department of Botany. University
    of Queensland. Brisbane.   B.S 2408

University of Queensland Papers. Department of Entomology.
    <u>See</u> Papers. Department of Entomology, University
    of Queensland.   E.S 2258 a

University of Queensland Papers. Department of Geology.
    <u>See</u> Papers. Department of Geology. University of
    Queensland. Brisbane.   P.S 1149

| TITLE | SERIAL No. |

University of Queensland Papers. Department of Zoology. Brisbane.
   See Papers. Department of Zoology, University of Queensland.   Z.S 2125

University of Queensland Papers. Great Barrier Reef Committee.
  Heron Island Research Station. Brisbane.
    Univ.Qd Pap.Gt.Barrier Reef Comm.   1966 →
    Replaces Report of the Great Barrier Reef Committee.
  Brisbane.   S. 2139 B

University of Rajasthan Studies. Jaipur.
    Univ.Rajasthan Stud. Biological Sciences 1958 →
    Formerly University of Rajputana Studies. Jaipur.   S. 1937 a

University of Rajputana Studies. Jaipur.
    Univ.Rajputana Stud. Biological Sciences 1951-1955
    Continued as University of Rajasthan Studies. Jaipur.   S. 1937 a

University Record. Chicago.
    Univ.Rec.Chicago 1896-1905.   S. 2330 a A

University of South Carolina Publications. Biology Ser.III.
  Columbia, S.C.
    Univ.S.Carol.Publs 1952-1960.   S. 2388

University Studies. University of Karachi. Karachi.
    Univ.Stud.Univ.Karachi 1964 →   S. 1939 a

University Studies of the University of Nebraska. Lincoln.
    Univ.Stud.Univ.Neb. 1888-1941.
    Continued as University of Nebraska Studies. Studies
  in Science & Technology. Lincoln.   S. 2473 A

University of Texas Bulletin. Austin.
    Univ.Tex.Bull. 1917-1937 (Geology only).
    Continued as University of Texas Publications. Austin.   P.S 1978

University of Texas Publications. Austin.
    Univ.Tex.Publs 1938 → (nat.hist.)
    Formerly University of Texas Bulletin. Austin. See Gen.Lbry Catalogue

University of Toronto Studies. Anatomical Series. Toronto.
    Univ.Toronto Stud.anat.Ser. 1900-1939.   S. 2650 A

University of Toronto Studies. Biological Series.
    Univ.Toronto Stud.biol.Ser. 1898-1956.   S. 2650 B

University of Toronto Studies. Contributions to Canadian
  Mineralogy. Toronto.
    Univ.Toronto Stud.Contr.Can.Miner. 1921-1947.   M.S 2704
                                          1921-1947.   S. 2650 C
  (Published in University of Toronto Studies Geological Series.)
    Continued as Contributions to Canadian Mineralogy.

University of Toronto Studies. Geological Series. Toronto.
    Univ.Toronto Stud.geol.Ser. 1900-1947.   S. 2650 C

TITLE	SERIAL No.

University of Toronto Studies. Pathological Series. Toronto.
    Univ.Toronto Stud.path.Ser. 1906-1931.   S. 2650 E

University of Toronto Studies. Physiological Series. Toronto.
    Univ.Toronto Stud.physiol.Ser. 1900-1928.   S. 2650 D

University of Udaipur Research Studies. Udaipur.
    Univ.Udaipur Res.Stud. 1963 →   S. 1909 b

University of Utah Biological Series. Salt Lake City.
    Univ.Utah biol.Ser. 1951 →
    Formerly Bulletin of the University of Utah.
    Salt Lake City. Biological Series.   S. 2398 A

University Vision. British Universities Film Council. London.
    Univ.Vision No.9 → 1972 →   S. 165 a

University of Washington Arboretum Bulletin. Seattle.
    See Arboretum Bulletin. University of Washington. Seattle.   B.S 4413 a

University of Washington Publications in Anthropology. Seattle.
    Univ.Wash.Publs Anthrop. 1920 →   P.A.S 805

University of Washington Publications in Biology. Seattle.
    Univ.Wash.Publs Biol. 1932 →
    Formerly University of Washington Publications in
    Fisheries. Seattle.   S. 2405 E

University of Washington. Publications in Botany. Seattle.
    Univ.Wash.Publs Bot. Vol.1, No.1, 1915.   B.S 4413

University of Washington Publications in Fisheries. Seattle.
    Univ.Wash.Publs Fish. 1925-1929.
    Continued as University of Washington Publications in
    Biology.   Z.S 2384

University of Washington Publications. Geology. Seattle.
    Univ.Wash.Publs Geol. 1916-1953.   P.S 1989 A

University of Washington Publications in Oceanography. Seattle.
    Univ.Wash.Publs Oceanogr. 1932-1960.
    Formerly Publications. Puget Sound Marine Biological
    Station of the University of Washington Seattle.   S. 2405 F a

University of Washington Publications in Oceanography. Seattle.
    Supplementary Series.
    Univ.Wash.Publs Oceanogr.Suppl.Ser. 1931-1938 (imp.)
    Continued as Contributions from the Department of
    Oceanography, University of Washington.   S. 2405 F b

University of Waterloo Biology Series. Waterloo.
    Univ.Waterloo Biol.Ser. 1971 →   S. 2680

TITLE SERIAL No.

University of Wyoming Publications. Laramie.
   Univ.Wyo.Publs Vol.2 → 1935 →
   Formerly University of Wyoming Publications in Science.
Botany and Geology.   S. 2471

University of Wyoming Publications in Science. Laramie.
   Univ.Wyo.Publs Sci. Botany 1922-1934.   B.S 4420
                         Geology 1929.   P.S 852
   Continued as University of Wyoming Publications.   S. 2471

Untersuchungen aus der Botanischen Laboratorium der Universität
   Göttingen. Berlin.
   Unters.Bot.Lab.Univ.Göttingen 1879–1883.   B.S.C 15

Untersuchungen aus dem Forstbotanischen Institut. München.
   Unters.forstbot.Inst.München 1880-1883.   B. 632.4 F.I.

Untersuchungen zur Naturlehre des Menschen und der Tiere.
   Frankfurt.
   Unters.Naturl.Mensch.Tiere 1856-1888.   Z.S 1490

Uppsala Universitets Arsskrift. Uppsala.
   Uppsala Univ.Arsskr. 1861 → (nat.hist.only)   S. 596 A

Uppsala Universitets Katalog.
   Uppsala Univ.Kat. 1839 →   S. 596 H

Uppsala Universitets Matrikel.
   Uppsala Univ.Matrikel 1595-1800 (publ. 1900-1946)
   (1595-1749 published in Uppsala Universitets Arsskrift.)   S. 596 E

Uppsatser i Praktisk Entomologi. Stockholm.
   Upps.prakt.Ent. 1891-1914.   E.S 502 a

Uragus. Zhurnal Sibirskogo Ornithologicheskogo Obshchestva.
   Tomsk.
   Uragus Nos.1-10, 1926-1929.   Z.S 1859

Uspekhi Eksperimental'noĭ Biologii. Moskva.
   Usp.eksp.Biol. Tom.1, Vyp.1.   Z.S 1812 a

Uspekhi Mikrobiologii. Veseoyuznoe Mikrobiologicheskoe Obshchestvo.
   Akademiya Nauk SSSR. Moskva.
   Usp.Mikrobiol.Moskva 1964 →   S. 1805 c

Uspekhi Sovremennoĭ Biologii, Moskva.
   Usp.sovrem.Biol. 1943 →
      (Wanting tom 24, nos. 2-4; tom 25, no. 4; tom 26
      nos.1 & 4; tom 32 nos.1 & 4.)   S. 1802 a A

TITLE	SERIAL No.

Utah Farm and Home Science. Logan, Utah.
   See Utah Science. Logan, Utah.   S. 2568

Utah Geological Association Publications. Salt Lake City.
   See Publications. Utah Geological Association. Salt Lake City.
                                                                P.S 844 B

Utah Science. Logan, Utah.
   Utah Sci. Vol.23 → 1962 →
   Formerly Farm and Home Science. Logan, Utah.   S. 2568

Utrecht Micropaleontological Bulletins. Utrecht.
   Utrecht micropaleont.Bull. 1969 →   P.S 299 a

Utrecht Micropaleontological Bulletins. Special Publication. Utrecht.
   Utrecht micropaleont.Bull.spec.Publs 1974 →   P.S 299 b

Uzbekskiĭ Biologicheskiĭ Zhurnal. Tashkent.
   Uzbek.biol.Zh. 1971 →   S. 1857 a B

Uzbekskiĭ Geologicheskiĭ Zhurnal. Tashkent.
   Uzbek.geol.Zh. 1965 →   P.S 554

V Pomoshch' Rabotayushchim po Zoologii v Pole i Laboratorii. Zoologicheskii Institut. Akademiya Nauk SSSR.
   V Pom.Rab.Zool.Pole Lab. 1956 →
      Nos.1-2 & 5.   Z. 72Q o L
      Nos.3-6.   E. 72Q o L

Vakblad voor Biologen. Helder.
   Vakbl.Biol. 1927 →   S. 630

Valda Avhandlingar av Carl von Linné i översättning utgivna av Svenska Linné-Sällskapet. Uppsala & Stockholm.
   Valda Avh.C.v.Linné 1921 →   S. 592 C

Van Gorcum's Geologische Reeks. Assen.
   Van Gorcum's geol.Reeks Deel 10-18, 1951-1954 (imp.)   P.S 296

Vanamon Kasvitieteellisen Seuran. Helsinki.
   See Suomalaisen Eläin-ja Kasvitieteellisen Seuran Vanamon Kasvitieteellisiä Julkaisuja. Helsinki.   B.S 1487

Vanasarn. Bangkok.
   Vanasarn 1970 →   S. 1913 d

Vår Fågelvärld. Stockholm.
   Vår Fågelvärld 1942 → Supplementum 1950 →   T.B.S 641

Variétés Scientifiques Recueillies par la Société des Sciences Naturelles et Physiques du Maroc.
   Var.scient.Soc.Sci.nat.phys.Maroc 1921-1960.   S. 2036 E
                                          6, 8-10; 1943-1960.   T.R.S 4305 D

TITLE	SERIAL No.

Varstvo Narave. Ljubljana.
    Varstvo Narave 1962 →   S. 1747 a

Varv. København.
    Varv 1964 →   P.S 406

Vasculum. Newcastle-upon-Tyne.
    Vasculum 1915 →   S. 451

Vasi Szemle. Szombathely.
    See Folia Sabariensia Vasi Szemle. Szombathely.   S. 1715

Växtskyddsnotiser. Statens Växtskyddsanstalt. Stockholm.
    Växtskyddsnotiser 1937 →   E.S 507

Věda Přírodní. Měsíčník pro Šíření a Pěstovani Věd Přírodních. Praha.
    Věda přír. Roč.8, 10-14, 1927-1933.   S. 1762 c

Vegetatio. Acta Geobotanica. Den Haag.
    Vegetatio 1948 →   B.S 306

Vegetation der Erde. Sammlung Pflanzengeographischer
    Monographien. Leipzig.
    Vegn Erde 1896-1923.   B.S 581.9

Vegetation of Siberia and the Far East. Current Index of
    Literature. Novosibirsk.
    See Rastitelnye Resury Sibiri i Dal'nego Vostoka.   B.S 1415 a

Vegetationsbilder. Jena.
    Vegetationsbilder 1903-1944.   B.S 1027

Veld. Newlands, C.P.
    Veld 1970.
    Continued as Veld & Flora. Newlands, C.P.   B.S 2265 a

Veld & Flora. Newlands, C.P.
    Veld Flora 1971 →
    Formerly Veld. Newlands, C.P.   B.S 2265 a

Veld & Vlei. Durban.
    Veld Vlei Vol.2 - Vol.3 No.11 1957-1958.
    Continued as Field & Tide.   Z.S 2095

Veliger. Berkeley, Cal.
    Veliger 1958 →   Z. Mollusca Section

Vellosia. Contribuições do Museu Botanico do Amazonas.
    Rio de Janeiro.
    Vellosia 1885-1888 Segunda edição. (1891-1892.)   S. 2212 & B.S 3062

Vellozia. Publicação do Centro de Pesquisas Florestais e
    Conservação da Natureza. Tijuca, Rio de Janeiro.
    Vellozia 1961 →   S. 2213 b

| TITLE | SERIAL No. |

Vem är Det ? Stockholm.
   <u>Vem är Det</u> ? 1912-1949.   REF.

Vem och Vad ? Helsingfors.
   <u>Vem och Vad</u> ? 1920, 1926-1948.   REF.

Vema Reports.
   <u>See</u> Vema Research Series.   Z. <u>See</u> Zool.Lbry Catalogue

Vema Research Series. New York.
   <u>Vema Res.Ser.</u> 1962 →   Z. <u>See</u> Zool.Lbry Catalogue

Venus. Kyoto.
   <u>Venus, Kyoto</u> 1928 →
   (From 1943-1948 Styled Japanese Journal of Malacology.)
                                                                       Z. Mollusca Section

Vereinsnachrichten der Badischen Entomologischen Vereinigung.
   Freiburg i.Br.
   <u>VerNachr.bad.ent.Verein</u> 1923-1924.   E.S 1337

Vereinsnachrichten. Entomologenverein Basel und Umgebung. Basel.
   <u>VerNachr.EntVer.Basel</u> 6-7, 1949-1950.
   <u>Continued as</u> Mitteilungen aus der Entomologischen
   Gesellschaft,Basel.   E.S 1203

Verhandeling van het Ryksinstituut voor Veldbiologisch Onderzoek
   ten Behoeve van het Natuurbehoud (RIVON).
   Nr.2, 1966 published in Mededeling van het Instituut voor
   Toegepast Biologisch Onderzoek in de Natuur. Arnhem.   S. 617

Verhandelingen van het Bataviaasch Genootschap van Kunsten
   en Wetenschappen. Batavia.
   <u>Verh.batav.Genoot.Kunst.Wet.</u> Deel 1-42, 1792-1881.
   (Deel 1 & 2, 3rd.impression, Deel 3-5 & 8 2nd.impression.)   S. 1951 B

Verhandelingen uitgegeven door de Commissie belast met het
   Vervaardigen eener Geologische Beschrijving en Kaart van
   Nederland te Haarlem.
   <u>Verh.Comm.Geol.Beschr.Kaart Ned.</u> 1853-1854.   P.S 1280

Verhandelingen van het Geologisch-Mijnbouwkundig Genostschap voor
   Nederland en Kolonien. 's Gravenhage. Geologische Serie.
   <u>Verh.geol-mijnb.Genoot.Ned.</u> 1912-1945
   <u>Continued as</u> Verhandelingen van het Nederlandsch
   Geologisch-Mijnbouwkundig Genootschap. 's Gravenhage.
   Geologische Serie.   P.S 1285

Verhandelingen uitgegeeven door de Hollandse Maatschappye
   der Wetenschappen te Haarlem.
   <u>Verh.holland.Maatsch.Wet.</u> 1759-1793.
   <u>Continued as</u> Natuurkundige Verhandelingen van de Bataafsche
   Hollandsche Maatschappye der Wetenschappen te Haarlem.   S. 621 B

| TITLE | SERIAL No. |

Verhandelingen der eerste Klasse van het Hollandsch
   (Koninklijk-Nederlandsche) Instituut van Wetenschappen,
   Letterkunde en Schoone Kunsten te Amsterdam.
   Verh.K.ned.Inst.Amst. 1812-1852.   S. 602 B

Verhandelingen. Koninklijk Belgisch Instituut voor Natuurwetenschappen.
   See Mémoires de l'Institut Royal des Sciences Naturelles
   de Belgique.   (2 series) S. 704 A a & S. 704 A b

Verhandelingen van het Koninklijk Nederlandsch (Nederlands)
   Geologisch Mijnbouwkundig Genootschap. 's Gravenhage.
   Geologische Serie.
   Verh.K.ned.geol.mijnb.Genoot. 1955 →
   Formerly Verhandelingen van het Nederlandsch
   Geologisch-Mijnbouwkundig Genootschap.   P.S 1285

Verhandelingen der Koninklijke Akademie van Wetenschappen.
   Amsterdam. Afdeeling Natuurkunde.
   Verh.K.Akad.Wet.,Amst. 1854-1890.
     Eerste Sectie: 1892-1935.
     Tweede Sectie: 1892-1937.
   Continued as Verhandelingen der Koninklijke Nederlandsche
   Akademie van Wetenschappen. Amsterdam.   S. 601 C & D

Verhandelingen der Koninklijke Nederlandsche (Nederlandse)
   Akademie van Wetenschappen. Amsterdam. Afdeeling Natuurkunde.
   Verh.K.ned.Akad.Wet. (1936 →)
     Eerste Sectie: 1936 →
     Tweede Sectie: 1938 →
   Formerly Verhandeling der Koninklijke Akademie van Wetenschappen.
   Amsterdam.   S. 601 C & D

Verhandelingen. Koninklyk Belgisch Koloniaal Instituut.
   See Mémoires. Institut Royal Colonial Belge.   S. 709 B

Verhandelingen. Koninklyk Natuurhistorisch Museum van België.
   See Mémoires du Musée Royal d'Histoire Naturelle de Belgique.
                        (2 series) S. 704 A a & S. 704 A b

Verhandelingen der Natuurkundige Vereeniging in Nederlandsch
   Indië. Batavia.
   Verh.natuurk.Ver.Ned.Ind. 1856-1860.   S. 1952 B & T.R.S 7603

Verhandelingen over de Natuurlijke Geschiedenis der Nederlandsche
   Overzeesche Bezittingen. Leiden.
   Verh.Natuurl.Gesch.Ned.Overzee.Bezitt. 1839-1845.   77.f.H

TITLE	SERIAL No.

Verhandelingen van het Nederlandsch Geologisch-Mijnbouwkundig
    Genootschap. 's Gravenhage. Geologische Serie.
    Verh.ned.geol.mijnb.Genoot. 1948-1954.
    Formerly Verhandelingen van het Geologisch-Mijnbouwkundig
    Genootschap voor Nederland en Kolonien.
    Continued as Verhandelingen van het K. Nederlandsch
    Geologisch-Mijnbouwkundig Genootschap.   P.S 1285

Verhandelingen en Rapporten uitgegeven door het
    Rijksinstitut voor Visscherijonderzoek. s'Gravenhage.
    Verh.Rapp.RijksInst.VisschOnderz. Deel.1 afl.1-3, 1920-1924.
    Formerly Rapport en Verhandelingen uitgegeven door
    het Rijksinstitut voor Visscherijonderzoek. s'Gravenhage.   Z. 22 q H

Verhandelingen. Rijksinstituut voor Natuurbeheer. Arnhem.
    Verh.rijksinst.Natuurbeh. 1970 →   S. 620

Verhandelingen uitgegeven door het Zeeuwsch Genootschap der
    Wetenschappen te Vlissingen. Middleburg.
    Verh.zeeuw.Genoot.Wet. 1769-1782.
    Continued as Nieuwe Verhandelingen van het Zeeuwsch
    Genootschap der Wetenschappen.   S. 657 A

Verhandlungen der Allgemeinen Schweizerischen Gesellschaft für
    die Gesamten Naturwissenschaften.
    Verh.schweiz.Ges. Sess. 11-22. 1825-1837.
    Formerly Kurze Übersicht der Verhandlungen der Allgemeinen
    Schweizerischen Gesellschaft für die Gesamten Naturwissenschaften.
    Continued as Verhandlungen der Schweizerischen Naturforschenden
    Gesellschaft.   S. 1201 A

Verhandlungen der Anatomischen Gesellschaft. Ergänzungsheft
    Anatomischer Anzeiger. Jena.
    Verh.anat.Ges.Jena 1887 →   Z.S 1520

Verhandlungen des Botanischen Vereins der Provinz
    Brandenburg. Berlin.
    Verh.bot.Ver.Prov.Brandenb. 1859 →   B.S 907

Verhandlungen des Comité du Néogène Méditerranéen.
    See Proceedings. Committee on Mediterranean Neogene
    Stratigraphy.   P.S 993 A

Verhandlungen der Deutschen Gesellschaft für Angewandte
    Entomologie. Berlin.
    Verh.dt.Ges.angew.Ent. 1918 →   E.S 1359 a

Verhandlungen der Deutschen Physikalischen Gesellschaft.
    Leipzig.
    Verh.dt.phys.Ges. 1899-1902
    Formerly Verhandlungen der Physikalischen
    Gesellschaft zu Berlin. Leipzig.   M.S 1427

| TITLE | SERIAL No. |

Verhandlungen des Deutschen Wissenschaftlichen Vereins zu Santiago.
de Chile. Valparaiso.
    Verh.dt.wiss.Ver.Santiago Chile 1885-1934.   S. 2235 A

Verhandlungen der Deutschen Zoologischen Gesellschaft.
Supplementband, Zoologischer Anzeiger. Leipzig.
    Verh.dt.zool.Ges. 1891 →   Z.S 1370
                       1902 →   T.R.S 1333

Verhandlungen der Geologischen Bundesanstalt. Wien.
    Verh.geol.Bundesanst.,Wien 1945 →
                           Sonderheft 1950 →
    Formerly Verhandlungen der Geologischen Reichsanstalt
(Staatsanstalt-Landesanstalt.) Wien.   P.S 1380

Verhandlungen der Geologischen Reichsanstalt
(Staatanstalt-Landesanstalt). Wien.
    Verh.geol.Reichsanst.(StAnst-Landesanst.), Wien 1867-1944.
    (From 1940-1944 See Berichte der Reichsstelle
     für Bodenforschung. Wien.)
    Continued as Verhandlungen der Geologischen
Bundesanstalt. Wien.   P.S 1380

Verhandlungen der Gesellschaft Deutscher Naturforscher und Arzte.
    Verh.Ges.dt.Naturf.Arzte 1890 →
    Formerly Tageblatt der Versammlung Deutscher
Naturforscher und Aerzte.   S. 1302 B

Verhandlungen der Gesellschaft für Erdkunde zu Berlin.
    Verh.Ges.Erdk.Berl. 1873-1888.   S. 1324 B

Verhandlungen der Gesellschaft Naturforschender Freunde zu Berlin.
    Verh.Ges.naturf.Freunde Berl. (1819)-1829.
    Formerly Magazin. Gesellschaft Naturforschender
Freunde zu Berlin.   S. 1326 A

Verhandlungen der Gesellschaft für Physische Anthropologie.
Stuttgart. Sonderheft des Anthropologischen Anzeigers.
    Verh.Ges.phys.Anthrop. 1926-1937.   P.A.S 45

Verhandlungen der Gesellschaft des Vaterländischen Museums
in Böhmen. Prag.
    Verh.Ges.vaterl.Mus.Prag 1823-1856.   S. 1761 A

Verhandlungen des Heil-und Naturwissenschaftlichen Vereins zu
Bratislava (Pressburg).
    See Poszonyi Orvos-Természettudományi Egyesület
Közleményei. Poszony.   S. 1763

Verhandlungen des Internationalen Geographischen
Kongresses in Berlin.
    Verh.Int.geogr.Kongr. 7th. 1899.   S. 2710

TITLE	SERIAL No.

Verhandlungen der Internationalen Vereinigung für
    Theoretische und Angewandte Limnologie. Stuttgart.
    Verh.int.Verein.theor.angew.Limnol. 1922 →   S. 1573 D

Verhandlungen des Internationalen Zoologen-Congresses.
    See International Congress of Zoology.   Z.S 2701

Verhandlungen der Kaiserlich-Königlich Geologischen Reichsanstalt
    (Staatsanstalt-Bundesanstalt). Wien.
    See Verhandlungen der Geologischen Reichsanstalt
    (Staatsanstalt-Landesanstalt.)   P.S 1380

Verhandlungen der Kaiserlich Königlichen Zoologisch-Botanischen
    Gesellschaft in Wien.
    See Verhandlungen der Zoologisch-Botanischen Gesellschaft
    in Wien.   S. 1779 A

Verhandlungen der Kaiserlichen Gesellschaft für die Gesammte
    Mineralogie zu St. Petersburg. St. Petersburg.
    Verh.K.Ges.Miner.St.Petersburg 1862
    (Contains the papers for 1859-1861)
    Formerly Verhandlungen der Russisch-Kaiserlichen
    Mineralogischen Gesellschaft zu St. Petersburg.
    Continued as Zapiski Imperatorskogo (S.-Peterburgskogo)
    Mineralogicheskogo Obshchestva.   M.S 1820 & P.S 500

Verhandlungen der Kaiserlichen Leopoldinisch-Carolinischen
    Akademie der Naturforscher.
    See Nova Acta Leopoldina, Halle a.S.   S. 1301 A

Verhandlungen. Kongress für Erforschung der Producktionskräfte
    und der Volkswirtschaft der Ukraine.
    See Pratsi. Z'yizd Doslidzhennya Produktsiinykh sil ta
    Narodn'oho Hospodarstva Ukrayiny.   P.S 548

Verhandlungen und Mitteilungen des Siebenbürgischen Vereins für
    Naturwissenschaften zu Hermannstadt.
    Verh.Mitt.siebenb.Ver.Naturw. 1850-1940.
    Continued as Mitteilungen der Arbeitsgemeinschaft für
    Naturwissenschaften Sibiu-Hermannstadt.   S. 1735

Verhandlungen der Naturforschenden Gesellschaft in Basel.
    Verh.naturf.Ges.Basel 1854 →   S. 1216 A
                               1854-1904.   T.R.S 1207
    Formerly Bericht über die Verhandlungen der
    Naturforschenden Gesellschaft in Basel.   S. 1216 A

Verhandlungen des Naturforschenden Vereins in Brünn.
    Verh.naturf.Ver.Brünn 1862-1944.   S. 1711

Verhandlungen des Naturhistorisch-Medizinischen
    Vereins zu Heidelberg.
    Verh.naturh.-med.Ver.Heidelb. 1874-1963.   S. 1458

TITLE	SERIAL No.

Verhandlungen des Naturhistorischen Vereins der
    Preussischen Rheinlande und Westfalens. Bonn.
    Verh.naturh.Ver.preuss.Rheinl. 1844-1933.
    <u>Continued as</u> Decheniana. Verhandlungen des Naturhistorischen
Vereins der Rheinlande und Westfalens. Bonn.   S. 1361 A

Verhandlungen des Naturwissenschaftlichen Vereins in
    Hamburg (Altona).
    Verh.naturw.Ver.Hamb. 1875-1880, 1893-1935.
    <u>Continued in</u> Abhandlungen und Verhandlungen des
Naturwissenschaften Vereins in Hamburg.   S. 1446 A

Verhandlungen des Naturwissenschaftlichen Vereins in Karlsruhe.
    Verh.naturw.Ver.Karlsruhe 1862-1935.   S. 1378

Verhandlungen der Ornithologischen Gesellschaft in Bayern.
    München.
    Verh.orn.Ges.Bayern Bd.4-22, 1903-1942.
    <u>Formerly</u> Jahresbericht des Ornithologischen Vereins
München.   T.R.S 1307 A & Z.S 1652

Verhandlungen der Physikalisch-Medizinischen Gesellschaft
    zu Würzburg.
    Verh.phys.-med.Ges.Wurzb. 1850-1935.
    <u>Continued as</u> Bericht der Physikalisch-Medizinischen
Gesellschaft zu Würzburg.   S. 1611 A

Verhandlungen der Physikalisch-Medizinischen Societät
    zu Erlangen.
    Verh.phys.-med.Soc.Erlangen 1865-1870.
    <u>Continued as</u> Sitzungsberichte der Physikalisch
Medizinischen Sozietät zu Erlangen.   S. 1405

Verhandlungen der Physikalischen Gesellschaft zu Berlin.
    Leipzig.
    Verh.phys.Ges.Berlin Vol.13-17; 1895-1898
    <u>Continued as</u> Verhandlungen der Deutschen
Physikalischen Gesellschaft. Leipzig.   M.S 1427

Verhandlungen der Russisch-Kaiserlichen Mineralogischen
    Gesellschaft zu St.Petersburg.   M.S 1820 &
    Verh.Russ.-Kaiser.Miner.Ges.St.Petersburg 1842-1858.   P.S 500
    <u>Continued as</u> Verhandlungen der Kaiserlichen Gesellschaft für
die Gesammte Mineralogie zu St. Petersburg.

Verhandlungen der Schweizerischen Naturforschenden Gesellschaft.
    Verh.schweiz.naturf.Ges. Sess. 23 → 1838 →   S. 1201 A
    Sess. 55-82, 1872-1900.
    <u>Formerly</u> Verhandlungen der Allgemeinen Schweizerischen   T.R.S 1210
Gesellschaft für die Gesamten Naturwissenschaften.   S. 1201 A

TITLE	SERIAL No.

Verhandlungen des Vereins für Natur-und Heilkunde zu Pressburg.
    Verh.Ver.Nat.-Heilk.Pressburg 1869-1872.
    Formerly Verhandlungen des Vereins für Naturkunde
    zu Pressburg.
    Continued as Pozsonyi Orvos-Természettudományi Egyesület
    Közleményei. Pozsony.   S. 1763

Verhandlungen des Vereins für Naturkunde zu Pressburg.
    Verh.Ver.Naturk.Pressburg 1856-1866.
    Continued as Verhandlungen des Vereins für Natur-und
    Heilkunde zu Pressburg.   S. 1763

Verhandlungen des Vereins für Naturwissenschaftliche
    Heimatforschung. Hamburg.
    Verh.Ver.naturw.Heimatforsch. 1930 →
    Formerly Verhandlungen des Vereins für Naturwissenschaftliche
    Unterhaltung zu Hamburg.   S. 1449

Verhandlungen des Vereins für Naturwissenschaftliche
    Unterhaltung zu Hamburg.
    Verh.Ver.naturw.Unterh.Hamb. 1871-1929.   S. 1449
                                    1871-1903.   T.R.S 1341
    Continued as Verhandlungen des Vereins für Naturwissenschaftliche
    Heimatforschung. Hamburg.

Verhandlungen der Zoologisch-Botanischen Gesellschaft in Wien.
    Verh.zool.-bot.Ges.Wien 1858 →   S. 1779 A & T.R.S 1805
                               1858-1910   E.S 1708
    Formerly Verhandlungen des Zoologisch-Botanischen Vereins
    in Wien.

Verhandlungen des Zoologisch-Botanischen Vereins in Wien.
    Verh.zool.-bot.Ver.Wien 1851-1857.
    Continued as Verhandlungen der Zoologisch-Botanischen
    Gesellschaft in Wien.   S. 1779 A & T.R.S 1805

Veröffentlichung der Haupt-Pilzstelle am Botanischen Museum
    der Universität in Berlin - Dahlem.
    Veröff.Haupt-Pilzst.bot.Mus.Univ.Berlin - Dahlem
    Nr.1, 1941.   B.S.M 38

Veröffentlichungen der Alexander Kohut Memorial Foundation. Wien.
    Veröff.Alex.Kohut meml Fdn Bd.2-4, 1924-1926.   B. Herbarium

Veröffentlichungen der Arbeitsgemeinschaft Donauforschung der
    Societas Internationalis Limnologiae. Stuttgart.
    See Archiv für Hydrobiologie. Supplement. Stuttgart.   S. 1573 B

Veröffentlichungen der Arbeitsgemeinschaft für Forschung des
    Landes Nordrhein-Westfalen. Köln.
    Veröff.ArbGem.Forsch.Landes Nordrhein-Westfal.
      Natur-, Ingenieur- und Gesellschaftswissenschaften.
      Heft 6-196, 1951-1969. (Natural History numbers only held.)
      Continued as Vorträge. Rheinisch-Westfälische Akademie
    der Wissenschaften. Opladen.   S. 1400

TITLE	SERIAL No.

Veröffentlichungen der Deutschen Akademischen Vereinigung
zu Buenos Aires.
   Veröff.dt.Akad.Verein.B.Aires Bd. 1, Hft 1 & 5.
    1899 & 1900.   S. 2227

Veröffentlichungen aus dem Deutschen Kolonial-und Ubersee-
Museum in Bremen.
   Veröff.dt.Kolon.u.Ubersee-Mus.Bremen 1935-1942.   S. 1372
                1935-1939.   T.R.S 1353
   Continued as Veröffentlichungen aus dem Museum für
Natur-Völker- und Handelskunde in Bremen. Reihe A.   S. 1372

Veröffentlichungen des Forschungsinstituts für Hydrobiologie
der Naturwissenschaftlichen Fakultät. Universität Istanbul.
   See Istanbul Universitesi fen Fakültesi Hidrobiologi.
Istanbul. Seri B.   S. 1887 C

Veröffentlichungen des Geobotanischen Institut Rübel. Zürich.
   Veröff.geobot.Inst.Zürich 1924 →   B.S 836

Veröffentlichungen aus dem Haus der Natur in Salzburg.
   Veröff.Haus Nat.Salzburg 1964 →   S. 1788 E

Veröffentlichungen des Instituts für Lagerställenforschung
der Türkei. Ankara.
   See Maden Tetkik ve Arama Enstitüsü Yayinlarindan. Ankara.   P.S 1468

Veröffentlichungen des Instituts für Meeresforschung in
Bremerhaven. Bremen.
   Veröff.Inst.Meeresforsch.Bremerh. 1952 →   S. 1373

Veröffentlichungen des Luxemburgischen Geologischen
Landesaufnahmedienstes.
   See Publications du Service de la Carte Géologique de
Luxembourg.   P.S 1279

Veroffentlichungen der Museen der Stadt Gera. Gera.
   Veröff.Mus.Stadt Gera
     Naturwiss.Reihe 1973 →   S. 1420

Veröffentlichungen des Museum Ferdinandeum in Innsbruck.
   Veröff.Mus.Ferdinandeum Innsb. 1922-1965.
   Continued as Veröffentlichungen des Tiroler Landesmuseum
Ferdinandeum. Innsbruck.   S. 1736 B

Veröffentlichungen aus dem Museum für Natur-Völker- und
Handelskunde in Bremen. Reihe A.
   Veröff.Mus.nat.-Völker- Handelsk.Bremen 1949.
   Formerly Veröffentlichungen aus dem Deutschen Kolonial- und
Uebersee- Museum in Bremen.
   Continued as Veröffentlichungen aus dem Ubersee-
Museum Bremen.   S. 1372

Veröffentlichungen des Museums für Völkerkunde zu Leipzig.
   Veröff.Mus.Völkerk.Leipzig Heft 14 → 1965 →   P.A.S 44 A
                        Heft 16, 1957.   B. 581.9(595)TRE

TITLE	SERIAL No.

Veröffentlichungen aus dem Naturhistorischen Museum in Basel.
   Veröff.naturh.Mus.Basel 1960 →   S. 1216 C

Veröffentlichungen aus dem Naturhistorischen Museum. Wien.
   Veröff.naturh.Mus.Wien Heft 1-5, 1924-1925.
      N.F. 1 → 1958 →   S. 1777 C

Veröffentlichungen des Naturwissenschaftlichen Vereins
   zu Osnabrück.
      Veröff.naturw.Ver.Osnabr. 1926-1970.
      Formerly Jahresbericht des Naturwissenschaftlichen Vereins
   zu Osnabrück.   S. 1566
      Replaced by Osnabrücker Naturwissenschaftliche
   Mitteilungen. Osnabrück.   S. 1566 B

Veröffentlichungen der Staatlichen Stelle für Naturschutz
   beim Württemburgischen Landesamt für Denkmalpflege. Stuttgart.
      Veröff.st.Stelle NatSchutz Württ Heft 1-11, 1924-1934.
   (For Heft 1 see Beilage zu den Jahresheften des Vereins
    für Vaterländische Naturkunde in Württemburg, Jahrg.80.
   and Heft 5 → Jahreshefte des Vereins für Vaterlandische
   Naturkunde in Württemburg. Jahrg.84 →)
      Continued as Veröffentlichungen der Württembergischen
   Landesstelle für Naturschutz.   S. 1591 C

Veröffentlichungen des Tiroler Landesmuseum Ferdinandeum. Innsbruck.
   Veröff.tirol.Landesmus.Ferdinandeum 1965 →
   Formerly Veröffentlichungen des Museums Ferdinandeum.
   Innsbruck.   S. 1736 B

Veröffentlichungen aus dem Ubersee-Museum Bremen. Reihe A.
   Veröff.Uberseemus.Bremen 1952 →
   Formerly Veröffentlichungen aus dem Museum für Natur-
   Völker- und Handelskunde in Bremen.   S. 1372

Veroffentlichungen der Universität Innsbruck. Innsbruck.
   Veröff.Univ.Innsbruck
      Alpenkundliche Studien 1968 →   P.S 386

Veröffentlichungen des Wissenschaftlichen Vereins in Zürich.
   See Monatsschrift des Wissenschaftlichen Vereins in Zürich.   S. 1258

Veröffentlichungen der Württembergischen Landesstelle für
   Landschaftspflege in Ludwigsburg und Tübingen (Baden -
   Württemberg und der Württembergischen Bezirksstellen in
   Stuttgart und Tübingen.)
      Veröff württ.Landest.LandschPfl. Hft.21-25, 1951-1957.
   (Published in Jahreshefte des Vereins für Väterlandische
   Naturkunde in Württemburg.)
      Formerly Veröffentlichungen der Württembergischen
   Landesstelle für Naturschutz.   S. 1591 A

| TITLE | SERIAL No. |

Veröffentlichungen der Württembergischen Landesstelle für
Naturschutz. Stuttgart.
Veröff.württ.Landest.NatSchutz Hft.12-16, 18-20, 1935-1950.
(Published in Jahreshefte des Vereins für Väterlandische
Naturkunde in Württemburg.)
Formerly Veröffentlichungen der Staatliche Stelle für
Naturschutz beim Württembergische Landesamt für Denkmalpflege.
Continued as Veroffentlichungen der Württembergischen
Landesstelle für Landschaftspflege in Ludwigsburg und
Tübingen (Baden - Württemberg und der Württembergischen
Bezirksstellen in Stuttgart und Tübingen.)   S. 1591 A

Veröffentlichungen der Zoologischen Staatssammlung München.
Veröff.zool.StSamml.Münch. 1950 →   Z.S 1565

Verslag.
See Verslagen.

Verslagen. Afdeeling Natuurkunde Koninklijke Nederlandsche
Akademie van Wetenschappen.
See Verslagen van de Gewone Vergadering der Afdeeling
Natuurkunde. Koninklijke (Nederlandse) Akademie van
Wetenschappen te Amsterdam.   S. 601 B a

Verslagen. Afdeeling Nederlandsch-Oost-Indië van de
Nederlandsche Entomologische Vereeniging.
See Verslagen van de Vergaderingen der Afdeeling
Nederlandsch-Oost-Indië van de Nederlandsche
Entomologische Vereeniging.   E.S 2034

Verslagen. Caraibisch Marien Biologisch Instituut. Curacao.
Versl.caraib.mar.biol.Inst. 1967/1968/1969.
Continued as Jaarverslag. Caraibisch Marien Biologisch
Instituut. Curacao.   S. 2281 B

Verslagen van de Gewone Vergadering der Afdeeling Natuurkunde.
Koninklijke (Nederlandse) Akademie van Wetenschappen
te Amsterdam.
Versl.gewone Vergad.Afd.Natuurk.K.ned.Akad. Deel. 38-43 & 52 →
    1929-1934 & 1943 →
(For previous Volumes see S. 601. B. Proceedings
for same year.)   S. 601 Ba

Verslagen van de Gewone Vergadering der Wis-en Natuurkundige
Afdeeling Koninklijke Akademie van Wetenschappen te Amsterdam.
Versl.gewone Vergad.wis-en natuurk.Afd.K.Akad.Wet.Amst. 1897-1898.
Formerly Verslagen der Zittingen van de Wis-en Natuurkundige
Afdeeling der Koninklijke Akademie van Wetenschappen. Amsterdam.
Continued as Proceedings, Section of Sciences of
Sciences. Koninklijke Akademie van Wetenschappen
te Amsterdam.   S. 601 B

TITLE   SERIAL No.

Verslagen en Mededeelingen Betreffende Indische Delfstoffen
    en hare Toepassingen. Dienst van den Mijnbouw in
    Nederlandsch Oost-Indië. Weltvreden.
    Versl.Meded.indische Delfstoffen   No.3, 6-19;
        1906-1926 (imp.)   M.S 1906

Verslagen en Mededeelingen. Commissie voor het Botanisch Onderzoek
    van de Zuiderzee en Omgeving. Bilthoven.
    Versl.Meded.Comm.bot.Onderz.Zuiderzee Omgev.
        No.2 - 4, 1927-1928.   B.S 304

Verslagen en Mededeelingen der Koninklijke Akademie van
    Wetenschappen (Afdeeling Natuurkunde). Amsterdam.
    Versl.Meded.K.Akad.wet.Amst. 1853-1892.   S. 601 B & T.R.S 757
    Continued as Verslagen der Zittingen van de Wis-en
    Natuurkundige Afdeeling der Koninklijke Akademie van
    Wetenschappen. Amsterdam.   S. 601 B

Verslagen en Mededeelingen. Nederlandsche Ornithologische
    Vereeniging. Wageningen.
    Versl.Meded.ned.orn.Vereen.   1904-1909.
    Continued as Jaarboek der Nederlandsche Ornithologische
    Vereeniging. Wageningen.   T.R.S 745

Verslagen. Nederlandsch-Indische Vereeniging tot Natuurbescherming.
    Buitenzorg.
    Versl.nederl.-ind.Vereen.Natuurbesch.   1920/1922, 1924-1934.   S. 658

Verslagen van de Openbare Vergadering der Eerste Klasse van het
    Koninklijk-Nederlandsch Instituut van Wetenschappen,
    Letterkunde en Schoone Kunsten. Amsterdam.
    Versl.Openb.Vergad.Kon.-Ned.Inst.Wet.Amst.   1817-1823.
    Formerly Verslag van de Werkzaamheden der Eerste Klasse
    van het Koninklijke Instituut van Wetenschappen, Letterkunde
    en Schoone Kunsten.   S. 602 A

Verslagen van het Proefstation voor de Java-Suikerindustrie.
    Soerabaia.
    Versl.Proefstn Javasuik.-Ind.   1928-1929.   E.S 2032

Verslagen van het Rijksherbarium. Leiden.
    Versl.Rijksherb.Leiden   1936-1938; 1959 →   B.S 312

Verslagen omtrent het Rijksmuseum van Natuurlijke Historie te Leiden.
    Versl.Rijksmus.nat.Hist.Leiden   1938 →   Z.S 601 C
                                     1938-1949.   T.R.S 749

Verslagen van de Vergaderingen der Afdeeling Nederlandsch-
    Oost-Indië van de Nederlandsche Entomologische Vereeniging.
    Amsterdam.
    Versl.Vergad.Afd.Ned.-Oost-Indië ned.ent.Vereen. 1931-1934.
    Continued as Entomologische Mededeelingen van
    Nederlandsch-Indië. Buitenzorg.   E.S 2034

| TITLE | SERIAL No. |

Verslagen van de Werkzaamheden der Eerste Klasse van het
Koninklijke Instituut van Wetenschappen Letterkunde en
Schoone Kunsten. Amsterdam.
Versl.Werkz.Kon.Inst.Wet.Amst. 1809-1816.
Continued as Verslagen van de Openbare Vergadering der
Eerste Klasse van het Koninklijk-Nederlandsch Instituut
van Wetenschappen, Letterkunde en Schoone Kunsten.
Amsterdam.   S. 602 A

Verslagen van de Werkzaamheden. Rijksinstituut voor Veldbiologisch
Onderzoek ten behoven van het Natuurbehoud. RIVON.
Bilthoven.
Versl.Werkzaamh.Rijksinst.Veldbiol. 1959-1961.
Continued as Jaarverslag. Rijksinstituut voor Veldbiologisch
Onderzoek ten behoeve van het Natuurbehoud. RIVON.   S. 619

Verslagen der Zittingen van de Wis-en Natuurkundige Afdeeling
der Koninklijke Akademie van Wetenschappen. Amsterdam.
Versl.Zitt.Wis-natuurk.Afd.K.Akad.wet.Amst. 1893-1896.
Formerly Verslagen en Mededeelingen der Koninklijke
Akademie van Wetenschappen (Afdeeling Natuurkunde). Amsterdam.
Continued as Verslagen van de Gewone Vergadering der
Wis-en Natuurkundige Afdeeling Koninklijke Akademie van
Wetenschappen te Amsterdam.   S. 601 B

Versuche und Abhandlungen der Naturforschenden Gesellschaft
in Danzig.
Vers.Abh.naturf.Ges.Danzig 1747-1756.
Continued as Neue Sammlung von Versuchen und Abhandlungen
der Naturforschenden Gesellschaft in Danzig.   S. 1384 A

Vertebrata Hungarica. Musei Historico-Naturalis Hungarici. Budapest.
Vertebr.hung. Vol.1 fasc.1 → 1959 →   Z.S 1744

Vertebrata Palasiatica. Peking.
Vertebr.palasiat. 1957 →   P.S 1793 A

Vertebratologické Zprávy. Brno.
Vertebr.Zprávy 1967 →   Z.S 1757 A

Verzameling der Wetenschappelijke Werken. Prins Leopold
Instituut voor Tropische Geneeskunde.
See Recueil des Travaux Scientifiques. Institut de Médecine
Tropicale Prince Leopold.   S. 776 B

Verzeichnis der Vorlesungen an den Königlichen Akademie zu
Braunsberg.
Verz.Vorles.K.Akad.Braunsberg 1912-1922.
Formerly Verzeichnis der Vorlesungen am Königlichen
Lyceum Hosianum zu Braunsberg.   B.S 940 b

Verzeichnis der Vorlesungen am Königlichen Lyceum Hosianum
zu Braunsberg.
Verz.Vorles.K.Lyceum hosian.Braunsberg 1906-1909.
Continued as Verzeichnis der Vorlesungen an der Königlichen
Akademie zu Braunsberg.   B.S 940 b

TITLE	SERIAL No.

Vesnik Geološkog Instituta Kraljevine Jugoslavije. Beograd.
    Vesn.geol.Inst., Jugosl. 1932-1940
    Continued as Geoloski Vesnik. Savezna Uprava za Geološka
    Istrazivanja. Beograd.   P.S 1463

Vesnik. Zavod za Geoloska i Geofizicka Istrazivanja. Beograd.
    Vesn.Zavod geol.geofiz.Istr. 1953 →
    Formerly Geoloski Vesnik. Savezna Uprava za Geoloska
    Istrazivanja. Beograd.   P.S 1463

Vestis. Latvijas PSR Zinatnu Akademijas. Riga.
    See Latvijas PSR Zinatnu Akademijas Vestis. Riga.   S. 1852 e

Vestnik Akademii Nauk Belorusskoĭ SSR.
    See Vestsi Akademii Navuk Belaruskaĭ SSR. Minsk.   S. 1814 a B

Vestnik Akademii Nauk SSSR. Moskva, Leningrad.
    Vest.Akad.Nauk, SSSR 1937-1940 (imp.) 1948 →   S. 1802 V

Vestnik Akademii Nauk Ukrainskoĭ SSR. Kiev.
    See Visnȳk Akademiyi Nauk Ukrayins'koyi RSR. Kyyiv.   S. 1834 a H

Vestnik. Akademiya Nauk Kirgizskoĭ SSR. Frunze.
    Vest.Akad.Nauk kirgiz.SSR 1961.   S. 1820 a E

Vestnik Belorusskogo Gosudarstvennogo Universiteta imeni
    V.I. Lenina. Minsk.
    Vest.belorussk.gos.Univ.
        Seriya II. 1972 →   S. 1814 B

Vestnik Botanicheskogo Obshchestva Gruzinskoĭ SSR. Tbilisi.
    Vest.bot.Obshch.gruz.SSR 1962 →   B.S 1432

Věstník Ceské Akademie Císaré Františka Josefa pro Vedy,
    Slovesnost a Umění v Praze.
    Věst.české Akad.Císaré Františka Josefa 1891-1920.
    Continued as Vestnik Ceske Akademie věd a Uméní v Praze.   S. 1760 B

Věstník Ceské Akademie Věd a Uměni v Praze.
    Věst.česke Akad.Věd.Uměni 1926-1952.
        (Wanting Vol. 28-29.)
    Formerly Věstník Cěske Akademie Císaré Františka
    Josefa pro Vedy, Slovesnost a Uměni v Praze.
    Continued as Věstník Ceskoslovenské Akademie věd. Praha.   S. 1760 B

Věstník Ceskoslovenské Akademie Věd. Praha.
    Věst.čsl.Akad.Věd. 1953 →
    Formerly Věstník Ceske Akademie věd a Uměni v Praze.   S. 1760 B

Věstník Ceskoslovenské Společnosti Zoologické. Praha.
    Věst.čsl.Spol.zool. 1952 →
    Formerly Věstnik Ceskoslovenské Zoologické
    Společnosti v Praze.   Z.S 1760

| TITLE | SERIAL No. |

Věstnik Ceskoslovenské Zoologické Společnosti v Praze.
    Věst.čsl.zool.Spol. 1934-1951.
    Continued as Věstnik Ceskoslovenské Společnosti
    Zoologické. Praha.   Z.S 1760

Věstnik Ceskoslovenského Zemědělského Museum. v Praze.
    Věst.čsl.zeměd.Mus. 1928-1931.   S. 1756 a

Vestnik du Comité Géologique. Leningrad.
    See Vestnik Geologicheskogo Komiteta. Leningrad.   P.S 1506 B

Vestnik Estestvoznaniya. S.-Peterburg.
    Vest.Estest.S.-Peterb. 1890-1893.   S. 1855 C

Vestnik Geologicheskogo Komiteta. Leningrad.
    Vest.geol.Kom. 1925-1928.   P.S 1506 B

Vestnik Gosudarstvennogo Geologicheskogo Komiteta ChSR. Praga.
    See Vestnik Státniho Geologického Ustavu Ceskoslovenské
    Republiky. Praha.   P.S 1354

Vestnik Gosudarstvennogo Muzeya Gruzii. Tbilisi.
    Vest.gos.Mus.Gruzii Ser.A & B. Vol.15 → 1951 →
    Formerly Bulletin du Muséum Géorgie. Tiflis.   S. 1858 C

Vestnik Khar'kovskogo Universiteta. Khar'kov.
    Vest.khar'kov.Univ. No.39, 55. 1970.
    Continued as Visnyk Kharkivskogo Universitetu. Kharkiv.   S. 1833 C

Vestník Klubu Přírodovědeckého v Prostějově.
    Vest.Klubu přír.Prostějove 1898-1940.   S. 1763 a

Věstnik Královské Ceské Společností Nauk Třída mat.-prír. Praha.
    See Sitzungsberichte der Königl.Böhmischen Gesellschaft
    der Wissenschaften. Prag.   S. 1703 A

Věstník Lekářsko-Přírodovědeckého Spolku v Bratislave.
    See Pozsonyi Orvos-Természettudományi Egyesület
    Közleményei Pozsony.   S. 1763

Vestnik Leningradskogo Gosudarstvennogo Universiteta. Leningrad.
    Ser.Biol., Geol.Geogr.
    Vest.Leningr.gos.Univ. 1955, 1957 → (imp.)   S. 1803 B

Vestnik Moskovskogo Universiteta. Moskva.
    Vest.Mosk.Univ.
      Ser. 4 Geologiya 1974 →   P.S 519 B
      Ser. 6 Biologiya, Pochvovedenie No.6 → 1963 →   S. 1844 F

Vestnik Penzenskogo Obshchestva Lyubiteleĭ
    Estestvoznaniya i Kraevedeniya. Penza.
    Vest.penz.Obshch.Lyub.Estest.Kraev. 1925.   S. 1848 B

TITLE	SERIAL No.

Věstnik Russkoĭ Flory. Yurêv.
    Vêst.russk.Flory 1916.   B.S 1450

Věstnik Russkoĭ Prikladnoĭ Entomologii. Kiev.
    Vêst.russk.prikl.Ent. 1914-1915.   E.S 1806

Vestnik i Sjezdu Ceskoslovenských Botaniku v Praze.
    See Bulletin du Congrès des Botanistes Tchécoslovaques.   B.S 1232

Věstník Státního Geologického Ustavu Ceskoslovenské
Republiky. Praha.
    Vést.st.geol.Ust.čsl.Repub. 1925-1950
    Continued as Vestník Ustředního Ustavu Geologického. Praha.   P.S 1354

Vestnik Tiflisskogo Botanicheskogo Sada. Tiflis.
    Vest.tiflis.bot.Sada 1905-1931.
    (wanting livr. 37-39, 1915)   B.S 1430 a

Věstník Ustředního Ustavu Geologického. Praha.
    Vest.ústred.Ust.geol. No.26 → 1951 →
    Formerly Vestník Státního Geologického Ustavu Ceskoslovenské
Republiky. Praha.   P.S 1354

Vestnik Zashchity Rasteniĭ. Leningrad.
    Vest.Zashch.Rast. 1939-1941.
    Formerly Zashchita Rasteniĭ. Leningrad.   E.S 1817

Vestnik Zoologii. Institut Zoologii. Akademiya Nauk Ukrainskoĭ
SSR. Kiev.
    Vest.Zool. 1967 →   Z.S 1827 E

Vestsi Akademii Navuk Belaruskaĭ SSR. Minsk.
    Vestsi Akad.Navuk BSSR
      Seriya Biol.Navuk 1956 → (imp.)
    Formerly Izvestiya Akademii Nauk Belorusskoĭ SSR. Minsk.   S. 1814 a B

Vetenskapliga och Praktiska Undersökningar i Lappland,
Stockholm.
    Vetensk.prakt.Unders.Lappl. 1910-1930.   S. 586

Veterinaria. Spojené Podniky pro Zdravotnickou Výrobu. Praha.
    Veterinaria Roc.10 → 1968 →   Z.S 1761

Veterinary and Animal Sciences. Ankara.
    See Key to Turkish Science. Veterinary and Animal Sciences.   Z. R69 o T

Veterinary Bulletin. Weybridge.
    Vet.Bull.Weybridge 1931 →   Z.B 69 V

Veterinary Research Report, New South Wales. Sydney.
    Vet.Res.Rep.N.S.W. 1927-1937.   E.S 2254 a

Veterinary Service Annual Report.
    See Report. Veterinary Service, Egypt.   Z. 74B o E

TITLE	SERIAL No.
Veteriner Fakültesi Dergisi, Ankara Universitesi. Vet.Fak.Derg.Ankara Univ. 1954 →	Z.S 1888
Viaggi di Studio ed Esplorazioni. Reale Accademia d'Italia. Viaggi Stud.Esplor. 1933-1938.	S. 1158 B
Victoria Naturalist. Victoria, B.C. Victoria Nat. 1944 →	S. 2655 a
Victorian Entomologist. (Melbourne.) Victorian Ent. 1972 →	E.S 2259 d
Victorian Naturalist. Melbourne. Victorian Nat. 1884 → 1894-1910.	S. 2114 A T.R.S 7207
Victoria's Resources. Springvale. Victoria's Resour. 1959 →	S. 2106
Vida Acuatica. Barcelona. Vida acuat. No. 4 → 1970 →	S. 1030
Vida Silvestre. Madrid. Vida silv. 1971 →	Z.S 1014
Videnskabelige Meddelelser fra Dansk Naturhistorisk Forening i Kjøbenhavn. Vidensk.Meddr dansk naturh.Foren. 1849 → 1853-1858.	S. 522 Z.S 580
Vidya. Ahmedabad. Vidya Vol.5 → 1962 →	S. 1910 d
Vie et Milieu. Bulletin du Laboratoire Arago, Université de Paris. Vie Milieu 1950 → from Vol.16, 1965 published in Series A - Biologie marine, B - Océanographie, C - Biologie terrestre.	S. 928 a A
Vieraea. La Laguna. Vieraea 1970 →	S. 1038
Vierteljahrsschrift der Naturforschenden Gesellschaft in Zürich. Vjschr.naturf.Ges.Zürich 1856 →	S. 1256 D
Viewpoint Series. Australian Conservation Foundation. Canberra. Viewpoint Ser.Aust.Conserv.Fdn 1967 →	S. 2122 a
Viewpoints in Biology. London. Viewpts Biol. 1962-1965.	S. 459 a

| TITLE | SERIAL No. |

Vijesnik, Vijesti.
　　See Vjesnik, Vjesti.

Vijnana Parishad Anusandhan Patrika. Allahabad.
　　See Research Journal of the Hindi Science Academy.　　S. 1911 b

Vikopni Fauni Ukraïni i Sumizhnikh Teritoriï. Akademiya Nauk
　　Ukrains'koi RSP. Institut Zoologii. Kiev.
　　Vikopni Fauni Ukr.Sumizh.Terit. 1 → 1962 →　　P.S 72 Q.o.K

Viltrevy. Jaktbiologisk Tidskrift Utgiven av Svenska
　　Jägareförbundet. Stockholm.
　　Viltrevy 1955 →　　Z.S 506

Vinatorul si Pescarul Sportiv. Bukarest.
　　Vinat.Pesc.Sportiv Vol.10 → 1957 → (imp.)　　Z.S 1892

Vintonia. Vinton, Iowa.
　　Vintonia 1941-1953.　　S. 2341 a

Virginia Journal of Science. Richmond.
　　Va J.Sci. Vol.1, Vol.2, No.6-8, Vol.3, No.2-7,
　　1940-1943; 1950 →　　S. 2440 B

Vision Research. Oxford.
　　Vision Res. Vol.10 → 1970 →　　S. 2745

Visnȳk Akademiyi Nauk Ukrayins'koyi RSR. Kȳyiv.
　　Visn.Akad.Nauk URSR 1950-1959, 1969 → (imp.)
　　Formerly Visti Akademiyi Nauk Ukrayins'koyi RSR. Kȳyiv.　　S. 1834 a H

Visnȳk Botanichnoho Sadu. Akademiya Nauk Ukrayins'koyi RSR. Kȳyiv.
　　Visn.bot.Sadu 1959-1962.　　B.S 1372 b

Visnyk Kharkivskogo Universitetu. Kharkiv.
　　Visn.Kharkiv.Univ. 1971 → (imp.)
　　Formerly Vestnik Khar'kovskogo Universiteta. Khar'kov.　　S. 1833 C

Visnȳk Kyyïvs'koho Botanichnoho Sadu.
　　Visn.kȳyiv.bot.Sadu 1925-1930.　　B.S 1372 a

Visnȳk Kȳyivs'koho Universȳtetu. Kȳyiv.
　　Visn.kȳyiv.Univ.
　　　Seriya Biolohiyi. 1958 →　　S. 1834 b C
　　　Seriya Heolohiyi. No.13 → 1971 →　　P.S 531 I

Visti Akademiyi Nauk Ukrayin'skoyi RSR. Kȳyiv.
　　Visti Akad.Nauk URSR 1938 (No.8) - 1946. (imp.)
　　Continued as Visnȳk Akademiyi Nauk Ukrayin'skoyi RSR.
　　Kȳyiv.　　S. 1834 a H

Vistnȳk.
　　See Visnȳk.

Vita degli Animali. Milano.
　　See Natura Viva. Milano.　　l. f. N

Vita Marina. Den Haag.
　　Vita Marina 1964 →　　S. 688 a

TITLE	SERIAL No.

Vivarium Darmstadt Informationen. Darmstadt.
    Vivar.Darmstadt Inf. 1973 →   Z.S 1386

Vjesnik Hrvatskog Državnog Geoložkog Zavode i Hrvatskog Državnog
    Geoložkog Muzesa. Zagreb.
    Vjesn.hrv.držav.geol.Zav. 1942-1944.   P.S 458

Vjesti Geološkoga Zavoda u Zagrebu. Zagreb.
    Vjesti geol.Zav.Zagr. 1925-1929.   P.S 399

Vodnye Bogatstva nedr Zemli na Sluzhbu Sotsialisticheskomu
    Stroitelstvu. Leningrad, etc.
    Vodn.Bogatstva Vyp. 4-7; 1933-34.   M.S 1814

Voeltzkow's Reise.
    See Wissenschaftliche Ergebnisse. Reise in Ostafrika
    1903-1905 von A. Voeltzkow. Stuttgart.   70.q.V

Vögel im Käfig und Voliere. Ein Handbuch für Vogelliebhaber.
    Aachen.
    Vögel im Käfig Voliere 1960 →
    Zweite Auflage 1963 →   T.B 70 S

Vogelwarte. Stuttgart.
    Vogelwarte 1948 →
    Formerly Vogelzug.   T.B.S 1310 & Z.S 1657

Vogelwelt. Berlin & München.
    Vogelwelt 1945 →   T.R.S 1321 A
        Vol.70 → 1949 →   Z.S 1654
    Formerly Deutsche Vogelwelt. Berlin.   T.R.S 1321 A

Vogelzug. Berlin.
    Vogelzug 1930-1943.
    Continued as Vogelwarte.   T.B.S 1310 & Z.S 1657

Volga. Akademiya Nauk SSSR. Tol'yatti.
    Volga 1968 →   Z. 72Q o L

Vollständiges Bücher-Lexicon. Leipzig.   REF.
    See Kayser's Vollständiges Bücher-Lexicon.   C.R.

Voprosy Antropologii. Moskva.
    Vop.Antrop. 1960 →
    Formerly Sovetskaya Antropologiya. Moskva Gosudarstvennyĭ
    Universitet im M.V. Lomonsova.   P.A.S 350

Voprosȳ Biostratigrafii Kontinental'nȳkh Tolshch.
    See Trudȳ... Sessii Vsesoyuznogo Paleontologicheskogo
    Obshchestva.   P.S 541

Voprosȳ Botaniki. Moskva.
    Vop.Bot. Vols. 1-2. 1954.   B.S 1403 h

| TITLE | SERIAL No. |

Voprosy Botaniki. Vilnius.
   See Botanikos Klausimai. Botanikos Institutas. Vilnius.   B.S 1516

Voprosy Ekologii. Kyyiv.
   Vop.Ekol. 1957-1962.   S. 1834 b B

Voprosy Ekologii i Biotsenologii. Leningrad.
   Vop.Ekol.Biotsen. Vyp.3, 5 → 1936 →
   (Suspended from 1940-1962.)   S. 1855 F

Voprosy Geologii Kuzbassa. Moskva.
   Vop.Geol.Kuzbassa 1956 →   P.S 1555

Voprosy Ikhtiologii. Moskva.
   Vop.Ikhtiol. Vyp.4 → 1955 → (imp.)   Z.S 1858

Voprosy Litologii i Petrografii. L'vov.
   Vip.Litol.Petrogr. 1969 →   M.S 1837

Voprosy Mikropaleontologii. Akademiya Nauk SSSR. Otdelenie
   Geologo-Geograficheskikh Nauk. Geologicheskii Institut. Moskva.
   Vop.Mikropaleont. 1956 →   P.S 514

Voprosy Mineralogii Osadochnykh Obrazovanii. (L'vov).
   Vop.Miner.osad.Obraz. Kn.2 → 1955 →   M.S 1835

Voprosy Obshchei i Meditsinskoi Mikrobiologii. Riga.
   See Trudy Instituta Mikrobiologii. Akademiya Nauk
   Latviiskoi SSR. Riga.   S. 1852 e D

Voprosy Paleobiologii i Biostratigrafii.
   See Trudy. Vsesoyuznoe Paleontologicheskoe Obshchestvo.   P.S 541

Voprosy Paleontologii. Leningrad.
   Vop.Paleont. 1950 →   P.S 520

Voprosy Pochvennoi Mikrobiologii. Riga.
   See Trudy Instituta Mikrobiologii. Akademiya Nauk
   Latviiskoi SSR. Riga.   S. 1852 e D

Voprosy Prudovogo Rybovodstva. Moskva.
   See Trudy. Vserossiiskii Nauchno-Issledovatel'skii
   Institut Prudovogo Rybnogo Khozyaistva. Moskva.   Z.S 1834

Voprosy Sel'skokhozyaistvennoi Mikrobiologii. Riga.
   See Trudy Instituta Mikrobiologii. Akademiya Nauk
   Latviiskoi SSR. Riga.   S. 1852 e D

Voprosy Stratigrafii. Leningrad.
   Vopr.Stratigr. 1974 →   P.S 520 A

TITLE	SERIAL No.

Voprosy̆ Virusologii. Riga.
    See Trudy̆ Instituta Mikrobiologii. Akademiya Nauk
    Latviĭskoĭ SSR. Riga.   S. 1852 e D

Voprozy̆ Paleobiogeografii i Biostratigrafii.
    See Trudy̆... Sessii Vsesoyuznogo Paleontologicheskogo
    Obshchestva.   P.S 541

Vorlesungen der Churpfälzischen Physikalisch-Ökonomischen
    Gesellschaft. Mannheim.
    Vorles.Churpfälz.phys.-ökon.Ges.   1784-1791.   S. 1457

Vorträge aus dem Gesamtgebiet der Botanik. Stuttgart.
    Vortr.GesGeb.Bot.   1962-1970.   B.S 901 a

Vorträge der Reichs-Zentrale für Pelztier-und
    Rauchwaren-Forschung. Leipzig.
    Vortr.Reichszent.Pelztier-u.RauchwarForsch.   1927.   Z.S 1357

Vorträge. Rheinisch-Westfälische Akademie der Wissenschaften.
    Opladen.
    Vortr.rhein.-westfäl.Akad.Wiss.
      Natur-, Ingenieur- und Wirtschaftswissenschaften.
      No.205 → 1970 → (Natural History numbers only held.)
      Formerly Veröffentlichungen der Arbeitsgemeinschaft für
      Forschung des Landes Nordrhein-Westfalen. Köln.   S. 1400

Vox Sanguinis. Basel.
    Vox Sang.   Vol.18 → 1970 →   P. SBG Unit

Vrule. Zadar.
    Vrule 1970 →   S. 1895 b

Vulkanologische en Seismologische Mededeelingen.
    Dienst van der Mijnbouw in Nederlandsch-Indië. Bandoeng.
    Vulkanolog.Meded.Bandoeng   No.9-11;   1928-1930.   M.S 1926

Výročni Zpráva Královske Céske Společnosti Nauk v Praze.
    Výr.Zpr.K.ceské Spol.Nauk   1918-1939.
    Formerly Jahresbericht der Königlich Böhmischen
    Gesellschaft der Naturwissenschaften. Prag.   S. 1703 B

Výroční Zpráva. Moravské Přírodovédecke Společností Brnó.
    Vyr.Zpr.morav.přír.Spol.   1925-1943.   S. 1713 B

Výroční Zpráva. Museum J.A.Komenského v Uherském Brodě.
    Uherský Brod.
    See Zprávy. Museum J.A. Komenského v Uherském Brodě.
    Uherský Brod.   S. 1747

Vytauto Didžiojo Universiteto Matematikos-Gamtos
    Fakulteto Darbai.
    Vytauto Didž.Univ.mat.gamtos Fak.Darb.   1930-1939   (imp.)
    Formerly Lietuvos Universiteto Matematikos Gamtos
    Fakulteto Darbai. Kaunas.   S. 1852 c

TITLE	SERIAL No.

WHO Chronicle. World Health Organization. Geneva.
    WHO Chron. Vol.16 → 1962 →   E.S 2563

WPRS Bulletin.
    See Bulletin. Section Regionale Ouest Palearctique,
    Organisation Internationale de Lutte Biologique.   E.S 2569

Waldhygiene. Würzburg.
    Waldhygiene 1954 →   E.S 1321

Walia. Addis Ababa.
    Walia 1969 →   S. 2031 a

Walker Museum Memoirs. Chicago.
    See Memoirs. Walker Museum. Chicago.   P.S 850

Walla Walla College Publications of the Department of Biological
    Sciences and the Biological Station. College Place, Washington.
    Walla Walla Coll.Publs Dep.biol.Sci. 1951-1962.   S. 2377 a

Wanderversammlung Deutscher Entomologen. Berlin-Dahlem.
    Wanderversamml.dt.Ent. 1926-1934: 1954 →
    1934-1953 Appeared in Entomologische Beihefte Berlin-Dahlem.E.S 1342 b

War Background Studies. Smithsonian Institution. Washington, D.C.
    War Backgr.Stud. 1942-1945.   S. 2426 D

Warren Spring Laboratory. Mineral Processing. Information Note.
    See Mineral Processing Information Note.   M.S 142

Wartime Pamphlets. Geological Survey of England and Wales. London.
    Wartime Pamphl.geol.Surv.Engl.Wales 1940-1949.   P.S 1013
        No.6-47, 1941-1946 (imp.)   M.S 155 A

Washington Geologic Newsletter. Olympia.
    Wash.geol.Newsl. 1973 →   P.S 1989 a

Washington University Record. St. Louis.
    Wash.Univ.Rec. 1905-1924. (imp.)   S. 2383 A

Washington University Studies. Science & Technology. St.Louis.
    Wash.Univ.Stud.Sci.Technol. 1928-1936.   S. 2383 Ba

Washington University Studies. Scientific Series. St.Louis, Mo.
    Wash.Univ.Stud.scient.Ser. 1913-1926.   S. 2383 B

Wasmann (formerly Club) Collector. San Francisco, Cal.
    Wasmann Club Collect. Vols. 3-7, 1939-1950. (imp.)
    Continued as Wasmann Journal of Biology. San Francisco.   S. 2536

Wasmann Journal of Biology. San Francisco.
    Wasmann J.Biol. Vol. 8 → 1950 →
    Formerly Wasmann (formerly Club) Collector. San Francisco.   S. 2536

| TITLE | SERIAL No. |

Water, Air, and Soil Pollution. Dordrecht.
    Wat.Air Soil Poll. 1971 →   S. 2750

Water Life. London.
    Wat.Life 1936-1941.
    (1940-1941 contained in "Animal & Zoo Magazine".)
    Continued as Water Life and Aquaria World.   Z.S 133

Water Life and Aquaria World. London.
    Wat.Life Aquar.Wld 1946-1957.
    Formerly Water Life.
    Continued as Fishkeeping and Water Life.   Z.S 133

Water Pollution Abstracts. London.
    Wat.Pollut.Abstr. 1949 →
    Formerly Summary of Current Literature. Water Pollution
    Research Board. London.   R.R.Abs 104

Water Pollution Research. Water Pollution Research Board
    D.S.I.R. London.
    Wat.Pollut.Res. 1957 →
    Formerly Report. Water Pollution Research Board
    D.S.I.R. London.   S. 205 W

Water Research. Oxford.
    Wat.Res. Vol.4 → 1970 →   S. 2752

Water Resources Bulletin. Louisiana Geological Survey. Baton Rouge.
    Wat.Resour.Bull.La geol.Surv. No.1-2, 1960-1961.
    Folio Series, No.1, 1960.   P.S 1926

Water Resources Pamphlet. Geological Survey. Somaliland Protectorate.
    Wat.Resour.Pamph.Somaliland No.1 1955.   P.S 1179 H

Water Resources Pamphlet. Louisiana Geological Survey. Baton Rouge.
    Wat.Resour.Pamph.La 1954 →   P.S 1926 D

Water Supply Paper. Geological Survey of Uganda. Entebbe.
    Wat.Supply Pap.Uganda 1941 →   P.S 1179 c

Water Supply Papers of the Geological Survey of Great Britain.
    London.
    Wat.Supply Pap.geol.Surv.Gt Br. 1963 →   P.S 1009 A

Watson's Microscope Record, London.
    Watson's Microsc.Rec. 1924-1939.   S. 433

Watsonia. Journal and Proceedings of the Botanical Society
    of the British Isles. Arbroath.
    Watsonia 1949 →
    Replaces Report. Botanical Society and Exchange Club
    of the British Isles.   B. British Herbarium

TITLE	SERIAL No.

Wealth of India. New Delhi.
    Wlth India 1948 →   REF.

Weather. London.
    Weather 1946 →   S. 210 B

Webbia. Raccolta di Scritti Botanici. Firenze.
    Webbia 1905-1923; 1948 →   B.S 702

Wecko-Skrift for Läkare och Naturforskare. Stockholm.
    Wecko-Skr. 1781-1785.
    Continued as Läkaren och Naturforskaren. Stockholm.   S. 580

Weekly-Bulletin. Paiforce Naturalists' Club.
    Wkly Bull.Paiforce Nat.Cl. Nos. 30-34, 36-48. 1945-1946.   S. 1919 a C

Weekly Entomologist. Altrincham & London.
    Wkly Ent. 1862-1863.   E.S 11 a

Weidmann. Wochenschrift für Jäger, Fischer und
    Naturfreunde. Bülach.
    Weidmann, Bülach 1919-1920.   Z.S 1285

Well Inventory Series. London.
    Well Inventory Ser. Metric Units. 1973 →   P.S 1009 B

Welsh Geological Quarterly.
    Welsh geol.Q. 1965 →   P.S 160 B

Welsh Regional Bulletin, Botanical Society of the British
    Isles. Aberystwyth.
    Welsh Reg.Bull.bot.Soc.Br.Isl. 1964 →   B.S. Europ. Room C

Wendigen. Utrecht Biohistorisch Instituut, Rijksuniversiteit.
    Utrecht.
    See U.B.I. Wendingen. Utrecht.   S. 678

Wentia. Koninglijke Nederlandse Botanische Vereniging. Amsterdam.
    Wentia 1959 →   B.S 302 a

Werdenda. Beiträge zur Pflanzenkunde. Bingen, Washington.
    Werdenda 1923-1931.   B.S 4405

Werken. Dienst Domaniale Natuurreservaten en Natuurbescherming.
    See Travaux. Service des Réserves Naturelles Domaniales
    et de la Conservation de la Nature.   S. 718

Werken van het Gennotschap ter Bervordering der Natuur-,
    Genees-en Heelkunde te Amsterdam.
    Werk.Genoot.Nat.-Genees-en Heelk. 1870-1871.
    Continued as Maandblad voor Natuurwetenschappen. Amsterdam.   S. 606

TITLE	SERIAL No.
Wesley Naturalist. London. **Wesley Nat.** 1887-1889.	S. 473
West American Scientist. San Diego, Cal. **W.Am.Scient.** 1884-1921.	S. 2444
West Australian Naturalist. **W.Aust.Nat.** 1939. Continued as Western Australian Naturalist. Perth.	S. 2158
West Indian Bulletin. Imperial Department of Agriculture. Barbados. **W.Indian Bull.** 1899-1922.	E.S 2385
West Midland Bird Report. Birmingham. **W.Midl.Bird Rep.** No.15 → 1949 → Formerly Report of the Birds of Warwickshire, Worcestershire and South Staffordshire.	T.B.S 258
West Virginia Geological Survey. Morgantown. **W.Va geol.Surv.** 1899-1951.	P.S 1986
Western Australian Naturalist. Perth. **W.Aust.Nat.** 1947 → Formerly West Australian Naturalist.	S. 2158
Western Australian Year Book. Perth. **W.Aust.Yb.** 1889-1891; 1893-1897; 1902-1904.	S. 2156 a
Western Birds. Del Mar (Ca.) **West.Birds** Vol.4 → 1973 → Formerly California Birds. San Diego.	T.B.S 5201
Western Naturalist. Paisley. **W. Nat.** 1972 →	S. 318 a
Westmorland Note-Book & Natural History Record. **Westmorland Note-Bk nat.Hist.Rec.** 1888-1889.	S. 134
Wetenschappelijke Mededeelingen. Dienst van de Mijnbouw in Nederlandsch Oost - Indië. Weltevreden. **Wet.Meded.Dienst Mijnb.Ned.Oost - Indië** 1924-1940 Continued as Publikasi Keilmuan. Djawatan Geologi. Kementerian Perekonomian. Republik Indonesia. Bandung.	P.S 1298
Wetenschappelijke Mededelingen. Koniklijke Nederlandse Natuurhistorische Vereniging. Amsterdam. **Wet.Meded.K.ned.natuurh.Veren.** No.4 → 1953 →	S. 700 G
Wetenschappelijke Uitkomsten der Snellius Expeditie. Leiden. See Snellius-Expedition in the Eastern Part of the Netherlands East-Indies. 1929-1930. Leiden.	77 A.q.A
Wetenskaplike Bydraes van die Potchefstroomse Universiteit vir C.H.O. Potchefstroom. **Wetensk.Bydr.potchefstroom.Univ.** Reeks B: Natuurwet. 1968 →	S. 2074
Wheat Information Service. Kyoto University. **Wheat Inf.Serv.Kyoto Univ.** 1954 →	B.S 1950

TITLE	SERIAL No.

Wiadomości Botaniczne. Krakow.
    Wiad.Bot. 1962 →   B.S 1300 d

Wiadomości Ekologiczne. Warsaw.
    Wiad.Ekol. 1970 →
    Formerly Ekologia Polska. Warsaw. Series B.   S. 1870 a B

Wiadomości Muzeum Ziemi. Warszawa.
    Wiad.Muz.Ziemi 1938-1952.   P.S 571

Wiadomości Parazytologiczne. Warszawa, etc.
    Wiad.parazyt. 1955 →   S. 1870 a G

Wichtigsten Lagerstätten der Erde. Berlin.
    Wicht.Lagerstätt.Erde No.2-19; 1940-1942 (imp.)   M.S 1317

Wiedemanns Archiv.
    See Archiv für Zoologie and Zootomie.   Z.S 1400

Wiedemanns Magazin.
    See Zoologisches Magazin.   Z.S 1540

Wiegmann's Archiv.
    See Archiv für Naturgeschichte.   Z.S 1390

Wielewaal. Turnhout.
    Wielewaal Jaarg. 36 → 1970 →   T.B.S 703

Wiener Botanische Zeitschrift. Wien.
    Wien.bot.Z. 1943-1944.
    Formerly and Continued as Österreichische Botanische
    Zeitschrift.   B.S 1212

Wiener Entomologische Monatschrift. Wien.
    Wien.ent.Monatschr. 1857-1864.   E.S 1705

Wiener Entomologische Rundschau. Wien.
    Wien.ent.Rdsch. 1949-1950.
    Continued in Entomologisches Nachrichtenblatt. Wien.   E.S 1710

Wiener Entomologische Zeitung. Wien.
    Wien.ent.Ztg 1882-1933.
    Amalgamated with Koleopterologische Rundschau 1934 →   E.S 1706

Wild Bird Protection in Norfolk. Norwich.
    Wild Bird Prot.Norfolk 1920-1952.   T.B.S 251

Wild Flower. Washington, D.C.
    Wild Flower Vol.27-31, 1951-1955.   B.S 4414

Wild Flower Magazine. Tunbridge Wells.
    Wild Flower Mag. 1931-1950. imp.   B.S 100

TITLE	SERIAL No.

Wildfowl. London.
    <u>Wildfowl</u> 1968 →
    <u>Formerly</u> Report of the Wildfowl Trust. London.   T.B.S 193

Wildfowl Trust Bulletin. Slimbridge (Gloucs.)
    <u>Wildf.Trust Bull.</u> No.68 → 1974 →   Z.S 78

Wild Life.
    <u>See</u> Wildlife.

Wildlife. Brisbane.
    <u>Wildlife, Brisbane</u> Vol.1, No.1, 1963.
    <u>Continued as</u> Wildlife in Australia. Wildlife Protection
    Society of Queensland. Brisbane.   S. 2104

Wild Life (M. Burton.) London.
    <u>Wild Life</u> (M. Burton) May 1957 → June 1957.
    <u>Continued as</u> Wild Life Observer.   S. 436 b

Wild Life. Nairobi.
    <u>Wild Life, Nairobi</u> 1959-1961.
    <u>Continued in</u> Africana.   Z.S 2052 A

Wildlife. Wellington.
    <u>Wildlife, Wellington</u> 1969 →   Z.S 2171

Wildlife in Australia. Wildlife Protection Society of Queensland.
    Brisbane.
    <u>Wildl.Aust.</u> Vol.1 → No.2 → 1963 →
    <u>Formerly</u> Wildlife. Brisbane.   S. 2104

Wildlife Bulletin of Nigeria. Ibadan.
    <u>Wildl.Bull.Nigeria</u> No.1 → 1973 →   Z.S 2026

Wildlife Circular. Fisheries and Game (Wildlife) Department,
    Victoria. Melbourne.
    <u>Wildl.Circ., Vict.</u> No.4 → 1957 → (imp.)   Z.S 2120 C

Wildlife Contributions. Fisheries & Wildlife Department,
    Victoria. Melbourne.
    <u>Wildl.Contr.Vict.</u> 1961 →
    <u>Formerly</u> Fauna Contributions. Fisheries and Game
    Department.   Z.S 2120 B

Wildlife and the Countryside. London.
    <u>Wildl.Ctryside</u> Oct.1966 - Feb.1970.
    <u>Formerly</u> Mainly about Wildlife and the Countryside. London.   S. 436 b

Wildlife Disease. Ames (Iowa)
    <u>Wildl.Dis.</u> 1959 → (lacks nos.22, 23 and 43 parts 1 and 4)
    (Microfiche)   Z.S 2326 B

Wildlife Management Bulletin. Canadian Wildlife Service. Ottawa.
    <u>Wildl.Mgmt Bull., Ottawa</u>
      Series 1, 1950-1966
      Series 2, 1950-1965
      Series 3, 1950-1957
    <u>Replaced by</u> Canadian Wildlife Service Report Series.   Z.S 2635

TITLE	SERIAL No.

Wildlife Monographs. A Publication of the Wildlife Society. Louisville, Kentucky.
    Wildl.Monogr. 1958 → (imp.) — Z.S 2538 A

Wild Life Observer. London.
    Wild Life Obsr. July 1957 - March 1966.
    Formerly Wild Life (M. Burton). London.
    Continued in Mainly about Wildlife and the Countryside. — S. 436 b

Wildlife Publication. New Zealand Department of Internal Affairs. Wellington.
    Wildl.Publ.,N.Z. No.8 → 1952 → (imp.). — Z.S 2181

Wildlife Research C.S.I.R.O.
    See C.S.I.R.O. Wildlife Research. — S. 2113 I

Wildlife Research Report. Washington, D.C.
    Wildl.Res.Rep. 1972 → — Z.S 2510 Q

Wildlife Review. Washington, D.C.
    Wildl.Rev. No.148 → 1973 → — Z. Rl o u

Wildlife Society News. Washington, D.C.
    Wildl.Soc.News No.116 → 1968 → — Z.S 2538 B

Wildlife Sound. Farnham, Surrey.
    Wildl.Sound No.3 → 1970 → — Z.S 275

Wild Life and Sport. Entebbe.
    Wild Life Sport Vol.2 - Vol.3, No.3, 1961-1963.
    Formerly Uganda Wild Life and Sport. — Z.S 2060

Wildlife Technical Bulletin. Alaska Department of Fish and Game. (Juneau.)
    See Game Technical Bulletin. Division of Game, Alaska Department of Fish and Game. (Juneau.) — Z.S 2596 A

Wilhelm Roux Archiv für Entwicklungsmechanik der Organismen. Berlin.
    Wilhelm Roux Arch.EntwMech.Org. Vol.105 → 1925 →
    Formerly Archiv für Mikroskopische Anatomie und Entwicklungsmechanik. — Z.S 1440

Willdenowia. Berlin.
    Willdenowia 1954 →
    Formerly Mitteilungen Botanischen Garten u. Museum. Berlin- Dahlem. — B.S 919 a

| TITLE | SERIAL No. |

Wilson Bulletin. Sioux City, Iowa.
    <u>Wilson Bull.</u> Vol.33 → 1921 →   Z.S 2488
                Vol.14 → 1902 →   T.R.S 5104

Wiltshire Archaeological and Natural History Magazine. Devizes.
    <u>Wilts.archaeol.nat.Hist.Mag.</u> 1854 →   S. 31
                          1854-1891.   T.R.S 168

Wiltshire Bird (and Plant) Notes. Devizes.
    <u>Wilts.Bird Pl.Notes</u> 1946.
    <u>Continued as</u> Report. Natural History Section. Wiltshire
    Archaeological and Natural History Society. Devizes.   S. 31 B

Wimbledon Museum Leaflet.
    <u>Wimbledon Mus.Leafl.</u> No.2 1917.   S. 398 C

Winchester College Bird Report. Winchester.
    <u>Winchester Coll.Bird Rep.</u> 1973 →
    <u>Formerly</u> Bird Report. Winchester College Natural
    History Society.   Z. 72 Aa f W

Wireless World. London.
    <u>Wireless Wld</u> Vol.80 No.1459 → 1974 →   Electronic Workshop

Wissenschaftliche Abhandlungen. Universität in Riga.
    <u>See</u> Latvijas Universitates Raksti. Riga.   S. 1851 a

Wissenschaftliche Arbeiten aus dem Burgenland. Eisenstadt.
    <u>Wiss.Arb.Burgenld</u> 1954 →   S. 1739

Wissenschaftliche Arbeiten. Landwirtschaftliche Hochschule
    "Wasil Kolarov". Plovdiv.
    <u>See</u> Nauchni Trudove. Vissh Selskostopanski Institut
    "Vasil Kolarov". Plovdiv.   B.S 1533

Wissenschaftliche Beiträge aus dem Osterlande. Altenburg.
    <u>Wiss.Beitr.Osterlande</u> 1934-1940.
    <u>Formerly</u> Mitteilungen aus dem Osterlande.
    <u>Continued as</u> Abhandlungen und Berichte des Naturkundlichen
    Museums "Mauritianum". Altenburg.   S. 1314

Wissenschaftliche Berichte. Institut für Geologie und Geographie.
    Vilnius.
    <u>See</u> Moksliniai Pranešimai. Geologijos ir Geografijos
    Institutas.   P.S 575 A

Wissenschaftliche Berichte der Moskauer Staats-Universität.
    Moskau.
    <u>See</u> Uchenye Zapiski Moskovskogo Gosudarstvennogo
    Universiteta.   S. 1844 C

Wissenschaftliche Berichte der Tomsker Staats Universität.
    <u>See</u> Trudȳ Biologicheskogo Fakul'teta Tomskogo
    Gosudarstvennogo Universiteta.   S. 1867 A

TITLE SERIAL No.

Wissenschaftliche Berichte. Ukrainisches Technisch
    Wirtschaftliches Institut. Augsburg.
    See Naukoviyī Byuleten. Ukrayins'kȳyi Tekhnichno-Hospodarskȳi
    Instȳtut. Augsburg, etc.   S. 1315

Wissenschaftliche Ergänzungshefte zur Zeitschrift des Deutschen
    und Osterreichischen Alpenvereins. Wien.
    Wiss.ErgänzHft.Z.dt.öst.Alpenver. 1897-1905.
    Continued as Wissenschaftlicher Veröffentlichungen des
    Deutschen und Osterreichischen Alpenvereins. Innsbruck.   S. 1738

Wissenschaftliche Ergebnisse der Altai-Pamir Expedition 1928. Berlin.
    Wiss.Ergebn.Altai-Pamir Exped. 1932.   73.q.B

Wissenschaftliche Ergebnisse der Deutschen Atlantischen
    Expedition auf dem Forschungs-und Vermessungsschiff "Meteor"
    1925-1927. Berlin.
    Wiss.Ergebn.dt.atlant.Exped."Meteor" 1932-1941.   78.q.B

Wissenschaftliche Ergebnisse der Deutschen Grönland-Expedition.
    Alfred Wegener 1929-1931. Leipzig.
    Wiss.Ergebn.dt.Grönland Exped.Alfred Wegener Bd.1-4 & 6.
    1933-1935.   71.o.B

Wissenschaftliche Ergebnisse der Deutschen Tiefsee-Expedition
    auf dem Dampfer "Valdivia" 1898-1899. Jena.
    Wiss.Ergebn.dt.Tiefsee-Exped."Valdivia" 1902-1940. 70.f.G & T.R.S 1366

Wissenschaftliche Ergebnisse des Deutschen Zentral-Afrika
    Expedition 1907-1908. Leipzig.
    Wiss.Ergebn.dt.ZentAfr.Exped. 1910-1925.   74 D.o.A & T.R.S 1364

Wissenschaftliche Ergebnisse der Finnischen Expeditionen
    nach der Halbinsel Kola. 1887-1892.
    Wiss.Ergebn.finn.Exped.Kola 1890-1894.   71.o.K

Wissenschaftliche Ergebnisse der Niederländischen Expeditionen
    in dem Karakorum. Leipzig.
    Wiss.Ergebn.niederl.Exped.Karakorum 1935-1940.   73 B.o.V

Wissenschaftliche Ergebnisse der Oldoway-Expedition 1913.
    Wiss.Ergebn.Oldoway Exped. N.F. Heft 2-3 1925-1928.   74 Db.q.R

Wissenschaftliche Ergebnisse. Reise in Ostafrika 1903-1905
    von A. Voeltzkow. Stuttgart.
    Wiss.Ergebn.Reise Ostafr. 1906-1923.   70.q.V

Wissenschaftliche Ergebnisse der Schwedischen-Expedition
    nach den Magellansländern 1895-1897. Stockholm.
    Wiss.Ergebn.schwed.Exped.Magellansländ. 1899-1907.   76.o.S

Wissenschaftliche Ergebnisse der Schwedischen Rhodesia-Kongo-
    Expedition 1911-1912. Stockholm.
    Wiss.Ergebn.schwed.Rhod.-Kongo-Exped. 1914-1921.   74.q.S

TITLE	SERIAL No.

Wissenschaftliche Ergebnisse der Schwedischen Südpolar-Expedition
   1901-1903. Stockholm.
    <u>Wiss.Ergebn.schwed.Südpolarexped</u>. 1905-1920.
    <u>Continued as</u> Further Zoological Results of the Swedish
   Antarctic Expedition 1901-1903. Stockholm.   80.q.N

Wissenschaftliche Ergebnisse der Dr. Trinkler'schen
   Zentralasien Expedition. 1927-28 Berlin.
    <u>Wiss.Ergebn.Trinkler.Zentralasien Exped</u>. 1932.   73 B.q.T

Wissenschaftliche Ergebnisse der Zweiten Deutschen Zentral-
   Afrika-Expedition 1910-1911.
    <u>See</u> Ergebnisse der Zweiten Deutschen Zentral-Afrika
   Expedition 1910-1911. Leipzig.   74 D.o.A

Wissenschaftliche Informationen für die Fischereipraxis.
   Hamburg.
    <u>Wiss.Inf.FischPrax</u>. Jg.5 - Jg.6 No.1 1958-1959.
    <u>Continued as</u> Informationen für die Fischwirtschaft
   des Auslandes. Hamburg.   Z.S 1305

Wissenschaftliche Meeresuntersuchungen der Kommission zur
   Wissenschaften Untersuchung der Deutschen Meere.
   Kiel & Leipzig. Neue Folge. Abteilung Kiel &
   Abteilung Helgoland, (after 1899.)
    <u>Wiss.Meeresunters</u>. (<u>Kiel</u> or <u>Helgol</u>.)
   1894 - (Helgol) 1935, (Kiel) 1936.
    <u>Formerly</u> Bericht der Commission zur Wissenschaft-
   lichen Untersuchungen der Deutschen Meere.
    <u>Continued as</u> Kieler Meeresforschungen <u>and as</u>
   Helgoländer Wissenschaftliche Meeresuntersuchungen.   Z.S 1360

Wissenschaftliche Mitteilungen aus Bosnien und Herzegovina.
   Wien.
    <u>Wiss.Mitt.Bosn.Herzeg</u>. 1893-1912.
    <u>Continued as</u> Wissenschaftliche Mitteilungen des Bosnisch-
   Herzegowinischen Landesmuseums. Sarajevo.   S. 1764 A

Wissenschaftliche Mitteilungen des Bosnisch-Herzegowninischen
   Landesmuseums. Sarajevo.
    <u>Wiss.Mitt.bosn.-herzeg.Landesmus</u>. Heft C: Naturw. 1971 →
    <u>Formerly</u> Wissenschaftliche Mitteilungen aus Bosnien
   und Herzegovina. Wien.   S. 1764 A

Wissenschaftliche Mitteilungen der Physikalisch-Medizinischen
   Societät zu Erlangen.
    <u>Wiss.Mitt.phys.-med.Soc.Erlangen</u> 1859.   S. 1405 B

Wissenschaftliche Mitteilungen. Universität Smolensk.
    <u>See</u> Nauchnye Izvestiya. Gosudarstvennyĭ Smolenskii
   Universitet.   S. 1809

Wissenschaftliche Mitteilungen des Vereins für Natur-und
   Heimatkunde in Köln a.Rh.
    <u>Wiss.Mitt.Ver.Natur-u.Heimatk.Köln</u> 1930-1935.   S. 1477

TITLE	SERIAL No.

Wissenschaftliche Monats Blätter. Königsberg.
    <u>Wiss.Mbl.Königsberg</u> 1873-1879.   S. 1643

Wissenschaftliche Veröffentlichungen des Deutschen und
    Osterreichischen Alpenvereins. Innsbruck.
    <u>Wiss.Veröff.dt.öst.Alpenver.</u> 1930-1931.
    <u>Formerly</u> Wissenschaftliche Ergänzungshefte zur Zeitschrift
    des Deutschen und Osterreichischen Alpenvereins. Wien.   S. 1738

Wissenschaftliche Zeitschrift der Ernst Moritz Arndt-Universität
    Greifswald. Mathematisch-Naturwissenschaftliche Reihe.
    <u>Wiss.Z.Ernst Moritz Arndt-Univ.Greifswald</u> 1954 →
    <u>Formerly</u> Wissenschaftliche Zeitschrift der Universität
    Greifswald. Mathematisch-Naturwissenschaftliche Reihe.   S. 1431

Wissenschaftliche Zeitschrift der Humboldt-Universität
    zu Berlin. Mathematisch-Naturwissenschaftliche Reihe.
    <u>Wiss.Z.Humboldt-Univ.Berl.</u> Jahrgang 2. 1952 →
                                 Sonderband 1965 →   S. 1323

Wissenschaftliche Zeitschrift der Martin-Luther Universität
    Halle-Wittenberg. Halle (Saale.) Mathematisch-
    Naturwissenschaftliche Reihe.
    <u>Wiss.Z.Martin-Luther-Univ.Halle-Wittenb.</u> 1951 →
    Sonderheft 1964 →   S. 1441 A

Wissenschaftliche Zeitschrift der Universität Greifswald.
    Mathematische-Naturwissenschaftliche Reihe.
    <u>Wiss.Z.Univ.Greifswald</u> 1951-1954.
    <u>Continued as</u> Wissenschaftliche Zeitschrift der Ernst Moritz
    Arndt-Universität Greifswald.   S. 1431

Wissenschaftliche Zeitschrift der Universität Rostock. Reihe.
    Mathematik-Naturwissenschaften.
    <u>Wiss.Z.Univ.Rostock</u> 1951 →   S. 1578

Wochenschrift für Aquarien und Terrarienkunde.
    (<u>later</u> Vereinigt mit Blätter für Aquarien- und
    Terrarienkunde.) Braunschweig.
    <u>Wschr.Aquar.-u.Terrarienk.</u> Vols.8-11, 36, 42-44;
    1911-1914, 1939, 1948-1950 (<u>imp.</u>).   Z.S 1477

Wombat. Geelong.
    <u>Wombat</u> Vol.1, No.1; Vol.3, No.3; Vol.5, No.4, 1895-1902.
    <u>See also</u> Geelong Naturalist.   S. 2120

Working Document. Council on Biological Sciences Information.
    Bethesda, Maryland.
    <u>Wkg Docum.Coun.biol.Sci.Inf.</u> 1970 →   S. 2318 a

Working for Nature. London.
    <u>See</u> Report. Council for Nature. London.   S. 229 C

TITLE	SERIAL No.

Working Paper. Tropical Fish Culture Research Institute. Malacca.
    Wkg Pap.trop.Fish.Cult.res.Inst. 1968 →   Z.S 1937

Works of Applied Entomology. Leningrad.
    See Trudȳ po Prikladnoi Entomologii.   E.S 1809

Works of the Bureau of Bird Banding. Moscow.
    See Trudȳ Byuro Kol'tsevaniya.   Z. 18 q R

Works issued by the Hakluyt Society. London.
    Wks Hakluyt Soc. 1847 →   70.o.H

World Animal Review. Rome.
    Wld Anim.Rev. 1972 →   Z.S 2713 Q

World Archaeology. London.
    Wld Archaeol. 1969 →   P.A.S 913

World Fishing. London.
    Wld Fishg Vol.17, No.12 → 1968 →   Z.S 2734

World Report on Palaeobotany. Utrecht.
    Wld Rep.Palaebot. 1950 → (1956 →)
    (Forms part of Regnum Vegetabile.)   P.S 50 o.I

World of Wildlife. London.
    Wld Wildl. 1971 →   Z.S 107

World Wildlife News. London.
    Wld Wildl.News Summer 1969 →   S. 248 a

Wrightia, a Botanical Journal. Dallas, Texas.
    Wrightia 1945 →   B.S 4387

Württembergische Naturwissenschaftliche Jahreshefte. Stuttgart.
    See Jahresheft des Vereins für Vaterländische Naturkunde
    in Württemberg. Stuttgart.   S. 1591 A

Würzburger Naturwissenschaftliche Zeitschrift.
    See Verhandlungen der Physikalisch-Medizinischen
    Gesellschaft zu Würzburg.   S. 1611 A

Wydawnictwa Luźne. Muzeum Ziemi. Warszawa.
    Wydaw.luźne 1960.   P.S 571 C

Wydawnictwa. Muzeum Imienia Dzieduszyckich we Lwowie.
    Wydaw.Muz.Dzieduszyck. 1880-1928.   S. 1791 A

Wydawnictwa Muzeum Slaskiego w Katowicach. Dzial III.
    Wydaw.Muz.slask.Katow. 1930-1935.   S. 1727 A

Wydawnictwa Okregowego Komitetu Ochrony Przyrody na Wielkopolske
    i Pomorze w Poznaniu. Poznań.
    Wydaw.okreg.Kom.Ochr.Przyr.Wielk.Pom.Poznań 1930-1937.   S. 1726 C

TITLE	SERIAL No.

Wydawnictwa Pomocnicze i Techniczno-Gospodarcze. Instytut Badawczy
    Lasów Państwowych. Warszawa. Ser. B.
    <u>Wydaw.pomoc.tech.-gosp.Inst.Badaw.Lasów pánst.</u> 1935-1939.   S. 1874 B

Wydawnictwa Popularnaukowe Kraków.
    <u>Wydaw.Pop.Kraków</u> No.10. 1955.   72 Q.o.S

Wydawnictwa Slaskie. Prace Geologiczne. Polska Akademja
    Umiejetności. Kraków.
    <u>See</u> Prace Geologiczne. Wydawnictwa Slaskie.   P.S 587

Wye College Reprints. Ashford.
    <u>Wye Coll.Repr.</u> 1952 →
    <u>Formerly</u> Reprints. Wye Agricultural College.   E.S 70

Wyoming Game and Fish Department Bulletin.
    <u>See</u> Bulletin. Wyoming Game & Fish Department. Laramie.   Z.S 2419

X-Ray Spectrometry. London.
    <u>X-ray Spectrom.</u> 1972 →   M.S 3017

X-Rays. Nakanoshima, Osaka.
    <u>X-rays</u> Vol.2-9, 1941-1956 (imp.)   M.S 1944

Y Gwyddonydd. Cardiff.
    <u>See</u> Gwyddonydd.   S. 450 a

Y.O.C. Newsletter. Sandy (Beds.)
    <u>YOC Newsl.</u> 1971 →   Z.S 476 B

YURT Report.
    <u>See</u> Report. Yorkshire Underground Research Team.

Yadoriga. Kyoto.
    <u>Yadoriga</u> No.21 → 1960 →   E.S 1909 a

Yale Conservation Studies. Yale Conservation Club. New Haven, Conn.
    <u>Yale Conserv.Stud.</u> 1952.   S. 2351 a

Yamaguchi Journal of Science. Yamaguchi.
    <u>Yamaguchi J.Sci.</u> 1950-1958.
    <u>Continued as</u> Science Reports of the Yamaguchi University.
    Yamaguchi.   S. 1985 e C

Year Book
    <u>See</u> Yearbook

Yearbook. The Academy of Natural Sciences of Philadelphia.
    <u>Yb.Acad.nat.Sci.Philad</u> 1923-1931.
    <u>Formerly</u> Report. The Academy of Natural Sciences
    of Philadelphia.
    <u>Continued as</u> Review. Academy of Natural Sciences of
    Philadelphia.   S. 2305 E

| TITLE | SERIAL No. |

Yearbook of Agriculture. United States Department of Agriculture.
Washington.
Yb.Agric.U.S.Dep.Agric. 1926 →
Formerly Agricultural Yearbook. Washington.   E.S 2451

Yearbook Alpine Garden Society. London.
Yb.alp.Gdn Soc. 1938-1939, 1946 →   B.S 81 a

Yearbook. American Amaryllis Society, Orlando, Fla.
Yb.Am.Amaryllis Soc. 1934-1935.
Continued as Herbertia, La Jolla.   B.S 4200

Yearbook. American Philosophical Society. Philadelphia. Pa.
Yb.Am.phil.Soc. 1937 →   S. 2304 G
1937-1942.   T.R.S 5113 B

Yearbook and Annual Report of the Canadian Institute. Toronto.
Yb.a.Rep.Can.Inst. 1912-1913.   S. 2605 D

Yearbook of Anthropology. Wenner-Gren Foundation for
Anthropological Research, Incorporated. New York.
Yb.Anthrop. 1955.   P.A.S 806 B

Yearbook of the Asiatic Society of Bengal, Calcutta.
Yb.Asiat.Soc.Beng. 1951 → (imp.)
Formerly Yearbook of the Royal Asiatic Society of Bengal.   S. 1902 C

Yearbook. Australian Academy of Science. Canberra.
Yb.aust.Acad.Sci. 1970 →   S. 2148 C

Yearbook. Botanical Society of the British Isles. Arbroath.
Yb.bot.Soc.Br.Isl. 1949-1953.   B.B.H.S 97

Yearbook of the Bureau of Entomology of Chekiang Province. Hangchow.
Yb.Bur.Ent.Chekiang Prov. 1931-1935.   E.S 1952

Yearbook and Calendar. Essex Field Club. Buckhurst Hill.
Yb.Essex Fld Cl. 1905-1912 (imp.)   S. 22 C

Yearbook. Canadian Institute. Toronto.
See Yearbook and Annual Report of the Canadian Institute.
Toronto.   S. 2605 D

Yearbook. Carnegie Institution of Washington.
Yb.Carnegie Instn Wash. 1902 →   S. 2417 A

Yearbook. Czestochowa Museum.
See Rocznik Muzeum w Czestochowie.   S. 1727 a

Yearbook. Department of Agriculture, Ceylon. Colombo.
Yb.Dep.Agric.Ceylon 1925-1927.   E.S 2019

Yearbook. Department of Agriculture, Madras.
Yb.Dep.Agric.Madras 1917-1929 (imp.)   E.S 2012

| TITLE | SERIAL No. |

Yearbook. Essex Field Club. Stratford.
    Yb.Essex Fld Cl. 1905-1912.   E.S 26

Yearbook. Faculty of Agriculture and Forestry of the University
   of Skopje.
    See Godisen Zbornik. Zemjudelski-Shumarski Fakultet na
   Universitetot, Skopje.   B.S 1329

Yearbook. Faculty of Agriculture, University of Ankara.
    Yb.Fac.Agric.Univ.Ankara 1958 → (imp.)   E.S 1887 b

Yearbook of Fishery Statistics. Food & Agriculture Organisation,
   United Nations. Washington, D.C. later Rome.
    Yb.Fish.Statist. 1947-1965.   Z.S 2713 B

Yearbook of the Geological Society. London.
    Yb.geol.Soc.Lond. 1971 →   P.S 100 C

Yearbook. Geological Society of America. Boulder, Colo.
    Yb.geol.Soc.Am. 1970 →   P.S 865 G

Yearbook. Geological Survey of Denmark. København.
    See Arbog. Danmarks Geologiske Undersøgelse.   P.S 1396

Yearbook. Indian National Science Academy. New Delhi.
    Yb.Indian natn.Sci.Acad. 1970 →
    Formerly Yearbook. National Institute of Sciences of
   India. New Delhi.   S. 1918 E

Yearbook. International Council of Scientific Unions.
   The Hague.
    Yb.int.Coun.scient.Un. 1959 →   S. 2701 C

Yearbook. International Dendrology Society. London.
    Yb.int.Dendrol.Soc. 1966 →   B.S 41

Yearbook. International Federation for Documentation. Hague.
    See FID Yearbook.   S. 637

Yearbook International Society of Plant Morphologists. Delhi.
    Yb.int.Soc.Plant Morph. 1962 →   B.S 4560 a

Yearbook. International Society for Tropical Ecology.
   Varanasi, India.
    Yb.Int.Soc.trop.Ecol. 1967 →   B.S 4604 a

Yearbook. Khedivial Agricultural Society. Cairo.
    See Yearbook Sultanieh Agricultural Society.   E.S 2154

Yearbook. Lamont-Doherty Geological Observatory. Palisades, N.Y.
    Yb.Lamont-Doherty geol.obs. 1973 →   M.S 2660 A

Yearbook. Malvern College Natural History Society.
    Yb.Malvern Coll.nat.Hist.Soc. 1903-1913. (imp.)
    Continued as Report. Malvern College Natural History
   Society.   S. 257

TITLE	SERIAL No.

Yearbook. National Institute of Sciences of India. New Delhi.
    Yb.natn.Inst.Sci.India   1960 - 1969.
    Continued as   Yearbook. Indian National Science Academy.
New Delhi.   S. 1918 E

Yearbook. National Museum of Canada.
    Yb.natn.Mus.Can.   1965.   S. 2632 D

Yearbook. National Museums of Victoria. Melbourne.
    Yb.natn.Mus.Vict.   1948.   S. 2112 C

Yearbook. National Trust for Scotland. Edinburgh.
    Yb.natn.Trust Scotland   1968 →   S. 430 a

Yearbook of New South Wales. Sydney.
    Yb.N.S.W.   1886   M.S 2405

Yearbook. New York Microscopical Society. New York.
    Yb.N.Y.microsc.Soc.   1952 →   S. 2509 C

Yearbook of Physical Anthropology. New York.
    Yb.phys.Anthrop.   1945 →   P.A.S 807

Yearbook. of the Public Museum of the City of Milwaukee.
    Yb.publ.Mus.Cy Milwaukee   1921-1930.
    Replaces   Report of the Public Museum of the City of
Milwaukee.   S. 2348 C

Yearbook. Queensland Government Mining Journal. Brisbane.
    Yb.Qd Govt Min.J.   1967 →   M.S 2402 B

Yearbook & Record. Royal Geographical Society. London.
    Yb.Rec.R.geogr.Soc.   1898-1914.   S. 211 H

Yearbook of the Rhododendron Association.
    Yb.Rhodod.Ass.   1929-1936.   B.S 65

Yearbook of the Royal Asiatic Society of Bengal. Calcutta.
    Yb.R.Asiat.Soc.Beng.   1935-1950, (imp.)   S. 1902 C
        1935-1949.   T.R.S 3003 B
    Continued as   Yearbook of the Asiatic Society of Bengal.

Yearbook. Royal Colonial Institute.
    Yb.R.colon.Inst.   1912-1916.   S. 209 B

Yearbook. Royal Geographical Society. London.
    See   Yearbook & Record. Royal Geographical Society. London.   S. 211 H

Yearbook of the Royal Society. London.
    Yb.R.Soc.   1896 →
    (1896-1958 at Tring.)   S. 3 G

TITLE	SERIAL No.

Yearbook of the Royal Society of Edinburgh.
    Yb.R.Soc.Edinb. 1940 →   S. 4 D

Yearbook of the Royal Society of Tropical Medicine & Hygiene.
    Yb.R.Soc.trop.Med.Hyg. 1922 →
    Formerly Year Book of the Society of Tropical
    Medicine and Hygiene.   S. 222 B

Yearbook Scottish Rock Garden Club. Edinburgh.
    Yb.Scott.Rock Gdn Club 1960 →   B.S 9 a

Yearbook of the Society of Tropical Medicine and Hygiene.
    Yb.Soc.trop.Med.Hyg. 1910-1911.
    Continued as Year Book of the Royal Society of
    Tropical Medicine & Hygiene.   S. 222 B

Yearbook of South Australia. London.
    Yb.S.Aust. 1898-1899.   77 Cb.o.S

Yearbook of the Sultanieh Agricultural Society. Cairo.
    Yb.sult.agric.Soc. 1905   E.S 2154

Yearbook of the United States Department of Agriculture. Washington.
    Yb.U.S.Dep.Agric. 1894-1922.
    Continued as Agricultural Yearbook. Washington.   E.S 2451

Yearbook. Worcestershire Nature Conservation Trust. Kidderminster.
    Yb.Worcs.Nat.Conserv.Trust 1972 →   S. 500 e A

Yearbook. World Wildlife Fund. Morges.
    Yb Wld Wildl.Fund 1968 →
    Formerly Report of the World Wildlife Fund.   Z.S 2703

Yearbook of the Zoological Society of India. Calcutta.
    Yb.zool.Soc.India 1956-1957 →   Z.S 1918 A

Year's Work in Librarianship. London.
    Yr's Wk Librship 1928-1947 (imp.)   S. 497 a E

Yedeoth. Proceedings of the Agricultural Experiment Station,
    Tel-Aviv, Palestine.
    Yedeoth 1936-1938   E.S 2036 c

Yorkshire Naturalists' Recorder. Manchester.
    Yorks.Nat.Rec. 1872-1873.   S. 474 a

Young Naturalist. London.
    Young Nat. 1879-1890.
    Continued as British Naturalist.   S. 470 a

Your Environment. London.
    Your Environ. 1969 →   S. 467 a

Yüksek Ziraat Enstitüsü Çalişmalari. Ankara.
    Yüks.Zir.Enst.Calism. 1934-1948.   E.S 1881 a
    No.10, 20-23, 27-30, 1936.   P.S 497

| TITLE | SERIAL No. |

Zabytki Przyrody Nieożywionej Ziem Rzeczypospolitej Polskiej.
　Warszawa.
　　Zabytki Przyr.nieożyw.Ziem Rzeczypos.pol. 1928 →   P.S 572

Zambia (Northern Rhodesia) Journal. Livingstone.
　　Zambia (Nth.Rhod.) J. Vol.6, 1965.
　　Formerly Northern Rhodesia Journal. Livingstone.   S. 2019 a

Zambia Museum Papers. Lusaka.
　　Zambia Mus.Pap. 1967 →   S. 2075 C

Zambia Museums Journal. Livingstone.
　　Zambia Mus.J. 1970 →   S. 2019 a

Zametki po Sistematike i Geografii Rastenii. Tbilisi.
　　Zametki Sist.Geogr.Rast. 1938 →   B.S 1430 c

Zapiski Akademii Nauk SSSR. Leningrad.
　　Zap.Akad.Nauk SSSR 1925-1930.
　　Formerly Zapiski Rossiĭskoĭ Akademii Nauk. Leningrad.   S. 1802 C

Zapiski Astrakhanskoĭ Stantsii Zashchitȳ Rastenii ot Vreditelei.
　　Astrakhan.
　　Zap.astrakh.Sta.Zashch.Rast.Vredit.
　　Vol.1, fasc.5-6 & Vol.2, fasc.1. 1927-1928.   E.S 1818 b

Zapiski Biologicheskoi Stantsii Obshchestva Lyubitelei
　　Estestvoznaniya, Antropologii i Etnografii. Moskva.
　　Zap.biol.Sta.Obshch.Lyub.Estest.Antrop.Etnogr. 1925-1930.   S. 1841 B

Zapiski Fizȳchno-Matematȳchnoho Viddilu. Vseukrayins'ka Akademiya
　　Nauk. Kyyiv.
　　Zap.fiz.-mat.Vidd.vseukr.Akad.Nauk 1923-1931.   S. 1834 a A

Zapiski Gornago Instituta Imperatritsȳ Ekaterinȳ II. S.-Peterburg.
　　Zap.gorn.Inst.imp.Ekaterinȳ 1907-1913.
　　Continued later as Zapiski Leningradskogo Ordena Lenina
　　i Trudogo Krasnogo Znameni Gornogo Instituta im
　　G.V. Plekhanova.   M.S 1801

Zapiski Gosudarstvennogo Nikitskogo Opȳtnogo Botanicheskogo
　　Sada. Leningrad, etc.
　　Zap.gos.nikit.opȳt.bot.Sada Vol.8-18, 1925-1939 (imp.)
　　Continued as Trudȳ Gosudarstvennogo Nikitskogo
　　Botanicheskogo Sada. Yalta.   B.S 1445

Zapiski Imperatorskago (S.Peterburgskogo) Mineralogicheskogo
　　Obshchestva.
　　Zap.imp.miner.Obshch. 1864-1915 (Series suspended 1864-1865.)
　　Formerly Verhandlungen der Kaiserlichen Gesellschaft
　　für die Gesammte Mineralogie zu St. Petersburg.
　　Continued as Zapiski Rossiiskogo Mineralogicheskogo
　　Obshchestva. Petrograd.   M.S 1820

TITLE	SERIAL No.

Zapiski Imperatorskoĭ Akademii Nauk po Fiziko-Matematicheskomu
   Otdeleniyu. S.-Peterburg.
   Zap.imp.Akad.Nauk 1894-1917.
   Formerly Mémoires de l'Academie Impériale des Sciences
   de St.Petersbourg.
   Continued as Zapiski Rossiĭskoĭ Akademii Nauk. Leningrad.   S. 1802 C

Zapiski Kavkazskago Muzeya. Tiflis.
   See Mémoires du Musée du Caucase. Tiflis.   S. 1858 D

Zapiski Kharbinskogo Obshchestva Estestvoispȳtateleĭ
   i Etnografov. Kharbin.
   Zap.kharbin.Obshch.Estest.Etnogr. 1946-1954.   S. 1986 f

Zapiski Kievskago Obshchestva Estestvoispȳtateleĭ. Kiev.
   Zap.kiev.Obshch.Estest. 1870-1929. (Wanting 1918-1925.)   S. 1834

Zapiski Kiĭvskago Tovaristva Prirodoznavtsiv. Kȳyiv.
   See Zapiski Kievskago Obshchestva Estestpoispytateleĭ. Kiev.   S. 1834

Zapiski Kirgizskogo Otdeleniya Vsesoyuznogo Mineralogicheskogo
   Obshchestva. Frunze.
   Zap.kirgiz.Otd.vses.miner.Obshch. 1959 →   M.S 1834

Zapiski Kommunisticheskogo Universiteta imeni Ya.M. Sverdlova.
   Moskva.
   Zap.kommun.Univ.Ya.M.Sverdlova Tom.2 1924.   S. 1837 A

Zapiski Krȳmskogo Obshchestva Estestvoispȳtateleĭ i Lyubiteleĭ
   Prirodȳ. Simferopol.
   Zap.krym.Obshch.Estest. nos. 1-3, 5-6, 8-12. 1911-1930.   S. 1813 A

Zapiski Leningradskogo Ordena Lenina i Trudogo Krasnogo Znameni
   Gornogo Instituta im G.V. Plekhanova.
   Zap.leningr.gorn.Inst. Tom.38, Vp.2 → 1961 → (imp.)
   Formerly (1907-1913) as Zapiski Gornago Instituta
   Imperatritsȳ Ekaterinȳ II. S.Peterburg.   M.S 1801

Zapiski Nauchno-Issledovatel'skogo Ob'edineniya. Russkii
   Svobodnȳi Universitet v Prage. Praga.
   Zap.nauchno-issled.Ob'ed.russk.svob.Univ.Praga 1935-1937.
   Formerly Nauchnye Trudy. Russkiĭ Narodnyĭ Universitet
   v Pragê.   S. 1762 a

Zapiski Nauchno-Prikladnȳkh Otdelov Tiflisskogo Botanicheskogo
   Sada. Tiflis.
   Zap.nauchno-prikl.Otd.tiflis.bot.Sada 1919-1930.   B.S 1430 b

Zapiski Naukovo-Doslidchoĭ Katedri Zoologiĭ. Kharkiv.
   See Trudy Khar'kovskogo Tovaristva Doslidnikiv
   Prirodi. Kharkiv.   S. 1833 A

Zapiski Novorossiĭskago Obshchestva Estestvoispȳtateleĭ. Odessa.
   Zap.novoross.Obshch.Estest. 1872-1912.   S. 1847 A

| TITLE | SERIAL No. |

Zapiski Obshchestva Ispȳtateleĭ Prirodȳ. Moskva.
　See Mémoires de la Société Impériale des Naturalistes
　de Moscou. 　　　　　　　　　　　　　　　S. 1838 B & T.R.S 2022

Zapiski Prȳrodnȳcho-Tekhnichnoho Viddilu. Vseukraĭns'ka Akademiya
　Nauk. Kȳyiv.
　　Zap.prȳrod.-tekh.Vidd.vseukr.Akad.Nauk　1931.　　　S. 1834 a

Zapiski Rossiĭskogo Mineralogicheskogo Obshchestva. Petrograd.
　　Zap.ross.miner.Obshch.　Vol.51, livr.2 - Vol.61, 1923-1932
　　(Wanting 1916-1918, Suspended 1919-1922.)
　　Formerly　Zapiski Imperatorskogo (S.-Peterburgskogo)
　　Mineralogicheskogo Obshchestvo.
　　Continued as　Zapiski Vserossiĭskogo Mineralogicheskogo
　　Obshchestva.　Moskva.　　　　　　　　　　　　　　　M.S 1820

Zapiski Rossiĭskoĭ Akademii Nauk. Leningrad. Otdel.Fiz.-Mat.Nauk.
　　Zap.ross.Akad.Nauk　1918-1924.
　　Formerly　Zapiski Imperatorskoĭ Akademii Nauk. S.-Peterburg.
　　Continued as　Zapiski Akademii Nauk SSR. Leningrad.　　S. 1802 C

Zapiski Russkogo Nauchnago Instituta v Bêlgradê.
　　Zap.russk.nauch.Inst.Bêlgr.　1930-1938.　　　　　　S. 1891 B

Zapiski Sverdlovskogo Otdeleniya Vsesoyuznogo Botanicheskogo
　Obshchestvo. Sverdlovsk.
　　Zap.sverdlov.Otd.vses.bot.Obshch.　Vȳp.3 → 1964 →　B.S 1403 b

Zapiski Ural'skago Obshchestva Lyubiteleĭ Estestvoznaniya.
　Ekaterinburg.
　　Zap.ural'.Obshch.Lyub.Estest.　1873-1927 (imp.).　　S. 1820 A

Zapiski Vladivostokskogo Otdela Gosudarstvennogo Russkogo
　Geograficheskogo Obshchestva. Vladivostok.
　　Zap.vladivost.Otd.gos.Russk.geogr.Obshch.　1928-1929.　S. 1830 D

Zapiski Vostochno-Sibirskago Otdela Imperatorskago Russkago
　Geograficheskago Obshchestva. Irkutsk.
　　Zap.vost.-sib.Otd.imp.russk.geogr.Obshch.　tom. 2, no. 1
　　　1892.　　　　　　　　　　　　　　　　　　　　　　S. 1830 A

Zapiski Vostochno-Sibirskogo Otdeleniya Vsesoyuznogo Mineralogicheskogo
　Obshchestva. Irkutsk.
　　Zap.vost.-sib.Otd.vses.Miner.Obshch.　vȳp.2-4, 1960-1962.　M.S 1820 A

Zapiski Vserossiĭskogo Mineralogicheskogo Obshchestva. Moskva.
　　Zap.vseross.miner.Obshch.　Vol.62-76, 1933-1947.
　　Formerly　Zapiski Rossiĭskogo Mineralogicheskogo Obshchestva.
　　Continued as　Zapiski Vsesoyuznogo Mineralogicheskogo
　　Obshchestva. Moskva.　　　　　　　　　　　　　　　　M.S 1820

Zapiski Vsesoyuznogo Mineralogicheskogo Obshchestva.
　Moskva.
　　Zap.vses.miner.Obshch.　Vol.78 → 1949 → (Wanting Vol.77, 1948.)
　　Formerly　Zapiski Vserossiĭskogo Mineralogicheskogo
　　Obshchestva. Moskva.　　　　　　　　　　　　　　　　M.S 1820

TITLE	SERIAL No.

Zapisnici Srpskog Geološkog Društva. Beograd.
   <u>Zapisnici srp.geol.Društ.</u> 1897-1907 (imp.) 1948 →
                                                   P.S 450 & P.S 450 A

Zapȳsky.
   <u>See</u> Zapiski.

Zashchita Rasteniĭ. Leningrad.
   <u>Zashch.Rast.</u> 1935-1939, 1967 →
   For 1940-1941 <u>See</u> Věstnik Zashchitȳ Rasteniĭ. Leningrad.   E.S 1817

Zaštita Bilja. Beograd.
   <u>Zašt.Bilja</u> 1950 →   E.S 1883

Zaštita Prirode. Beograd.
   <u>Zašt.Prir.</u> 1950-1951; 1964 →   S. 1890 a

Zavod za Ratarstvo. Zagreb.
   <u>Zavod Ratarst.</u> 1948 → (imp.)   B.S 1336

Zbirnyk Biolohichnoho Fakul'tetu. Odes'kȳĭ Derzhavnȳĭ Universȳtet
   im. I.I. Mechnȳkova. Kȳyiv.
   <u>See</u> Sbornik Biologicheskogo Fakul'teta. Odesskiĭ Gosudarstvennȳĭ
   Universitet im. I.I. Mechnikova. Kiev.   S. 1847 b

Zbirnȳk. Poltavskiĭ Derzhavniĭ Muzei im. V.G. Korolenka. Poltava.
   <u>Zbirn.poltav.derzh.Muz.V.G.Korolenka</u> 1927-1928.   S. 1876.

Zbirnȳk Prats' Dniprovs'koyĭ Biolohichnoyĭ Stantsiyi. Kȳyiv.
   <u>Zbirn.prats'dniprov.biol.Sta.</u> No.2 1927.   S. 1834 a M

Zbirnȳk Prats' z Henetȳkȳ Instytutu Zoolohiyi i Biolohiyi.
   Akademiya Nauk URSR. Kȳyiv.
   <u>Zbirn.prats' Henet.</u> Nos.2-3, 1938-1939.   S. 1834 a L

Zbirnȳk Prats' z Morfolohiyi Tvarȳn. Kȳyiv.
   <u>Zbirn.Prats' Morf.Tvarȳn</u> 1935-1939.   Z.S 1827 B

Zbirnȳk Prats' z Parazȳtologiyi. Pratsi Naukovoyi Konferentsiyi
   Parazȳtologyiv URSR. Kȳyiv.
   <u>Zbirn.Prats'Parazȳt.Pratsi nauk.Konf.Parazit.URSR</u> 1947.
   <u>Continued as</u> Problemȳ Parazitologii. Trudȳ Nauchnoĭ
   Konferentsii Parazitologov USSR. Kiev.   Z.S 1827 F

Zbirnȳk Prats Zoolohichnoho Muzeyu. Kȳyiv.
   <u>Zbirn.Prats zool.Mus.</u> 1926 →   Z.S 1827
                 1926-1938.   T.R.S 2007
   <u>Forms part of</u> Trudȳ, Ukraïns'ka Akademiya Nauk,
   Fizicho-Matematichnoho Viddilu.

Zbornik Geološkog i Rudarskog Fakulteťa, Tehnička Velika
   Skola. Beograd.
   <u>Zborn.geol.rud.Fak.teh.Velika Sk.Beogr.</u> 1952-1956.   P.S 457

TITLE	SERIAL No.

Zbornik Matice Srpske. Novi Sad. Serija Prirodnih Nauka.
  Zborn.Matice srp. Vol. 2. 1952.
  Formerly Naučni Sbornik Matice Srpske. Novi Sad. Serija
  Prirodnih Nauka.   S. 1707 B

Zbornik. Muzej na Grad Skopje.
  Zborn.Muz.Grad Skopje 1964 →   P.A.S 112

Zborník. Pedagogická Fakulta Univerzity Komenského. Bratislava.
  See Sborník Pedagogického Inštitútu v Trnave.   S. 1706 b

Zborník Pedagogickej Fakulty Univerzity Komenského v Bratislave.
  Zborn.pedag.Fak.Univ.komensk.Bratislave
    Prírodné Vedy. 1966 →
    Formerly Sborník Pedagogického Inštitútu v Trnave.
    Prírodné Vedy.   S. 1740 B
    Prírodné Vedy: Geografia. 1970 →   S. 1740 C

Zbornik Prirodoslovnega Društva. Ljubljana.
  Zborn.prir.Društ. 1939-1946.   S. 1746 a B

Zbornik na Rabotite. Hidrobiološki Zavod. Ohrid. Prirodno-
  Matematički Fakultet na Universitetot. Skopje.
  Zborn.Rab.hidrobiol.Zav.Ohrid. 1953 →   S. 1764 b A

Zbornik Radova. Biološki Institut N.R. Srbije. Beograd.
  Zborn.Rad.biol.Inst.N.R.Srb. 1957-1961.   S. 1888 b B

Zbornik Radova. Geoloshki Institut N.R. Srbije. Beograd.
  Zborn.Rad.geol.Inst. 1950-1953.
  Continued as Zbornik Radova. Geoloshkog Instituta
  "Jovan Zhujovich". Beograd.   P.S 456 A

Zbornik Radova Geoloshkog Instituta "Jovan Zhujovich". Beograd.
  Zborn.Rad.geol.Inst.J.Zhujovich Knjiga 7 → 1954 →
  Formerly Zbornik Radova. Geoloshki Institut N.R. Srbije.
  Beograd.   P.S 456 A

Zbornik Radova Geoloshkog i Rudarskog Fakulteta. Beograd.
  See Zbornik Geološkog i Rudarskog Fakulteta, Tehnička
  Velika Skola. Beograd.   P.S 457

Zbornik Radova. Institut za Ekologiju i Biogeografiju.
  Srpska Akademija Nauka. Beograd.
  Zborn.Rad.Inst.Ekol.Biogeogr. 1950-1956.   S. 1888 H

Zbornik Radova. Institut za Oceanografiju i Ribarstvo. Split.
  Zborn.Rad.Inst.Oceanogr.Ribarst. 1968/1969 →   S. 1704 a F

Zbornik Radova Pol'oprivrednog Fakulteta, Universitet u Beogradu.
  Zborn.Rad.pol'opriv.Fak.Univ.Beogr. 1953 →   B.S 1333

TITLE	SERIAL No.

Zborník Slovenského Národného Múzea. Bratislava.
    Zborn.slov.národ.Muz. Prírodné Vedy 1968 →
    Formerly Sborník Slovenského Národného Muzea. Bratislava.   S. 1754 a A

Zbornik Východoslovenského Múzea. Košice.
    Zborn.vychodoslov.Múz.   Séria A, 1967-1972.
                           Séria B, 1966-1972.
    Formerly Sborník Východoslovenského Múzea. Košice.
    Continued as   Seria AB, Vol.Xl (A) → Vol.XII (B) → 1973 → S. 1763 d A-B

Zeepaard, Het. (Den Haag.)
    Zeepaard 1941 →   S. 699 B

Zeitschrift für Acclimatisation. Berlin.
    Z.Acclim. 1858-1874.   Z.S 1420

Zeitschrift für Allgemeine Erdkunde. Berlin.
    Z.allg.Erdk.Berl. 1853-1865.
    Formerly Monatsberichte über die Verhandlungen der
    Gesellschaft für Erdkunde. Berlin.
    Continued as   Zeitschrift der Gesellschaft für
    Erdkunde zu Berlin.   S. 1324 A

Zeitschrift für Analytische Chemie. Wiesbaden, etc.
    Z.analyt.Chem. 1862 →   M.S 1362

Zeitschrift für Anatomie und Entwickelungsgeschichte. Leipzig.
    Z.Anat.EntwGesch., Leipzig 1875-1877.
    Continued as   Archiv für Anatomie und Entwickelungsgeschichte. Z.S 1590

Zeitschrift für Angewandte Entomologie. Berlin.
    Z.angew.Ent. 1914 →   E.S 1359

Zeitschrift für Angewandte Geologie. Berlin.
    Z.angew.Geol.   Bd.1, Hft.2-3: Bd.2, Hft.2-12: Bd.3: Bd.4,
                    Hft. 1,4,6-12: Bd.5 → 1955 →   P.S 1339

Zeitschrift für Angewandte Mikroskopie und Klinische Chemie. Leipzig.
    Z.angew.Mikrosk.klin.Chem. 1895-1908.   S. 1694

Zeitschrift für Angewandte Mineralogie. Berlin.
    Z.angew.Miner.   Bd.1, Hft.1-4, Bd.2, Hft.1. 1937-1939.   M.S 1314

Zeitschrift für Angewandte Zoologie. Berlin.
    Z.angew.Zool.   Vol.41 → 1954 →   Z.S 1435

Zeitschrift für Anorganische (und Allgemeine) Chemie.
    Leipzig. etc.
    Z.anorg.allg.Chem.   1892-1934   M.S 1361

Zeitschrift für Aquarien- und Terrarien-Vereine. Berlin.
    Z.Aquar.-u.TerrarVer. 1937 No.34 - 1939 No.15
    Formerly Nachrichtenblatt fur Aquarien- und
    Terrarien-Vereine.   Z.S 1315 A

| TITLE | SERIAL No. |

Zeitschrift der Arbeitsgemeinschaft Osterreichischer
   Entomologen. Wien.
   Z.ArbGem.öst.Ent. 1960 →
   Formerly Entomologisches Nachrichtenblatt. Wien.   E.S 1711

Zeitschrift für Biologie. Moskau.
   See Biologicheskiĭ Zhurnal. Moskva.   S. 1845 a

Zeitschrift für Botanik. Jena, Stuttgart.
   Z.Bot. 1909-1965.
   Continued as Zeitschrift für Pflanzenphysiologie. Stuttgart.   B.S 1024

Zeitschrift der Bulgarischen Geologischen Gesellschaft.
   See Spisanie na Bŭlgarskoto Geologichesko Druzhestvo.   P.S 490

Zeitschrift der Deutschen Geologischen Gesellschaft. Berlin.
   Z.dt.geol.Ges. 1849 →   P.S 340
                1849-1882.   M.S 1429

Zeitschrift der Deutschen Mikrologischen Gesellschaft. München.
   See Kleinwelt.   S. 1589 a

Zeitschrift für Entomologie. Breslau.
   Z.Ent. 1847-1907. 1927-1944.
   From 1908-1924 Appeared as Jahresheft des Vereins für
   Schlesische Insektenkunde zu Breslau.   E.S 1320

Zeitschrift für Entomologie (Germar). Leipzig.
   Z.Ent.(Germar) 1839-1844.
   Continued as Linnaea Entomologica. Berlin.   E.S 1311

Zeitschrift für Ethnologie und ihre Hülfswissenschaften als
   Lehre from Menschen in seinen Beziehungen zur Natur und zur
   Geschichte. Berlin.
   Z.Ethnol. 1869-1887.   P.A.S 43

Zeitschrift des Ferdinandeums für Tirol und Vorarlberg. Innsbruck.
   Z.Ferdinand.Tirol 1852-1920.
   Formerly Neue Zeitschrift des Ferdinandeums für Tirol
   und Vorarlberg. Innsbruck.   S. 1736 A

Zeitschrift für Fischerei und deren Hilfswissenschaften. Berlin.
   Z.Fisch. Bd.1-19, 1893-1917.
   Neue Folge 1952-1970.   Z.S 1312

Zeitschrift für Garten-und Blumenfreunde, Kunst-und
   Handelsgärtner. Hamburg.
   See Neue Allgemeine Deutsche Garten-und Blumenzeitung.
   Hamburg.   B.S 1005

Zeitschrift für Geologie und Geographie. Bucarest.
   See Revue de Géologie et de Géographie.   P.S 464

TITLE	SERIAL No.

Zeitschrift für Geologische Wissenschaften. Berlin.
    Z.geol.Wiss. 1973 →
    Replaces Geologie. Zeitschrift für das Gesamtgebiet der
    Geologischen Wissenschaften.   P.S 1338
    Bericht der Deutschen Gesellschaft für Geologische
    Wissenschaften.   P.S 342
    Paläontologische Abhandlungen. Deutsche Gesellschaft für
    Geologische Wissenschaften.   P.S 342 A

Zeitschrift für die Gesammten Naturwissenschaften Halle.
    Z.ges.naturw.Halle 1853-1881.
    Formerly Jahresbericht des Naturwissenschaftlichen
    Vereins in Halle.
    Continued as Zeitschrift für Naturwissenschaften. Halle.   S. 1437 A

Zeitschrift für Geschiebeforschung. Berlin, etc.
    Z.Geschiebeforsch. 1925-1935.
    Continued as Zeitschrift für Geschiebeforschung und
    Flachlandsgeologie.   P.S 343

Zeitschrift für Geschiebeforschung und Flachlandsgeologie. Leipzig.
    Z.Geschiebeforsch.Flachldgeol. Band 12, 1936-1944.
    Formerly Zeitschrift für Geschiebeforschung.   P.S 343

Zeitschrift der Gesellschaft für Erdkunde zu Berlin.
    Z.Ges.Erdk.Berl. 1866-1888; 1904.   S. 1324 A
        Bd. 15 1880.   T.R.S 1331
    Formerly Zeitschrift für Allgemeine Erdkunde. Berlin.

Zeitschrift für Gletscherkunde, für Eiszeitforschung und Geschichte
    des Klimas. Berlin.
    Z.Gletscherk.Eiszeitforsch.Gesch.Klimas 1906-1912.   P.S 345

Zeitschrift für Heimatkunde. Laibach.
    See Carniola.   S. 1746 B

Zeitschrift für Hundeforschung. Leipzig.
    Z.Hundeforsch. Bd.12 → 1938 →
    Bd.1-11. Issued as part of Zentralblatt für Kleintierkunde
    und Pelztierkunde. Leipzig.   Z. Mammal Section

Zeitschrift für Hydrologie (Hydrographie, Hydrobiologie,
    Fischereiwissenschaft.) Aarau.
    Z.Hydrol.Hydrogr.Hydrobiol. 1941-1948.
    Formerly Revue Hydrologie (Hydrographie, Hydrobiologie,
    Pisciculture). Aarau.
    Continued as Schweizerische Zeitschrift für
    Hydrologie. Basel.   S. 1201 F

Zeitschrift für Induktive Abstammungs-und Vererbungslehre. Berlin.
    Z.indukt.Abstamm.u.VererbLehre 1908-1957.
        Supplement 1-2, 1928.
    Continued as Zeitschrift für Vererbungslehre.   S. 1645

TITLE	SERIAL No.

Zeitschrift des Internationalen Afrikanischen Instituts.
   See Africa. Journal of the International African
   Institute. London.   P.A.S 8 B

Zeitschrift für Karst-und Höhlenkunde. Berlin.
   Z.Karst.-u.Höhlenk. 1941.
   Formerly Mitteilungen über Höhlen-und Karstforschung. Berlin.   S. 1652

Zeitschrift für Kleintierkunde und Pelztierkunde. Leipzig.
   Z.Kleintierk.Pelztierk. Jahrg.12, Hft.1-4, 1936.
   Formerly Kleintier und Pelztier. Leipzig.
   Continued as Zentralblatt für Kleintierkunde und
   Pelztierkunde ("Kleintier und Pelztier".) Leipzig.   Z. Mammal Section

Zeitschrift des Kölner Zoo. Köln.
   Z.Kölner Zoo Jahrg.14 → 1971 →
   Formerly Freunde des Kölner Zoo.   Z.S 1309

Zeitschrift für Kristallographie, Kristallgeometrie,
   Kristallphysik, Kristallchemie. Leipzig, etc.
   Z.Kristallogr.Kristallgeom. 1921 →
   Formerly Zeitschrift für Kristallographie und Mineralogie.
   Leipzig.   M.S 1306

Zeitschrift für Kristallographie und Mineralogie. Leipzig.
   Z.Kristallogr.Miner. 1877-1920.
   Vol.1-51; 1877-1913 duplicated.
   Continued as Zeitschrift für Kristallographie,
   Kristallgeometrie, Kristallphysik, Kristallchemie. Leipzig.   M.S 1306

Zeitschrift für Kristallographie, Mineralogie und Petrographie.
   Leipzig. Abt.B. Vols 40-45 (imp.); 1930-1943.
   Included in Tschermaks Mineralogische und Petrographische
   Mitteilungen. Wien.   M.S 1310

Zeitschrift für Lepidopterologie. Krefeld.
   Z.Lepid. 1950-1955.   E.S 1341 a

Zeitschrift des Mährischen Landesmuseums. Brünn.
   Z.mähr.Landesmus. 1901-1919.   S. 1712 A

Zeitschrift für Malakozoologie. Hannover.
   Z.Malakozool. 1844-1853.
   Continued as Malakozoologische Blaetter.   Z. Mollusca Section

Zeitschrift für Mikrosokopische-Anatomische Forschung. Leipzig.
   Z.mikrosk.-anat.Forsch. 1924 →   Z.S 1580 B

Zeitschrift für Mineralogie, Geologie und Paläontologie. Stuttgart.
   Z.Miner.geol.Paläont. 1907-1909.
   Formerly Monatsschrift für Mineralien-Gesteins-und
   Petrifaktensammler. Rochlitz i. Sa.   M.S 1306 A

| TITLE | SERIAL No. |

Zeitschrift für Mineralogie, Taschenbuch. Frankfurt am Main.
    Z.Miner.Taschenb. 1825-1829.   M.S 1301
              No.4 & 9, 1825 (imp.)   P.S 311
    Formerly Taschenbuch für die Gesammte Mineralogie.
    Continued as Jahrbuch für Mineralogie, Geognosie, Geologie
    und Petrefaktenkunde.

Zeitschrift für Morphologie und Anthropologie. Stuttgart.
    Z.Morph.Anthrop. 1899-1911; 1959 →   P.A.S 40

Zeitschrift für Morphologie und Okologie der Tiere. Berlin.
    Z.Morph.Okol.Tiere 1924-1967
    Continued as Zeitschrift für Morphologie der Tiere.   Z.S 1535

Zeitschrift für Morphologie der Tiere. Berlin.
    Z.Morph.Tiere 1967 →
    Formerly Zeitschrift für Morphologie und Okologie
    der Tiere. Berlin.   Z.S 1535

Zeitschrift für Naturforschung. Wiesbaden.
    Z.Naturf. 1946.
      Teil a. 1947-1973.
      Teil b. 1947-1973.
      Teil c. 1973 →   S. 1602

Zeitschrift für Naturwissenschaften. Halle a.S.
    Z.Naturw. 1882-1941.   S. 1437 A
             1890-1937.   T.R.S 1361
    Formerly Zeitschrift für die Gesammten Naturwissenschaften. S. 1437 A

Zeitschrift für Oologie und Nidologie. Stuttgart.
    Z.Ool.Nidol. 1911.   Z.S 1655

Zeitschrift für Ornithologie und Praktische Geflügelzucht. Stettin.
    Z.Orn.prakt.Geflugelz. 1883-1892.   T.B.S 1303

Zeitschrift des Osterreichischen Entomologen-Vereins, Wien.
    Z.öst.EntVer. 1916-1939.
    Continued as Zeitschrift des Wiener Entomologen-Vereins.
    Wien.   E.S 1707

Zeitschrift für Parasitenkunde. Berlin.
    Z.Parasitenk. 1928 →
        (Wanting Bd. 13, Nos. 3-6.)   S. 1660

Zeitschrift für Parasitenkunde, Jena.
    Z.Parasitenk.Jena 1869-1875.   S. 1675

Zeitschrift für Pflanzenkrankheiten (afterwards Pflanzenpathologie
und Pflanzenschutz). Stuttgart.
    Z.Pflkrankh.PflPath.PflSchutz 1891-1944: 1948-1951.   B.M.S 30

Zeitschrift für Pflanzenphysiologie. Stuttgart.
    Z.PflPhys. Vol.55 → 1965 →
    Formerly Zeitschrift für Botanik. Jena, Stuttgart.   B.S 1024

TITLE	SERIAL No.

Zeitschrift für Physikalische Chemie, Stöchiometrie und
Verwandtschaftlehre. Leipzig. etc.
Z.phys.Chem. 1887-1914   M.S 1430

Zeitschrift für Physiologie. Heidelberg.
Z.Physiol. 1824-1835.   Z.S 1510

Zeitschrift für Pilzfreunde. Dresden.
Zeit.Pilzfr. 1883-1885.   B.M.S 25

Zeitschrift für Pilzkunde. Heilbronn, Karlsruhe.
Z.Pilzk. 1948 →   B.M.S 28

Zeitschrift für Praktische Geologie. Berlin.
Z.prakt.Geol. 1893-1942   M.S 1309

Zeitschrift der Rheinischen Naturforschenden Gesellschaft. Mainz.
Z.rhein.naturf.Ges.Mainz 1961-1966.
Continued as Mainzer Naturwissenschaftliches Archiv.   S. 1527

Zeitschrift der Rumänischen Geologischen Gesellschaft.
See Buletinul Societátii Romane de Geologie.   P.S 462

Zeitschrift für Säugetierkunde. Berlin.
Z.Säugetierk. 1926 →   Z. Mammal Section
1926-1972.   T.R.S 1355

Zeitschrift für Seefahrt-und Meereskunde. Hamburg.
See Annalen der Hydrographie u. Maritime Meteorologie. Berlin. M.S 1381

Zeitschrift für Sukkulentenkunde. Berlin.
Z.SukkulKde 1923-1928.
Formerly Monatsschrift für Kakteenkunde.
Continued as Monatsschrift der Deutschen Kakteen-
Gesellschaft.   B.S 906

Zeitschrift für Systematische Hymenopterologie und Dipterologie.
Leipzig.
Z.syst.Hymenopt.Dipterol. 1901-1908.   E.S 1350

Zeitschrift für Tierpsychologie. Berlin.
Z.Tierpsychol. 1937 →   Z.S 1595

Zeitschrift für Tirol und Vorarlberg. Innsbruck.
See Beitrage zur Geschichte, Statistik, Naturkunde und
Kunst von Tirol und Vorarlberg. Innsbruck.   S. 1736 A

Zeitschrift der Ungarischen Geologischen Gesellschaft.
See Földtani Közlöny. A Magyar Földtani Társulat Folyóirata.   P.S 365

Zeitschrift des Vereins der Naturbeobachter und Sammler. Wien.
Z.Ver.NatBeob.Sammler Wien 1926-1934.   S. 1774
1926-1932.   E.S 1712

Zeitschrift für Vererbungslehre. Berlin.
Z.VererbLehre 89-98. 1958-1966.
Formerly Zeitschrift für Induktive Abstammungs-und
Vererbungslehre.
Continued as Molecular & General Genetics. Berlin.   S. 1645

TITLE	SERIAL No.

Zeitschrift für Vergleichende Physiologie. Berlin.
    Z.vergl.Physiol. Bd.50-76, 1965-1972.
    Continued as Journal of Comparative Physiology. Berlin.   S. 1684

Zeitschrift für Vogelschutz und andere Gebiete des
    Naturschutzes. Berlin.
    Z.Vogelschutz 1920-1921.
    Continued as Naturschutz.   S. 1654

Zeitschrift für Vulkanologie. Berlin.
    Z.Vulk. 1914-1938.   M.S 1308

Zeitschrift des Wiener Entomologen-Vereins. Wien.
    Z.wien EntVer. 1939-1942.
    Formerly Zeitschrift des Osterreichischen Entomologen-
    Vereins, Wien.
    Continued as Zeitschrift der Wiener Entomologischen
    Gesellschaft. Wien.   E.S 1707

Zeitschrift der Wiener Entomologischen Gesellschaft. Wien.
    Z.wien.ent.Ges. 1943 →
    Formerly Zeitschrift des Wiener Entomologen-Vereins. Wien.   E.S 1707

Zeitschrift für Wissenschaftliche Biologie.
    Abt.A; See Zeitschrift für Morphologie und
        Okologie der Tiere.   Z.S 1535
    Abt.D; See Wilhelm Roux Archiv für Entwicklungs-
        Mechanik der Organismen.   Z.S 1440
        See Archiv für Wissenschaftliche Botanik.   B.S 903
    Abt.F; See Zeitschrift für Parasitenkunde. Berlin.   S. 1660

Zeitschrift für Wissenschaftliche Insektenbiologie. Berlin.
    Z.wiss.InsektBiol. 1905-1937.
    Formerly Allgemeine Zeitschrift für Entomologie. Berlin.   E.S 1338

Zeitschrift für Wissenschaftliche Mikroskopie und Mikroskopische
    Technik. Leipzig.
    Z.wiss.Mikrosk. 1884-1971.
      (Wanting Bd.59, Hft.4)
    Continued as Microscopica Acta.   S. 1690

Zeitschrift für Wissenschaftliche Zoologie. Leipzig.
    Z.wiss.Zool. 1848 →
    (Wanting Bd.154-155 & Bd.156, Hft.1. 1941-1944.)   Z.S 1600

Zeitschrift für Zellforschung und Mikroskopische Anatomie. Berlin.
    Z.Zellforsch.mikrosk.Anat. Bd.76 → 1967 →   S. 1692

Zeitschrift für Zoologische Systematik und Evolutionsforschung.
    Frankfurt am Main.
    Z.Zool.Syst.EvolForsch. 1963 →   Z.S 1438

| TITLE | SERIAL No. |

Zeitung für Geognosie, Geologie und Innere Naturgeschichte
  der Erde. Weimar.
  Ztg.geog.geol.Erde 1826-1831.   P.S 301

Zeitung für Zoologie, Zootomie, und Palaeozoologie. Leipzig.
  Ztg Zool.Zoot.Palaeoz.Leipzig 1848-1849.   S. 1671
                                                              Z.S 1610 & Tweeddale

Zemlevedenie. Sbornik Moskovskogo Obshchestva Ispȳtateleĭ
  Prirodȳ. Moskva.
  Zemlevedenie N.S.1 (41) → 1940 →   P.S 537 A

Zentralblatt für Bakteriologie, Parasitenkunde, Infektionskrankheiten
  (und Hygiene). Jena.
  Zentbl.Bakt.ParasitKde 1887-1894.
    Abt.I, Med.-Hyg.Bakt. 1895-1901.
      Abt.I, Originale. 1902-1971. Reihe A. 1971 →
                                   Reihe B. 1971 →
    Abt.I, Referate. 1902 →
    Abt.II, Bakt.Gärungsphys.PflPath. 1895 →
    (Wanting Bd.106, Hft.25-26.)   S. 1676 A - B

Zentralblatt für Geologie und Paläontologie. Stuttgart.
  Zentbl.Geol.Paläont. Teil 1, 1950 →
  Formerly Zentralblatt für Mineralogie, Geologie und
  Paläontologie. Teil 4. 1943.   P.S 319

Zentralblatt für Geologie und Paläontologie. Stuttgart. Teil 2.
  Zentbl.Geol.Paläont. Teil 2. 1950 →
  Formerly Zentralblatt für Mineralogie, Geologie und
  Paläontologie. Teil 3. 1943-1949.   P.S 319

Zentralblatt für das Gesamtgebiet der Entomologie. Lienz.
  Zentbl.Gesamtgeb.Ent. 1946-1949.   E.S 1702

Zentralblatt für Kleintierkunde und Pelztierkunde
  ("Kleintier und Pelztiere"), Leipzig.
  Zentbl.Kleintierk. Jahrg.12, Hft.5 - Jahrg.14,
  Hft.8, 1936-1938.
  Formerly Zeitschrift für Kleintierkunde und
  Pelztierkunde. Leipzig.
  Continued in Beiträge zur Tierkunde und Tierzucht;
  Monographien der Wildsäugetiere & Zeitschrift für
  Hundeforschung.   Z. Mammal Section

Zentralblatt für Mineralogie, Stuttgart.
  Zentbl.Miner. 1950 →
    Teil I. Kristallographie, Mineralogie.
    Teil II. Gesteinskunde, Technische Mineralogie, Geochemie,
        Lagerstättenkunde, 1950-1958; Petrographie,
        Technische Mineralogie, Geochemie und
        Lagerstättenkunde, 1958 →
  Formerly Zentralblatt für Mineralogie, Geologie und
  Paläontologie, Stuttgart. Teil I-II, 1943-1949.   M.S 1302

| TITLE | SERIAL No. |

Zentralblatt für Mineralogie, Geologie und Paläontologie. Stuttgart.
   Zentbl.Miner.Geol.Paläont.
     1900-1924; Abt.B. Geologie und Paläontologie, 1925-1942.   P.S 316
     1900-1924; Abt.A. Mineralogie und Petrographie, 1925-1942.   M.S 1305
    Continued as Neues Jahrbuch für Mineralogie, Geologie und
    Paläontologie. Monatshefte Abt.A & B. 1943-1949.

Zentralblatt für Mineralogie, Geologie und Paläontologie, Stuttgart.
   Zentbl.Miner.Geol.Paläont. Teil I-II, 1943-1949.
     Teil I. Kristallographie und Mineralogie, 1943-1949.
     Teil II. Gesteinskunde, Lagerstättenkunde, Allgemeine
         und Angewandte Geologie, 1946-1949.
    Formerly Neues Jahrbuch für Mineralogie, Geologie und
    Paläontologie. Stuttgart. Referate, Teil I. 1928-1942.
    Continued as Zentralblatt für Mineralogie, Stuttgart.
    Teil I-II, 1950 →   M.S 1302

Zentralblatt für Mineralogie, Geologie und Paläontologie.
   Stuttgart. Teil 3.
   Zentbl.Miner.Geol.Paläont. Teil 3, 1943-1949.
    (1945-1947 not published.)
   Formerly Neues Jahrbuch für Mineralogie, Geologie und
   Paläontologie. Referate 3. 1928-1942.
   Continued as Zentralblatt für Geologie und Paläontologie.
   Teil 2, 1950 →   P.S 319

Zentralblatt für Mineralogie, Geologie und Paläontologie.
   Stuttgart. Teil 4.
   Zentbl.Miner.Geol.Paläont. Teil 4. Tom.18, Hft.1-3, 1943.
   Formerly Neues Jahrbuch für Mineralogie, Geologie
   und Paläontologie. Referate Teil 2. 1928-1942.
   Continued as Zentralblatt für Geologie und Paläontologie.
   Teil 1. 1950 →   P.S 319

Zentralblatt für Naturwissenschaften und Anthropologie. Leipzig.
   Zentbl.Naturw.Anthrop. 1853-1854.   Z.S 1555

Zentralblatt für Zoologie, Allgemeine und Experimentelle Biologie.
   Leipzig.
   Zentbl.Zool.allg.exp.Biol. 1912-1918.   T.R.S 1349

Zephyrus. Fukuoka.
   Zephyrus 1929-1947.   E.S 1904

Zeszyty Naukowe Uniwersytetu Jagiellońskiego. Krakow.
   Zesz.nauk.Uniw.jagiellonsk.
   Seria nauk Biologicznych. Zoologia 1957-1958.
   Also numbered as part of the Zeszyty.
   Continued as Prace Zoologiczne. (Zeszyty Naukowe
   Uniwersytetu Jagiellońskiego) Uniwersytet
   Jagiellonski. Krakow.   Z.S 1800

Zeszyty Naukowe Uniwersytetu Jagiellonskiego: Prace Botaniczne.
   Crakow.
   Zesz.nauk.Uniw.Jagiellon., Bot. 1973 →   B.S 1305

TITLE	SERIAL No.

Zeszyty Naukowe Uniwersytetu Łódzkiego. Łódź.
  Zesz.nauk.Uniw.Łódz. Seria II 1955 →   S. 1726 b

Zeszyty Naukowe Uniwersytetu Mikołaja Kopernika w Toruniu.
  Nauki Matematyczno-Przyrodnicze.
  Zesz.nauk.Uniw.Mikołaja Kopernika Torun No.1-30, 1956-1972.
  Continued as Acta Universitatis Nicolai Copernici. Torun.   S. 1874 b

Zeszyty Problemowe "Kosmosu". Warszawa.
  Zesz.probl."Kosmosu" Nos. 2 & 4 → 1956 →   S. 1789 B

Zeszyty Problemowe "Nauki Polskiej". Warszawa.
  Zesz.probl.Nauki pol. 1954-1966.   S. 1870 a F

Zhivotnyĭ Mir SSSR. Zoologicheskiĭ Institut, Akademiya
  Nauk SSSR. Moskva & Leningrad.
  Zhivotn.Mir SSSR Tom.1, 3-5. 1936-1958.   Z. 72Q q Z

Zhiwuxue Zazhi.
  See Botanical Magazine. Peking.   B.S 1890 a

Zhurnal Bio-Zoolohichnoho Tsȳklu. Kȳyiv.
  Zh.bio-zool.Tsȳklu Kȳyiv No.1-3, 1932.   Z.S 1827 A

Zhurnal Bolêzneĭ Rasteniĭ.
  Zh.bolêz.Rast. 1907-1915.
  Continued as Bolezni Rasteniĭ. S.-Peterburg.   B.M.S 62

Zhurnal Eksperimental'noĭ Biologii i Meditsinȳ. Moskva.
  Zh.eksp.Biol.Med. vols. 2-7. 1926-1931.
  (Vols. 2-5 are Seriya A.)
  Continued as Biologicheskiĭ Zhurnal. Moskva.   S. 1845 a

Zhurnal Evolyutsionnoĭ Biokhimii i Fiziologii. Leningrad.
  Zh.evol.Biokhim.Fiziol. Tom 6 → 1970 →   S. 1804 c

Zhurnal Geologii i Geografii. Bucarest.
  See Revue de Géologie et de Géographie.   P.S 464

Zhurnal Geologo-Geografichnogo Tsikla.
  See Zhurnal Heoloho-Heohrafichnoho Tsȳklu.   P.S 532

Zhurnal Heoloho-Heohrafichnoho Tsȳklu. Vseukrayins'ka
  Akademiya Nauk. Kiev.
  Zh.heol.-heohr.Tsȳklu 1932.
    (Wanting: 5-8 1933-1934)
  Continued as Heolohichnyi Zhurnal.   P.S 532

Zhurnal Instȳtutu Botanikȳ. Akademiya Nauk URSR. Kȳyiv.
  Zh.Inst.Bot.Kȳyiv
  No.4 (12), 6 (14) - 23 (31), 1935-1940.
  Continued as Botanichniyi Zhurnal. Kȳyiv.   B.S 1371

Zhurnal Obshcheĭ Biologii, Moskva.
  Zh.obshch.Biol. vol. 4 → 1943 →
    (Wanting vol. 7, no. 6; vol. 8 & 14.)   S. 1802 b

TITLE	SERIAL No.

Zhurnal Russkago Fiziko-Khimicheskago Obshchestva Imperatovskom
   St.-Peterburgskom Universitete. St. Petersburg.
   Zh.russk.fiz.-khim.Obshch. 1869-1899.   M.S 1876

Ziemleviedieniie. Recueil de la Société des Naturalistes
   de Moscou.
   See Zemlevedenie. Sbornik Moskovskogo Obshchestva
   Ispytateleĭ Prirodȳ. Moskva.   P.S 537 A

Zinruigaku Zassi. Tokyo.
   See Journal of the Anthropological Society of Nippon.
   Tokyo.   P.A.S 284

Ziraat Fakültesi Yilligi. Ankara.
   See Ankara Universitesi Ziraat Fakultesi Yilliği.   E.S 1887

Zirai Mücadele Arastirma Yilligi. Ankara.
   See Plant Protection Research Annual. Ankara.   E.S 1887 a

Zitteliana. München.
   Zitteliana 1969 →   P.S 338 a A

Ziva. Praha.
   Ziva Roc.1-9; 13, Sv.1-3; 14, Sv.1, 1853-1867.
   (N.S.) Roc.18 (56) → 1970 →   S. 1754

Zoe. A Biological Journal. San Francisco, Cal.
   Zoe 1890-1908.   S. 2442

Zoo. Antwerp.
   Zoo, Antwerp Ann.14 → 1949 → (imp.)   Z.S 710
                   No.21 → 1955 →   T.R.S 704

Zoo. London.
   Zoo Lond. 1936-1938.
   Continued as Animal & Zoo Magazine.   Z. 1 C

Zoo. Barcelona.
   Zoo, Barcelona 1962 →   T.R.S 1109 & Z.S 1007

Zoo Federation News. London.
   Zoo Fed.News 1971 →   Z.S 2

Zoo Life. London.
   Zoo Life 1946-1957.   Z. 1 M

Zoogeografica. Jena.
   Zoogeografica 1932-1942.   Z.S 1345

Zooleo. Bulletin de la Société de Botanique et de Zoologie
   Congolaise. Léopoldville.
   Zooleo 1938-1940: 1949-1960. (imp.)
   (From 1940-1942 See Bulletin de la Société de Botanique
   et de Zoologie Congolaise.)   S. 2057

| TITLE | SERIAL No. |

Zoologia Platense. Berazategui, Buenos Aires.
    Zool.platense 1967 →   Z.S 2254

Zoologica. New York.
    Zoologica, N.Y. 1907 →   T.R.S 5134 C & Z.S 2465 C

Zoologica. Seoul.
    Zoologica, Seoul 1962 →   Z.S 1981

Zoologica. Stuttgart.
    Zoologica, Stuttg. 1897 →
    Formerly Bibliotheca Zoologica. Cassel & Stuttgart.   Z.S 1480

Zoologica Africana. Cape Town.
    Zoologica afr. 1965 →   Z.S 2033 A

Zoologica Gothoburgensia. Göteborg.
    Zool.gothoburg. 1965 →   Z.S 504

Zoologica Palaearctica. Dresden.
    Zoologica palaearct. 1923-1924.
    Continued as Pallasia, Dresden.   Z.S 1341

Zoologica Poloniae. Wrocław.
    Zoologica Pol. 1935 →   Z.S 1802

Zoologica Scripta. Stockholm.
    Zoologica Scr. 1971 →   Z.S 501 A
    Replaces Arkiv för Zoologi. Stockholm.   Z.S 501 & T.R.S 646

Zoological Bulletin. Boston.
    Zool.Bull. 1897-1899.   Z.S 2525

Zoological Bulletin. Division of Zoology, Pennsylvania Department
    of Agriculture. Harrisburg.
    Zool.Bull.Div.Zool.Pa Dep.Agric. Vol.5 - Vol.7,
    No. 1-2, 4-12, 1907-1910.
    Formerly Monthly Bulletin. Division of Zoology, Pennsylvania
    Department of Agriculture.
    Continued as Bi-monthly Zoological Bulletin. Pennsylvania
    Department of Agriculture.   E.S 2489 a

Zoological Journal. London.
    Zool.J.Lond. 1824-1835.   E.S 24
         T.R.S 162 & Z.S 280 & Tweeddale

Zoological Journal of the Linnean Society of London.
    Zool.J.Linn.Soc. Vol. 48 → 1969 →   T.R.S 132 B &
    Formerly Journal of the Linnean Society of London. Zoology.   Z.S 20 B

| TITLE | SERIAL No. |

Zoological Leaflet. Chicago Natural History Museum.
   See Leaflet. Field Museum of Natural History. Chicago.
   Zoology.   Z.S 2330 L

Zoological Magazine. London.
   Zool.Mag.Lond. 1833.   Z.S 290 & S. 457

Zoological Magazine. Tokyo Zoological Society.
   Zool.Mag.Tokyo 1889 → (imp.).   Z.S 1950 B

Zoological Memoirs. University of Bombay.
   Zool.Mem.Univ.Bombay 1948 →   Z.S 1924

Zoological Memoirs. University of Travancore Research
   Institute. Trivandrum.
   Zool.Mem.Univ.Travancore 1956.   Z.S 1922

Zoological Papers. Biological Institute. Erevan.
   See Zoologicheskiĭ Sbornik. Biologicheskiĭ (Zoologicheskiĭ)
   Institut, Akademiya Nauk Armyanskoĭ SSR. Erevan.   Z.S 1852

Zoological Record. Kiev.
   See Vestnik Zoologii. Institut Zoologii. Akademiya
   Nauk Ukrainskoĭ SSR. Kiev.   Z.S 1827 E

Zoological Record. London.
   Zool.Rec. 1870 →   REF.ABS.105, Z. B 1 Z & E. REF.
            1870-1900.   T.B.S 9999
            1901 → (Aves Section only)   T.B.S 9999
            1870-1966.   T.R.REF.
   (1906-1914 Also formed Section N of International Catalogue
   of Scientific Literature.)
   Formerly Record of Zoological Literature. London.

Zoological Society Bulletin. New York Zoological Society.
   Zool.Soc.Bull.N.Y. No.6 - Vol.30, 1901-1926.
   Formerly News Bulletin. New York Zoological Society.   Z.S 2465 A
   Continued as Bulletin. New York Zoological Society.
                                Z.S 2465 A & T.R.S 5134 B

Zoologicheskiĭ Sbornik. Biologicheskiĭ (Zoologicheskiĭ)
   Institut, Akademiya Nauk Armyanskoĭ SSR. Erevan.
   Zool.Sb.Erevan Vyp.1-4. 9 → 1939 → (imp.)   Z.S 1852

Zoologicheskii Vêstnik. Petrograd.
   Zool.Vêst. 1916-1918.   Z.S 1849

Zoologicheskiĭ Zhurnal. Moskva.
   Zool.Zh. Tom 11 → 1932 →
   Formerly Russkiĭ Zoologicheskiĭ Zhurnal.   T.R.S 2030 & Z.S 1850

Zoologické a Entomologické Listy. Brno.
   Zool.ent.Listy 1952-1955.
   Formerly Entomologické Listy. Brno.
   Continued as Zoologické Listy. Brno.   E.S 1724

Zoologické Listy. Brno.
   Zool.Listy 1956 →   Z.S 1757 &
   Formerly Zoologické a Entomologické Listy. Brno. E.S 1724

| TITLE | SERIAL No. |

Zoologijas Muzeja Biletens. P.Stuckas Latvijas Valsts
    Universitate. Riga.
    Zool.Muz.Bilet.P.Stuckas latvij.Valsts Univ. 1967 →   Z.S 1806

Zoologijas Muzeja Raksti. P.Stuckas Latvijas Valsts Universitate.
    Riga.
    Zool.Muz.Rak.P.Stuckas latvij.Valst Univ. Nr.2 → 1970 →   Z.S 1806 A

Zoologische Abhandlungen Staatliches Museum für Tierkunde in Dresden.
    Leipzig.
    Zool.Abh.st.Mus.Tierk.,Dresden Bd.26 → 1961 →
    Formerly Abhandlungen und Berichte aus dem Staatlichen
    Museum für Tierkunde in Dresden.   T.R.S 1350 A & Z.S 1320

Zoologische Annalen. Weimar.
    Zool.Annln Weimar 1794.   Z. Tweeddale

Zoologische Annalen. Würzburg.
    Zool.Annln 1904-1919.   T.R.S 1347 & Z.S 1640

Zoologische Beiträge. Breslau. (Berlin).
    Zool.Beitr. 1883-1892.
        Neue Folge 1950 →   Z.S 1470

Zoologische Beiträge aus Uppsala.
    See Zoologiska Bidrag från Uppsala.   Z.S 525

Zoologische Bijdragen. Leiden.
    Zool.Bijdr. 1955 →   T.R.S 748, E.S 608 & Z.S 601 E

Zoologische Documentatie.
    See Documentation Zoologique.   Z.S 760 F

Zoologische Garten. Frankfurt.
    Zool.Gart., Frankf. 1859-1905.   T.R.S 1332 &
    Continued as Zoologischer Beobachter. Frankfurt.   Z.S 1340

Zoologische Garten. Leipzig.
    Zool.Gart.Lpz. 1928 →
    Formerly Zoologischer Beobachter. Frankfurt.   T.R.S 1332 & Z.S 1340

Zoologische Jahrbücher. Jena.
    Zool.Jb. 1886 →
        Abteilung für Systematik Bd.3 → 1887 →   Z.S 1530 & T.R.S 1360
        Abteilung für Anatomie Bd.3 → 1888 →   Z.S 1530
        Abteilung für Allgemeine Zoologie Bd.30 → 1910 →   Z.S 1530 & T.R.S 1360

Zoologische Mededeelingen. Leiden.
    Zool.Meded.Leiden 1915 →   E.S 606 & Z.S 601 B
    Formerly Notes from the Leyden Museum.   T.R.S 746

Zoölogische Monographieën van het Rijksmuseum van Natuurlijke
    Historie. Leiden.
    Zool.Monogrn Rijksmus.nat.Hist. 1973 →   Z.S 601 F

TITLE	SERIAL No.

Zoologische Verhandelingen. Leiden.
    Zool.Verh.Leiden 1948 →   T.R.S 747, E.S 607 & Z.S 601 D

Zoologische Vorträge. Leipzig.
    Zool.Vortr. 1889-1902.   Z. 1 o M

Zoologischer Anzeiger. Leipzig.
    Zool.Anz. 1878 →   T.R.S 1339 & Z.S 1620
              1878-1910.   E.S 1331
    Diario...adnexa. See Bibliographia Zoologica.
    Supplementband. See Verhandlungen der Deutschen
    Zoologischen Gesellschaft.

Zoologischer Beobachter. Frankfurt.
    Zool.Beob. 1906-1911.
    Formerly Zoologische Garten. Frankfurt.
    Later continued as Zoologische Garten. Leipzig.   T.R.S 1332 & Z.S 1340

Zoologischer Bericht. Jena.
    Zool.Ber. 1922-1944.   Z.S 1625

Zoologischer Jahresbericht. Zoologische Station zu Neapel. Berlin.
    Zool.Jber.Neapel 1879-1913.   T.R.S 1334 A & Z.S 1131
                  1879-1905.   T.R.S 1334 A

Zoologisches Archiv, von F.A.A.Meyer. Leipzig.   Z. 4 o M &
    Zool.Arch. 1796.   Tweeddale

Zoologisches Centralblatt.
    See Zoologisches Zentralblatt.   T.R.S 1348 & Z.S 1630

Zoologisches Magazin. Kiel.
    Zool.Mag.Kiel 1817-1823.   Z.S 1540

Zoologisches Zentralblatt. Leipzig.
    Zool.Zentbl. 1894-1912.   T.R.S 1348 & Z.S 1630

Zoologisk Revy. Stockholm.
    Zool.Revy Vol.18 → 1956 →
    Formerly Svensk Faunistisk Revy.   Z.S 512

Zoologiska Bidrag fran Uppsala. Stockholm.
    Zool.Bidr.Upps. 1911/12 - 1969.
      Suppl. 1920-1972.   Z.S 525
    Replaced by Zoon. Uppsala.   Z.S 525 A

Zoologist. London.
    Zoologist 1843-1916.   T.R.S 131, E.S 25 & Z.S 260

Zoology of the Faroes...edited by Ad.S.Jensen,
    W.Lundebeck &c. Copenhagen.
    Zoology Faroes 1928 →   Z. 71 o J

Zoology of Iceland...editors: A.Fridriksson & S.L.Tuxen.
    Copenhagen & Reykjavik.
    Zoology Iceland 1937 → (Excluding entomology.)   Z. 71 o F
              Vol.3 → 1957 → (Entomology only.)   E. 71 o F

TITLE	SERIAL No.

Zoology Leaflet. British Museum (Natural History). London.
   <u>Zoology Leafl.Br.Mus.nat.Hist.</u> 1971 →   S.B.M 10 & Z. 072 Aa o B

Zoology Publications. Los Angeles County Museum.
   <u>See</u> Publications. Los Angeles County Museum.
   (Science Series.) Zoology.   Z.S 2570

Zoology Publications from Victoria University College. Wellington.
   <u>Zoology Publs Vict.Univ.Coll.</u> 1949-1957.
   <u>Continued as</u> Zoology Publications from Victoria
   University of Wellington.   Z.S 2175

Zoology Publications from Victoria University of Wellington.
   <u>Zool.Publs Vict.Univ.Wellington</u> No.22 → 1957 →
   <u>Formerly</u> Zoology Publications from Victoria
   University College.   Z.S 2175

Zoön. Pretoria.
   <u>Zoön, Pretoria</u> 1963 →   Z.S 2035

Zoon. Uppsala.
   <u>Zoon, Uppsala</u> 1973 →
      Suppl. 1973 →   Z.S 525 A
   <u>Replaces</u> Zoologiska Bidrag från Uppsala.   Z.S 525

Zoonooz. San Diego.
   <u>Zoonooz</u> Vol.4, No.29-30; Vol.17, No.2 → 1938-1944 →   Z.S 2502 A

Zoopathologica. New York.
   <u>Zoopathologica</u> 1916-1928.   Z.S 2465 D

Zoophysiology and Ecology. Berlin.
   <u>Zoophysiol.Ecol.</u> 1971 →   Z.S 2737

Zprava.
   <u>See</u> Zprávy.

Zpravodaj Anthropologické Spolěcnosti. Brno.
   <u>Zprav.anthrop.Spol.,Brno.</u> 5 → 1952 →   P.A.S 76

Zpravodaj. Vysoká Skola Zemědělska v Brně. Knižní a Dokumentační.
Brno.
   <u>See</u> Knižní a Dokumentační Zpravodaj. Vysoká Skola Zemědělska
   v Brně.   S. 1710 a G

TITLE	SERIAL No.

Zprávy Ceskoslovenské Botanické Společnosti pří CSAV. Praha.
    Zpr.čsl.bot.Spol.ČSAV 1966 →   B.S 1232 a

Zpravy Ceskoslovenské Společnosti Entomologické při CSAV. Praha.
    Zpr.čsl.Spol.ent.CSAV 1965 →   E.S 1725

Zprávy Geografického Ustavu. CSAV. Opava.
    Zpr.geogr.Ust.CSAV, Opava 1 (130-B) → 1964 →
    Formerly Zprávy Slezského Ustavu CSAV v Opavě.
    Přírodní Vědi.   S. 1711 b

Zprávy o Geologických Vyskumoch. Praha and Bratislava.
    Zpr.geol.Vysk. 1961, 1964.   P.S 1356

Zprávy. Museum J.A. Komenského v Uherském Brodě. Uherský Brod.
    Zpr.Mus.J.A.Komenského Uh.Brodě 1956-1966.   S. 1747

Zprávy. Odbor Přírodních Věd. Vlastivědný Ustav v Olomouci.
    Zpr.Odb.Přír.Věd No.1-3, 1963.
    Continued as Práce Odboru Přírodních věd Vlastivědného
    Ustava v Olomouci.   S. 1748 a A

Zprávy Slezského Ustavu CSAV v Opavě. Přírodní Vědi.
    Zpr.slezsk.Ust.CSAV, Opave No. 124 B—129 B, 1963.
    Continued as Zprávy Geografického Ustavu CSAV. Opava.   S. 1711 b

Zpravy Vlastivedného Ústavu v v Olomouci.
    Zpr.Vlast.Úst. Čislo 100 → 1962 →   S. 1748 a B

Zprávy Vyzkumného Ustavu pro Minerály v Turnově. Turnova.
    Zpr.vyzk.Ust.Miner.,Turnove Rada IV. No.2; 1955   M.S 1733

Zprávy o Zasedání Královské Ceské Společnosti Náuk v Praze.
    See Sitzungsberichte der Königl. Böhmischen Gesellschaft
    der Wissenschaften. Prag.   S. 1703 A

Zuid-Afrikaan. Cape Town.
    Zuid-Afrik. 1830-1836.   Z. 74G ff Z

Zürcherische Jugend. Naturforschende Gesellschaft. Zürich.
    Zürch.Jug.naturf.Ges.Zürich 1799-1872.
    Continued as Neujahrsblatt Naturforschende
    Gesellschaft in Zürich.   S. 1256 E

Zweite Deutsche Nordpolarfahrt. 1869-1870. Leipzig.
    Zweite dt.Nordpolarfahrt 1873-1874.   71.o.B